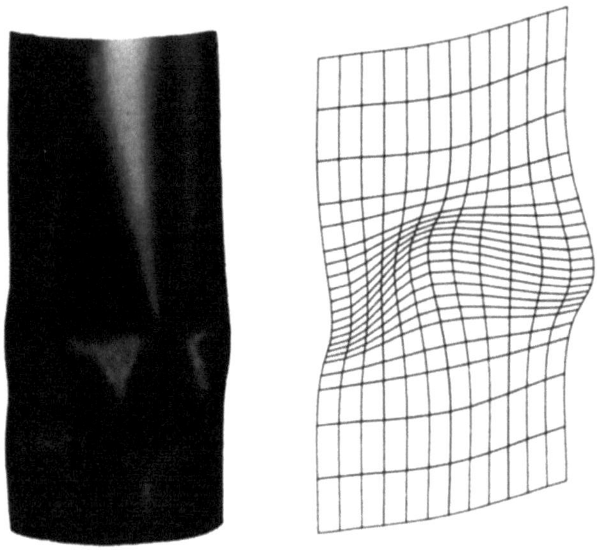

Buckling of Shells

Proceedings of a State-of-the-Art Colloquium
Universität Stuttgart, Germany, May 6–7, 1982

Editor: E. Ramm

With 388 Figures and 16 Tables

Springer-Verlag
Berlin Heidelberg New York 1982

Prof. Dr.-Ing. E. Ramm
Institut für Baustatik, Universität Stuttgart
Pfaffenwaldring 7, D-7000 Stuttgart 80, Germany

ISBN-13: 978-3-642-49336-2 e-ISBN-13: 978-3-642-49334-8
DOI: 10.1007/978-3-642-49334-8

Library of Congress Cataloging in Publication Data.
Main entry under title: Buckling of shells.

"Sponsored by the Deutsche Forschungsgemeinschaft (DFG), the Universität Stuttgart, the International Association for Shell and Spatial Structures (IASS), and the Control Data GmbH."--Pref.
Bibliography: p. Includes index.
1. Shells (Engineering)--Congresses. 2. Buckling (Mechanics)--Congresses. I. Ramm, E. (Ekkehard), (1940 - 1951). II. Deutsche Forschungsgemeinschaft TA660.S5B825 1982 624.1'7762 82-10710

Softcover reprint of the hardcover 1st edition 1982

2061/3020-543210

Dedicated to

Professor Dr.-Ing.
Friedrich Wilhelm Bornscheuer

PREFACE

Thin shells are very popular structures in many different branches of engineering. There are the domes, water and cooling towers, the containments in civil engineering, the pressure vessels and pipes in mechanical and nuclear engineering, storage tanks and platform components in marine and offshore engineering, the car bodies in the automobile industry, planes, rockets and space structures in aeronautical engineering, to mention only a few examples of the broad spectrum of application. In addition there is the large applied mechanics group involved in all the computational and experimental work in this area.

Thin shells are in a way optimal structures. They play the role of the "primadonnas" among all kinds of structures. Their performance can be extraordinary, but they can also be very sensitive. The susceptibility to buckling is a typical example. David Bushnell says in his recent review paper entitled "Buckling of Shells - Pitfall for Designers": "To the layman buckling is a mysterious, perhaps even awe inspiring phenomenon that transforms objects originally imbued with symmetrical beauty into junk". For the engineer the buckling of shells is not mysterious. In contrary it is very real; but it is definitely a very complex phenomenon. Furthermore, it is a very important subject because it is closely related to the safety of a structure. Y. C. Fung and E. E. Sechler wrote over twenty years ago: "In a sense the instability of thin elastic shells remains the most challenging of all classical problems of the theory of elasticity". Despite a remarkable progress made in the last years it is felt that the statement is still valid. Many problems are still unsolved, design procedures have to be developed or improved. The increasing number of publications strongly indicates the interest of many engineers in science and practice in the subject. This, in turn, makes it necessary to condense once in a while all the information available in such an important area. Therefore, the meeting had been termed a state-of-the-art colloquium and it was the main purpose to compile the current status of the instability problem of shells in research and practice. In twenty-three invited lectures theoretical concepts, analytical and numerical methods, experimental investigations and the practical implication are discussed. In addition the main structural systems like cylindrical, conical, spherical, torispherical and hyperboloidal shells are reviewed.

It should be mentioned that the idea for this colloquium is closely related to the 65th anniversary of Professor Dr.-Ing. F. W. Bornscheuer on May 2, 1982 who has been the head of the organizing "Institut für Baustatik" of the University of Stuttgart since nearly 25 years. Dr. Bornscheuer was involved in the design of shell structures for many years when he worked in the aeronautical and civil engineering industry. In 1958 he was appointed to the Chair of Structural Analysis and Theory of Elasticity. In the last years Professor Bornscheuer was the chairman of the German committee (Schalenbeul-Ausschuß des Deutschen Ausschusses für Stahlbau) which recently published the "DASt-Richtlinie 013" on shell buckling. He is also a member of task group 9 of the European Convention for Constructional Steelwork (ECCS) working also on this subject.

The colloquium was sponsored by the Deutsche Forschungsgemeinschaft (DFG), the Universität Stuttgart, the International Association for Shell and Spatial Structures (IASS) and the Control Data GmbH. This support is very much appreciated. In particular I like to thank all authors for their cooperation and effort in preparing the papers and lectures in time. Thanks are extended also to Springer-Verlag for a speedy publication of these proceedings. Last but not least I am especially grateful to all members of the "Institut für Baustatik" for the enormous assistance in organizing the colloquium and during the preparation of the final publication.

Stuttgart, May 21, 1982

Ekkehard Ramm

CONTENTS

CONTRIBUTING AUTHORS

Paper 1	W. T. Koiter	Delft University of Technology Mechanical Engineering Depart. Mekelweg 2 NL - 2628 CD Delft Netherlands
Paper 2	W. B. Krätzig Y. Basar U. Wittek	Ruhr-Universität Bochum Institut für konstruktiven Ingenieurbau Postfach 10 21 48 D - 4630 Bochum 1 F. R. Germany
Paper 3	J. H. Argyris H. Balmer J. Bühlmeier M. Haase H. -P. Mlejnek P. K. Schmolz	Universität Stuttgart Institut für Statik und Dynamik Postfach 80 11 40 D - 7000 Stuttgart 80 F. R. Germany
Paper 4	E. Stein A. Berg W. Wagner	Technische Universität Hannover Lehrstuhl für Baumechanik Callinstraße 32 D - 3000 Hannover F. R. Germany
Paper 5	W. Wunderlich H. J. Rensch H. Obrecht	Ruhr - Universität Bochum Institut für konstruktiven Ingenieurbau Postfach 10 21 48 D - 4630 Bochum 1 F. R. Germany
Paper 6	B. -H. Kröplin	Technische Universität Braunschweig Lehrstuhl und Institut für Baustatik Beethovenstraße 51 D - 3300 Braunschweig F. R. Germany
Paper 7	E. Ramm H. Stegmüller	Universität Stuttgart Institut für Baustatik Pfaffenwaldring 7 D - 7000 Stuttgart 80 F. R. Germany

Paper 8	P. J. Dowling J. E. Harding N. Agelidis W. Fahy	Imperial College of Science and Technology Depart. of Civil Engineering London SW7 2BU United Kingdom
Paper 9	A. C. Walker Y. Segal S. McCall	University of Surrey Depart. of Mechanical Engineering Guildford - Surrey GU2 5XH United Kingdom
Paper 10	F. Resinger R. Greiner	Technische Universität Graz Institut für Stahlbau, Holzbau und Flächentragwerke Rechbauerstraße 12 A - 8010 Graz Austria
Paper 11	H. Saal	Bismarckstraße 19 D - 6111 Lengfeld (Odenwald) F. R. Germany
Paper 12	M. Esslinger	Deutsche Forschungs- und Versuchsanstalt für Luft- und Raumfahrt e. V. Postfach 3267 D - 3300 Braunschweig F. R. Germany
	J. Ciprian	Hoechst Aktiengesellschaft Abteilung K 801 Postfach 80 03 20 D - 6230 Frankfurt (Main) 80 F. R. Germany
Paper 13	D. Vandepitte J. Rathé B. Verhegghe R. Paridaeus C. Verschaeve	Rijksuniversiteit Gent Lab. voor Modelonderzoek Grote steenweg Noord 12 B - 9710 Zwijnaarde Belgium
Paper 14	L. Kollár	Karap u. 9 H - 1122 Budapest XII Hungary
Paper 15	G. D. Galletly	The University of Liverpool Depart. of Mechanical Engineering P.O. Box 147 Liverpool L69 3BX United Kingdom

Paper 16	W. Zerna I. Mungan	Ruhr - Universität Bochum Institut für konstruktiven Ingenieurbau Postfach 10 21 48 D - 4630 Bochum 1 F. R. Germany
Paper 17	G. Hees	Technische Universität Berlin Lehrgebiet Statik der Baukonstruktionen Straße des 17. Juni 135 D - 1000 Berlin 12 F. R. Germany
Paper 18	J. Singer	Technion Israel Institut of Technology Depart. of Aeronautical Engineering Technion City Haifa 32000 Israel
Paper 19	J. Arbocz	TH Delft Lucht- en Ruimevaarttechniek Kluyverweg 1 NL - 2629 HS Delft Netherlands
Paper 20	D. F. Fischer F. G. Rammerstorfer	Voest - Alpine Postfach 2 A - 4010 Linz Austria
Paper 21	F. W. Bornscheuer	Universität Stuttgart Institut für Baustatik Pfaffenwaldring 7 D - 7000 Stuttgart 80 F. R. Germany
Paper 22	L. A. Samuelson	IFM Akustikbyran AB Warfinger Vaeg 26 S - 11251 Stockholm Sweden
Paper 23	H. Isler	Lyssachschachen Postfach CH - 3400 Burgdorf Switzerland

Part I :

Theoretical Concepts - Numerical Methods

THE APPLICATION OF THE INITIAL POST-BUCKLING ANALYSIS TO SHELLS

W. T. KOITER

Delft University of Technology
Laboratory for Engineering Mechanics
Mechanical Engineering Department
Delft, The Netherlands

Abstract

The energy approach to elastic stability is employed to describe characteristic differences in buckling behaviour between bars in compression, plates loaded in their plane, and shells, in particular in the absence of inextensional deformation. The initial stage of post-buckling behaviour of shells is discussed from the stand-point of shallow shell theory and it is applied to more or less localized buckling patterns.

1. Introduction

The German term "Ausweichprobleme" for buckling problems, coined by BIEZENO and GRAMMEL in their famous treatise "Technische Dynamik", contains a good description of the phenemenon in question, the evasion of an increase of loads on a structure by more or less suddenly occurring deflections. There are two basic types of such phenomena, viz. the limit point type of singularity where the load attains a maximum in the load-deflection diagram, and the bifurcation point where two (or more) load-deflection paths intersect. The latter branching behaviour is in a mathematical sense the exception rather than the rule, but it plays a dominant role in buckling theory. One reason is the far greater ease with which bifurcation problems may be attacked, in particular if the pre-buckling behaviour may be described with adequate accuracy by linear theory. Limit point singularities, however, are only open to a non-linear analysis. Fortunately there is also a more factual reason for the popularity of bifurcation problems. The most efficient structural elements are one-dimensional or two-dimensional, with approximately constant stresses over the cross-section of the member or over the thickness of the plate or shell. In the idealized cases of a truly constant stress dis-

Buckling of Shells, Ed.: E. Ramm
Proceedings of a State-of-the-Art Colloqium

tribution over the cross-section of a bar, or of a membrane state of stress in a plate or shell, buckling if it occurs is necessarily of the bifurcation type. The irregularities present in actual structures will disturb the pre-buckling membrane state of stress and will thus result in limit point behaviour (if buckling occurs at all). In the case of small irregularities, however, this modified behaviour is then amenable to a simple perturbation type analysis of the idealized branching behaviour. This was indeed the basic philosophy of our attack on buckling problems some forty years ago.

2. The energy criterion of elastic stability

Buckling under conservative loads and in the elastic domain may be described completely in terms of the potential energy of the structure and its external loads. Let x_i $(i = 1,2,3)$ denote the Cartesian components of the radius vector $\underset{\sim}{x}$ of a generic point of the structures in the fundamental state of equilibrium I to be investigated as to its stability. An arbitrary kinematically admissible displacement field from this state I to some adjacent state II is described by the vector field $\underset{\sim}{u}(\underset{\sim}{x})$. The additional deformation due to this displacement field is described by the (additional) strain tensor

$$\gamma_{ij} = \frac{1}{2}\,(u_{i,j}+u_{j,i}+u_{h,i}u_{h,j}), \tag{2.1}$$

where an additional subscript j preceded by a comma denotes partial differentiation with respect to the coordinate x_j, and where the summation convention has been employed. Due to the (additional) deformation the elastic strain energy density per unit volume in the fundamental state I is increased by

$$\left(\frac{\partial W}{\partial \gamma_{ij}}\right)_I \gamma_{ij} + \frac{1}{2}\left(\frac{\partial^2 W}{\partial \gamma_{ij}\partial \gamma_{k\ell}}\right)_I \gamma_{ij}\gamma_{k\ell} + \,..\, =$$

$$= S_{ij}\gamma_{ij} + \frac{1}{2}\,E^I_{ijk\ell}\gamma_{ij}\gamma_{k\ell} + \,..\,, \tag{2.2}$$

where S_{ij} is the CAUCHY stress tensor and $E^I_{ijk\ell}$ is the tensor of elastic moduli in the fundamental state. In the case of dead loading the potential energy functional which describes the increment in potential energy due to the displacement field $\underset{\sim}{u}(\underset{\sim}{x})$ is now given by

$$P[\underset{\sim}{u}(\underset{\sim}{x})] = \int \left[\frac{1}{2} S_{ij} u_{h,i} u_{h,j} + \frac{1}{2} E^{I}_{ijk\ell} \gamma_{ij} \gamma_{k\ell} + .. \right] dv, \qquad (2.3)$$

where the linear term in $\underset{\sim}{u}(\underset{\sim}{x})$ occurring in (2.2) has been cancelled by the linear potential energy functional of the dead external loads. The equilibrium in the fundamental state I is now stable if and only if the potential energy functional $P[\underset{\sim}{u}(\underset{\sim}{x})]$ (2.3) is positive definite.

Expansion of the energy functional with respect to the displacement field leads immediately to the criterion that the second variation

$$P_2[\underset{\sim}{u}(\underset{\sim}{x})] = \int \left[\frac{1}{2} S_{ij} u_{h,i} u_{h,j} + \frac{1}{2} E^{I}_{ijk\ell} \theta_{ij} \theta_{k\ell} \right] dv \qquad (2.4)$$

must be non-negative for all admissible $\underset{\sim}{u}(\underset{\sim}{x})$, where the linearized additional strain tensor θ_{ij} is defined by

$$\theta_{ij} = \frac{1}{2}(u_{i.j} + u_{j,i}) . \qquad (2.5)$$

We shall also employ the criterion of a positive definite second variation as a sufficient condition for stability, and we shall not dwell on the mathematical difficulties involved in its justification.

The foregoing discussion based on the Cartesian coordinates of a particle in the fundamental state I is indeed the simplest possible description, if we have to consider only a single fundamental state I. This simplicity is lost completely if we wish to consider a family of fundamental states, e.g. corresponding to various load factors in loading conditions specified by the product of a unit loading system and a scalar load factor λ. In these cases it is far more convenient to employ Cartesian coordinates in some fixed reference configuration, e.g. the undeformed structure, as independent variables. The price we may have to pay here is a considerably more complicated expression for the potential energy functional.

In many problems of buckling of structures we may ignore the difference between the Cartesian coordinates in the undeformed structure and in the fundamental state as independent variables. In the case of a uniform state of stress in the fundamental state the relative error of this approximation is of the order of the maximum principal strain in the fundamental state, and thus negligible in the elastic range of most structural materials. In this case we have an approximately linear pre-buckling state where $S_{ij} = \lambda S^0_{ij}$ and the tensor of elastic moduli $E^I_{ijk\ell}$ is identified with the tensor of the classical linear theory of elasticity $E_{ijk\ell}$. The energy functional and its second variation now take the form

$$P[\underset{\sim}{u}(\underset{\sim}{x})] = \int \left[\frac{1}{2}\lambda S^0_{ij} u_{h,i} u_{h,j} + \frac{1}{2} E_{ijk\ell}\gamma_{ij}\gamma_{k\ell} + \dots\right] dv, \qquad (2.6)$$

$$P_2[\underset{\sim}{u}(\underset{\sim}{x})] = \int \left[\frac{1}{2}\lambda S^0_{ij} u_{h,i} u_{h,j} + \frac{1}{2} E_{ijk\ell}\theta_{ij}\theta_{k\ell}\right] dv, \qquad (2.7)$$

where the integrations are extended over the volume of the undeformed body.

One of the most significant aspects of the simplification achieved in (2.6) and (2.7) is that the second terms in the integrands may be interpreted as the elastic energy density of the physically linear and geometrically non-linear theory of elasticity in the case of (2.6), and as the elastic energy density of the classical linear theory of elasticity in the case of (2.7). In the case of various types of structures the appropriate simplified energy expressions of linear theory are thus immediately available for the discussion of stability. We shall employ this aspect first in a qualitative characterisation of buckling problems for bars, plates and shells.

3. Characteristic differences between buckling problems for bars, plates, and shells

A slender bar under a compressive load may buckle in the plane of least flexural rigidity by bending in this plane. The second terms in (2.6) and (2.7) integrated over the cross-section then result in a flexural energy term per unit length $\frac{1}{2}B\kappa^2$ in the case of (2.6) and $\frac{1}{2}Bw^2_{;xx}$ in the case of (2.7), where B is the flexural rigidity, κ is the curvature of the axis of the bar, and $w_{,xx}$ is the second derivative of the deflection along the axis. The critical case of neutral equilibrium occurs when the second variation has a minimum equal to zero, and this occurs for a compressive stress $\sigma = \pi^2 E i^2/\ell^2$, where E is YOUNG's modulus, i is the radius of gyration of the cross-section and ℓ is the length of the bar, supposedly supported in hinged ends. Note that the critical stress is proportional to the square of the ratio of a characteristic dimension of the cross-section to the length of the bar. Since $w_{,xx}$ is a good approximation to κ up to fairly large deflections, the complete energy expression (2.6) is approximated by the second variation (equal to zero!) up to quite large deflections. The post-buckling behaviour is governed by stationary values of the complete energy expression (2.6), and it is therefore approximated by the zero value of the second variation up to quite large deflections. In other words, the post-buckling behaviour is neutral to a good approximation. This is indeed confirmed by a rigorous analysis which results in a slightly stable post-buckling behaviour of the bar. The basic physical reason for the approximately neutral behaviour is that the axial load on the bar remains constant along the axis and does not change appreciably in the initial post-buckling range.

A quite different behaviour is displayed by the buckling of flat plates loaded in their plane. Retaining only quadratic terms in the normal deflection w(x,y) in the mid-plane strain components, and no non-linear terms in the changes of curvature of the mid-plane, we have

$$\gamma_{xx} = u_{,x} + \tfrac{1}{2} w_{,x}^2,\quad \gamma_{yy} = v_{,y} + \tfrac{1}{2} w_{,y}^2,\quad \gamma_{xy} = \tfrac{1}{2}(u_{,y} + v_{,x} + w_{,x} w_{,y})\,; \tag{3.1}$$

$$\kappa_{xx} = w_{,xx},\quad \kappa_{yy} = w_{,yy},\quad \kappa_{xy} = w_{,xy}. \tag{3.2}$$

The complete energy functional (2.6) and the second variation (2.7) are now evaluated on the assumption of an approximately plane state of stress in the plate. We obtain

$$P[\underset{\sim}{u}] = \int\Big[\tfrac{1}{2}\,\lambda h(\sigma_x^0 w_{,x}^2+\sigma_y^0 w_{,y}^2+2\tau_{xy}^0 w_{,x}w_{,y}) + $$
$$+ \frac{Eh}{2(1-\nu^2)}\Big\{\gamma_{xx}^2+\gamma_{yy}^2+2\nu\gamma_{xx}\gamma_{yy}+2(1-\nu)\gamma_{xy}^2\Big\} +$$
$$+ \frac{Eh^3}{24(1-\nu^2)}\Big\{w_{,xx}^2+w_{,yy}^2+2\nu w_{,xx}w_{,yy}+2(1-\nu)w_{,xy}^2\Big\}\Big]dA, \tag{3.3}$$

$$P_2[\underset{\sim}{u}] = \int\Big[\tfrac{1}{2}\,\lambda h(\sigma_x^0 w_{,x}^2+\sigma_y^0 w_{,y}^2+2\tau_{xy}^0 w_{,x}w_{,y}) + $$
$$+ \frac{Eh}{2(1-\nu^2)}\Big\{u_{,x}^2+v_{,y}^2+2\nu u_{,x}v_{,y}+\tfrac{1}{2}(1-\nu)(u_{,y}+v_{,x})^2\Big\} +$$
$$+ \frac{Eh^3}{24(1-\nu^2)}\Big\{w_{,xx}^2+w_{,yy}^2+2\nu w_{,xx}w_{,yy}+2(1-\nu)w_{,xy}^2\Big\}\Big]dA, \tag{3.4}$$

where E is YOUNG's modulus, ν is POISSON's ratio and h is the thickness of the plate.

Since the in-plane displacement components u,v only occur in positive definite form in the second line of the second variation (3.4) these displacements will be zero in the buckling mode. Buckling occurs when the sum of the first and third lines in (3.4) has a minimum equal to zero. From dimensional analysis it is obvious that the critical stresses σ_x, σ_y, τ_{xy} are of order Eh^2/b^2 where b is a characteristic dimension in the mid-plane. This result is comparable to the critical stress of a bar in compression. However, the behaviour after buckling is completely different. The second line in the complete energy functional (3.3) is positive definite, and it implies a stable post-buckling behaviour since the sum of the first and third lines is zero in the buckling mode. The deflections grow indeed very slowly when the load is increased beyond its critical value. In the initial post-buckling range they are of order $h(\lambda/\lambda_{cr}-1)^{\frac{1}{2}}$, in contrast with the bar in compression where the deflections in the initial post-buckling range are of order $\ell(\lambda/\lambda_{cr}-1)^{\frac{1}{2}}$. The physical reason for this difference in behaviour is that the second line in (3.3) entails a redistribution of in-plane stresses in the post-buckling range from the original, usually more or less uniform distribution of initial stresses σ_x^0, σ_y^0, τ_{xy}^0.

The most characteristic buckling behaviour of shells occurs in shallow buckling modes whose minimum wave length L is small in comparison with the smallest principal radius of curvature of the middle surface. The simplified non-linear theory of shells for shallow deformation patterns is connected with the names DONNELL, MARGUERRE, MUSHTARI and VLASOV. Formulated in terms of Cartesian coordinates x,y in the directions of the principal curvatures $1/R_1$, $1/R_2$ in the tangent plane, the middle surface strains γ_{xx} and γ_{yy} are now given by

$$\gamma_{xx} = u,_x - \frac{w}{R_1} + \frac{1}{2} w,_x^2 , \quad \gamma_{yy} = v,_y - \frac{w}{R_2} + \frac{1}{2} w,_y^2 , \tag{3.5}$$

whereas the other formulae (3.1) and (3.2) remain valid. The sign convention in (3.5) is in accordance with the convention that a curvature is positive if its center lies on the positive normal. The potential energy functional (3.3) remains valid for shallow buckling modes in a shell in a region whose area is small in comparison with the square of the minimum principal radius of curvature and large in comparison with L^2, provided of course that γ_{xx} and γ_{yy} are evaluated from (3.5). The second variation (3.4) is now replaced by

$$\begin{aligned} P_2[\underset{\sim}{u}] = \int \Bigg[& \frac{1}{2} \lambda h (\sigma_x^0 w,_x^2 + \sigma_y^0 w,_y^2 + 2\tau_{xy}^0 w,_x w,_y) + \\ & + \frac{Eh}{2(1-\nu^2)} \Big\{ (u,_x - \frac{w}{R_1})^2 + (v,_y - \frac{w}{R_2})^2 + \\ & + 2\nu (u,_x - \frac{w}{R_1})(v,_y - \frac{w}{R_2}) + \frac{1}{2}(1-\nu)(u,_y + v,_x)^2 \Big\} + \\ & + \frac{Eh^3}{24(1-\nu^2)} \Big\{ w,_{xx}^2 + w,_{yy}^2 + 2\nu w,_{xx} w,_{yy} + 2(1-\nu) w,_{xy}^2 \Big\} \Bigg] dA. \end{aligned} \tag{3.6}$$

The essential modification in (3.6) is that the first and last lines in terms of the normal deflection w are now coupled with the middle lines which represent the energy due to stretching of the middle surface in the (linear) buckling mode. A characteristic critical stress σ now contains a contribution proportional to EL^2/R^2 in addition to the contribution Eh^2/L^2 reminiscent of plate theory. The minimum value of the critical stress is obtained for $L^2 = O(hR)$, and its value is of order Eh/R. This result is of course subject to the restriction that the stretching energy in (3.6) does not vanish in the buckling mode. In other words, critical stresses of order Eh/R occur in shells which cannot exhibit (infinitesimal) inextensional bending. As we shall see sich high critical stresses occur in particular in shells of positive Gaussian curvature, and also in some cases in cylindrical shells. The wave length L of order $\sqrt{hR}$ then ensures the validity of the basic assumptions of shallow shell theory.

A further consequence of the stretching of the middle surface in the buckling mode is that we are no longer ensured of stability of the equilibrium at the critical load. The nonlinear terms in the strain components of the middle surface may result in a decrease of the stretching energy in the post-buckling range. This situation of an unstable post-buckling behaviour occurs in the most serious form, if the cubic part of the energy functional

$$\begin{aligned} P_3[\underset{\sim}{u}(\underset{\sim}{x})] = \frac{Eh}{2(1-\nu^2)} \iint \Big[& \Big\{ u,_x - \frac{w}{R_1} + \nu (v,_y - \frac{w}{R_2}) \Big\} w,_x^2 + \\ & + \Big\{ v,_y - \frac{w}{R_2} + \nu (u,_x - \frac{w}{R_1}) \Big\} w,_y^2 + \\ & + (1-\nu)(u,_y + v,_x) w,_x w,_y \Big] dA \end{aligned} \qquad (3.7)$$

is non-zero in a buckling mode. The analysis is more complicated when (3.7) happens to be zero in all critical modes. In that case we have to investigate the modification of the deflection pattern in the initial post-buckling range. The result may be a stable rising post-buckling behaviour, as in the case of flat plates, or an unstable steeply descending post-buckling path.

Up to now we have ignored the fact that our discussion of shell buckling in terms of Cartesian coordinates in a tangent plane can only be a valid approximation in a restricted area in the vicinity of the point of contact, of dimensions small in comparison with the minimum principal radius of curvature. This is also the reason why the approximate analysis can only hold for small wave lengths L of the buckling pattern. In the case of a periodic array of buckles over the entire surface of the shell, however, as may occur in cylindrical or spherical shells, the ave r a g e d energy expressions, obtained from (3.3), (3.6) and (3.7) by a division by the area of the shell portion under consideration, remain valid as averages for the entire shell. In the case of more localized buckling modes a suitable modification of the analysis is discussed in the next section.

4. Localized buckling of shells

The equations of neutral equilibrium associated with the second variation (3.6) are written down in the form

$$\left.\begin{aligned}
&u_{,xx}+\tfrac{1}{2}(1-\nu)u_{,yy}+\tfrac{1}{2}(1-\nu)v_{,xy}-\left(\frac{1}{R_1}+\frac{\nu}{R_2}\right)w_{,x} = 0,\\
&\tfrac{1}{2}(1+\nu)u_{,xy}+\tfrac{1}{2}(1-\nu)v_{,xx}+v_{,yy}-\left(\frac{1}{R_2}+\frac{\nu}{R_1}\right)w_{,y} = 0,\\
&\frac{h^2}{12}\Delta\Delta w-\left(\frac{1}{R_1}+\frac{\nu}{R_2}\right)u_{,x}-\left(\frac{1}{R_2}+\frac{\nu}{R_1}\right)v_{,y}+\left(\frac{1}{R_1^2}+\frac{2\nu}{R_1R_2}+\frac{1}{R_2^2}\right)w -\\
&\qquad -\lambda\frac{1-\nu^2}{E}(\sigma_x^0 w_{,xx}+\sigma_y^0 w_{,yy}+2\tau_{xy}^0 w_{,xy}) = 0,
\end{aligned}\right\} \tag{4.1}$$

where $\Delta = \frac{\partial^2}{\partial x^2}+\frac{\partial^2}{\partial y^2}$ is the Laplacian operator. In the derivation of (4.1) free use is made of the approximation that the curvatures $1/R_1$ and $1/R_2$, and the stresses σ_x, σ_y, τ_{xy} in the fundamental state may be dealt with as constants in the case of shallow buckling modes. Eliminating u and v we obtain a single eight-order equation in the normal deflection

$$\begin{aligned}
&\frac{h^2}{12}\Delta^4 w+(1-\nu^2)\left[\frac{1}{R_2^2}w_{,xxxx}+\frac{2}{R_1R_2}w_{,xxyy}+\frac{1}{R_1^2}w_{,yyyy}\right] -\\
&\qquad -\lambda\frac{1-\nu^2}{E}\left[\sigma_x^0\Delta^2 w_{,xx}+\sigma_y^0\Delta^2 w_{,yy}+2\tau_{xy}^0\Delta^2 w_{,xy}\right] = 0.
\end{aligned} \tag{4.2}$$

For a cylindrical shell $1/R_1 = 0$, $R_2 = R$, this equation reduces to DONNELL's well-known stability equation [1].

Equation (4.2) has periodic solutions of the form $\cos px/R_2 \cos py/R_2$, if $\tau_{xy} = 0$ in the fundamental state, provided that the wave numbers p and q satisfy the equation

$$\frac{h^2}{4c^2R^2}(p^2+q^2)^4+ (p^2+\alpha q^2)^2+ \frac{\lambda}{E}(p^2+q^2)^2(p^2\sigma_x^0+q^2\sigma_y^0) = 0, \tag{4.3}$$

where $\alpha = R_2/R_1$ and $c = [3(1-\nu^2)]^{\frac{1}{2}}$. In the case of shear loading in the fundamental state a periodic solution is of the form $\cos(px+qy)/R_2$, if the wave numbers p and q satisfy the equation

$$\frac{h^2}{4c^2R_2^2}(p^2+q^2)^4+(p^2+\alpha q^2)^2+\frac{\lambda}{E}(p^2+q^2)^2 2pq\tau^0_{xy} = 0. \qquad (4.4)$$

In the case of a single periodic buckling mode of the types considered above the cubic functional (3.7) vanishes because the average value of the cube of a sine or cosine is zero. The nonlinear analysis in the case of a simple buckling mode thus requires the evaluation of the modification of the buckling mode in the initial post-buckling range. This analysis has been carried out for many examples, cf. [1,2] for appropriate references. Multiple buckling modes do occur, however, in significant problems, and they lead to a far simpler analysis, if the cubic functional (3.7) does not vanish identically for a linear combination of buckling modes. This situation arises whenever the pairs of wave numbers of three buckling modes p_1, q_1; p_2, q_2; p_3, q_3 satisfy the conditions $p_1 \pm p_2 \pm p_3 = 0$, $q_1 \pm q_2 \pm q_3 = 0$. It is well known that this situation indeed occurs in the case of cylindrical shells under axial compression [3] and spherical shells under external pressure [4].

In the case of cylindrical shells under axial compression we shall restrict our attention, for the sake of brevity, to the case of only two simultaneous buckling modes which may occur if $m = \sqrt{cR/2h}$ is an integer. The linear combination of periodic buckling modes in question is

$$w = h\left[b_0 \cos p_o x/R + c_m \cos mx/R \cos my/R\right], \qquad (4.5)$$

where an outward deflection is considered to be positive and $p_0 = \sqrt{2cR/h} = 2m$. For a more general discussion we refer to [5], and in particular to [3]. Taking $\sigma^0_x = -Eh/cR$, the critical load factor λ is unity, and the energy functional, i.e. the sum of the second variation (3.6) and the cubic functional (3.7), is evaluated at the average value per unit area of the middle surface

$$\frac{Eh^3}{R^2}\left[\frac{1}{2}(1-\lambda)(b_0^2+\frac{1}{8}c_m^2) + \frac{3c}{32}b_0 c_m^2\right] \qquad (4.6)$$

The post-buckling paths are obtained by putting equal to zero the partial derivatives of (4.6) with respect to b_0 and c_m. We obtain

$$b_0 = -2(1-\lambda)/3c, \quad c_m = \pm 8(1-\lambda)/3c, \tag{4.7}$$

and the deflection in the unstable post-buckling paths is given by

$$w = -2(1-\lambda)h/3c\left[\cos p_0 x/R \pm 4 \cos mx/R \cos my/R\right], \tag{4.8}$$

Geometric imperfections in the shape of the periodic post-buckling path are described by

$$w_0 = \kappa h\left[\cos p_0 x/R \pm 4 \cos mx/R \cos my/R\right], \tag{4.9}$$

and they result for $\kappa<0$ in a limit point λ^* given by

$$(1-\lambda^*)^2 = -6c\kappa\lambda^*. \tag{4.10}$$

In the case of a s p h e r i c a l s h e l l under uniform external pressure $\sigma_x^0 = \sigma_y^0 = -p^0R/2h$ the critical stress is again $\sigma_x^0 = \sigma_y^0 = -Eh/cR$, and the critical load factor is again unity. A linear combination of two buckling modes is doubly periodic

$$w = h\left[a_0\cos p_0 x/R + a_1\cos\tfrac{1}{2}p_0 x/R \cos\tfrac{1}{2}p_0\sqrt{3}\, y/R\right]. \tag{4.11}$$

The sum of the second variation (3.6) and the cubic functional (3.7) is again evaluated per unit area, and the result is *)

$$\frac{Eh^3}{R^2}\left[\frac{1}{2}(1-\lambda)(a_0^2+\tfrac{1}{2}a_1^2) - \frac{9c}{32}a_0a_1^2\right]. \tag{4.12}$$

From the equilibrium conditions we obtain

$$a_0 = \frac{8}{9c}(1-\lambda), \quad a_1 = \pm\frac{16}{9c}(1-\lambda), \tag{4.13}$$

---

*) In the spherical shell the deflection is taken to be positive inwards

and the deflection in the unstable post-buckling paths is given by [4]

$$w = \frac{8}{9c}(1-\lambda)h\left[\cos p_0 x/R \pm 2\cos\tfrac{1}{2}p_0 x/R \cos\tfrac{1}{2}\sqrt{3}\, p_0 y/R\right]. \tag{4.14}$$

Geometric imperfections in the shape of the periodic post-buckling paths are here described by

$$w_0 = \kappa h\left[\cos p_0 x/R \pm 2\cos\tfrac{1}{2}p_0 x/R \cos\tfrac{1}{2}\sqrt{3}\, p_0 y/R\right], \tag{4.15}$$

and they result in a reduced limit point λ^* for $\kappa>0$ given by [4]

$$(1-\lambda^*)^2 = \frac{9c}{2}\kappa\lambda^*. \tag{4.16}$$

Up to here our analysis has been confined to p e r i o d i c buckling modes such as described by (4.5) in the case of cylindrical shells and by (4.11) in the case of spherical shells. More or less l o c a l i z e d b u c k l i n g m o d e s are obtained from the periodic modes by a multiplication of all displacement components by a modulating factor

$$E^*(x,y) = \exp\left[-\tfrac{1}{2}\mu^2(x^2+y^2)/R^2\right], \tag{4.17}$$

where $\mu^2 << p_0^2$. In differentiations of a displacement component with respect to x and y such a modulating factor may be treated as a constant factor. The second and third variations of the energy functional in the modulated, more or less localized modes are thus obtained from the similar variations in the periodic modes by a multiplication of their integrands by the square and cube of the modulating factor E^* (4.17) respectively. In view of the decay of the exponential factor the integrations may now be carried out in both x- and y-directions from $-\infty$ to $+\infty$. In both directions the quadratic integrals are all of the type

$$\int_{-\infty}^{\infty} e^{-\mu^2\alpha^2}\cos n\alpha\, d\alpha = \frac{\sqrt{\pi}}{\mu}\, e^{-n^2/4\mu^2}, \tag{4.18}$$

where n is either a large number, of order p_0, or zero. It follows that we need only consider integrals of type (4.18) with $n = 0$, with the value $\sqrt{\pi}/\mu$. The cubic integrals are of the same type with μ^2 replaced by $(3/2)\mu^2$ and their value for $n = 0$ is thus $\sqrt{2\pi}/\mu\sqrt{3}$, a factor $\sqrt{2/3}$ smaller.

The essential part of the energy expressions (4.6) and (4.12) is the expression between the brackets. The factor in front is just a constant dimensional factor. The single modification of the expressions between brackets in the case of modulated modes is that the coefficient of the cubic term is decreased by a factor 2/3 in comparison with the quadratic term. The critical load factor remains unity, and the post-buckling paths are described for the cylindrical shell by

$$b_0 = -(1-\lambda)/c, \quad c_m = \pm 4(1-\lambda)/c, \tag{4.19}$$

and for the spherical shell by

$$a_0 = \frac{4}{3c}(1-\lambda), \quad a_1 = \pm \frac{8}{3c}(1-\lambda). \tag{4.20}$$

More or less localized imperfections are taken into account by the same modulating factor $E^*(x,y)$ applied to the periodic imperfection pattern (4.9) or (4.15). The result is that the coefficients in the right-hand members of (4.10) and (4.16) are reduced by a factor 2/3. The imperfection sensitivity to more or less localized geometric imperfections for cylindrical shells under axial compression is therefore given by the formula [6]

$$(1-\lambda^*)^2 = -4c\kappa\lambda^*. \tag{4.21}$$

The similar formula for the spherical shell under external pressure is

$$(1-\lambda^*)^2 = 3c\kappa\lambda^*. \tag{4.22}$$

The foregoing discussion is, of course, quite approxomate in character, but it is believed to give a good qualitative picture of imperfection sensitivity. One of the most doubtful aspects of the analysis might appear to be the treatment of the modulating factor $E^{*}(x,y)$ defined by (4.17) as a constant factor for the purpose of differentiation of displacement components. The influence of derivatives of $E^{*}(x,y)$ has been investigated by GRISTCHAK [7] in the case of a cylindrical shell under axial compression. For $\mu^2/m^2 = 0.1$, resulting in a reasonable representation of a more or less localized dimple, he has established that our first approximation has an error of no more than about 11 percent.

5. Concluding remarks

We have given preference to a rather extensive discussion of the cases of multiple simultaneous buckling modes with a resulting non-zero cubic functional (3.7) because they are so characteristic for the explosive buckling behaviour of some shells. Moreover, the analysis of the cases with simple buckling modes in which the cubic functional vanishes, at least approximately, is less attractive for lecturing purposes. It seems appropriate, however, to end with a word of caution which refers to many discussions of only the initial range of post-buckling behaviour. The range of validity of this analysis is in some cases extremely small. We have no question about the approximate validity of HUTCHINSON's results for the spherical shell [4], is spite of the somewhat rough approach through the equations of shallow shell theory. The hexagonal buckling pattern described by (4.14) is also confirmed by experiment [9]. Curiously enough, the more exact approach employed in [8] leads to far less satisfactory results. The explanation for this unexpected and a little alarming situation is to be found in the cluster of critical load factors just above the buckling load factor equal to unity. It is the nonlinear interaction between the associated higher modes with the critical modes which is responsible for the failure of the initial stage of post-buckling behaviour to describe the actual behaviour adequately.

References

1. Brush, D.O. and Almroth, B.O.: Buckling of bars, plates and shells. McGraw-Hill, New York (1975).

2. Hutchinson, J.W. and Koiter, W.T.: Postbuckling theory. Appl.

Mech. Revs. 23 (1970), 1353-1366.

3. Koiter, W.T.: On the stability of elastic equilibrium (in Dutch). Thesis Delft University of Technology (1945). Engl. transl. NASA TT F 10,833 (1967) and AFFDL, TR 70-25 (1970).

4. Hutchinson, J.W.: Imperfection-sensitivity of externally pressurized spherical shells. J. Appl. Mech. 34 (1967), 49-55.

5. Koiter, W.T.: General theory of shell stability. In "Thin shell theory, New trends and applications", edited by W. Olszak. CISM courses and lectures Nr.240, Springer Verlag, Wien, New York (1980), 65-87.

6. Koiter, W.T.: The influence of more or less localized imperfections on the buckling of circular cylindrical shells under axial compression. In "Complex analysis and its applications", the I.N. Vekua 70-th Anniversary volume. Akademia Nauk USSR, Moscow (1978), 242-244.

7. Gristchak, V.Z.: Asymptotic formula for the buckling stress of axially compressed circular cylindrical shells with more or less localized short-wave imperfections. Report WTHD-88, Dept. Mech. Engng, Delft University of Technology (1976).

8. Koiter, W.T.: The nonlinear buckling problem of a complete spherical shell under external pressure. Proc. Kon. Ned. Ak. Wet. B72, 40-123 (1969).

9. Carlson, R.L., Sendelbeck, R.L. and Hoff, N.J.: Experimental studies of the buckling of complete spherical shells. Exper. Mech. 7, 281-288 (1967).

Nonlinear Behavior and Elastic Stability of Shells
- Theoretical Concepts - Numerical Computations - Results

W.B. Krätzig, Y. Basar, U. Wittek

Institut für Konstruktiven Ingenieurbau
Ruhr-Universität Bochum
Bochum, West-Germany

Summary

The paper sketches aspects of nonlinear response and instability behavior of general shell structures, loaded time-invariantly. Starting from their basic mechanical formulations suitable finite element algorithms lead to the computation of highly nonlinear phenomena. Unity of theoretical and numerical concepts is emphasized for a successful tackling.

1. Introduction

Thin shell structures are able to sustain heavy loads by virtue of their curvature. On the other hand the curved middle surface produces a large variety of different response phenomena especially in the geometrically nonlinear range. Stretching and bending modes interact in a considerably more capricious manner than in linear problems. Often, the 1. and 2. strain tensor change their participations during a certain loading process, giving rise to surprising response phenomena. The investigation of thin shell structures, particularly with unknown response behavior, thus requires:

- a precise description of their basic kinematic and dynamic equations;
- a discretization, modelling the undeformed and the deformed configuration both with particular care;
- numerical algorithms, stable, convergent and adequately sensible for the intended phenomena.

The price for the inherent stiffness of thin shell structures is their sensitiveness to initial imperfections, which influences any nonlinear response. The ingnorance of shape and magni-

Buckling of Shells, Ed.: E. Ramm
Proceedings of a State-of-the-Art Colloqium

tude of future imperfections at the state of design adds a fourth point to the listed requirements: predictions of the stiffness-loss under estimated imperfection bounds.

By no means, the aim of this paper can be a survey over the whole field of nonlinear shell problems. Its objective rather is to portray the topic from the point of view of three individual authors, emphasizing the equal importance of mechanical, analytical and computational aspects. In addition the paper presents some of the authors' contributions to the wide field of geometrically nonlinear shell problems.

2. Energy-consistent nonlinear theory of elastic shells

2.1 Description of the shell space

Any nonlinear or buckling analysis of thin shells should be based on energy-consistent shell theories. This attribute specifies shell theories, which have been derived under regard of thermodynamic conservation laws [1, 2]. Energy-consistent theories prove the kinematic and the equilibrium operator, e.g. connected in a principle of virtual work, to be adjoint operators [5, 9]. Such theories have been developed in certain levels of approximation [3, 4] within the last two decades; in the following we repeat their basic equations for large displacements and moderate rotations [6, 7, 8, 11].

According to fig. 1 points P of the deformed middle surface F are described by a set of convected curvi-linear coordinates Θ^α $(\alpha = 1,2)$

$$\underset{\sim}{r} = \underset{\sim}{r}(\Theta^1,\Theta^2) = \underset{\sim}{r}(\Theta^\alpha) , \tag{2.1}$$

which define the deformed frame of reference $\underset{\sim}{a}_\alpha$, $\underset{\sim}{a}_3$ through

$$\underset{\sim}{a}_\alpha = \frac{\partial \underset{\sim}{r}}{\partial \Theta^\alpha} = \underset{\sim}{r}_{,\alpha} , \quad \underset{\sim}{a}_3 = \frac{\underset{\sim}{a}_1 \times \underset{\sim}{a}_2}{\sqrt{a}} . \tag{2.2}$$

By virtue of (2.2) metric- and curvature-tensor of F

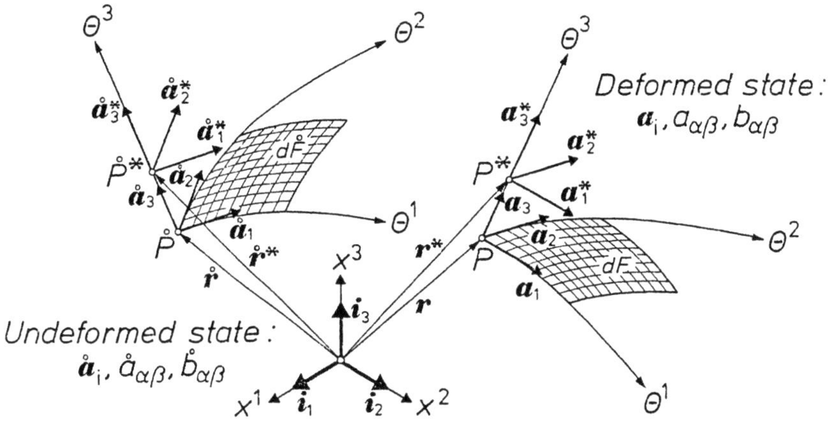

Fig. 1: Middle surface F and shell space in the undeformed and the deformed configuration

$$a_{\alpha\beta} = \underset{\sim}{a}_\alpha \cdot \underset{\sim}{a}_\beta , \qquad b_{\alpha\beta} = -\frac{1}{2}\left(\underset{\sim}{a}_\alpha \cdot \underset{\sim}{a}_{3,\beta} + \underset{\sim}{a}_\beta \cdot \underset{\sim}{a}_{3,\alpha}\right) \tag{2.3}$$

as well as further geometrical functions can be introduced in the usual manner [5, 12]. Christoffelsymbols of the 2. kind

$$\Gamma^\lambda_{\alpha\beta} = \underset{\sim}{a}^\lambda \cdot \underset{\sim}{a}_{\alpha,\beta} = \underset{\sim}{a}^\lambda \cdot \underset{\sim}{a}_{\beta,\alpha} , \qquad \Gamma^\lambda_{\lambda\alpha} = \frac{(\sqrt{a})_{,\alpha}}{\sqrt{a}} \tag{2.4}$$

admit the definition of covariant derivatives $(\ldots)//_\alpha$ with respect to the deformed middle surface, denoted by a double stroke.

The distance between any arbitrary point P^* of the shell space and its orthogonal projection P on F is measured by the straight coordinate Θ^3. As shown in fig. 1 the position vector $\underset{\sim}{r}^*$ of P^* then can be represented by

$$\underset{\sim}{r}^* = \underset{\sim}{r} + \Theta^3 \underset{\sim}{a}_3 , \tag{2.5}$$

which defines the frame of reference of P^* as

$$\underset{\sim}{a}^{*}_{\alpha} = \underset{\sim}{r}^{*}_{,\alpha} = \underset{\sim}{a}_{\alpha} + \Theta^{3}\underset{\sim}{a}_{3,\alpha} = \mu^{\rho}_{\alpha}\,\underset{\sim}{a}_{\rho}\;, \qquad \underset{\sim}{a}^{*}_{3} = \underset{\sim}{r}^{*}_{,3} = \underset{\sim}{a}_{3}\;,$$

$$\mu^{\rho}_{\alpha} = (\delta^{\rho}_{\alpha} - \Theta^{3} b^{\rho}_{\alpha}). \tag{2.6}$$

In the following all further functions related to a point P^{*} outside F will be starred.

Analogous relations will be established for all geometrical elements of the undeformed middle surface F and undeformed shell space: they will be marked by a circle, e.g. $\overset{\circ}{\underset{\sim}{a}}_{\alpha}$, $\overset{\circ}{\underset{\sim}{a}}_{3}$, $\overset{\circ}{\underset{\sim}{a}}{}^{*}_{\alpha}$, $\overset{\circ}{a}_{\alpha\beta}$, $\overset{\circ}{b}_{\alpha\beta}$. Covariant derivatives with respect to F will be denoted by a single stroke $(\ldots)/_{\alpha}$.

2.2 External kinematic variables for large displacements and moderate rotations

The state of deformation of the shell, illustrated in fig. 2, is governed by the Kirchhoff-Love-Hypothesis, under which the undeformed unit normal $\overset{\circ}{\underset{\sim}{a}}_{3}$ deforms into a vector $\underset{\sim}{a}_{3}$ preserving its normality and unit length. Thus the displacement vector $\underset{\sim}{v}^{*}$ of an arbitrary point P^{*} of the shell space can be represented by the linear form

$$\underset{\sim}{v}^{*} = \underset{\sim}{r}^{*} - \overset{\circ}{\underset{\sim}{r}}{}^{*} = (\underset{\sim}{r} - \overset{\circ}{\underset{\sim}{r}}) + \Theta^{3}\,(\underset{\sim}{a}_{3} - \overset{\circ}{\underset{\sim}{a}}_{3}) = \underset{\sim}{v} + \Theta^{3}\underset{\sim}{w}\;, \tag{2.7}$$

which defines the displacement vector $\underset{\sim}{v}$ and the associated difference vector $\underset{\sim}{w}$ as functions of the middle surface:

$$\underset{\sim}{v} = \underset{\sim}{r} - \overset{\circ}{\underset{\sim}{r}} = v_{\alpha}\,\overset{\circ}{\underset{\sim}{a}}{}^{\alpha} + v_{3}\,\overset{\circ}{\underset{\sim}{a}}{}^{3} = v^{\alpha}\overset{\circ}{\underset{\sim}{a}}_{\alpha} + v^{3}\overset{\circ}{\underset{\sim}{a}}_{3}\;,$$

$$\underset{\sim}{w} = \underset{\sim}{a}_{3} - \overset{\circ}{\underset{\sim}{a}}_{3} = w_{\alpha}\,\overset{\circ}{\underset{\sim}{a}}{}^{\alpha} + w_{3}\overset{\circ}{\underset{\sim}{a}}{}^{3} = w^{\alpha}\overset{\circ}{\underset{\sim}{a}}_{\alpha} + w^{3}\overset{\circ}{\underset{\sim}{a}}_{3}\;. \tag{2.8}$$

In all subsequent computations it will turn out to be convenient to decompose both vectors with respect to the undeformed frame of reference.

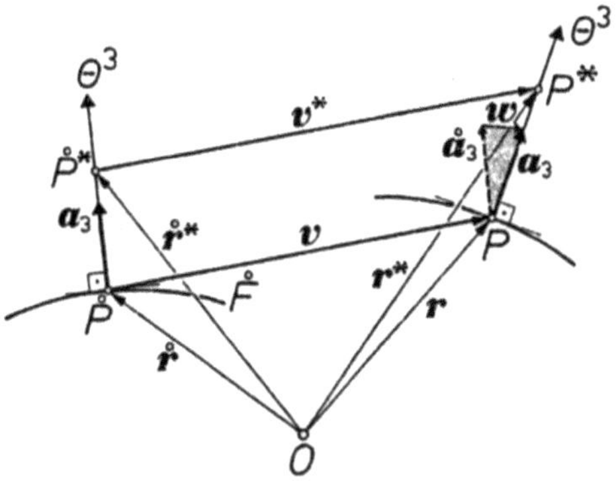

Fig. 2: State of the deformation

The partial derivative of $\underset{\sim}{v}$

$$\underset{\sim}{v}_{,\alpha} = (v_\rho|_\alpha - v_3 \mathring{b}_{\rho\alpha})\,\underset{\sim}{\mathring{a}}^\rho + (v_{3,\alpha} + v_\lambda \mathring{b}^\lambda_\alpha)\,\underset{\sim}{\mathring{a}}_3 = \varphi_{\alpha\rho}\,\underset{\sim}{\mathring{a}}^\rho + \varphi_{\alpha 3}\,\underset{\sim}{\mathring{a}}^3 \tag{2.9}$$

defines the deformation gradients

$$\varphi_{\alpha\rho} = (v_\rho|_\alpha - v_3 \mathring{b}_{\rho\alpha}), \qquad \varphi_{\alpha 3} = v_{3,\alpha} + v_\lambda \mathring{b}^\lambda_\alpha ,$$
$$\varphi_{\alpha\cdot}^{\ \beta} = \varphi_{\alpha\rho}\,\mathring{a}^{\rho\beta}, \qquad \varphi^{\beta}_{\cdot 3} = \varphi_{\alpha 3}\,\mathring{a}^{\alpha\beta} , \tag{2.10}$$

from which the transformation of the undeformed frame of reference into the deformed one can be derived:

$$\underset{\sim}{a}_\alpha = (\delta^\rho_\alpha + \varphi_{\alpha\cdot}^{\ \rho})\,\underset{\sim}{\mathring{a}}_\rho + \varphi_{\alpha 3}\,\underset{\sim}{\mathring{a}}_3 , \quad \underset{\sim}{a}_3 = w_\alpha\,\underset{\sim}{\mathring{a}}^\alpha + (1 + w_3)\,\underset{\sim}{\mathring{a}}_3 . \tag{2.11}$$

Using these transformations the Kirchhoff-Love-Hypothesis

$$\underset{\sim}{a}_\alpha \cdot \underset{\sim}{a}_3 = 0 , \quad \underset{\sim}{a}_3 \cdot \underset{\sim}{a}_3 - \underset{\sim}{\mathring{a}}_3 \cdot \underset{\sim}{\mathring{a}}_3 = 0 \tag{2.12}$$

describes the coupling of $\underset{\sim}{v}$ and $\underset{\sim}{w}$ for arbitrary large displacements:

$$w_\rho(\delta^\rho_\alpha + \varphi_\alpha{}^{\rho}) + \varphi_{\alpha 3}(1 + w_3) = 0 ,$$

$$w_3(2 + w_3) + w_\alpha w^\alpha = 0 , \qquad (2.13)$$

These equations are satisfied up to terms of third order by the approximations [13]:

$$w_\alpha = -\varphi_{\alpha 3} + \varphi_\alpha{}^{\lambda} \varphi_{\lambda 3} , \qquad w_3 = -\frac{1}{2} \varphi_{\alpha 3} \varphi^{\alpha}_{\cdot 3} . \qquad (2.14)$$

Remembering the basic restriction of small strains

$$\{\varphi_{\alpha\rho} , \varphi_{\alpha 3}\} \leq O(\eta) , \qquad 1 + \eta^2 \approx 1 \qquad (2.15)$$

for long-wavy deformations $L_w = O(R)$ [5], in which η denotes the norm of $\varphi_{\alpha\beta}$ or $\gamma_{(\alpha\beta)}$ (2.26), R the smallest radius of curvature of F and L_w the wave-length of v_i, (2.13) can be simplified by:

$$w_\rho(\delta^\rho_\alpha + \varphi_\alpha{}^{\rho}) + \varphi_{\alpha 3} = 0 , \qquad w_3 = -\frac{1}{2} w_\alpha w^\alpha . \qquad (2.16)$$

For geometrically linear theories we find from this naturally

$$w_\alpha = -\varphi_{\alpha 3} , \quad w_3 = 0 . \qquad (2.17)$$

Recent nonlinear shell theories [3, 4, 6, 7] have advantageously introduced a vector of rotation $\underset{\sim}{\omega}$ by

$$\underset{\sim}{w} = \underset{\sim}{\omega} \times \overset{\circ}{\underset{\sim}{a}}_3 + w_3 \overset{\circ}{\underset{\sim}{a}}_3 , \quad \underset{\sim}{\omega} = \overset{\circ}{\underset{\sim}{a}}_3 \times \underset{\sim}{w} = \omega_\alpha \overset{\circ}{\underset{\sim}{a}}{}^\alpha = \omega^\alpha \overset{\circ}{\underset{\sim}{a}}_\alpha , \qquad (2.18)$$

due to fig. 3 orthogonal to $\overset{\circ}{\underset{\sim}{a}}_3$ and $\underset{\sim}{a}_3$. By virtue of (2.8) we find the mutual transformations between $\underset{\sim}{\omega}$ and $\underset{\sim}{w}$:

$$\omega_\alpha = \overset{\circ}{\varepsilon}_{\beta\alpha} w^\beta , \qquad w_\alpha = \overset{\circ}{\varepsilon}_{\alpha\beta} \omega^\beta . \qquad (2.19)$$

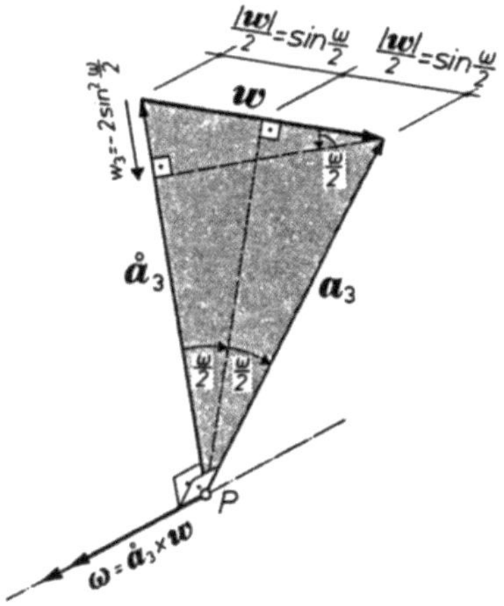

Fig. 3: Vector of rotation $\underset{\sim}{\omega}$ and difference vector $\underset{\sim}{w}$

As can be seen from fig. 3 too the length of $\underset{\sim}{\omega}$

$$|\underset{\sim}{\omega}| = |\underset{\sim}{\mathring{a}}_3 \times \underset{\sim}{w}| = |\underset{\sim}{\mathring{a}}_3||\underset{\sim}{w}| \cos\frac{\omega}{2} = 2\sin\frac{\omega}{2}\cos\frac{\omega}{2} = \sin\omega \qquad (2.20)$$

corresponds to the sine of the angle of rotation:

$$w_3 = -2\sin^2\frac{\omega}{2} = -\frac{1}{2}\underset{\sim}{w}\cdot\underset{\sim}{w} = -\frac{\sin^2\omega}{2\cos^2\frac{\omega}{2}} = -\frac{1}{2\cos^2\frac{\omega}{2}}\underset{\sim}{\omega}\cdot\underset{\sim}{\omega} \; ,$$

$$\underset{\sim}{w} = \underset{\sim}{a}_3 - \underset{\sim}{\mathring{a}}_3 = \underset{\sim}{\omega}\times\underset{\sim}{\mathring{a}}_3 - \frac{1}{2\cos^2\frac{\omega}{2}}(\underset{\sim}{\omega}\cdot\underset{\sim}{\omega})\,\underset{\sim}{\mathring{a}}_3$$

$$= \underset{\sim}{\omega}\times\underset{\sim}{\mathring{a}}_3 + \frac{1}{2\cos^2\frac{\omega}{2}}\underset{\sim}{\omega}\times(\underset{\sim}{\omega}\times\underset{\sim}{\mathring{a}}_3) \; , \qquad (2.21)$$

which clearly demonstrates the properties of $\underset{\sim}{\omega}$ as a vector of rotation. In conclusion we find from (2.15, 2.18) for the order of magnitude of $\underset{\sim}{\omega}$

$$\omega = |\underset{\sim}{\omega}| = \sqrt{\omega_\alpha \omega^\alpha} = \sqrt{w_\alpha w^\alpha} = O(\eta) \tag{2.22}$$

which proves the validity of the derived relations for moderate rotations [7]. Approximating (2.21) with the help of

$$2\cos^2\frac{\omega}{2} = 2 - \frac{1}{2!}\omega^2 + \ldots \approx 2 \tag{2.23}$$

we finally breake down this expression as

$$\underset{\sim}{w} = \underset{\sim}{\omega} \times \underset{\sim}{\mathring{a}}_3 - \frac{1}{2}(\underset{\sim}{\omega}\cdot\underset{\sim}{\omega})\mathring{a}_3 = \underset{\sim}{\omega} \times [\underset{\sim}{\mathring{a}}_3 + \frac{1}{2}(\underset{\sim}{\omega} \times \underset{\sim}{\mathring{a}}_3)]. \tag{2.24}$$

2.3 Kinematic relations

In order to define 2-dimensional strain measures we apply Green's strain tensor to the shell space (i, j = 1, 2, 3):

$$\gamma_{(ij)} = \frac{1}{2}(a^*_{ij} - \mathring{a}^*_{ij}) = \frac{1}{2}(\underset{\sim}{a}^*_i \cdot \underset{\sim}{a}^*_j - \underset{\sim}{\mathring{a}}^*_i \cdot \underset{\sim}{\mathring{a}}^*_j). \tag{2.25}$$

Considering (2.3, 2.6) as well as (2.10, 2.11) $\gamma_{(\alpha\beta)}$ can be approximated as a linear form

$$\gamma_{(\alpha\beta)} = \alpha_{(\alpha\beta)} + \Theta^3 \omega_{(\alpha\beta)}, \tag{2.26}$$

which defines the 1. and 2. strain tensor of the middle surface [5, 8]:

$$\alpha_{(\alpha\beta)} = \frac{1}{2}(a_{\alpha\beta} - \mathring{a}_{\alpha\beta}) = \frac{1}{2}(\varphi_{\alpha\beta} + \varphi_{\beta\alpha} + \varphi_{\alpha\lambda}\varphi_{\beta\cdot}^{\lambda} + \varphi_{\alpha 3}\varphi_{\beta 3}), \tag{2.27}$$

$$\omega_{(\alpha\beta)} = -(b_{\alpha\beta} - \mathring{b}_{\alpha\beta}) = \frac{1}{2}[(w_\alpha|_\beta + w_\beta|_\alpha - \mathring{b}^\lambda_\alpha \varphi_{\beta\lambda} - \mathring{b}^\lambda_\beta \varphi_{\alpha\lambda}) + (\mathring{b}_{\alpha\beta} w_\lambda w^\lambda$$

$$w^\lambda|_\alpha \varphi_{\beta\lambda} + w^\lambda|_\beta \varphi_{\alpha\lambda} + \mathring{b}^\lambda_\alpha w_\lambda \varphi_{\beta 3} + \mathring{b}^\lambda_\beta w_\lambda \varphi_{\alpha 3})]. \tag{2.28}$$

Additionally we have substituted w_3 in $\omega_{(\alpha\beta)}$ according to (2.16) and neglected terms of third order. Using (2.16) $\omega_{(\alpha\beta)}$ also can be expressed in terms of the displacement gradients:

$$\omega_{(\alpha\beta)} = -\frac{1}{2}[\varphi_{\alpha 3}|_{\beta} + \varphi_{\beta 3}|_{\alpha} + \overset{\circ}{b}{}^{\lambda}_{\beta}\varphi_{\alpha\lambda} + \overset{\circ}{b}{}^{\lambda}_{\alpha}\varphi_{\beta\lambda} - \varphi^{\lambda}_{\cdot 3}(\varphi_{\alpha\lambda}|_{\beta} + \varphi_{\beta\lambda}|_{\alpha}$$

$$+ \overset{\circ}{b}_{\alpha\beta}\varphi_{\lambda 3} - \overset{\circ}{b}_{\lambda\beta}\varphi_{\alpha 3} - \overset{\circ}{b}_{\lambda\alpha}\varphi_{\beta 3})] . \tag{2.29}$$

With those previously defined functions further geometrical elements of the deformed middle surface F can be retransformed into the fundamental state $\overset{\circ}{F}$ neglecting quadratic deformations. We find for

$$a^{\alpha\beta} \approx \overset{\circ}{a}{}^{\alpha\beta} - 2\alpha^{(\alpha\beta)} \approx \overset{\circ}{a}{}^{\alpha\beta} - \varphi^{\alpha\beta} - \varphi^{\beta\alpha} ,$$

$$\sqrt{\frac{a}{\overset{\circ}{a}}} \approx 1 + \alpha^{\lambda}_{\lambda} \approx 1 + \varphi_{\lambda}{}^{\lambda}_{\cdot} ,$$

$$b^{\beta}_{\alpha} = b_{\alpha\rho}\, a^{\rho\beta} = \overset{\circ}{b}{}^{\beta}_{\alpha} - L^{\beta}_{\alpha} , \tag{2.30}$$

$$\sqrt{\frac{a}{\overset{\circ}{a}}}\, L^{\lambda}_{\lambda\alpha} = \left(\sqrt{\frac{a}{\overset{\circ}{a}}}\right)_{,\alpha} = \left(\sqrt{\frac{a}{\overset{\circ}{a}}}\right)\Big|_{\alpha} = \left(\sqrt{\frac{a}{\overset{\circ}{a}}}\right)\Big\|_{\alpha}$$

with the difference-curvature-tensor L^{α}_{β} and the difference-Christoffel-symbol $L^{\alpha}_{\beta\gamma}$ [11]:

$$L^{\beta}_{\alpha} \approx w^{\beta}|_{\alpha} + \overset{\circ}{b}{}^{\lambda}_{\alpha}\varphi_{\lambda}{}^{\beta}_{\cdot} \approx -\varphi^{\beta}_{\cdot 3}|_{\alpha} + \overset{\circ}{b}{}^{\lambda}_{\alpha}\varphi_{\lambda}{}^{\beta}_{\cdot} ,$$

$$L^{\alpha}_{\beta\gamma} = \Gamma^{\alpha}_{\beta\gamma} - \overset{\circ}{\Gamma}{}^{\alpha}_{\beta\gamma} \approx \varphi_{\beta}{}^{\alpha}_{\cdot}|_{\gamma} - \varphi_{\beta 3}\,\overset{\circ}{b}{}^{\alpha}_{\gamma} + \varphi^{\alpha}_{\cdot 3}\,\overset{\circ}{b}_{\beta\gamma} . \tag{2.31}$$

2.4 Nonlinear conditions of equilibrium

In fig. 4 we consider an element of the deformed middle surface F under external loads. The load vector $\underset{\sim}{p}$ is related to surface elements dF, the physical components of the stress-resultant-vectors $\underset{\sim}{n}^{\alpha}$ and the moment-vector $\underset{\sim}{m}^{\alpha}$ to line elements ds_{α} of F [5, 12]:

$$\begin{aligned} \underset{\sim}{n}^{\alpha} \sqrt{a}\, d\theta^{\beta} &= \underset{\sim}{n}^{\langle\alpha\rangle} \sqrt{a_{\beta\beta}}\, d\theta^{\beta} , \\ \underset{\sim}{m}^{\alpha} \sqrt{a}\, d\theta^{\beta} &= \underset{\sim}{m}^{\langle\alpha\rangle} \sqrt{a_{\beta\beta}}\, d\theta^{\beta} . \end{aligned} \tag{2.32}$$

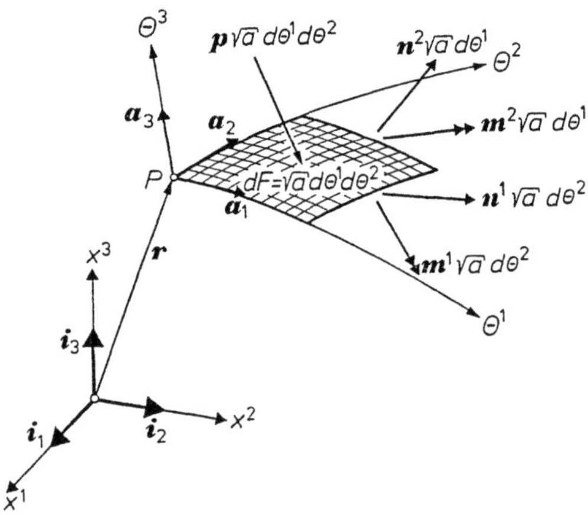

Fig. 4: Element of the deformed middle surface with internal and external force vectors

We start with the well-known equations of equilibrium, formulated in the deformed configuration:

$$\underset{\sim}{n}^{\alpha}\big\|_{\alpha} + \underset{\sim}{p} = 0 , \quad \underset{\sim}{m}^{\alpha}\big\|_{\alpha} + \underset{\sim}{a}_{\alpha} \times \underset{\sim}{n}^{\alpha} = 0 . \tag{2.33}$$

Transformation of their covariant derivatives to the undeformed middle surface $\overset{\circ}{F}$ by virtue of (2.30, 2.31) leads to

$$\left(\sqrt{\frac{a}{\overset{\circ}{a}}}\, \underset{\sim}{n}^{\alpha}\right)\bigg|_{\alpha} + \sqrt{\frac{a}{\overset{\circ}{a}}}\, \underset{\sim}{p} = 0 , \quad \left(\sqrt{\frac{a}{\overset{\circ}{a}}}\, \underset{\sim}{m}^{\alpha}\right)\bigg|_{\alpha} + \underset{\sim}{a}_{\alpha} \times \sqrt{\frac{a}{\overset{\circ}{a}}}\, \underset{\sim}{n}^{\alpha} = 0 . \tag{2.34}$$

Multiplication of $\underset{\sim}{p}$ with the scalar $\sqrt{a/\mathring{a}} = \sqrt{F/\mathring{F}}$ relates this vector to the undeformed middle surface; its decomposition shall be carried out in two different ways:

$$\sqrt{\frac{a}{\mathring{a}}}\,\underset{\sim}{p} = \mathring{p}^{\alpha}\mathring{\underset{\sim}{a}}_{\alpha} + \mathring{p}^{3}\mathring{\underset{\sim}{a}}_{3} = p^{\alpha}\underset{\sim}{a}_{\alpha} + p^{3}\underset{\sim}{a}_{3} \tag{2.35}$$

connected by

$$\mathring{p}^{\alpha} = p^{\rho}(\delta^{\alpha}_{\rho} + \varphi_{\rho}{}^{\alpha}_{\cdot}) + p^{3}w^{\alpha}, \quad \mathring{p}^{3} = p^{\alpha}\varphi_{\alpha 3} + p^{3}(1 + w_{3}). \tag{2.36}$$

For all conservative loads $\mathring{p}^{i}$ advantageously remains independent of the deformation pattern.

The internal stress-vectors in (2.34) shall be decomposed into Piola-Kirchhoff-quantities of the 2. kind:

$$\sqrt{\frac{a}{\mathring{a}}}\,\underset{\sim}{n}^{\alpha} = n^{\alpha\beta}\underset{\sim}{a}_{\beta} + q^{\alpha}\underset{\sim}{a}_{3}\,, \qquad \sqrt{\frac{a}{\mathring{a}}}\,\underset{\sim}{m}^{\alpha} = m^{\alpha\rho}\,\underset{\sim}{a}_{3}\times\underset{\sim}{a}_{\rho} = m^{\alpha\rho}\varepsilon_{\rho\beta}\underset{\sim}{a}^{\beta}. \tag{2.37}$$

Their use is of great advantage in all further computations because of their easy transformation - by multiplication with $\sqrt{\mathring{a}/a}$ - into Cauchy-stresses, in which restrictions of the shell strength are formulated in a natural manner. Introducing now (2.11, 2.14) and (2.35, 2.37) into (2.34) we obtain after neglect of quadratic deformation terms, the following nonlinear equations of equilibrium:

$$N^{\alpha\beta}|_{\alpha} - Q^{\alpha}\mathring{b}^{\beta}_{\alpha} + \mathring{p}^{\beta} = 0,$$

$$N^{\alpha\beta}\mathring{b}_{\alpha\beta} + Q^{\alpha}|_{\alpha} + \mathring{p}^{3} = 0, \tag{2.38}$$

$$M^{\alpha\beta}|_{\alpha} - q^{\alpha}(\delta^{\beta}_{\alpha} + \varphi_{\alpha}{}^{\beta}_{\cdot}) - \mathring{b}_{\alpha\rho}(m^{(\alpha\beta)}\varphi^{\rho}_{\cdot 3} - m^{(\alpha\rho)}\varphi^{\beta}_{\cdot 3}) = 0.$$

Herein the subsequent nonlinear stress components have been used

$$N^{\alpha\beta} = n^{\alpha\rho}(\delta^{\beta}_{\rho} + \varphi_{\rho}{}^{\beta}_{\cdot}) + q^{\alpha}w^{\beta},$$

$$\approx \tilde{n}^{(\alpha\rho)}(\delta^{\beta}_{\rho} + \varphi_{\rho}{}^{\beta}_{\cdot}) - m^{(\alpha\rho)}(\mathring{b}^{\beta}_{\rho} - w^{\beta}|_{\rho}) + q^{\alpha}w^{\beta},$$

$$\approx \tilde{n}^{(\alpha\rho)}(\delta^{\beta}_{\rho} + \varphi_{\rho}{}^{\beta}_{\cdot}) - m^{(\alpha\rho)}(\mathring{b}^{\beta}_{\rho} + \varphi^{\beta}_{\cdot 3}) + q^{\alpha}\varphi^{\beta}_{\cdot 3},$$

$$Q^{\alpha} = q^{\alpha} + n^{\alpha\beta}\varphi_{\beta 3} \approx q^{\alpha} + \tilde{n}^{(\alpha\rho)}\varphi_{\rho 3} - \mathring{b}^{\lambda}_{\rho} m^{(\alpha\rho)}\varphi_{\lambda 3} \quad ,$$

$$M^{\alpha\beta} = m^{(\alpha\rho)}(\delta^{\beta}_{\rho} + \varphi_{\rho}{}^{\beta}_{\cdot}) \ . \qquad (2.39)$$

From the equation of moment equilibrium around the a_3-axis the symmetric pseudo-stress-tensor $\tilde{n}^{(\alpha\beta)}$ has been defined in the usual way:

$$\tilde{n}^{(\alpha\beta)} = n^{\alpha\beta} + m^{(\alpha\rho)} b^{\beta}_{\rho} \approx n^{\alpha\beta} + m^{(\alpha\rho)}(\mathring{b}^{\beta}_{\rho} - w^{\beta}|_{\rho} - \mathring{b}^{\lambda}_{\rho}\ \varphi_{\lambda}{}^{\beta}_{\cdot}) \ . \qquad (2.40)$$

2.5 Constitutive equations

Hyperelastic shells are characterized by the existence of a density function $\Pi_i(\Theta^{\alpha})$ of deformation energy, from which general constitutive equations under the Kirchhoff-Love-Hypothesis can be derived as [1, 2]:

$$\tilde{n}^{(\alpha\beta)} = \frac{1}{2}\left(\frac{\partial\pi_i}{\partial\alpha_{(\alpha\beta)}} + \frac{\partial\pi_i}{\partial\alpha_{(\beta\alpha)}}\right), \quad m^{(\alpha\beta)} = \frac{1}{2}\left(\frac{\partial\pi_i}{\partial\omega_{(\alpha\beta)}} + \frac{\partial\pi_i}{\partial\omega_{(\beta\alpha)}}\right) \ . \qquad (2.41)$$

Their special form for Hookean and isotropic material (under small strains, but large displacements) reads

$$\tilde{n}^{(\alpha\beta)} = D H^{\alpha\beta\rho\lambda}\alpha_{(\rho\lambda)} \ , \qquad m^{(\alpha\beta)} = B H^{\alpha\beta\rho\lambda}\omega_{(\rho\lambda)} \ , \qquad (2.42)$$

abbreviating the stretching-stiffness, the bending-stiffness and the tensor of elasticity as:

$$D = \frac{Eh}{1-\nu^2} \ , \quad B = \frac{Eh^3}{12(1-\nu^2)} \ .$$

$$H^{\alpha\beta\rho\lambda} = \frac{1-\nu}{2}(\mathring{a}^{\alpha\lambda}\mathring{a}^{\beta\rho} + \mathring{a}^{\alpha\rho}\mathring{a}^{\beta\lambda} + \frac{2\nu}{1-\nu}\mathring{a}^{\alpha\beta}\mathring{a}^{\lambda\rho}) \ , \qquad (2.43)$$

2.6 Boundary variables and boundary conditions

Any deformed middle surface F under consideration will be bounded by a closed and continuous curve C, along which the force-vector $\underset{\sim}{n}$ and the moment-vector $\underset{\sim}{m}$, both related to the unit length of C, act as force-variables. $\underset{\sim}{v}$ (2.8) and $\underset{\sim}{\omega}$ (2.18) are defined as corresponding displacement-variables. For $\mathring{C}$, bounding the undeformed middle surface $\mathring{F}$, we evaluate the unit frame of reference $(\mathring{\underset{\sim}{u}}, \mathring{\underset{\sim}{t}}, \mathring{\underset{\sim}{a}}_3)$

$$\mathring{\underset{\sim}{u}} = \mathring{\underset{\sim}{t}} \times \mathring{\underset{\sim}{a}}_3 = \mathring{u}_\alpha \mathring{\underset{\sim}{a}}^\alpha , \quad \mathring{\underset{\sim}{t}} = \frac{d\mathring{\underset{\sim}{r}}}{d\mathring{s}} = \mathring{t}_\alpha \mathring{\underset{\sim}{a}}^\alpha , \quad \mathring{\underset{\sim}{a}}_3 = \mathring{\underset{\sim}{a}}^3 ,$$

$$\mathring{u}_\beta = \mathring{\varepsilon}_{\beta\alpha} \frac{d\Theta^\alpha}{d\mathring{s}} = \mathring{\varepsilon}_{\beta\alpha} \mathring{t}^\alpha , \quad \mathring{t}^\alpha = \frac{d\Theta^\alpha}{d\mathring{s}} = \mathring{\varepsilon}^{\beta\alpha} \mathring{u}_\beta . \tag{2.44}$$

in which $\mathring{s}$ represents the undeformed length of $\mathring{C}$.

We now relate $\underset{\sim}{n}$, $\underset{\sim}{m}$ to $\mathring{s}$ and decompose both with respect to $(\mathring{\underset{\sim}{u}}, \mathring{\underset{\sim}{t}}, \mathring{\underset{\sim}{a}}_3)$:

$$\frac{ds}{d\mathring{s}} \underset{\sim}{n} = n_t \mathring{\underset{\sim}{t}} + n_u \mathring{\underset{\sim}{u}} + n_3 \mathring{\underset{\sim}{a}}_3 , \quad \frac{ds}{d\mathring{s}} \underset{\sim}{m} = m_t \mathring{\underset{\sim}{t}} + m_u \mathring{\underset{\sim}{u}} + m_3 \mathring{\underset{\sim}{a}}_3 . \tag{2.45}$$

Observing the coupling to (2.32) [5]

$$\underset{\sim}{n} = \underset{\sim}{n}^\alpha u_\alpha , \quad \underset{\sim}{m} = \underset{\sim}{m}^\alpha u_\alpha \tag{2.46}$$

as well as (2.11), (2.37) and the abbreviations (2.39), we obtain for the components:

$$n_t = N^{\alpha\beta} \mathring{u}_\alpha \mathring{t}_\beta , \quad n_u = N^{\alpha\beta} \mathring{u}_\alpha \mathring{u}_\beta , \quad n_3 = Q^\alpha \mathring{u}_\alpha ,$$

$$m_t = M^{\alpha\beta} \mathring{u}_\alpha \mathring{u}_\beta , \quad m_u = -M^{\alpha\beta} \mathring{u}_\alpha \mathring{t}_\beta , \quad m_3 = m^{\alpha\beta} w^\rho \varepsilon_{\rho\beta} \mathring{u}_\alpha = m^{\alpha\beta} \omega_\beta \mathring{u}_\alpha . \tag{2.47}$$

From an analogous decomposition of $\underset{\sim}{v}$ and $\underset{\sim}{\omega}$

$$\underset{\sim}{v} = v_t \mathring{\underset{\sim}{t}} + v_u \mathring{\underset{\sim}{u}} + v_3 \mathring{\underset{\sim}{a}}_3 , \quad \underset{\sim}{\omega} = \omega_t \mathring{\underset{\sim}{t}} + \omega_u \mathring{\underset{\sim}{u}} \tag{2.48}$$

dual boundary deformations will be defind:

$$v_t = v_\alpha \overset{\circ}{t}{}^\alpha , \qquad v_u = v_\alpha \overset{\circ}{u}{}^\alpha , \qquad v_3 = v^3 ,$$
$$\omega_t = \omega^\alpha \overset{\circ}{t}_\alpha = w_\alpha \overset{\circ}{u}{}^\alpha , \quad \omega_u = \omega^\alpha \overset{\circ}{u}_\alpha = - w_\alpha \overset{\circ}{t}{}^\alpha . \tag{2.49}$$

It should be emphasized, that m_3 of course vanishes in a decomposition with respect to the deformed frame ($\underset{\sim}{u}$, $\underset{\sim}{t}$, $\underset{\sim}{a}_3$) putting ω_3 into the role of an immaterial variable.

In order to enable the prescription of boundary quantities we formulate the virtual work expression of the stresses along the boundary during a virtual displacement $\delta\underset{\sim}{v}^* = \delta\underset{\sim}{v} + \Theta^3\delta\underset{\sim}{w}$. Eliminating the dependant rotational component ω_u we find for the slightly changed Kirchhoff-Love-type boundary conditions of nonlinear shells:

$$\tilde{n}_t = N^{\alpha\beta}\overset{\circ}{u}_\alpha \overset{\circ}{t}_\beta + \overset{\circ}{b}{}^\lambda_\beta \overset{\circ}{t}{}^\beta \overset{\circ}{t}_\lambda m_u , \qquad v_t = v_\alpha \overset{\circ}{t}{}^\alpha ,$$
$$\tilde{n}_u = N^{\alpha\beta}\overset{\circ}{u}_\alpha \overset{\circ}{u}_\beta + \overset{\circ}{b}{}^\lambda_\beta \overset{\circ}{t}{}^\beta \overset{\circ}{u}_\lambda m_u , \qquad v_u = v_\alpha \overset{\circ}{u}{}^\alpha ,$$
$$\tilde{n}_3 = Q^\alpha \overset{\circ}{u}_\alpha - \frac{\partial m_u}{\partial \overset{\circ}{s}} , \qquad v_3 = v^3 , \tag{2.50}$$
$$m_t = M^{\alpha\beta}\overset{\circ}{u}_\alpha \overset{\circ}{u}_\beta , \qquad \omega_t = \omega_\alpha \overset{\circ}{u}{}^\alpha = - w_\beta \overset{\circ}{t}{}^\beta .$$

2.7 Principle of virtual displacements

The energy-consistency of our nonlinear shell theory can be proved by formulating a principle of virtual displacements from all previously derived formulae and receiving an identical expression for the kinematic field (2.7), starting from a 3-dimensional context. This has been carried out in detail in [5, 8], from which we summarize the results.

Let $\overset{\circ}{C} = \overset{\circ}{C}_r + \overset{\circ}{C}_t$ again be the closed and smooth shell boundary curve, along which the following boundary conditions have been presribed:

$$\underset{\sim}{v} = \overset{\circ}{\underset{\sim}{v}}, \underset{\sim}{\omega}_t = \overset{\circ}{\underset{\sim}{\omega}}_t \text{ on } \overset{\circ}{C}_r \; ; \; \underset{\sim}{n} = \overset{\circ}{\underset{\sim}{n}}, \quad \underset{\sim}{m}_t = \overset{\circ}{m}_t \text{ on } \overset{\circ}{C}_t . \tag{2.51}$$

The principle of virtual displacements can be cast into the simple form:

$$\delta^* A = \delta^* A_a + \delta^* A_i = \iint_{\overset{\circ}{F}} (\overset{\circ}{p}{}^{\alpha} \delta v_\alpha + \overset{\circ}{p}{}^3 \delta v_3) d\overset{\circ}{F} + \int_{\overset{\circ}{C}_t} (\tilde{n}^\circ_t \delta v_t + \tilde{n}^\circ_u \delta v_u +$$
$$+ \tilde{n}^\circ_3 \delta v_3 + m^\circ_t \delta \omega_t) d\overset{\circ}{s} - \iint_{\overset{\circ}{F}} (\tilde{n}^{(\alpha\beta)} \delta \alpha_{(\alpha\beta)} + m^{(\alpha\beta)} \delta \omega_{(\alpha\beta)}) d\overset{\circ}{F} = 0 , \tag{2.52}$$

in which $\delta \underset{\sim}{v}$ is coupled with (2.27, 2.28) and $\delta\omega_t$ restricted to the nonlinear Kirchhoff-Love-Hypothesis (2.14) due to (2.49). In order to release this latter restriction (2.52) can be transformed into:

$$\delta^* \tilde{A} = \delta^* \tilde{A}_i + \delta \tilde{A}_a$$
$$= \iint_{\overset{\circ}{F}} (\overset{\circ}{p}{}^{\alpha} \delta v_\alpha + \overset{\circ}{p}{}^3 \delta v_3) d\overset{\circ}{F} + \int_{\overset{\circ}{C}_t} (n^\circ_t \delta v_t + n^\circ_u \delta v_u + n^\circ_3 \delta v_3 + m^\circ_t \delta \omega_t + m^\circ_u \delta \omega_u) d\overset{\circ}{s}$$
$$- \iint_{\overset{\circ}{F}} (\tilde{n}^{(\alpha\beta)} \delta \alpha_{(\alpha\beta)} + m^{(\alpha\beta)} \delta \omega_{(\alpha\beta)}) d\overset{\circ}{F} - \iint_{\overset{\circ}{F}} q^\alpha [(\delta^\rho_\alpha + \varphi_\alpha{}^{\rho}{}_{\cdot}) \delta w_\rho +$$
$$+ (\delta v^\rho|_\alpha - \delta v_3 \overset{\circ}{b}{}^\rho_\alpha) w_\rho + (\delta v_{3,\alpha} + \overset{\circ}{b}{}^\lambda_\alpha \delta v_\lambda)] d\overset{\circ}{F} - \iint_{\overset{\circ}{F}} \delta q^\alpha [(\delta^\rho_\alpha + \varphi_\alpha{}^{\rho}{}_{\cdot}) w_\rho +$$
$$+ \varphi_{\alpha 3}] d\overset{\circ}{F} = 0 . \tag{2.53}$$

In this principle v_i, w_α and q^α can be varied independently, because the orthogonality condition (2.14) is contained in the last surface integrals. It can be shown by standard methods [5, 8], that the Euler-equations of both principles correspond to the conditions of equilibrium (2.38) and to the natural boundary conditions (2.51).

2.8 Review of variables and operators

Kinematic (2.27, 2.28) and dynamic (2.38, 2.39) field equations, constitutive relations (2.42) and boundary conditions (2.50, 2.51) form a complete set of energy-consistent nonlinear equations for hyperelastic shells. All external variables of this set are related to the undeformed middle surface F and decomposed with respect to $\mathring{\underset{\sim}{a}}_i$, $\mathring{\underset{\sim}{a}}^i$. The internal force variables are related to line elements of F and decomposed with respect to $\underset{\sim}{a}_i$, additionally multiplied by the scalar $\sqrt{a/\mathring{a}}$.

In order to discover essential properties of the joined operators we now group external and internal variables due to fig. 5.

External variables:

$$\mathring{\boldsymbol{p}} = \begin{bmatrix} \mathring{\boldsymbol{p}}^{\alpha} \\ \hline \mathring{\boldsymbol{p}}^{3} \end{bmatrix} = \begin{bmatrix} \mathring{p}^{1} \\ \mathring{p}^{2} \\ \hline \mathring{p}^{3} \end{bmatrix} \qquad \boldsymbol{u} = \begin{bmatrix} \boldsymbol{v}_{\alpha} \\ \hline \boldsymbol{v}_{3} \end{bmatrix} = \begin{bmatrix} v_1 \\ v_2 \\ \hline v_3 \end{bmatrix}$$

Internal variables:

$$\boldsymbol{\sigma} = \begin{bmatrix} \tilde{\boldsymbol{n}}^{(\alpha\beta)} \\ \hline \boldsymbol{m}^{(\alpha\beta)} \end{bmatrix} \qquad \boldsymbol{\epsilon} = \begin{bmatrix} \boldsymbol{\alpha}_{(\alpha\beta)} \\ \hline \boldsymbol{\omega}_{(\alpha\beta)} \end{bmatrix}$$

force variables *deformation variables*

Fig. 5: Grouping of variables

Using its abbreviations the different relations can be broken down as follows [5, 11].

Equations of equilibrium (2.38, 2.39):

$$-\underline{\mathring{p}} = \underline{D}_e\, \underline{\sigma} = (\underline{D}_{eL} + \underline{D}_{eN}(\underline{u}))\cdot\underline{\sigma} = \left(\begin{bmatrix} \underline{D}_{eLD} \\ \hline \underline{D}_{eLB} \end{bmatrix} + \begin{bmatrix} \underline{D}_{eND}(\underline{u}) \\ \hline \underline{D}_{eNB}(\underline{u}) \end{bmatrix} \right)\cdot\underline{\sigma} \tag{2.54}$$

with $\underline{D}_{el}$ linear differential operator;

$\underline{D}_{eN}(\underline{u})$ linear differential operator, beeing a linear functional of $\underline{u}$ and its derivatives.

Kinematic relations (2.27, 2.28):

$$\underline{\varepsilon} = \underline{D}_k\, \underline{u} = (\underline{D}_{kL} + \underline{D}_{kN}(\underline{u})\cdot\underline{u}) = \left(\begin{bmatrix} \underline{D}_{kLD} \\ \hline \underline{D}_{kLB} \end{bmatrix} + \begin{bmatrix} \underline{D}_{kND}(\underline{u}) \\ \hline \underline{D}_{kNB}(\underline{u}) \end{bmatrix} \right)\cdot\underline{u} \tag{2.55}$$

with $\underline{D}_{kl}$ linear differential operator, adjoint to $\underline{D}_{el}$;
$\underline{D}_{kN}(\underline{u})$ nonlinear differential operator, beeing a linear functional of $\underline{u}$ and its derivatives.

$$\underline{D}_{kN}(\underline{u})\cdot\underline{u} = \underline{u}^T\,\underline{D}^*_{kN}\,\underline{u} \tag{2.56}$$

$\underline{D}_e$ and $\underline{D}_k$ are mutually adjoint operators.

Constitutive relations (2.42):

$$\underline{\sigma} = \underline{D}\,\underline{E}\,\underline{\varepsilon} = D\left[\begin{array}{c|c} \underline{E}_{11} & \underline{0} \\ \hline \underline{0} & \frac{h^2}{12}\underline{E}_{11} \end{array}\right]\cdot\underline{\varepsilon} \tag{2.57}$$

with $\underline{E}$, $\underline{E}_{11}$ purely algebraic operators.

Displacement boundary conditions (2.50, 2.51):

$$\underline{r}^\circ = \underline{R}_r\,\underline{u} \tag{2.58}$$

with $\underline{R}_r$ purely algebraic operator.

Force boundary conditions (2.50, 2.51):

$$\underline{t}^\circ = (\underline{R}_{tL} + \underline{R}_{tN}(\underline{u}))\cdot\underline{\sigma} \tag{2.59}$$

with $\underline{R}_{tL}$ linear differential operator;
$\underline{R}_{tN}(\underline{u})$ linear differential operator, beeing linear in $\underline{u}$ and its derivatives.

A survey over all operator equations is offerend by fig. 6, where most operators are seperated into their inplane (D)- and bending (B)-contributions.

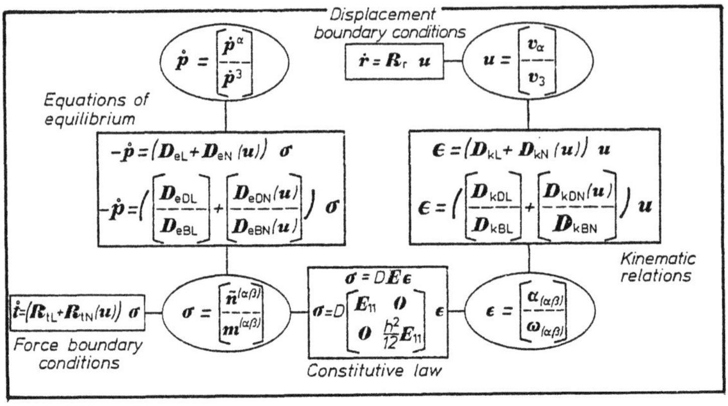

Fig. 6: Structure of nonlinear shell theory

3. Computational concepts and algorithms

3.1 Incrementation of the potential energy

In the last decade several numerical tools have been developed to tackle successfully nonlinear response phenomena of rather general kind. All these tools are based on incremental procedures; their basic equations shall be derived subsequently for displacement finite element algorithms.

We start with the total potential energy (see section 2.7) for geometrically nonlinear shells

$$\Pi = \Pi_i + \Pi_a = \frac{1}{2} \iint_{\mathring{F}} \left(D H^{\alpha\beta\lambda\mu} \alpha_{(\alpha\beta)} \alpha_{(\lambda\mu)} + B H^{\alpha\beta\lambda\mu} \omega_{(\alpha\beta)} \omega_{(\lambda\mu)} \right) d\mathring{F}$$

$$- \iint_{\mathring{F}} \left(\mathring{p}^{\alpha} v_{\alpha} + \mathring{p}^{3} \delta v_3 \right) d\mathring{F}$$

$$- \int_{\mathring{C}_t} \left(\mathring{\tilde{n}}_t v_t + \mathring{\tilde{n}}_u v_u + \mathring{n}_3 v_3 + \mathring{m}_t \omega_t \right) d\mathring{s}$$

$$= \frac{1}{2} \iint_{\overset{\circ}{F}} \underline{\varepsilon}^T D \underline{E} \underline{\varepsilon} \, d\overset{\circ}{F} - \iint_{\overset{\circ}{F}} \underline{u}^T \overset{\circ}{\underline{p}} \, d\overset{\circ}{F} - \int_{\overset{\circ}{C}_t} \underline{r}^T \underline{t}^{\circ} \, d\overset{\circ}{s} \, . \tag{3.1}$$

in which variables and operators, as defined in section 2.8, have been used as abbreviations. In order to increment the total potential energy we introduce three different states of deformation:

- the initial state $(\overset{\circ}{\ldots})$, representing the unloaded, stress-free and undeformed configuration of the ideal shell,
- the fundamental state $(\bar{\ldots})$ as a linear or nonlinear state of equilibrium,
- the adjacent state $(\overset{+}{\ldots})$, incrementally neighboured to the fundamental state and reached from it by a first variation of the displacement field.

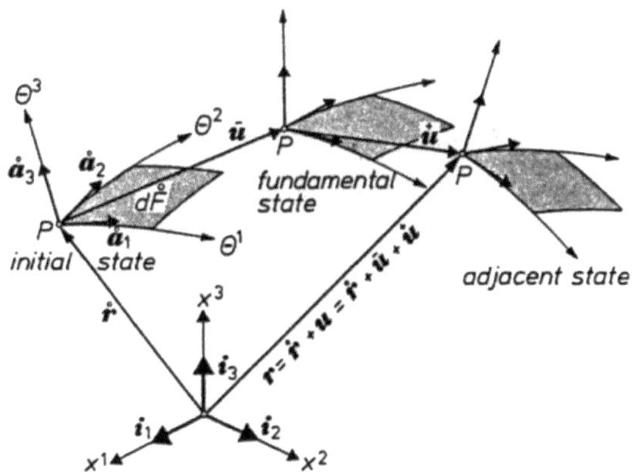

Fig. 7: Different states of deformation

Due to fig. 7 any displacement field shall be decomposed according to

$$\underline{u} = \bar{\underline{u}} + \overset{+}{\underline{u}} \, , \tag{3.2}$$

where $\bar{\underline{u}}$ describes an arbitrary large displacement (see fig. 5) and $\overset{+}{\underline{u}}$ its first variation. From (2.55) we consequently find

$$\underline{\varepsilon} = \underline{D}_k\,\underline{u} = (\underline{D}_{kL} + \underline{D}_{kN}(\bar{\underline{u}} + \overset{+}{\underline{u}}))\,(\bar{\underline{u}} + \overset{+}{\underline{u}})$$

$$= \underbrace{\underline{D}_{kL}\bar{\underline{u}}} + \underbrace{\underline{D}_{kN}(\bar{\underline{u}})\bar{\underline{u}}} + \underbrace{\underline{D}_{kL}\overset{+}{\underline{u}}} + \underbrace{\underline{D}_{kN}(\bar{\underline{u}})\overset{+}{\underline{u}} + \underline{D}_{kN}(\overset{+}{\underline{u}})\bar{\underline{u}}} + \underbrace{\underline{D}_{kN}(\overset{+}{\underline{u}})\overset{+}{\underline{u}}} \tag{3.3}$$
$$= \bar{\underline{\varepsilon}} + \bar{\bar{\underline{\varepsilon}}} + \overset{+}{\underline{\varepsilon}} + \overset{-+}{\underline{\varepsilon}} + \overset{++}{\underline{\varepsilon}}\,.$$

By the help of (2.58, 2.59) the following incrementations can be carried out:

$$\underline{r} = \underline{R}_r\,\underline{u} = \underline{R}_r\,\bar{\underline{u}} + \underline{R}_r\,\overset{+}{\underline{u}} = \bar{\underline{r}} + \overset{+}{\underline{r}} \tag{3.4}$$

$$\overset{\circ}{\underline{t}} = \underline{R}_t\,\underline{\sigma} = (\underline{R}_{tL} + \underline{R}_{tN}(\bar{\underline{u}} + \overset{+}{\underline{u}}))\,\sigma\,. \tag{3.5}$$

Substituting these abbreviations (3.2) - (3.5) in (3.1) and suppressing all displacement depending load terms we finally find

$$\begin{aligned} \Pi = \frac{1}{2}\iint_{\overset{\circ}{F}} &(\bar{\underline{\varepsilon}} + \bar{\bar{\underline{\varepsilon}}} + \overset{+}{\underline{\varepsilon}} + \overset{-+}{\underline{\varepsilon}} + \overset{++}{\underline{\varepsilon}})^T D\,\underline{E} \\ &(\bar{\underline{\varepsilon}} + \bar{\bar{\underline{\varepsilon}}} + \overset{+}{\underline{\varepsilon}} + \overset{-+}{\underline{\varepsilon}} + \overset{++}{\underline{\varepsilon}})\,d\overset{\circ}{F} \\ &- \iint_{\overset{\circ}{F}} (\bar{\underline{u}} + \overset{+}{\underline{u}})^T \overset{\circ}{\underline{P}}\,d\overset{\circ}{F} - \int_{\overset{\circ}{c}_t} (\bar{\underline{r}} + \overset{+}{\underline{r}})^T \overset{\circ}{\underline{t}}\,d\overset{\circ}{s} \\ = \bar{\Pi} + \overset{+}{\Pi} + \overset{++}{\Pi} + \ldots \end{aligned} \tag{3.6}$$

where $\overset{+}{\Pi}$ and $\overset{++}{\Pi}$ denote potential terms linear and quadratic in the displacement increments $\overset{+}{\underline{u}}$.

3.2 Equilibriums of the adjacent state

The condition of stationarity of (3.6)

$$\delta\Pi = \delta\Pi_i + \delta\Pi_a = 0 \longrightarrow \Pi = stat \tag{3.7}$$

under the kinematic restrictions

$$\underline{\varepsilon} = \underline{D}_k \underline{u}\,, \quad \underline{r} = \underline{r}^\circ \text{ along } \overset{\circ}{C}_r \tag{3.8}$$

represents a weak formulation of the equations of equilibrium (2.54) of the adjacent state for linear elastic behavior. Postulationg the fundamental state to be in equilibrium ($\delta\bar{\Pi}=0$) we find from (3.6):

$$\begin{aligned}\delta\Pi = \delta\Big[\tfrac{1}{2}\Big\{\iint_{\overset{\circ}{F}} \Big(&\overset{+}{\underline{\varepsilon}}^T \underline{D}\underline{E}\,\overset{+}{\underline{\varepsilon}} + \overset{++}{\underline{\varepsilon}}^T \underline{D}\underline{E}\,\bar{\underline{\varepsilon}} + \bar{\underline{\varepsilon}}^T \underline{D}\underline{E}\,\overset{++}{\underline{\varepsilon}} \\ &+ \overset{++}{\underline{\varepsilon}}^T \underline{D}\underline{E}\,\overset{--}{\underline{\varepsilon}} + \overset{--}{\underline{\varepsilon}}^T \underline{D}\underline{E}\,\overset{++}{\underline{\varepsilon}} + \overset{+}{\underline{\varepsilon}}^T \underline{D}\underline{E}\,\overset{-+}{\underline{\varepsilon}} + \overset{-+}{\underline{\varepsilon}}\,\underline{D}\underline{E}\,\overset{+}{\underline{\varepsilon}} + \overset{-+}{\underline{\varepsilon}}^T \underline{D}\underline{E}\,\overset{-+}{\underline{\varepsilon}} \\ &+ \bar{\underline{\varepsilon}}^T \underline{D}\underline{E}\,\overset{+}{\underline{\varepsilon}} + \overset{+}{\underline{\varepsilon}}^T \underline{D}\underline{E}\,\bar{\underline{\varepsilon}} + \overset{--}{\underline{\varepsilon}}^T \underline{D}\underline{E}\,\overset{+}{\underline{\varepsilon}} + \overset{+}{\underline{\varepsilon}}^T \underline{D}\underline{E}\,\overset{--}{\underline{\varepsilon}} \\ &+ \bar{\underline{\varepsilon}}^T \underline{D}\underline{E}\,\overset{-+}{\underline{\varepsilon}} + \overset{-+}{\underline{\varepsilon}}\,\underline{D}\underline{E}\,\bar{\underline{\varepsilon}} + \overset{--}{\underline{\varepsilon}}^T \underline{D}\underline{E}\,\overset{-+}{\underline{\varepsilon}} + \overset{-+}{\underline{\varepsilon}}^T \underline{D}\underline{E}\,\bar{\underline{\varepsilon}}\Big)\, d\overset{\circ}{F}\Big\} \\ &- \iint_{\overset{\circ}{F}} \overset{+}{\underline{u}}^T \overset{\circ}{\underline{p}}\, d\overset{\circ}{F} - \int_{\overset{\circ}{C}_t} \overset{+}{\underline{r}}^T \overset{\circ}{\underline{t}}\, d\overset{\circ}{s}\Big] = 0\end{aligned} \tag{3.9}$$

In this condition the symbol δ operates on the displacements $\overset{+}{\underline{u}}$ respectively $\overset{+}{\underline{r}}$ of the adjacent state. Carrying out these variations, (3.9) turnes out - due to the properties of the joined operators in section 2.8 - to be a linear functional in $\overset{+}{\underline{u}}$, containing linear and nonlinear terms of the displacement field $\bar{\underline{u}}$ of the fundamental state.

3.3 Discretization and tangent stiffness relation

For all following considerations the reader should again keep in mind, that for the Lagrangian formulation used all external kinematic variables of the different states of deformation are related to the base vectors of the known middle surface $\overset{\circ}{F}$ of the initial state.

In order to derive discrete algorithms we now apply the conditions (3.7, 3.8) to an arbitrary finite element $\overset{\circ}{F}{}^p$ of the middle surface $\overset{\circ}{F}$. Its displacement field

$$\underline{u}^p = \underline{\Omega}^p \, \underline{v}^p \tag{3.10}$$

shall be approximated by the nodal displacement $\underline{v}^p$ and the matrix $\underline{\Omega}^p$ of the interpolation functions.

After introduction of these approximations for both displacement fields

$$\underline{\bar{u}}^p = \underline{\Omega}^p \, \underline{\bar{v}}^p \tag{3.11}$$

$$\underline{\overset{+}{u}}^p = \underline{\Omega}^p \, \underline{\overset{+}{v}}^p \tag{3.12}$$

into (3.9) and after variation - now with respect to $\underline{\overset{+}{v}}^p$ - we find the following increment equation of the p-th element:

$$\underline{k}_T^p \, \underline{\overset{+}{v}}^p = (\underline{k}_e + \underline{k}_{gL} + \underline{k}_{gN} + \underline{k}_{uL} + \underline{k}_{uN}) \, \underline{\overset{+}{v}}^p = \underline{p} - \underline{p}_i \; . \tag{3.13}$$

The tangent stiffness matrix $\underline{k}_T^p$ can be devided into the following submatrices:

$$\underline{k}_e \, \underline{\overset{+}{v}}^p = \frac{\partial}{\partial \underline{\overset{+}{v}}^p} \left[\frac{1}{2} \iint_{\overset{\circ}{F}^p} (\underline{\overset{+}{\varepsilon}}^T \underline{D} \underline{E} \, \underline{\overset{+}{\varepsilon}}) \, d\overset{\circ}{F}^p \right] \tag{3.14}$$

with $\underline{k}_e$: linear elastic stiffness matrix,

$$\underline{k}_{gL} \, \underline{\overset{+}{v}}^p = \frac{\partial}{\partial \underline{\overset{+}{v}}^p} \left[\frac{1}{2} \iint_{\overset{\circ}{F}^p} (\underline{\overset{++}{\varepsilon}}^T \underline{D} \underline{E} \, \underline{\bar{\varepsilon}} + \underline{\bar{\varepsilon}}^T \underline{D} \underline{E} \, \underline{\overset{++}{\varepsilon}}) \, d\overset{\circ}{F}^p \right] \tag{3.15}$$

with $\underline{k}_{gL}$: geometric stiffness matrix (linear in $\underline{\bar{v}}$),

$$\underline{k}_{gN} \, \underline{\overset{+}{v}}^p = \frac{\partial}{\partial \underline{\overset{+}{v}}^p} \left[\frac{1}{2} \iint_{\overset{\circ}{F}^p} (\underline{\overset{++}{\varepsilon}}^T \underline{D} \underline{E} \, \underline{\bar{\bar{\varepsilon}}} + \underline{\bar{\bar{\varepsilon}}}^T \underline{D} \underline{E} \, \underline{\overset{++}{\varepsilon}}) \, d\overset{\circ}{F}^p \right] \tag{3.16}$$

with $\underline{k}_{gN}$: geometric stiffness matrix (nonlinear in $\underline{\bar{v}}$),

$$\underline{k}_{uL} \, \underline{\overset{+}{v}}^p = \frac{\partial}{\partial \underline{\overset{+}{v}}^p} \left[\frac{1}{2} \iint_{\overset{\circ}{F}^p} (\underline{\overset{+}{\varepsilon}}^T \underline{D} \underline{E} \, \underline{\overset{-+}{\varepsilon}} + \underline{\overset{-+}{\varepsilon}}^T \underline{D} \underline{E} \, \underline{\overset{+}{\varepsilon}}) \, d\overset{\circ}{F}^p \right] \tag{3.17}$$

with k_{uL}: initial displacement matrix (linear in $\underline{\bar{v}}$),

$$\underline{k}_{uN} \, \underline{\overset{+}{v}}^p = \frac{\partial}{\partial \underline{\overset{+}{v}}^p} \left[\frac{1}{2} \iint_{\overset{\circ}{F}^p} (\underline{\overset{-+}{\varepsilon}}^T \underline{D} \underline{E} \, \underline{\overset{-+}{\varepsilon}}) \, d\overset{\circ}{F}^p \right] \tag{3.18}$$

with $\underline{k}_{uN}$: initial displacement matrix (nonlinear in $\underline{\bar{v}}$).

The terms on the right-hand side of (3.9) can be devided into the following vectors:

$$\underline{p}_i = \frac{\partial}{\partial \underline{\overset{+}{v}}^p} \left[\frac{1}{2} \iint_{\overset{\circ}{F}} \left\{ (\underline{\bar{\varepsilon}} + \underline{\bar{\bar{\varepsilon}}})^T \underline{DE} (\underline{\overset{+}{\varepsilon}} + \underline{\overset{+}{\bar{\varepsilon}}}) + (\underline{\overset{+}{\varepsilon}} + \underline{\overset{+}{\bar{\varepsilon}}})^T \underline{DE} (\underline{\bar{\varepsilon}} + \underline{\bar{\bar{\varepsilon}}}) \right\} d\overset{\circ}{F}^p \right] \tag{3.19a}$$

with $\underline{p}_i$: vector of internal nodal forces,

$$\underline{p} = \iint_{\overset{\circ}{F}^p} \underline{\Omega}^T \underline{\overset{\circ}{p}}\, dF^{\circ p} + \int_{\overset{\circ}{c}_t} \underline{\Omega}^T \underline{R}^T \underline{\overset{\circ}{t}}'\, d\overset{\circ}{s} \tag{3.19b}$$

with $\underline{p}$: vector of external loads.

Any subsequent standard assemblage process of these element matrices creates the corresponding tangent stiffness relation and the right-hand side vectors of the complete shell structure according to (3.13)

$$\underline{\overset{+}{v}} = \underline{a}\, \underline{\overset{+}{V}} :$$

$$\underline{K}_T \underline{\overset{+}{V}} = [\underline{K}_e + \underline{K}_{gL} + \underline{K}_{gN} + \underline{K}_{uL} + \underline{K}_{uN}]\, \underline{\overset{+}{V}} = \underline{P} - \underline{P}_i \,. \tag{3.20}$$

3.4 Computational algorithms

Algorithms for nonlinear responses in finite element computations [19] are usually based on combined incremental and iterative techniques: during a step-wise load incrementation the nonlinear load-displacement relation is linearized and the linearization error is corrected by additional equilibrium iterations.

The central equation of this technique is (3.20), where the out of balance forces are computed from the difference of the total applied loads $\underline{P}$ and those internal equilibrium forces $\underline{P}_i$, which can be evaluated from the reached state of deformation (n: last number of iterations):

$$\underline{P}_i = \underline{P}_i(\underline{\bar{u}}_n) = \underline{P}_i(\underline{\bar{u}}_{n-1} + \underline{\overset{+}{u}}_n)\,. \tag{3.21}$$

Through this interpretation the above equation changes into an iteration process: (3.20) will be satisfied only, if the applied loads $\underline{P}$ equalize the internal forces $\underline{P}_i$; hence the out of balance forces have reached the zero level.

This general concept has been detailed in two different ways:

- the standard and the modified Newton-Raphson algorithms [19];
- the standard and the modified Riks-Wempner-Wessels algorithms [21].

Newton-Raphson algorithms are inefficient or fail completely in the neighbourhood of limit points:

$$\overset{+}{\underline{V}}^T \underline{K}_T \overset{+}{\underline{V}} \mathrel{\hat{=}} \det \underline{K}_T = 0 . \tag{3.22}$$

Algorithms of the Riks-Wempner-kind on the opposite easily overcome those difficulties, if negative load incrementation at positive displacement increments has been implemented for decreasing load paths, where $\underline{K}_T$ is negative definite:

$$\overset{+}{\underline{V}}^T \underline{K} \overset{+}{\underline{V}} < 0 . \tag{3.23}$$

The completion of Wessels [21] to this algorithm removes the originally destroyed symmetry of the tangent stiffness matrix.

During the computation of nonlinear load-displacement paths without or with existence of limit points - indicating snap-through buckling - secondary equilibrium paths have to be expected at many points. To find these points of neutral equilibrium we replace the original decomposition (3.2) of the displacement field by

$$\underline{u} = \lambda \bar{\underline{u}} + \overset{+}{\underline{u}} , \tag{3.24}$$

in terms of discrete quantities by

$$\underline{V} = \lambda \bar{\underline{V}} + \overset{+}{\underline{V}} . \tag{3.25}$$

If the variation (3.7) is fulfilled under the restriction, that the load does not change during the transition from the fundamental to the adjacent state, the equilibrium condition yields

$$\delta\Pi = \delta\overset{++}{\Pi} = 0 , \qquad (3.26)$$

as the energetic criterion for neutral equilibrium. After a similar way of derivation as given in section 3.2 and 3.3 we end up with a quadratic eigenvalue problem

$$\underline{K}_T \overset{+}{\underline{V}} = [\underline{K}_e + \lambda(\underline{K}_{gL} + \underline{K}_{uL}) + \lambda^2(\underline{K}_{gN} + \underline{K}_{uN})]\overset{+}{\underline{V}} = 0 \qquad (3.27)$$

$$\longrightarrow \lambda_c , \overset{+}{\underline{V}}_c . \qquad (3.28)$$

The most common approximation of this nonlinear eigenvalue problem in structural stability analysis neglects all stiffness submatrices, expect $\underline{K}_{gL}$, in comparison to $\underline{K}_e$. By this simplification we end up with the linear eigenvalue problem of the so-called classical shell stability:

$$[\underline{K}_e + \lambda \underline{K}_{gL}]\overset{+}{\underline{V}} = 0 \longrightarrow \lambda_c , \overset{+}{\underline{V}}_c . \qquad (3.29)$$

The next improvement step considers also the linear part $\underline{K}_{uL}$ of the initial displacement matrix, which again leads to a linear eigenvalue problem:

$$[\underline{K}_e + \lambda(\underline{K}_{gL} + \underline{K}_{uL})]\overset{+}{\underline{V}} = 0 \longrightarrow \lambda_c , \overset{+}{\underline{V}}_c . \qquad (3.30)$$

The consideration of all stiffness-submatrices of the linear and nonlinear displacement of the fundamental state ($\lambda\underline{\bar{u}}$, see fig. 8) finally requires the solution of the complete quadratic eigenvalue problem (3.28).

To localize points of neutral equilibrium at a completely nonlinear deformation path (fig. 8) we again start from (3.28):

$$\underline{K}_T \overset{+}{\underline{V}} = [\underline{K}_e + \lambda(\underline{K}_{gL} + \underline{K}_{uL}) + \lambda^2(\underline{K}_{gN} + \underline{K}_{uN})]\overset{+}{\underline{V}} = 0 . \qquad (3.31)$$

$\underline{\bar{u}}$ respectively $\underline{\bar{V}}$, $\underline{\bar{v}}^p$ (remember 3.15 - 3.18) will now be identified with the unknown displacement field $\underline{\bar{u}}^*$, $\underline{\bar{V}}^*$ of the fundamental state, at which bifurcation occurs. At this bifurcation point the eigenvalue $\lambda = \lambda_c^* = 1$ - as defined value of the bifurcation load - is known and the quadratic eigenvalue problem (3.31) may be replaced by its linearization:

$$[\underline{K}_e + \lambda(\underline{K}_{gL} + \underline{K}_{uL} + \underline{K}_{gN} + \underline{K}_{uN})]\,\overset{+}{\underline{V}} = 0 \,. \qquad (3.32)$$

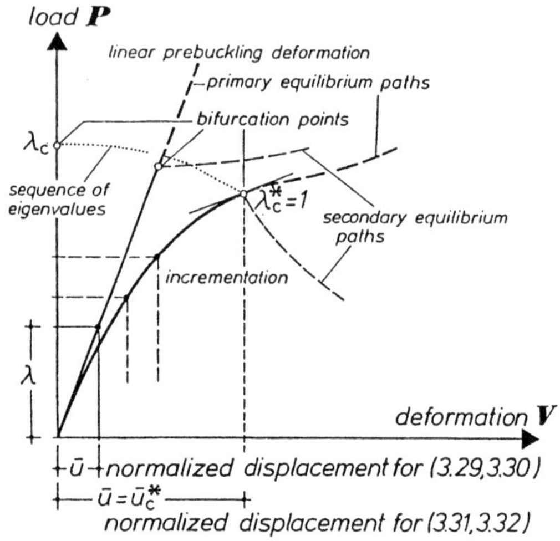

Fig. 8: Nonlinear responses and eigenvalue problems

Only for $\lambda = \lambda_c^* = 1$ both solution fields $\overset{+}{\underline{V}}$ of (3.31) and (3.32) will be identical, they will differ for all other values of λ.

We start the computation process at any arbitrary point, e.g. with the classical solution λ_c for $\bar{\underline{u}} = \bar{\underline{V}} = 0$ and solve the linear eigenvalue problem (3.32) for step-wise increasing displacements $\bar{\underline{u}}$, $\bar{\underline{V}}$. As the sequence of uncorrect values $\lambda \neq 1$ approaches the real path of deformation (see fig. 8), the linear solution (3.32) approaches that one of (3.31). For $\lambda = \lambda_c^* = 1$ we find from (3.32) the desired instability point and the eigenvector $\overset{+}{\underline{V}}$. Starting from this point of bifurcation additional nonlinear computation yield higher equilibrium paths in the postbuckling range.

3.5 Nonlinear triangular finite shell elements

Only relatively few elements for arbitrarily curved shells are published in the literature. The existing ones are often members of the isoparametric family. This element concept is advantageously able to fulfil all three convergence criteria

- completeness
- compatibility
- capability of strain-free rigid body modes

in an exact sense. Its great disadvantage lies in the equivalent approximation of shell geometry and displacement field by polynomials of the same degree. Hence the curvature tensor b^{α}_{β}, the most significantly geometrical influence upon the response, is often evaluated approximately.

Based on our experiences with high precision elements [15] we have developed in the last years a new family of arbitrary curved triangular shell elements (ACS), geometrically linear and nonlinear, in which the middle surface $\overset{\circ}{F}$ of the initial state and the applied loads have been described exactly. As it is evident from fig. 9, the tensorial displacement components v_i (and their partial derivatives) with respect to the individual nodal point base vectors are interpolated directly by polynominals up to the 5th degree. The stiffness submatrices are automatically formed by the computer for various shell theories in a very efficient algorithm [14].

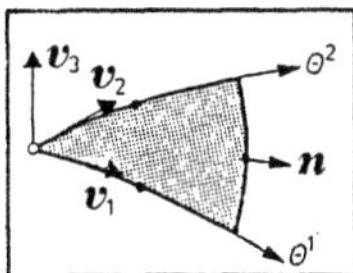	*Name of nonlinear element:*				
	NACS 15	NACS 27	NACS 36	NACS 54	NACS 63
Polynomial approximation	$v_\lambda: P^1$ $v_3: P^3$	$v_\lambda: P^3$ $v_3: P^3$	$v_\lambda: P^3$ $v_3: P^5_R$	$v_\lambda: P^5_R$ $v_3: P^5_R$	$v_\lambda: P^5$ $v_3: P^5$
	P^n: *Polynomials of degree n* P^n_R: *Reduced polynomials of degree n*				
Nodal parameters	v_λ v_3 $v_{3,\alpha}$	v_λ $v_{\lambda,\alpha}$ v_3 $v_{3,\alpha}$	v_λ $v_{\lambda,\alpha}$ v_3 $v_{3,\alpha}$ $v_{3,\alpha\beta}$	v_λ $v_{\lambda,\alpha}$ $v_{\lambda,\alpha\beta}$ v_3 $v_{3,\alpha}$ $v_{3,\alpha\beta}$	v_λ $v_{\lambda,\alpha}$ $v_{\lambda,\alpha\beta}$ v_3 $v_{3,\alpha}$ $v_{3,\alpha\beta}$ $v_{\lambda,n}$ $v_{3,n}$
	15	27	36	54	63
Points of integration	12	12	21	21	21

Fig. 9: Family of nonlinear finite elements [14]

No doubt, the elements fulfil the requirements of completeness up to the highest order (expection NACS 54). For the compatibility requirements at least C^1-continuity is guaranteed. A weak point - in principle - of any curved element is its ability to describe arbitrary rigid body motions without producing strain energy, e.g. nodal forces, especially for geometrically nonlinear problems. It has been proved, that - if the interpolation matrix $\underline{\Omega}^p$ in (3.10) is able to perform 6 linearly independent rigid body modes for $b^\alpha_\beta \rightarrow 0$ - the curved element ($b^\alpha_\beta \neq 0$) will converge to zero-strain modes as the element length L^p approaches zero. The quality of convergence, depending on the applied shell theory and its discretization, can be checked by the evaluation of eigenvalues of the element. Postulating namely a free-free vibration with a unit mass matrix for a single element p

$$\underline{k}^p_T \underline{v}^p = \underline{p}^p = \lambda \underline{v}^p \longrightarrow (\underline{k}^p_T - \lambda \underline{I}) \underline{v}^p = 0 \tag{3.33}$$

under the normality condition

$$\underline{v}^{p^T} \underline{v}^p = 1 \tag{3.34}$$

for the eigenvectors, each eigenvalue λ describes the strain-energy of the respective mode:

$$\lambda = \underline{v}^{p^T} \underline{k}^p_T \underline{v}^p = 2\Pi^p_i \,. \tag{3.35}$$

Fig. 10 relates, for the linear stiffness matrix $\underline{k}_e$ and some selected middle surface geometries, the maximum λ_6 of the 6 smallest eigenvalues (beeing all zero in the ideality) to the first non-zero eigenvalue λ_7. The ratio is given for the derived shell theory (section 2) and various element-lenghts L^p. Thus fig. 10 demonstrates the high quality of fulfil of this - especially for nonlinear responses - important condition.

All later examples in section 4 have been computed on the basis of these elements.

Circular cylindrical shell - R/h = 100:

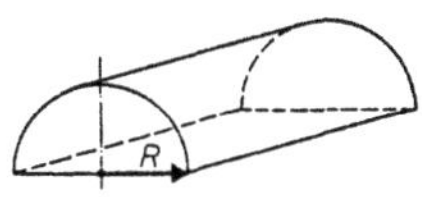

R/L^p	λ_6	λ_7	$\lvert\lambda_6/\lambda_7\rvert$
2,0	$0{,}72 \cdot 10^{-1}$	$0{,}79 \cdot 10^{2}$	$9{,}1 \cdot 10^{-4}$
3,3	$0{,}55 \cdot 10^{-3}$	$0{,}55 \cdot 10^{2}$	$1{,}0 \cdot 10^{-5}$
10,0	$-0{,}57 \cdot 10^{-6}$	$0{,}21 \cdot 10^{2}$	$2{,}7 \cdot 10^{-8}$

Spherical shell - R/h = 100:

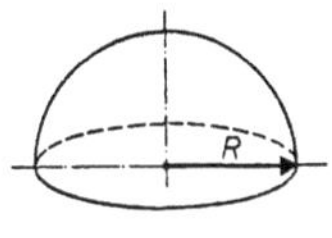

R/L^p	λ_6	λ_7	$\lvert\lambda_6/\lambda_7\rvert$
2,0	$0{,}15 \cdot 10^{1}$	$0{,}87 \cdot 10^{2}$	$1{,}7 \cdot 10^{-2}$
3,3	$0{,}15 \cdot 10^{-2}$	$0{,}36 \cdot 10^{2}$	$4{,}2 \cdot 10^{-5}$
5,0	$0{,}11 \cdot 10^{-4}$	$0{,}17 \cdot 10^{2}$	$6{,}7 \cdot 10^{-7}$
10,0	$-0{,}24 \cdot 10^{-6}$	$0{,}43 \cdot 10^{1}$	$5{,}6 \cdot 10^{-8}$

Fig. 10: Pseudo-zero-strain rigid body modes for NACS 54 (L^p denotes a typical length of the element)

4. Response phenomena and numerical examples

4.1 Survey of instability phenomena

Previously we have described certain numerical tools for a successful treatment of nonlinear response phenomena. Elastic shell structures may suffer two classes of instability phenomena (see fig. 11):

- loss of stability due to bifurcation-buckling,
- loss of stability due to snap-through-buckling.

The simplest treatment of bifurcation-buckling is the classical eigenvalue problem (3.29) of shell stability, in which all prebuckling deformations are neglected (No. 1 in fig. 11). Extended eigenvalue problems (3.30) or (3.31, 3.32) have to be solved, if the prebuckling deformations become of more severe influence. The influence might be linear (No. 2) or nonlinear (No. 3).

Once bifurcation points have been found the behavior of the shell in the postbuckling region has to be investigated. For perfect shells three different cases may appear. There might be an increase of the load bearing capacity (No. 3a in fig. 11) or a desastreous decrease (No. 3b) along the secondary equilibrium

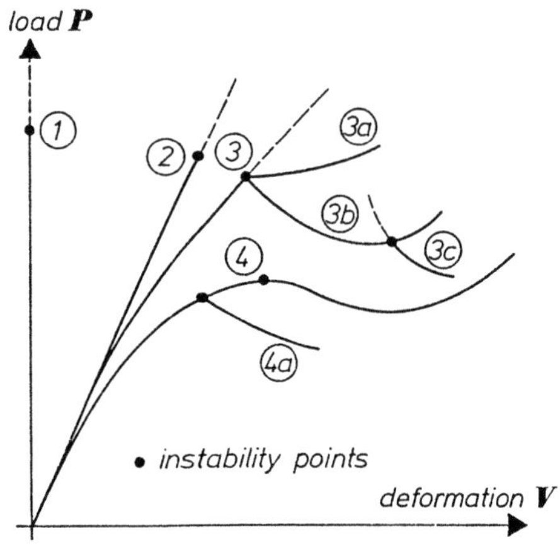

Fig. 11: Buckling phenomena

path. In both cases the behavior can be symmetric or nonsymmetric with important influences on the imperfection sensitivity [23, 24]. In the course of the path No. 3b secondary bifurcation points with higher equilibrium paths (No. 3c) may occur. All these phenomena can be treated by using the Riks-Wempner iteration technique, simultaneously combined with a bifurcation point detect due to (3.31, 3.32).

The second class of phenomena of loss of stability is connected to snap-through-buckling (No. 4). Once the shell has reached a limit load snapping occurs as a dynamic process until a new stable equilibrium configuration has gained. This phenomenon is typical for perfect shells under certain load conditions and for shells with imperfection patterns, it can be treated without severe problems by the Riks-Wempner algorithm. But even such a correct tracing of the nonlinear load-displacement path may be insufficient, if the shell suffers bifurcation buckling before having reached its limit point (No. 4a); thus again a simultaneous eigenvalue treatment (3.31, 3.32) has to support the nonlinear load-displacement iteration.

If we summarize our recognitions from fig. 11 we find, that any software program for the treatment of the complete nonlinear and instability response of elastic shell structures should at least contain

- a nonlinear algorithm solving the problems near limit points with positive (for $\overset{+}{\underline{V}}^T \underline{K}_T \overset{+}{\underline{V}} > 0$) and negative (for $\overset{+}{\underline{V}}^T \underline{K}_T \overset{+}{\underline{V}} < 0$) load incrementation, e.g. the Riks-Wempner algorithm;
- a general bifurcation point detect-algorithm, e.g. as described in section 3.4, with the possibility of evaluation of the eigenvector $\overset{+}{\underline{V}}$ as starting vector for the secondary (or tertiary) load-displacement path.

Our modular software system MISS [27], especially its subsytem FEMAS [30], is equipped with these algorithms and thus admits a treatment of the instability phenomena of fig. 11.

On the other side, difficulties in handling nonlinear algorithms should not be underestimated In the surrounding of sharp curves of the (many dimensional) load-displacement path manual load incrementation often is required in order to avoid overshooting, and the searching time for nonlinear instability points of unknown shell responses may exceed the most difficult puzzles by far. In addition, the severe influence of geometric imperfections on the load bearing capacity and their ignorance at the state of design of the shell have motivated the construction of simpler, but also realistic approximation concepts. The most important ones, the concept of initial-post-buckling [22, 25] and the concept of reduced-membrane-buckling [24, 26] should be mentioned here. Especially the latter one is considered to be a simple and - at least for shells with unique buckling modes - sufficiently exact tool to evaluate lower bounds of critical loads.

4.2 Linear eigenvalue analysis

The mentioned algorithms shall now be demonstrated on the basis of three examples. Fig. 12 shows a very large and wide cooling tower with lower and upper ring beam. The critical load-factor λ_c, computed as lowest eigenvalue from (3.29) for dead-weight, wind and internal suction (all three assumed to act with rotational symmetry), yields to an unsufficiently low value. Thus

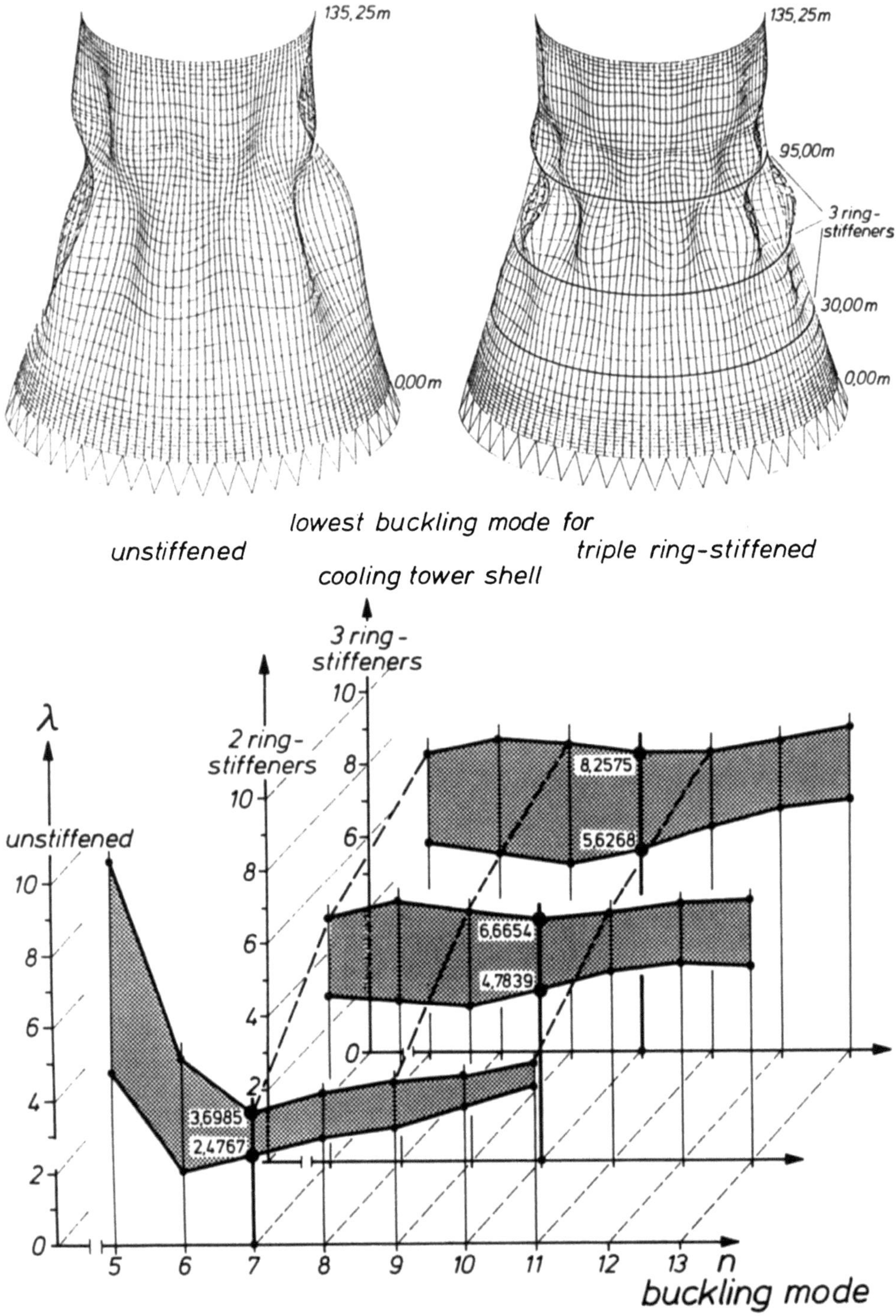

Fig. 12: Linear stability and imperfection sensitivity for unstiffened and ring-stiffened hyperboloidal shells

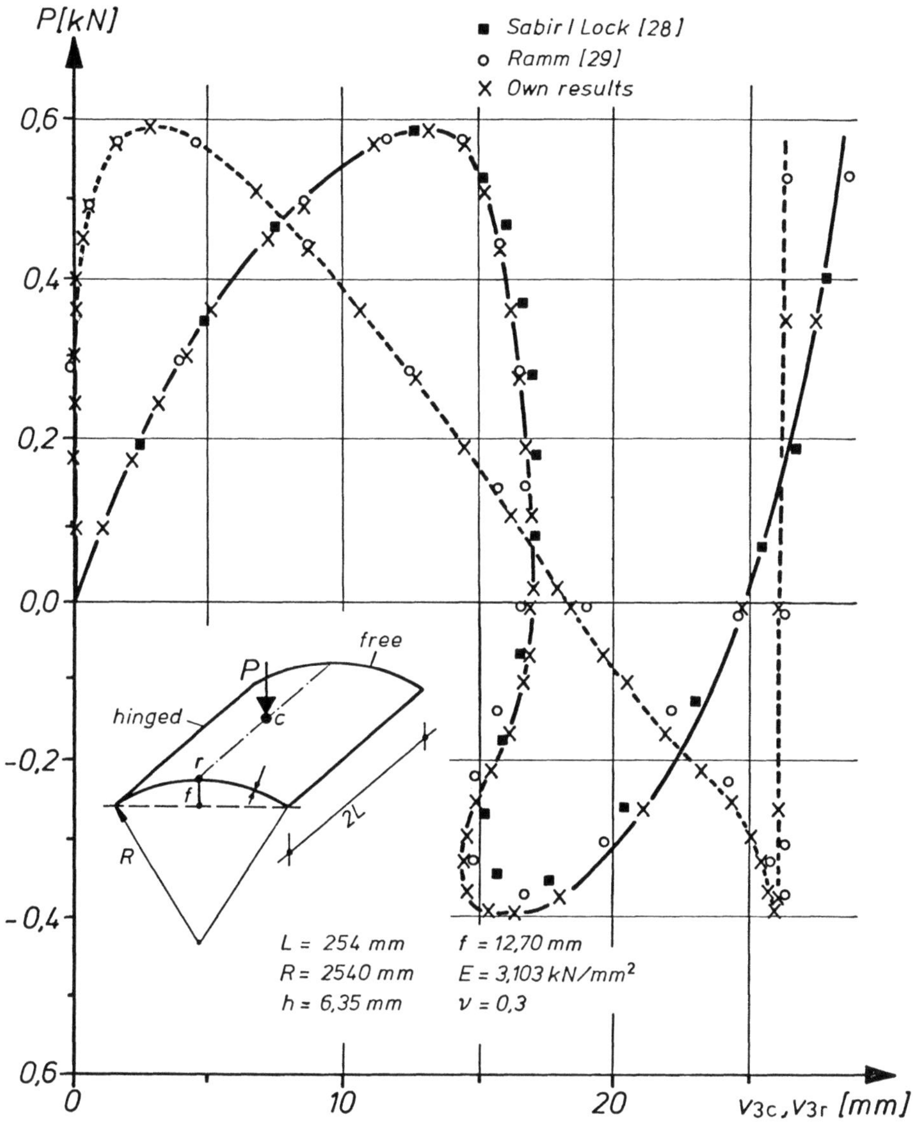

Fig. 13: Snap through of shallow cylindrical shell

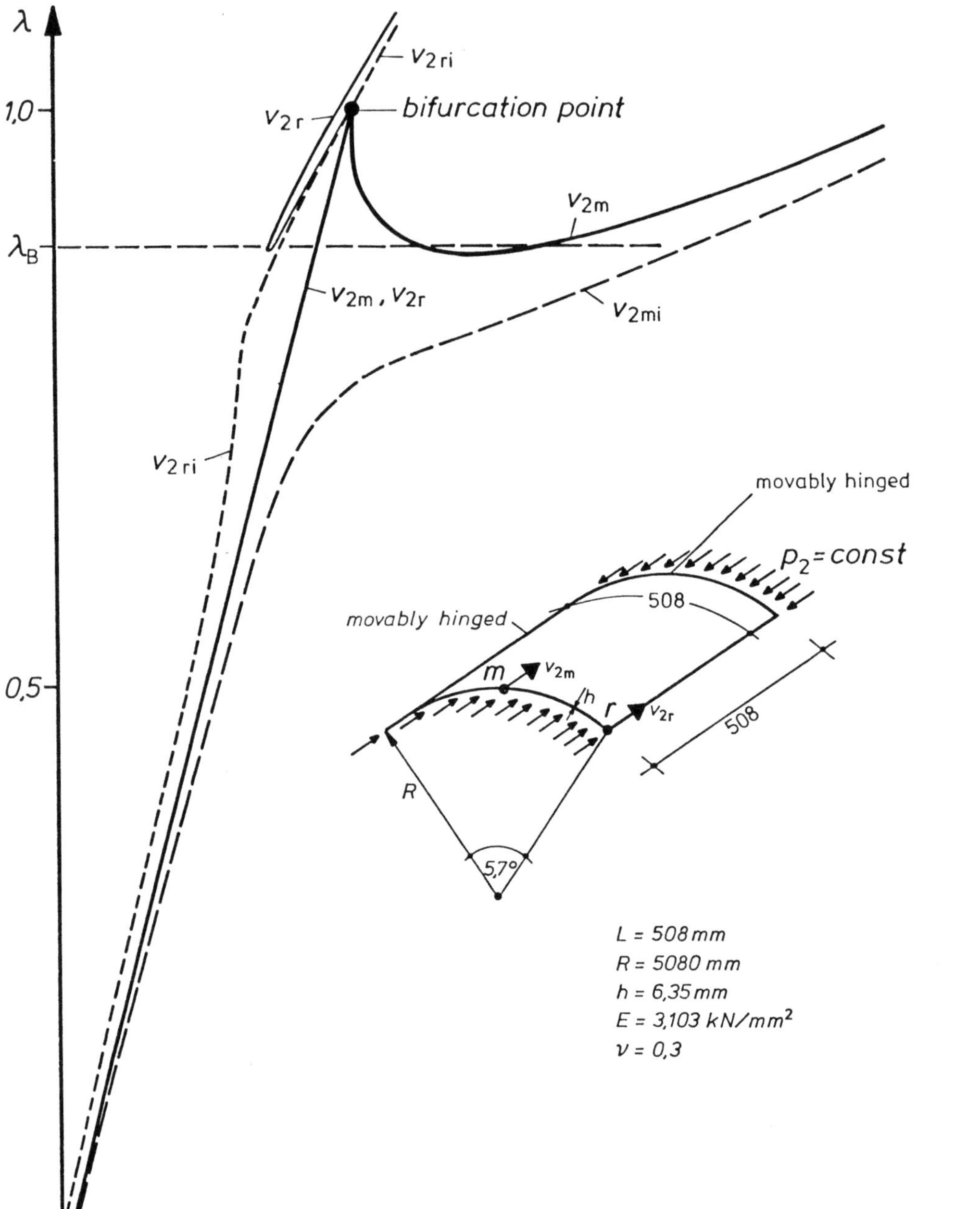

Fig. 14: Bifurcation of axially loaded shallow cylindrical shell

additional ring-stiffeners have to be applied increasing λ_c considerably. The lowest bound of the imperfection sensitivity has been estimated by the concept of reduced-membrane-buckling [26] taking only strain energy due to bending of the buckled configuration into account.

4.3 Snap-through of shallow cylindrical shell

The snap-through of the cylindrical pannel under a central load P, as given in fig. 13, has been computed originally by Sabir and Lock [28] to demonstrate the advantage of changing from load-to displacement-incrementation and vice versa using the Newton-Raphson-technique during a complicated response. We compare our results with those of them and Ramm [29].

As can be seen from the different responses v_{3c} and v_{3r}, the center c at the beginning suffers increasing deflections, while the edges r remain nearly undeflected. Having reached the limit point - close to a flat plate - the whole structure snaps through, nearly into its opposite position. The new system behaves considerably stiffer; especially the arches along the free edges suffer no further vertical displacements. The automatic computation has been carried out for a quarter of the structure using 8 triangular elements NACS 54. The total number of degrees of freedom counts, after fulfil of all boundary- and symmetry-conditions, 91 [14].

4.4 Nonlinear bifurcation-buckling of shallow cylindrical shell

Fig. 14 shows an axially loaded circular cylindrical pannel with load-displacement paths for two given points m and r. All edges are supported in a tangentially movable way. For the perfect shell both meridional displacements v_{2m} and v_{2r} increase linearly up to the bifurcation point, where the shell buckles with one half-wave in each direction. v_{2m} bifurcates into the secondary path, exceeding the post-buckling minimum (close to λ_B according to the concept of reduced-membrane-buckling [26]) and increasing again in the deeper post-buckling range. The boundary displacement v_{2r} after bifurcation briefly decreases, then - after a very sharp curve - increases in a much steeper ascend than v_{2m}.

Applying a pre-imperfection-pattern-affin to the buckling mode of v_{3mi} = -0.50 mm of amplitude, both response variables behave nearly linearly up to 70% of the critical load. There, after the nonlinear terms in the tangential stiffness relation (3.19) become dominant, v_{2mi} as well as v_{2ri} start their approach to the respective ideal paths. 8 elements NACS 27 with a total of 50 degrees of freedom have been used for a quarter of the structure [17].

References

1. Green, A.E.; Naghdi, P.M.: Non-Isothermal Theory of Rods, Plates and Shells. Int. J. Solids Struct. 6 (1970), p. 209.

2. Krätzig, W.B.: Allgemeine Schalentheorie beliebiger Werkstoffe und Verformungen. Ing.-Archiv 40 (1971), p. 311.

3. Koiter, W.T.: On the nonlinear theory of thin shells. Proc. Kon. Ned. Akad. Wet., 1966, Ser. B, Vol. 66, No. 1, p. 1.

4. Reissner, E.: Linear and Nonlinear Theory of Shells. Thin-Shell Structures, edited by Fung, Y.C.; Sechler, E.E., Prentice-Hall Inc., Englewood Cliffs, New Jersey 1974.

5. Basar, Y.; Krätzig, W.B.: Mechanik der Flächentragwerke. Vieweg Verlag (in preparation).

6. Simmonds, J.G.; Danielson, D.A.: Non-linear Shell Theory with Finite Rotations and Stress-Function Vectors. J. Appl. Mech., Trans. ASME Ser. E, 39 (1972), p. 1085.

7. Pietraszkiewicz, W.: Introduction to the Non-Linear Theory of Shells. Mitt. Inst. Mech. No. 10, Ruhr-University Bochum 1977.

8. Basar, Y.: Eine geometrisch nicht-lineare Schalentheorie. konstruktiver ingenieurbau-berichte 38/39, Vulkan-Verlag, Essen 1981.

9. Mason, J.: Variational, Incremental and Energy Methods in Solid Mechanics and Shell Theory. Elsevier Publ. Comp., Amsterdam 1980.

10. Naghdi, P.M.: The Theory of Shells and Plates. Handbuch der Physik, Vol. IV, A2. Springer, Berlin 1972.

11. Basar, Y.; Krätzig, W.B.: Struktur konsistenter Grundgleichungen für das Beul- und Nachbeulverhalten allgemeiner Flächentragwerke. Der Stahlbau 5 (1977), p. 138.

12. Green, A.E.; Zerna, W.: Theoretical Elasticity. 2.ed., Clarendin Press, Oxford 1968.

13. Basar, Y.: Eine allgemeine Schalentheorie endlicher Verformungen. Dr.-Ing.-Thesis, University of Hannover 1964.

14. Harte, R.: Doppelt gekrümmte Dreieckselemente für lineare und geometrisch nichtlineare Berechnungen allgemeiner Flächentragwerke. Dr.-Ing.-Thesis, Ruhr-University Bochum (in preparation).

15. Eckstein, U., Krätzig, W.B., Wittek, U.: Finite-Element-Berechnungen zur Grenztragfähigkeit der Rotationsschalen. Techn.-wiss. Mitt. No. 80-4 of the Institut für Konstr. Ingenierbau, Bochum 1980.

16. Wittek, U.: Die Stabilitätsgleichungen der Schalentheorie und ihre Anwendung zur Beurteilung des Schwingungs- und Stabilitätsverhaltens der Rotationsschalen. Festschrift W. Zerna und KIB, Werner-Verlag, Düsseldorf 1976.

17. Eckstein, U.: Nichtlineare Stabilitätsberechnung elastischer Schalentragwerke. Dr.-Ing.-Thesis, Ruhr-University Bochum (in preparation).

18. Brendel, B.: Zur geometrischen Elastostabilität. Dr.-Ing.-Thesis, University of Stuttgart 1979.

19. Bathe, K.-J., Ramm, E., Wilson, E.L.: Finite Element Formulations for Large Deformation Dynamic Analysis. Int. J. Num. Meth. Eng. 9 (1975), p. 353.

20. Ramm, E.: Geometrisch nichtlineare Elastostatik und Finite Elemente. Habilitationsschrift, Universität Stuttgart 1976.

21. Wessels, M.: Das statische und dynamische Durchschlagsproblem der imperfekten flachen Kugelschale. Mitt. d. Inst. f. Statik der T.U. Hannover No. 23, 1977.

22. Budiansky, B., Hutchinson, J.W.: Buckling-Progress and Challenge. In: Trends in Solid Mechanics. Delft University Press 1979.

23. Koiter, W.T.: General equations of elastic stability for thin shells. Proc. Symp. of the Theory of Thin Shells, University of Houston, Texas 1967.

24. Wittek, U.: The Concept of Reduced-Membrane-Buckling. Proc. 2nd IUTAM-Symp. on Stability of the Mechanics of Continua, Nümbrecht 1981.

25. Koiter, W.T.: Over de stabiliteit van het elastisch evenwicht. Dr.ir.-Thesis, Delft 1945. NASA TT F-10, 833, 1967.

26. Wittek, U.: Beitrag zum Tragverhalten der Strukturen bei endlichen Verformungen unter besonderer Beachtung des Nachbeulverhaltens dünner Flächentragwerke. Techn.-wiss. Mitt. No. 80-1 of the Institut für Konstr. Ingenieurbau, Bochum 1980.

27. Krätzig, W.B., Weber, B., et al.: MISS - Modulares Interaktives Software-System. Interner Arbeitsbericht KIB III 81-3, Institut f. Konstr. Ingenieurbau, Bochum 1981.

28. Sabir, A.B., Lock, A.C.: The Application of Finite Elements to the Large Deflection - Geometrically Non-linear Behavior of Cylindrical Shells. Variational Methods in Engineering 7/66, Southampton University Press 1973.

29. Ramm, E.: Strategies for Tracing the Nonlinear Response Near Limit Points. Europe-US-Workshop: Nonlinear Finite Element Analysis in Structural Mechanics, Springer-Verlag, Berlin 1981.

30. Beem, H. et al.: FEMAS-FINITE-ELEMENT-MODULN ALLGEMEINER STRUKTUREN. Interne Arbeitsberichte KIB III, Institut für Konstr, Ingenieurbau, Bochum 1981, 1982.

THE NATURAL METHOD: SIMPLE AND ELEGANT

J.H. ARGYRIS, H. BALMER, J. BÜHLMEIER, M. HAASE, H.-P. MLEJNEK, P.K. SCHMOLZ

Institut für Statik und Dynamik der Luft- und Raumfahrtkonstruktionen
Universität Stuttgart

The paper presents a survey of the natural mode technique with special emphasis on plate and shell type structures. First, the natural method is briefly recapitulated on the basis of the plane triangular simplex element (TRIM3). Such an element with its simple natural deformation modes (elongations of the sides) may then be used in a differential sense as building block for higher order and curved membrane elements which are not necessarily of triangular shape. This is the basic idea of the subelement concept.

Subsequently, the method is extended to general plate and shell elements. Here, it is essential that not only the relevant mechanical quantities (stress resultants and corresponding strain measures) but also the geometry of the middle surface of the shell is adapted to the triangular form of the subelement! The methodology is then applied to a number of shell models.

Considering next nonlinear geometric effects, it is shown that the subelement concept is also particularly useful for the construction of the geometrical stiffness of complex shell elements. Reference to plasticity effects is also made.

The final part of the paper deals with selected applications to the buckling of realistic shell structures.

List of References

1. Argyris, J.H.: Recent advances in matrix methods of structural analysis, Progress in Aeronautical Sciences 4 (Pergamon Press, Oxford, 1964).
2. Argyris, J.H.: Continua and discontinua, Matrix methods in structural mechanics, Opening address, Proceedings of the Conference on Matrix Methods held at Wright-Patterson Air Force Base, Ohio (Oct. 26-28, 1965).

Buckling of Shells, Ed.: E. Ramm
Proceedings of a State-of-the-Art Colloqium

3. Argyris, J.H.; Scharpf, D.W.: Some general considerations on the natural mode technique. Part I, Small displacements, Part II, Large displacements, Aeron. J. Roy. Aeron. Soc. 73 (1969) 219-226, 361-368.

4. Argyris, J.H.; Scharpf, D.W.: A sequel to technical note 13: The curved tetrahedronal and triangular elements TEC and TRIC for the matrix displacement method, Aeron. J. Roy. Aeron. Soc. 73 (1969) 55-65.

5. Argyris, J.H.; Scharpf, D.W.: The SHEBA family of shell elements for the matrix displacement method. Part I, Natural definition of geometry and strains, Part II, Interpolation scheme and stiffness matrix, Aeron. J. Roy. Aeron. Soc. 72 (1968) 873-883.

6. Argyris, J.H.; Scharpf, D.W.: The SHEBA family of shell elements for the matrix displacement method. Part III, Large displacements, Aeron. J. Roy. Aeron. Soc. 73 (1969) 423-426.

7. Argyris, J.H.; Haase, M.; Malejannakis, G.A.: Natural geometry of surfaces with specific reference to the matrix displacement analysis of shells, Proc. Kon. Nederl. Akad. Wet. ser. B, 5, 76 (1973) 361-410.

8. Argyris, J.H.; Balmer, H.; Doltsinis, J.St.; Dunne, P.C.; Haase, M.; Kleiber, M.; Malejannakis, G.A.; Mlejnek, H.-P.; Müller, M.; Scharpf, D.W.: Finite element method - the natural approach, Comp. Meths. Appl. Mech. Eng. 17/18 (1979) 1-106.

9. Argyris, J.H.; Dunne, P.C.; Malejannakis, G.A.; Schelkle, E.: A simple triangular facet shell element with applications to linear and nonlinear equilibrium and elastic stability problems, Comp. Meths. Appl. Mech. Eng. 10 (1977) 371-403, 11 (1977) 97-131.

10. Argyris, J.H.; Haase, M.; Mlejnek, H.-P.: On an unconventional but natural formation of a stiffness matrix, Comp. Meths. Appl. Mech. Eng. 22 (1980) 1-22.

11. Argyris, J.H.; Balmer, H.; Kleiber, M.; Hindenlang, U.: Natural description of large inelastic deformations for shells of arbitrary shape - application of TRUMP element, Comp. Meths. Appl. Mech. Eng. 22 (1980) 361-389.

12. Argyris, J.H.; Haase, M.; Mlejnek, H.-P.: Some considerations on the natural approach, Comp. Meths. Appl. Mech. Eng., to be published.

13. Bühlmeier, J.: Ein Beitrag zur natürlichen Formulierung von Platten- und Schalenelementen beliebiger Dicke, Dr.-Ing. thesis, to be submitted to the Faculty of Aerospace Engineering (University of Stuttgart, 1982).

14. Argyris, J.H.; Mlejnek, H.-P.: Einführung in die Methode der finiten Elemente, Elementare Strukturmechanik, Band I Statik (Vieweg, Braunschweig 1982), to be published.

15. Argyris, J.H.; Balmer, H.; Bühlmeier, J.; Haase, M.; Mlejnek, H.-P.; Schmolz, P.K.: The natural method: simple and elegant, Comp. Meths. Appl. Mech. Eng., to be published.

16. Argyris, J.H.; Balmer, H.; Doltsinis, J.St.; Haase, M.; Hindenlang, U.; Mlejnek, H.-P.: A natural engineering approach to the nonlinear behaviour of shells, paper presented at the ASME WAM 81 Symposium on nonlinear finite element analysis of shells, Washington D.C., to be published.

PLANE GEOMETRY AND COORDINATES

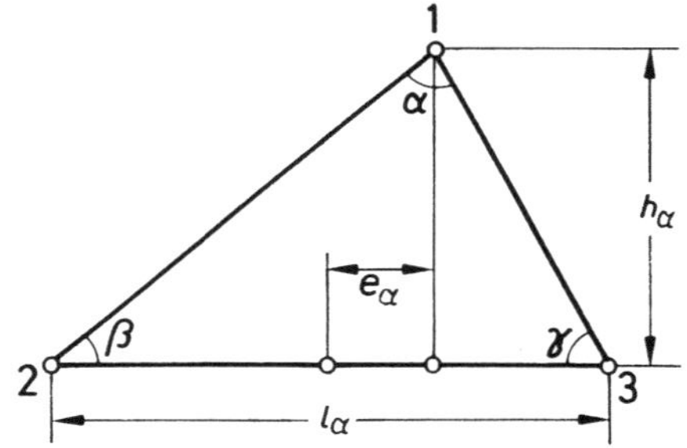

SIDE LENGTH $\boldsymbol{l} = \lceil l_\alpha \quad l_\beta \quad l_\gamma \rfloor$

HEIGHT $\boldsymbol{h} = \lceil h_\alpha \quad h_\beta \quad h_\gamma \rfloor$

ECCENTRICITY $\boldsymbol{\mu} = \lceil \mu_\alpha \quad \mu_\beta \quad \mu_\gamma \rfloor$

$$\mu_\alpha = \frac{2e_\alpha}{l_\alpha} = \frac{l_\gamma^2 - l_\beta^2}{l_\alpha^2}$$

$$\boldsymbol{s} = \lceil s_\alpha \quad s_\beta \quad s_\gamma \rfloor \quad , \quad \boldsymbol{c} = \lceil c_\alpha \quad c_\beta \quad c_\gamma \rfloor$$

$$s_\alpha = \sin\alpha \qquad c_\alpha = \cos\alpha$$

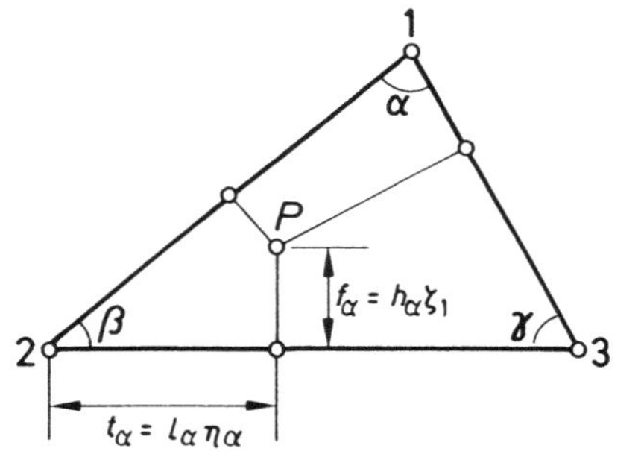

COORDINATES OF POINT P $\boldsymbol{f} = \{ f_\alpha \quad f_\beta \quad f_\gamma \}$

$\boldsymbol{t} = \{ t_\alpha \quad t_\beta \quad t_\gamma \}$

NORMALISATION $\boldsymbol{\zeta} = \boldsymbol{h}^{-1}\boldsymbol{f} = \{ \zeta_1 \quad \zeta_2 \quad \zeta_3 \}$

$\boldsymbol{\eta} = \boldsymbol{l}^{-1}\boldsymbol{t} = \{ \eta_\alpha \quad \eta_\beta \quad \eta_\gamma \}$

NOTE: $\mathbf{e}_3^t \boldsymbol{\zeta} = \zeta_1 + \zeta_2 + \zeta_3 = 1 \quad , \quad \mathbf{e}_3 = \{1 \quad 1 \quad 1\}$

DIFFERENTIATION

SCALAR FUNCTION $\quad F(\zeta_1, \zeta_2, \zeta_3) = F(\boldsymbol{\zeta})$

TOTAL DIFFERENTIAL

$$dF = \frac{\partial F}{\partial \zeta_1} d\zeta_1 + \frac{\partial F}{\partial \zeta_2} d\zeta_2 + \frac{\partial F}{\partial \zeta_3} d\zeta_3$$

DIRECTIONAL DERIVATIVE

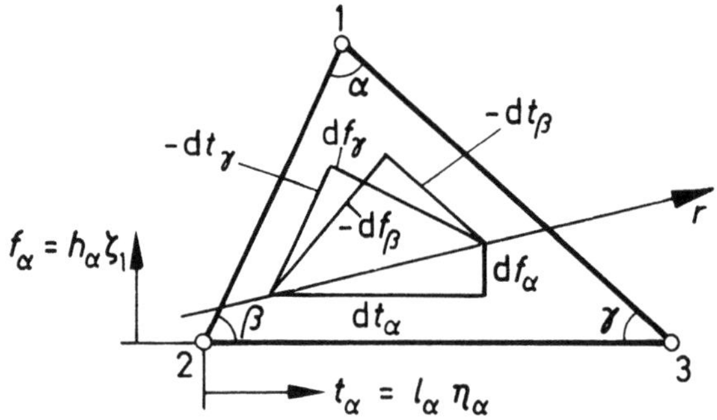

$$\frac{dF}{dr} = \frac{\partial F}{\partial \zeta_1} \frac{d\zeta_1}{dr} + \frac{\partial F}{\partial \zeta_2} \frac{d\zeta_2}{dr} + \frac{\partial F}{\partial \zeta_3} \frac{d\zeta_3}{dr}$$

NOTE: $\partial F / \partial \zeta$ ONLY FORMAL DERIVATIVES

$$\frac{d}{dr}\left(\zeta_1 + \zeta_2 + \zeta_3 \right) = 0$$

EXAMPLE: DERIVATIVES IN DIRECTION t_α AND f_α

$$\frac{dF}{dt_\alpha} = \frac{1}{l_\alpha}\left(\frac{\partial F}{\partial \zeta_1} \times 0 + \frac{\partial F}{\partial \zeta_2} \times -1 + \frac{\partial F}{\partial \zeta_3} \times 1 \right) = \frac{1}{l_\alpha}\left(-\frac{\partial F}{\partial \zeta_2} + \frac{\partial F}{\partial \zeta_3} \right) = \frac{1}{l_\alpha} \frac{\partial F}{\partial \eta_{\alpha t}}$$

$$\frac{dF}{df_\alpha} = \frac{1}{h_\alpha}\left(\frac{\partial F}{\partial \zeta_1} + \frac{\mu_\alpha - 1}{2} \frac{\partial F}{\partial \zeta_2} - \frac{\mu_\alpha + 1}{2} \frac{\partial F}{\partial \zeta_3} \right) = \frac{1}{h_\alpha} \frac{\partial F}{\partial \zeta_{1t}}$$

STRESS AND STRAIN

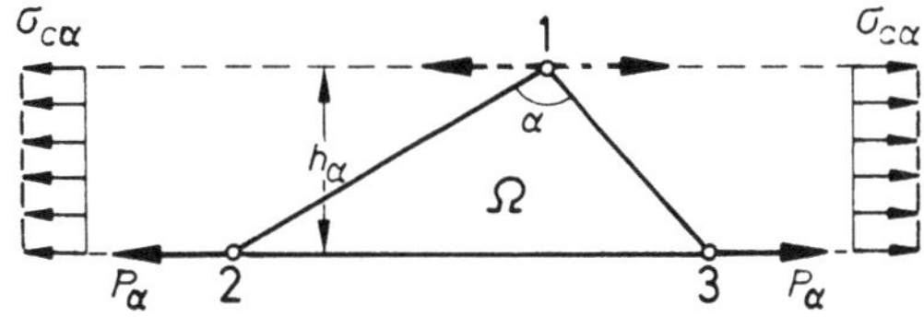

$$P_\alpha = \frac{h_\alpha}{2} t\,\sigma_{c\alpha} = \frac{\Omega}{l_\alpha} t\,\sigma_{c\alpha}$$

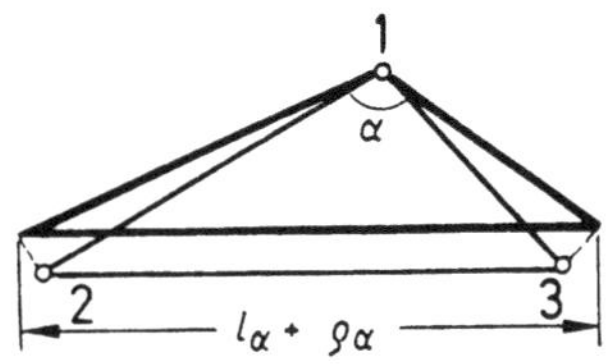

$$\varepsilon_{t\alpha} = \rho_\alpha / l_\alpha$$

COMPONENT STRESS

$$\boldsymbol{\sigma}_c = \{\, \sigma_\alpha \quad \sigma_\beta \quad \sigma_\gamma \,\}_c$$

TOTAL STRAIN

$$\boldsymbol{\varepsilon}_t = \{\, \varepsilon_\alpha \quad \varepsilon_\beta \quad \varepsilon_\gamma \,\}_t$$

ELASTIC STRESS - STRAIN RELATION

$$\boldsymbol{\sigma}_c = \boldsymbol{\varkappa}_{ct}\,\boldsymbol{\varepsilon}_t$$

ISOTROPIC $\boldsymbol{\varkappa}_{ct} = 2G\mathcal{A}^{-1}\left(\mathcal{A} + \frac{\nu}{1-\nu}\,\mathbf{e}_3\mathbf{e}_3^t\right)\mathcal{A}^{-1}$, $\mathcal{A} = \begin{bmatrix} 1 & c_\gamma^2 & c_\beta^2 \\ & 1 & c_\alpha^2 \\ \text{sym} & & 1 \end{bmatrix}$

NATURAL FORCES

$$\mathbf{P}_N = \{\, P_\alpha \quad P_\beta \quad P_\gamma \,\}$$

NATURAL DISPLACEMENTS

$$\boldsymbol{\rho}_N = \{\, \rho_\alpha \quad \rho_\beta \quad \rho_\gamma \,\}$$

FORCE - DISPLACEMENT RELATION

$$\mathbf{P}_N = \mathbf{k}_N\,\boldsymbol{\rho}_N$$

NATURAL STIFFNESS

$$\mathbf{k}_N = \Omega t\,\mathbf{l}^{-1}\boldsymbol{\varkappa}_{ct}\,\mathbf{l}^{-1}$$

SUBELEMENT CONCEPT

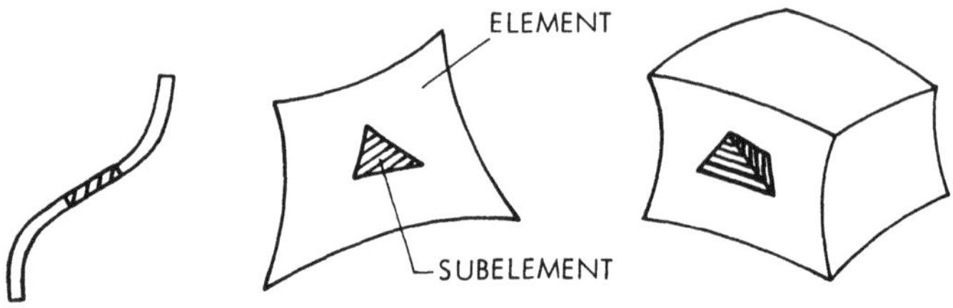

- RESOLUTION OF SUBELEMENT DISPLACEMENT INTO RIGID BODY AND DEFORMATION (NATURAL) PARTS

$$\bar{\boldsymbol{\rho}}_L = \begin{bmatrix} \boldsymbol{\rho}_{L0} \\ \boldsymbol{\rho}_{LN} \end{bmatrix} = \begin{bmatrix} \boldsymbol{a}_{L0} \\ \boldsymbol{a}_{LN} \end{bmatrix} \boldsymbol{\rho}$$

WHERE $\boldsymbol{\rho}$ CARTESIAN ELEMENT DISPLACEMENT

- FORCE - DISPLACEMENT RELATION FOR SUBELEMENT

$$\boldsymbol{P}_{LN} = \boldsymbol{k}_{LN} \boldsymbol{\rho}_{LN}$$

WHERE $\boldsymbol{k}_{LN}$ SUBELEMENT STIFFNESS CORRESPONDING TO URELEMENT (CONSTANT STRAIN) STIFFNESS

- FORCE - DISPLACEMENT RELATION FOR A FINITE NUMBER OF SUBELEMENTS

$$\boldsymbol{P} = \sum \boldsymbol{a}_{LN}^{it}(\boldsymbol{\zeta}_i) \boldsymbol{P}_{LN}^{i}(\boldsymbol{\zeta}_i) = \left[\sum \boldsymbol{a}_{LN}^{it}(\boldsymbol{\zeta}_i) \boldsymbol{k}_{LN}(\boldsymbol{\zeta}_i) \boldsymbol{a}_{LN}^{i}(\boldsymbol{\zeta}_i) \right] \boldsymbol{\rho} = \boldsymbol{k}\boldsymbol{\rho}$$

- ELEMENT STIFFNESS IN THE LIMIT

$$\boldsymbol{k} = \int_V \boldsymbol{a}_{LN}^{t}(\boldsymbol{\zeta}) \boldsymbol{k}_{LN}(\boldsymbol{\zeta}) \boldsymbol{a}_{LN}(\boldsymbol{\zeta})$$

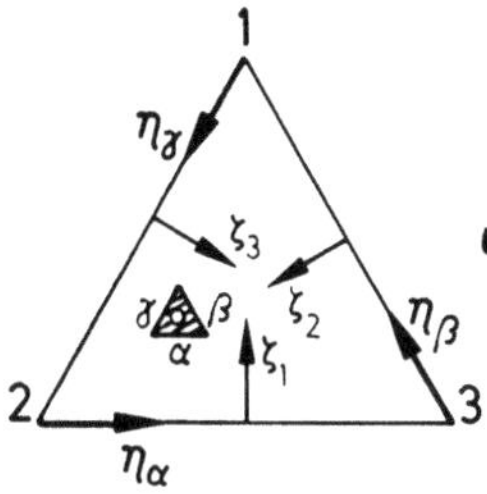

PARAMETER PLANE

MAPPING $\boldsymbol{x}(\boldsymbol{\zeta})$

INTERPOLATION OF CARTESIAN DISPL.

$$\boldsymbol{u}(\boldsymbol{\zeta}) = \boldsymbol{\omega}(\boldsymbol{\zeta})\boldsymbol{\rho}$$

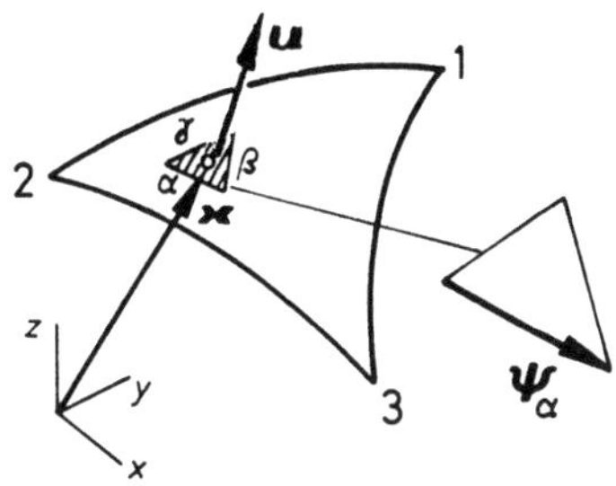

REAL SPACE

SIDE VECTORS OF SUBELEMENT $\qquad \boldsymbol{\psi}_\alpha = \dfrac{\partial \boldsymbol{x}}{\partial \eta_{\alpha t}} \, d\eta_\alpha$

SIDE LENGTHS

$$l_{L\alpha} = \left(\boldsymbol{\psi}_\alpha^t \, \boldsymbol{\psi}_\alpha\right)^{\frac{1}{2}} = m_\alpha \, d\eta_\alpha \qquad \text{WHERE} \qquad m_\alpha = \left(\frac{\partial \boldsymbol{x}^t}{\partial \eta_{\alpha t}} \frac{\partial \boldsymbol{x}}{\partial \eta_{\alpha t}}\right)^{\frac{1}{2}}$$

RELATIVE DISPLACEMENT IN DIRECTION α

$$d\boldsymbol{u}_\alpha = \frac{\partial \boldsymbol{u}}{\partial \eta_{\alpha t}} \, d\eta_\alpha = \frac{\partial \boldsymbol{\omega}}{\partial \eta_{\alpha t}} \, d\eta_\alpha \boldsymbol{\rho}$$

PROJECTION ONTO DIRECTION α

$$\rho_{L\alpha} = \frac{1}{m_\alpha} \frac{\partial \boldsymbol{x}^t}{\partial \eta_{\alpha t}} \frac{\partial \boldsymbol{\omega}}{\partial \eta_{\alpha t}} \, d\eta_\alpha \boldsymbol{\rho} = \boldsymbol{a}_{L\alpha} \boldsymbol{\rho}$$

SUBELEMENT STIFFNESS

$$\boldsymbol{k}_{LN} = \Omega_L \, t \, \boldsymbol{l}_L^{-1} \boldsymbol{\varkappa}_{ct} \boldsymbol{l}_L^{-1}$$

$$\text{WHERE} \quad \boldsymbol{l}_L = \left\lceil l_\alpha \;\; l_\beta \;\; l_\gamma \right\rfloor_L \, , \quad \Omega_L = \tfrac{1}{2} l_{L\alpha} l_{L\beta} \sin \gamma$$

ELEMENT STIFFNESS BY AREA INTEGRATION IN PARAMETER PLANE

ANALOGOUS PROCEDURE FOR QUADRILATERAL

MAPPING $\boldsymbol{x}(\xi_1, \xi_2)$

$\boldsymbol{u}(\xi_1, \xi_2)$

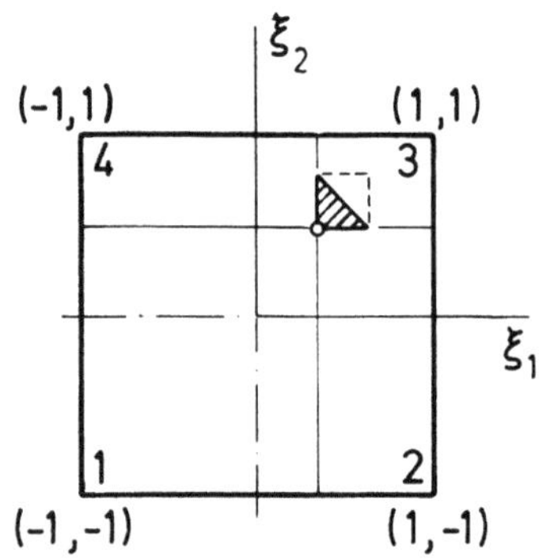

PARAMETER PLANE

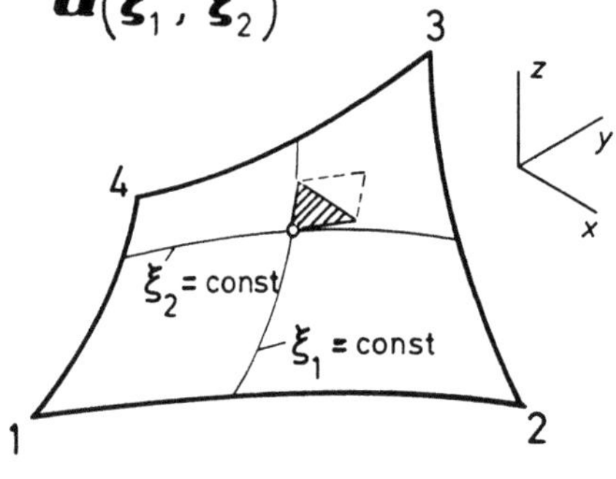

REAL SPACE

RELATION BETWEEN PARAMETERS, E.G. FOR η_α

$$\frac{\partial \boldsymbol{x}}{\partial \eta_{\alpha t}} = \sum_{i=1}^{2} \frac{\partial \boldsymbol{x}}{\partial \xi_i} \frac{\partial \xi_i}{\partial \eta_{\alpha t}} = -\frac{\partial \boldsymbol{x}}{\partial \xi_1} + \frac{\partial \boldsymbol{x}}{\partial \xi_2}$$

$$\frac{\partial \boldsymbol{u}}{\partial \eta_{\alpha t}} = -\frac{\partial \boldsymbol{u}}{\partial \xi_1} + \frac{\partial \boldsymbol{u}}{\partial \xi_2}$$

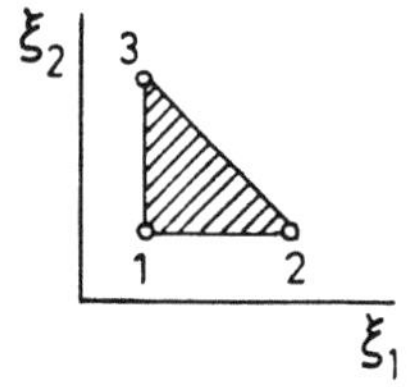

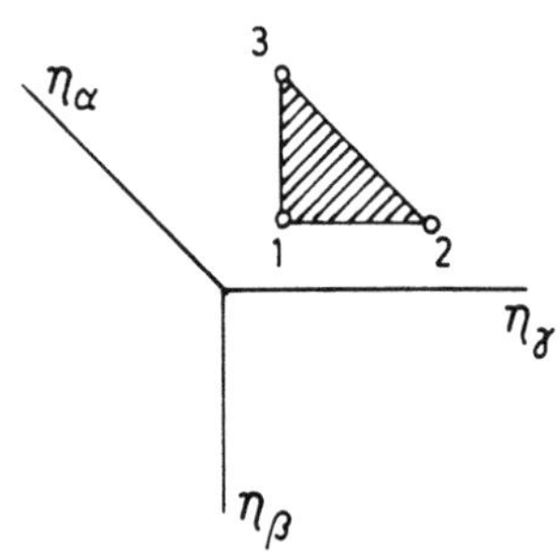

a) PROBLEM

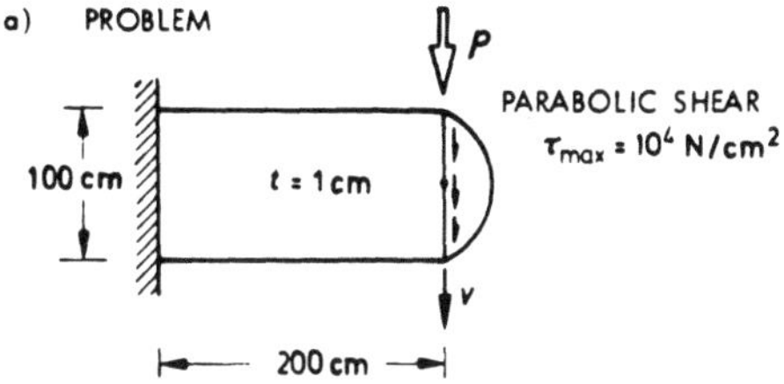

MODULUS OF ELASTICITY

$E = 2.0 * 10^7$ N/cm^2

POISSON'S RATIO

$\nu = 0.2$

COMPARISON SOLUTION

$\frac{Etv}{P} = 37.16$

b) TESTRESULTS

GRID	$\frac{Etv}{P}$ (FEM) QUAM$_N$ (SUBELEMENT)	QUAM$_C$ (CONVENTIONAL)	PERCENTAGE DIFFERENCE
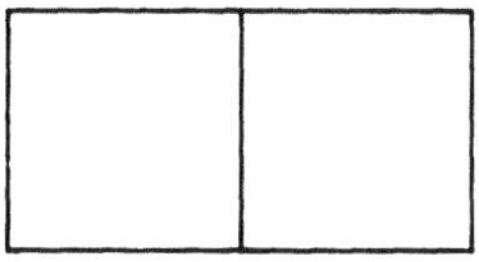	25.37	25.37	—
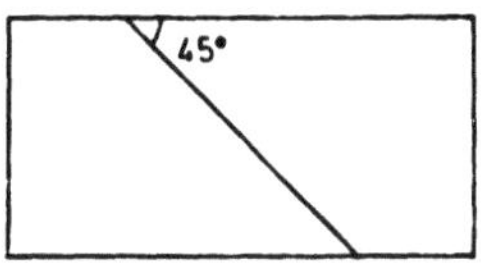	18.56	17.91	3.63
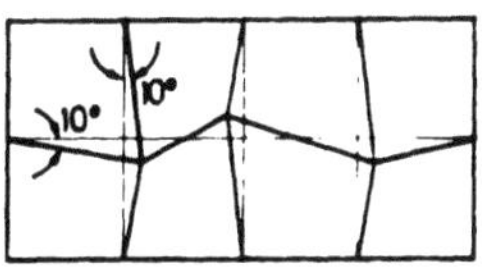	31.85	31.82	0.09
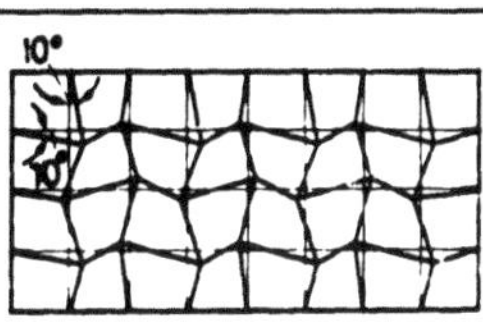	35.29	35.27	0.06

COMPARISON BETWEEN QUADRILATERAL MEMBRANE ELEMENTS QUAM$_N$ AND QUAM$_C$

THIN SHELL (KIRCHHOFF), ISOPARAMETRIC, FULLY COMPATIBLE

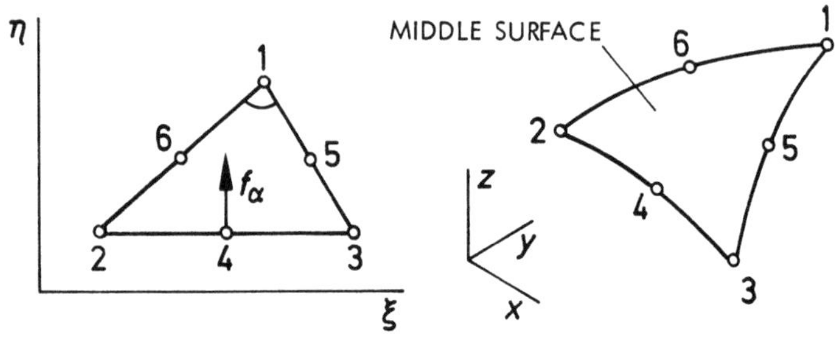

● INTERPOLATION: COMPLETE 5TH ORDER POLYNOMIAL IN $\boldsymbol{\zeta}$

DEGREES OF FREEDOM: 21 INCLUDING 3 RIGID BODY

SET OF 21 MODAL FUNCTIONS ω OF TYPE: $\zeta_1^i \zeta_2^j \zeta_3^k$, $i + j + k = 5$

$$\boldsymbol{\omega}_p = \lceil \bar{\boldsymbol{\omega}} \;\; \bar{\boldsymbol{\omega}} \;\; \bar{\boldsymbol{\omega}} \rfloor , \quad \bar{\boldsymbol{\omega}} = [\; \omega_1 \quad \omega_2 \cdots \; \omega_{21} \;]$$

● CARTESIAN (GENERALISED) NODAL DISPLACEMENT

$$\boldsymbol{\rho} = \{ \boldsymbol{\rho}_1 \;\; \boldsymbol{\rho}_2 \;\; \boldsymbol{\rho}_3 \;\; \boldsymbol{\rho}_4 \;\; \boldsymbol{\rho}_5 \;\; \boldsymbol{\rho}_6 \}$$

VERTICES: $\boldsymbol{\rho}_i = \{ \boldsymbol{u} \;\; \boldsymbol{u}_{,\xi} \;\; \boldsymbol{u}_{,\eta} \;\; \boldsymbol{u}_{,\xi\xi} \;\; \boldsymbol{u}_{,\xi\eta} \;\; \boldsymbol{u}_{,\eta\eta} \}_i$, $\boldsymbol{u}_{,\xi} = \dfrac{\partial \boldsymbol{u}}{\partial \xi}$

MIDPOINTS: NORMAL SLOPES, E.G. $\boldsymbol{\rho}_4 = \mathrm{d}\boldsymbol{u}/\mathrm{d}f_\alpha\big|_4$

● CARTESIAN DISPLACEMENT FIELD

$$\boldsymbol{u} = \boldsymbol{\omega}_p \boldsymbol{p}_u = \boldsymbol{\omega} \boldsymbol{\rho} \quad \text{WHERE } \boldsymbol{p}_u \text{ MODAL AMPLITUDES}$$

● GEOMETRY ANALOGOUSLY

$$\boldsymbol{x} = \boldsymbol{\omega}_p \, \boldsymbol{p}_x = \boldsymbol{\omega} \boldsymbol{x}_I$$

ELASTIC ELEMENT STIFFNESS

RESOLUTION INTO MEMBRANE AND BENDING PARTS

NATURAL SUBELEMENT DISPLACEMENT

$$\boldsymbol{\rho}_{LM} = \boldsymbol{a}_{LM}\,\boldsymbol{\rho} \qquad\qquad \boldsymbol{\rho}_{LB} = \boldsymbol{a}_{LB}\,\boldsymbol{\rho}$$

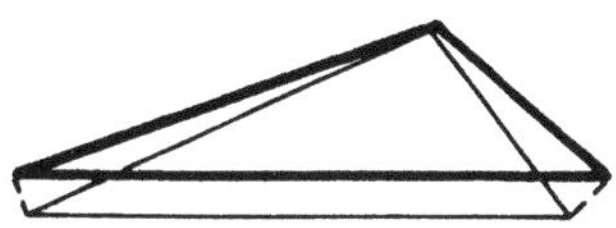

URELEMENT: TRIM3

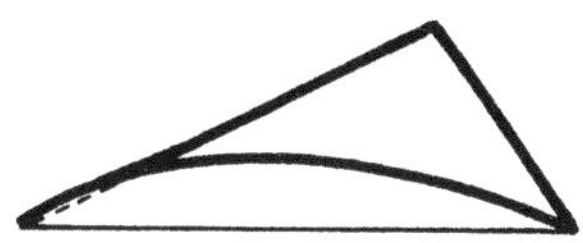

CONSTANT CURVATURE

ELEMENT STIFFNESS

$$\boldsymbol{k} = \boldsymbol{k}_M + \boldsymbol{k}_B = \int_{\Omega} \boldsymbol{a}^t_{LM}\,\boldsymbol{k}_{LM}\,\boldsymbol{a}_{LM} + \int_{\Omega} \boldsymbol{a}^t_{LB}\,\boldsymbol{k}_{LB}\,\boldsymbol{a}_{LB}$$

NATURAL SUBELEMENT STIFFNESS

$$\boldsymbol{k}_{LM} = \Omega_L\, t\, \boldsymbol{l}_L^{-1} \boldsymbol{\varkappa}_{ct}\, \boldsymbol{l}_L^{-1} \quad , \quad \boldsymbol{k}_{LB} = \Omega_L \frac{t^3}{12} \boldsymbol{l}_L^{-1} \boldsymbol{\varkappa}_{ct}\, \boldsymbol{l}_L^{-1}$$

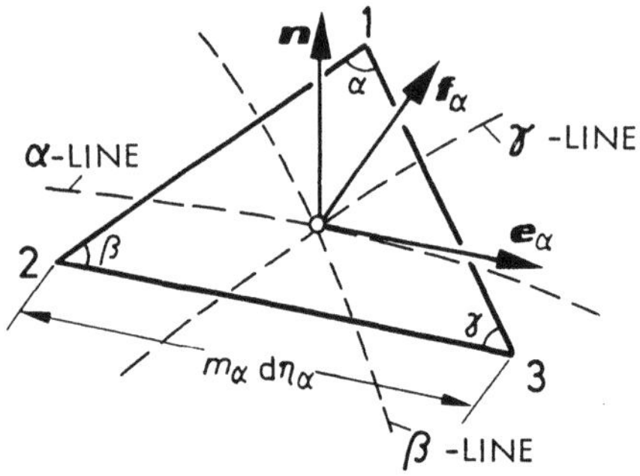

SUBELEMENT ON SHELL MIDDLE SURFACE

- SET OF 3 MOVING TRIPEDS ASSOCIATED WITH NATURAL DIRECTIONS, E.G.

$$\boldsymbol{D}_\alpha = [\, \boldsymbol{e}_\alpha \; \boldsymbol{f}_\alpha \; \boldsymbol{n} \,]$$

WHERE

$$\boldsymbol{e}_\alpha = m_\alpha^{-1} \frac{\partial \boldsymbol{x}}{\partial \eta_{\alpha t}} \;, \qquad \boldsymbol{n} = \frac{\boldsymbol{e}_\alpha \times \boldsymbol{e}_\beta}{|\boldsymbol{e}_\alpha \times \boldsymbol{e}_\beta|} \;, \qquad \boldsymbol{f}_\alpha = \boldsymbol{n} \times \boldsymbol{e}_\alpha$$

- VARIATION OF TRIPED ORIENTATION ALONG α-LINE

$$\frac{d\boldsymbol{D}_\alpha}{dt_\alpha} = m_\alpha^{-1} \frac{\partial \boldsymbol{D}_\alpha}{\partial \eta_{\alpha t}} = \boldsymbol{D}_\alpha \begin{bmatrix} 0 & -\varkappa_g & -\varkappa_n \\ \varkappa_g & 0 & -\tau_g \\ \varkappa_n & \tau_g & 0 \end{bmatrix}^\alpha$$

GEODESIC CURVATURE $\quad \varkappa_g^\alpha = m_\alpha^{-1} \, \boldsymbol{f}_\alpha^t \, \dfrac{\partial \boldsymbol{e}_\alpha}{\partial \eta_{\alpha t}}$

NORMAL CURVATURE $\quad \varkappa_n^\alpha = m_\alpha^{-1} \, \boldsymbol{n}^t \, \dfrac{\partial \boldsymbol{e}_\alpha}{\partial \eta_{\alpha t}}$

GEODESIC TORSION $\quad \tau_g^\alpha = -m_\alpha^{-1} \, \boldsymbol{f}_\alpha^t \, \dfrac{\partial \boldsymbol{n}}{\partial \eta_{\alpha t}}$

DETERMINATION OF $\boldsymbol{a}_{LM}$ AND $\boldsymbol{a}_{LB}$, KINEMATICS

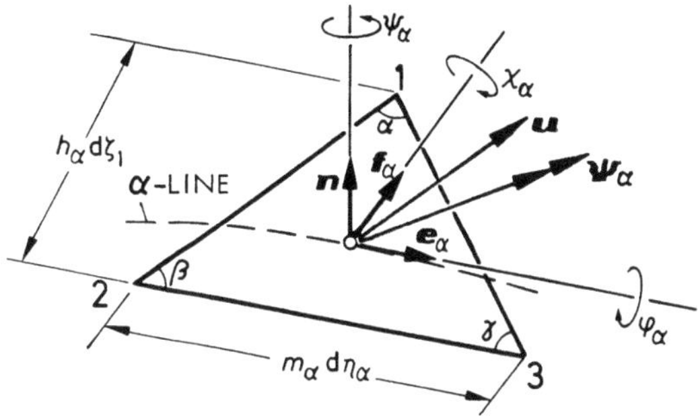

KINEMATICS OF LINE ELEMENT α

- MEMBRANE PART AS PREVIOUSLY:

$$\rho_{M\alpha} = \boldsymbol{e}_\alpha^t \frac{\partial \boldsymbol{u}}{\partial \eta_{\alpha t}} d\eta_\alpha \qquad \text{AND} \qquad \boldsymbol{u} = \boldsymbol{\omega\rho} \longrightarrow \rho_{M\alpha} = \boldsymbol{a}_{M\alpha}\boldsymbol{\rho}$$

- BENDING PART:

ROTATION VECTOR OF LINE ELEMENT α

$$\boldsymbol{\psi}_\alpha = \varphi_\alpha \boldsymbol{e}_\alpha + \chi_\alpha \boldsymbol{f}_\alpha + \psi_\alpha \boldsymbol{n}$$

$$\varphi_\alpha = h_\alpha^{-1}\boldsymbol{n}^t \frac{\partial \boldsymbol{u}}{\partial \zeta_{1t}} \quad , \quad \chi_\alpha = -m_\alpha^{-1}\boldsymbol{n}^t \frac{\partial \boldsymbol{u}}{\partial \eta_{\alpha t}} \quad , \quad \psi_\alpha = m_\alpha^{-1}\boldsymbol{f}_\alpha^t \frac{\partial \boldsymbol{u}}{\partial \eta_{\alpha t}}$$

RELATIVE ROTATION IN DIRECTION α

$$d\boldsymbol{\psi}_\alpha = \frac{\partial \boldsymbol{\psi}_\alpha}{\partial \eta_{\alpha t}} d\eta_\alpha$$

PROJECTION ONTO $\boldsymbol{f}_\alpha$

$$\rho_{B\alpha} = \boldsymbol{f}_\alpha^t \frac{\partial \boldsymbol{\psi}_\alpha}{\partial \eta_{\alpha t}} d\eta_\alpha$$

$$= \Bigg[\left(x_n^\alpha \boldsymbol{e}_\alpha^t + m_\alpha^{-2}\frac{\partial m_\alpha}{\partial \eta_{\alpha t}}\boldsymbol{n}^t\right)\frac{\partial \boldsymbol{u}}{\partial \eta_{\alpha t}} + \cot\beta\, x_g^\alpha \boldsymbol{n}^t \frac{\partial \boldsymbol{u}}{\partial \eta_{\beta t}}$$

$$-\cot\gamma\, x_g^\alpha \boldsymbol{n}^t \frac{\partial \boldsymbol{u}}{\partial \eta_{\gamma t}} - m_\alpha^{-1}\boldsymbol{n}^t \frac{\partial^2 \boldsymbol{u}}{\partial \eta_{\alpha t}^2}\Bigg] d\eta_\alpha$$

$$\text{AND} \qquad \boldsymbol{u} = \boldsymbol{\omega\rho} \longrightarrow \rho_{B\alpha} = \boldsymbol{a}_{B\alpha}\boldsymbol{\rho}$$

ISOPARAMETRIC TRIANGULAR BENDING ELEMENT INCLUDING TRANSVERSE SHEAR (TRBSC6)

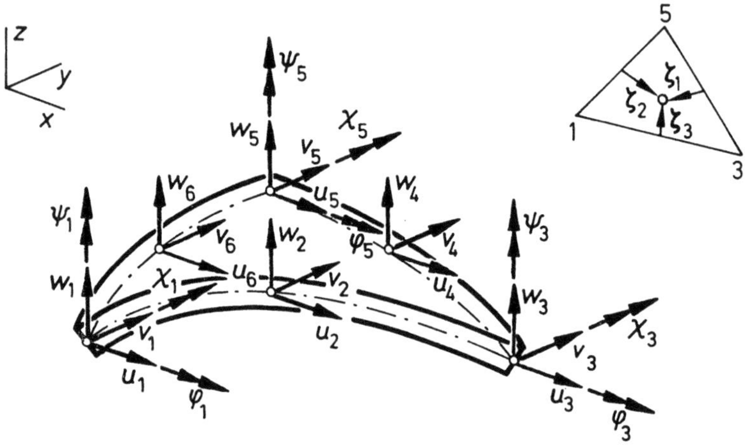

NODAL DISPLACEMENTS $\boldsymbol{\rho} = \left\{ \boldsymbol{\rho}_T \quad \boldsymbol{\rho}_R \right\}$

TRANSLATION $\boldsymbol{\rho}_T$, ROTATION $\boldsymbol{\rho}_R$

DISPLACEMENT FIELD

TRANSLATION $\boldsymbol{u} = \boldsymbol{\omega}_n \boldsymbol{\rho}_T$

ROTATION $\boldsymbol{\varphi} = \boldsymbol{\omega}_m \boldsymbol{\rho}_R$

WHERE: $\boldsymbol{\omega}_n = \left\lceil \bar{\boldsymbol{\omega}}_n \quad \bar{\boldsymbol{\omega}}_n \quad \bar{\boldsymbol{\omega}}_n \right\rfloor$

$\boldsymbol{\omega}_m = \left\lceil \bar{\boldsymbol{\omega}}_m \quad \bar{\boldsymbol{\omega}}_m \quad \bar{\boldsymbol{\omega}}_m \right\rfloor$

$$\bar{\boldsymbol{\omega}}_n = \left[\zeta_1(2\zeta_1-1) \quad 4\zeta_1\zeta_2 \quad \zeta_2(2\zeta_2-1) \quad 4\zeta_2\zeta_3 \quad \zeta_3(2\zeta_3-1) \quad 4\zeta_3\zeta_1 \right]$$

$$\bar{\boldsymbol{\omega}}_m = \left[\zeta_1 \quad \zeta_2 \quad \zeta_3 \right]$$

NATURAL DISPLACEMENTS OF SUBELEMENT IN TERMS OF CARTESIAN NODAL POINT DISPLACEMENTS

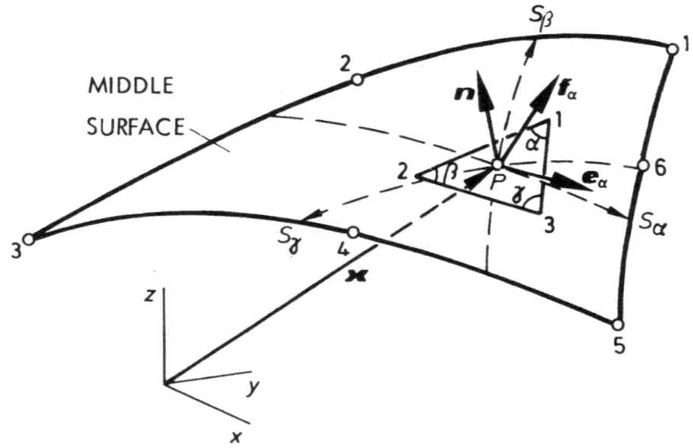

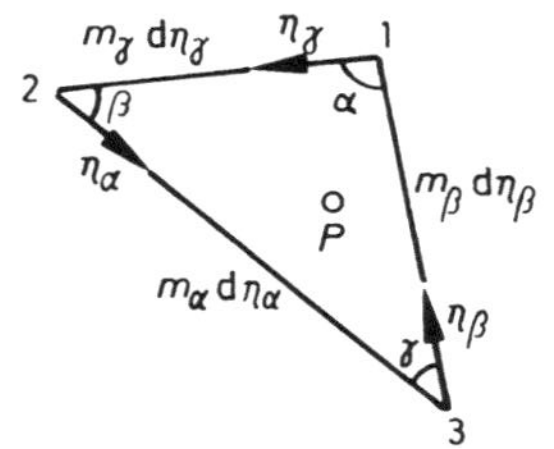

SUBELEMENT
(POINT P)

E.G. SIDE $\boldsymbol{\alpha}$:

● MEMBRANE PART:

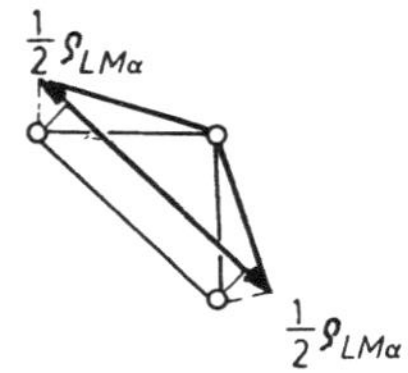

$$\varrho_{LM\alpha} = \boldsymbol{e}_\alpha^t \, d\boldsymbol{u}_\alpha = \boldsymbol{e}_\alpha^t \frac{\partial \boldsymbol{\omega}_n}{\partial \eta_{\alpha t}} d\eta_\alpha \boldsymbol{\varrho}_T = \boldsymbol{a}_{LM\alpha} \boldsymbol{\varrho}_T$$

● BENDING PART:

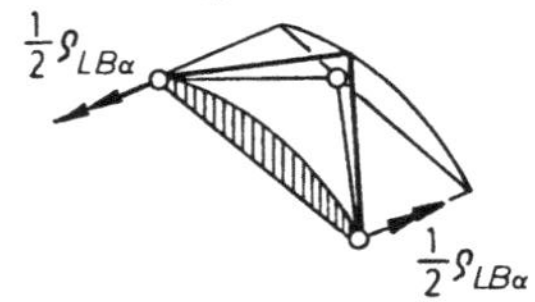

$$\varrho_{LB\alpha} = \boldsymbol{f}_\alpha^t \, d\boldsymbol{\varphi}_\alpha = \boldsymbol{f}_\alpha^t \frac{\partial \boldsymbol{\omega}_m}{\partial \eta_{\alpha t}} d\eta_\alpha \boldsymbol{\varrho}_R = \boldsymbol{a}_{LB\alpha} \boldsymbol{\varrho}_R$$

● SHEAR PART:

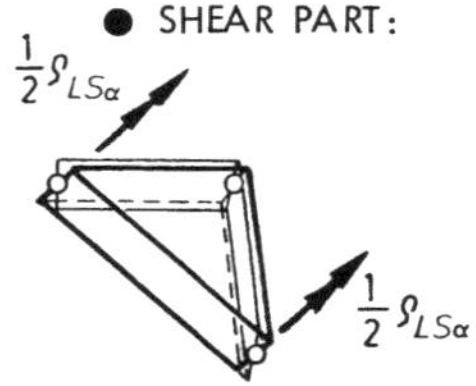

$$\varrho_{LS\alpha} = \frac{2}{m_\alpha d\eta_\alpha} \boldsymbol{n}^t d\boldsymbol{u}_\alpha + 2\boldsymbol{f}_\alpha^t \boldsymbol{\varphi}_\alpha = \frac{2}{m_\alpha d\eta_\alpha} \boldsymbol{n}^t \frac{\partial \boldsymbol{\omega}_n}{\partial \eta_{\alpha t}} d\eta_\alpha \boldsymbol{\varrho}_T + 2\boldsymbol{f}_\alpha^t \boldsymbol{\omega}_m \boldsymbol{\varrho}_R = \boldsymbol{a}_{LS1\alpha} \boldsymbol{\varrho}_T + \boldsymbol{a}_{LS2\alpha} \boldsymbol{\varrho}_R$$

ELASTIC STIFFNESS MATRIX OF FINITE ELEMENT

$$\boldsymbol{k} = \int_V \boldsymbol{a}_{LN}^t \, \boldsymbol{k}_{LN} \, \boldsymbol{a}_{LN}$$

- SUBELEMENT KINEMATICS

$$\boldsymbol{\rho}_{LN} = \begin{bmatrix} \boldsymbol{\rho}_M \\ \boldsymbol{\rho}_B \\ \boldsymbol{\rho}_S \end{bmatrix}_L = \left[\begin{array}{c:c} \boldsymbol{a}_{LM} & \\ \hdashline & \boldsymbol{a}_{LB} \\ \hdashline \boldsymbol{a}_{LS1} & \boldsymbol{a}_{LS2} \end{array}\right] \begin{bmatrix} \boldsymbol{\rho}_T \\ \boldsymbol{\rho}_R \end{bmatrix} = \boldsymbol{a}_{LN} \, \boldsymbol{\rho}$$

- SUBELEMENT STIFFNESS

$$\boldsymbol{k}_{LN} = \left\lceil \boldsymbol{k}_{LN} \; \boldsymbol{k}_{LB} \; \boldsymbol{k}_{LS} \right\rfloor$$

WHERE: $\boldsymbol{k}_{LN}$, $\boldsymbol{k}_{LB}$ AS PREVIOUSLY

$\boldsymbol{k}_{LS} = 2G \mathcal{A}^{-1} \Omega_L \, t$ (ISOTROPIC MATERIAL)

ELIMINATION OF SHEAR LOCKING EFFECT FOR THIN ELEMENTS

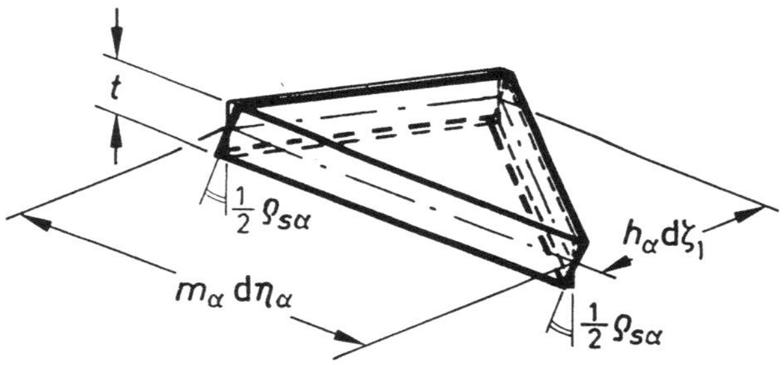

ELEMENT KINEMATICS

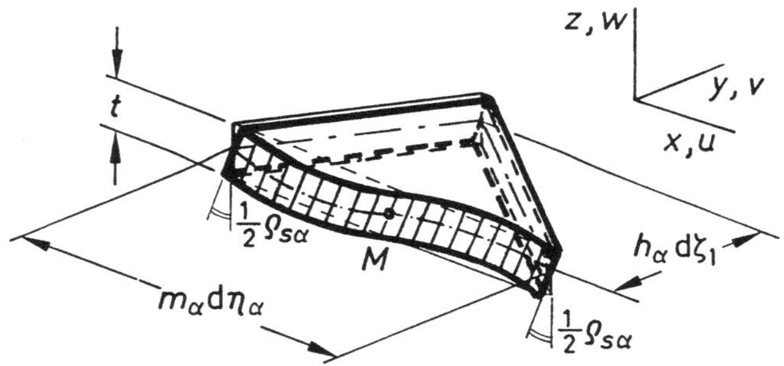

DESIRED SHAPE

SHEAR ENERGY FOR $\varrho_{s\alpha} = 1$ $\quad u_{s\alpha} = \frac{1}{2} G t\, m_\alpha h_\alpha$

BENDING ENERGY FOR $\varrho_{s\alpha} = 1$ $\quad u_{B\alpha} = \int_0^1 u_B(\zeta_1)\, d(\zeta_1)$

$$f_\alpha = \frac{u_{B\alpha}}{u_{s\alpha}}$$

REDUCED SHEAR STIFFNESS MATRIX

$$\left(2G\mathcal{A}^{-1}\right)^{red} = 2G\,\mathbf{f}^t \mathcal{A}^{-1} \mathbf{f} \qquad \text{WHERE} \qquad \mathbf{f} = \left\lceil f_\alpha \quad f_\beta \quad f_\gamma \right\rfloor^{1/2}$$

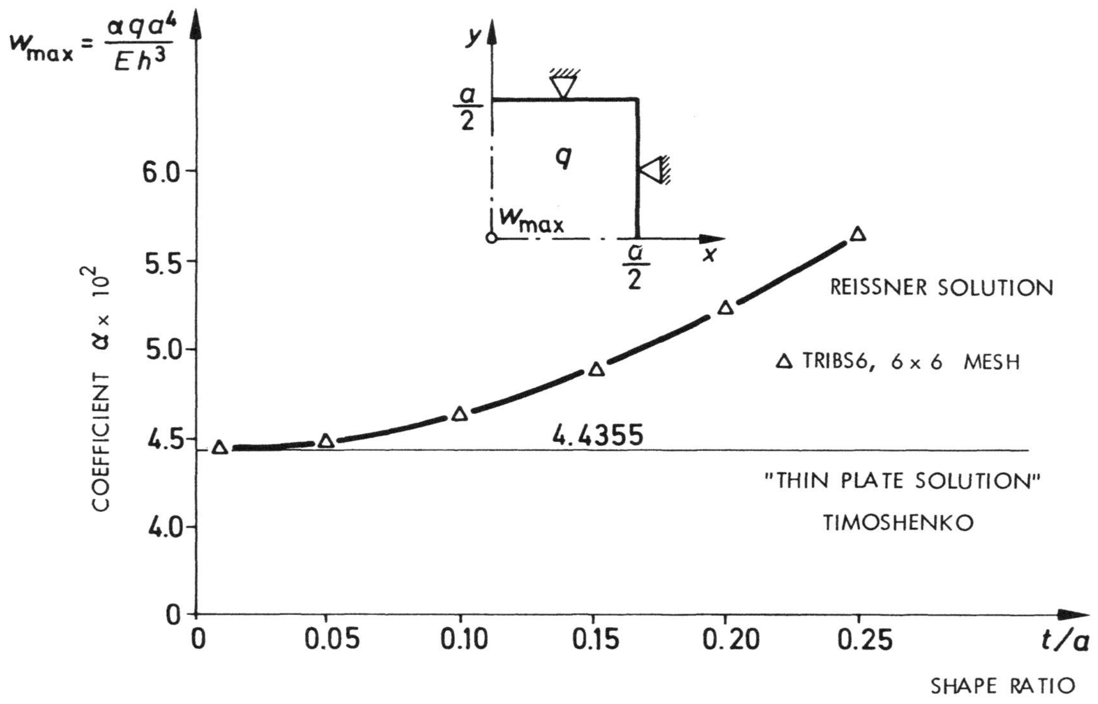

INFLUENCE OF TRANSVERSE SHEAR EFFECTS ON THE MAXIMUM DISPLACEMENT OF A PLATE UNDER UNIFORMLY DISTRIBUTED PRESSURE

TANGENT STIFFNESS OF FINITE ELEMENT

NONLINEAR NATURAL DEFORMATIONS $\tilde{\rho}_{Ni}$ ENTAILING THE SAME STRAIN FIELDS AS CORRESPONDING ρ_{Ni} OF LINEAR THEORY

$$\tilde{\rho}_{Ni} = \boldsymbol{a}_{Ni}\boldsymbol{\rho} + \frac{1}{2}\boldsymbol{\rho}^t \boldsymbol{\Lambda}_i \boldsymbol{\rho}$$

$$= \rho_{Ni} + \frac{1}{2}\bar{\boldsymbol{\rho}}^t \begin{bmatrix} \boldsymbol{\Lambda}_{i00} & \boldsymbol{\Lambda}_{i0N} \\ \boldsymbol{\Lambda}_{iN0} & \boldsymbol{\Lambda}_{iNN} \end{bmatrix} \bar{\boldsymbol{\rho}} \qquad \text{WITH} \quad \bar{\boldsymbol{\rho}} = \{\boldsymbol{\rho}_o \ \boldsymbol{\rho}_N\}$$

DERIVATIVES OF NONLINEAR NATURAL DEFORMATIONS

$$\frac{\partial \tilde{\rho}_{Ni}}{\partial \boldsymbol{\rho}^t} = \boldsymbol{a}_{Ni} + \boldsymbol{\rho}^t \boldsymbol{\Lambda}_i \quad ; \quad \frac{\partial^2 \tilde{\rho}_{Ni}}{\partial \boldsymbol{\rho} \partial \boldsymbol{\rho}^t} = \boldsymbol{\Lambda}_i$$

STRAIN ENERGY

$$\bar{U} = \frac{1}{2}\tilde{\boldsymbol{\rho}}_N^t \boldsymbol{k}_{NE} \tilde{\boldsymbol{\rho}}_N$$

TANGENTIAL STIFFNESS

$$\boldsymbol{k}_T = \frac{\partial^2 \bar{U}}{\partial \boldsymbol{\rho} \partial \boldsymbol{\rho}^t}$$

$$\boldsymbol{k}_T = \boldsymbol{a}_N^t \boldsymbol{k}_{NE} \boldsymbol{a}_N + \sum_i P_{Ni} \boldsymbol{\Lambda}_i = \boldsymbol{k}_E + \boldsymbol{k}_G$$

DERIVATION OF $\boldsymbol{\Lambda}_{i00}$ AND $\boldsymbol{\Lambda}_{i0N}$

ELEMENT SUBJECTED TO RIGID BODY ROTATION $\boldsymbol{\rho}_0$ DESCRIBED BY $\boldsymbol{T}$

$$\boldsymbol{T}(\boldsymbol{\rho}_0) = \boldsymbol{I}_3 + \boldsymbol{S} + \frac{1}{2}\boldsymbol{S}^2 \quad \text{WITH} \quad \boldsymbol{S} = \begin{bmatrix} 0 & -\rho_{oz} & \rho_{oy} \\ \rho_{oz} & 0 & -\rho_{ox} \\ -\rho_{oy} & \rho_{ox} & 0 \end{bmatrix}$$

DISPLACEMENT VECTOR AFTER ROTATION

$$\boldsymbol{\rho}' = \boldsymbol{T}(\boldsymbol{\rho}_0)\,\boldsymbol{\rho}$$

NATURAL DEFORMATIONS ARISING FROM $d\boldsymbol{\rho}$

$$d\tilde{\rho}_{Ni} = \boldsymbol{a}_{Ni}\,\boldsymbol{T}(\boldsymbol{\rho}_0)\,d\boldsymbol{\rho}$$

DIFFERENTIATION

$$\frac{\partial^2 \tilde{\rho}_{Ni}}{\partial \boldsymbol{\rho}_0\, \partial \boldsymbol{\rho}^t} = \frac{\partial}{\partial \boldsymbol{\rho}_0}\left[\boldsymbol{a}_{Ni}\,\boldsymbol{T}(\boldsymbol{\rho}_0)\right]$$

APPLICATION OF THE RELATION $\boldsymbol{\rho} = \left[\boldsymbol{A}_0 \quad \boldsymbol{A}_N\right]\bar{\boldsymbol{\rho}}$ AND THE CHAIN RULE LEADS TO

$$\boldsymbol{\Lambda}_{i00} = \frac{\partial}{\partial \boldsymbol{\rho}_0}\left[\boldsymbol{a}_{Ni}\,\boldsymbol{T}(\boldsymbol{\rho}_0)\right]\boldsymbol{A}_0$$

$$\boldsymbol{\Lambda}_{i0N} = \frac{\partial}{\partial \boldsymbol{\rho}_0}\left[\boldsymbol{a}_{Ni}\,\boldsymbol{T}(\boldsymbol{\rho}_0)\right]\boldsymbol{A}_N$$

GEOMETRICAL STIFFNESS OF SUBELEMENT

NONLINEAR NATURAL DISPLACEMENT RELATION FOR INFINITESIMAL SUBELEMENT

$$d\tilde{\rho}_{LNi} = d\rho_{LNi} + \frac{1}{2}\begin{bmatrix} \boldsymbol{\rho}_{L0}^{t} & d\boldsymbol{\rho}_{LN}^{t} \end{bmatrix} \begin{bmatrix} \boldsymbol{\Lambda}_{Li00} & \boldsymbol{\Lambda}_{Li0N} \\ \boldsymbol{\Lambda}_{LiN0} & \mathbf{0} \end{bmatrix} \begin{bmatrix} \boldsymbol{\rho}_{L0} \\ d\boldsymbol{\rho}_{LN} \end{bmatrix}$$

REMARKS: 1) FOR SMALL STRAINS $d\boldsymbol{\rho}_{LN}^{t}\,\boldsymbol{\Lambda}_{LiNN}\,d\boldsymbol{\rho}_{LN}$ MAY BE NEGLECTED

2) FOR CONSERVATIVE SYSTEMS $\boldsymbol{\Lambda}_{LiN0} = \boldsymbol{\Lambda}_{Li0N}^{t}$

TANGENT STIFFNESS OF SUBELEMENT

$$\boldsymbol{k}_{LT} = \boldsymbol{k}_{LE} + \boldsymbol{k}_{LG} = \begin{bmatrix} \mathbf{0} & \mathbf{0} \\ \mathbf{0} & \boldsymbol{k}_{LNE} \end{bmatrix} + \begin{bmatrix} \boldsymbol{k}_{LG00} & \boldsymbol{k}_{LG0N} \\ \boldsymbol{k}_{LGN0}^{t} & \mathbf{0} \end{bmatrix}$$

WITH

$$\boldsymbol{k}_{LG00} = \sum_i P_{LNi}\,\boldsymbol{\Lambda}_{Li00}$$

$$\boldsymbol{k}_{LG0N} = \sum_i P_{LNi}\,\boldsymbol{\Lambda}_{Li0N}$$

$\boldsymbol{\Lambda}_{Li00}$ AND $\boldsymbol{\Lambda}_{Li0N}$ ARE OBTAINED FROM $\boldsymbol{\Lambda}_{i00}$ AND $\boldsymbol{\Lambda}_{i0N}$ BY LIMITING PROCEDURE

INTEGRATION OVER SUBELEMENTS LEADS TO GEOMETRICAL STIFFNESS

$$\boldsymbol{k}_{G} = \int_V \begin{bmatrix} \boldsymbol{a}_{L0}^{t}(\boldsymbol{\zeta}) & \boldsymbol{a}_{LN}^{t}(\boldsymbol{\zeta}) \end{bmatrix} \boldsymbol{k}_{LG}(\boldsymbol{\zeta}) \begin{bmatrix} \boldsymbol{a}_{L0}(\boldsymbol{\zeta}) \\ \boldsymbol{a}_{LN}(\boldsymbol{\zeta}) \end{bmatrix}$$

INELASTIC BEHAVIOUR

LAYER MODEL

- VARIATION OF MATERIAL PROPERTIES ALONG THICKNESS

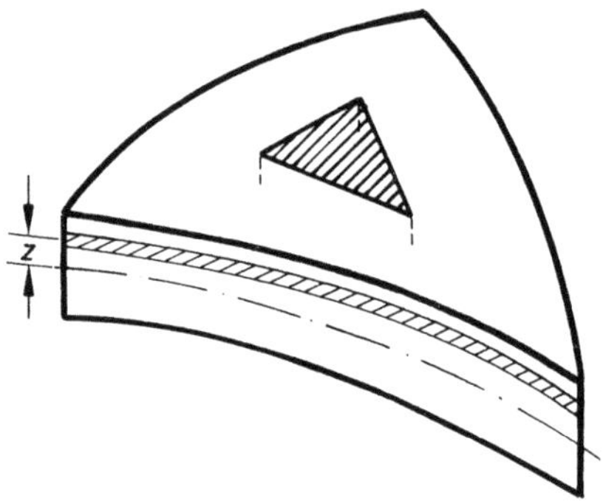

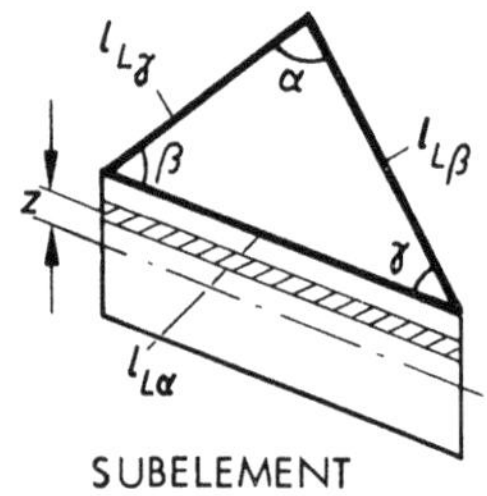

INDIVIDUAL LAYER

- KINEMATICS

TOTAL NATURAL STRAIN $\quad \boldsymbol{\gamma}_t = \boldsymbol{l}_L^{-1} (\boldsymbol{\rho}_{LM} + z\boldsymbol{\rho}_{LB})$

WHERE: $\boldsymbol{\gamma}_t = \{ \gamma_\alpha \quad \gamma_\beta \quad \gamma_\gamma \}_t \,, \quad \boldsymbol{l}_L = \lceil l_\alpha \quad l_\beta \quad l_\gamma \rfloor_L$

- CONSTITUTIVE ASSUMPTIONS

$$\boldsymbol{\gamma}_t = \boldsymbol{\varepsilon}_t + \boldsymbol{\eta}_t$$

ELASTIC PART: $\quad \boldsymbol{\varepsilon}_t = \boldsymbol{\kappa}_{ct}^{-1} \, \boldsymbol{\sigma}_c$

INELASTIC PART: $\dot{\boldsymbol{\eta}}_t$ FROM PLASTICITY, VISCOPLASTICITY

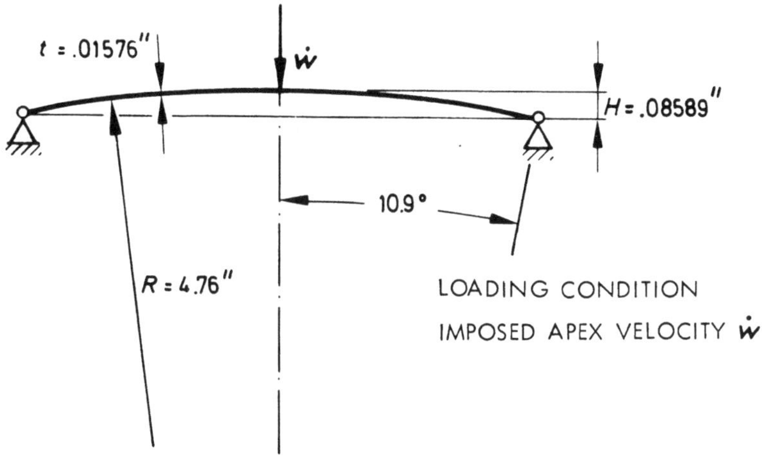

(a) GEOMETRY AND LOADING

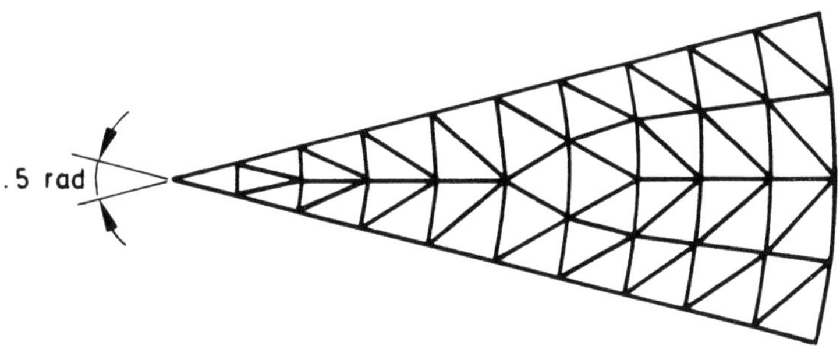

(b) FINITE ELEMENT MESH (TRUMP, 10 LAYERS)

SHALLOW SPHERICAL CAP SUBJECTED TO
IMPOSED VELOCITY AT THE APEX

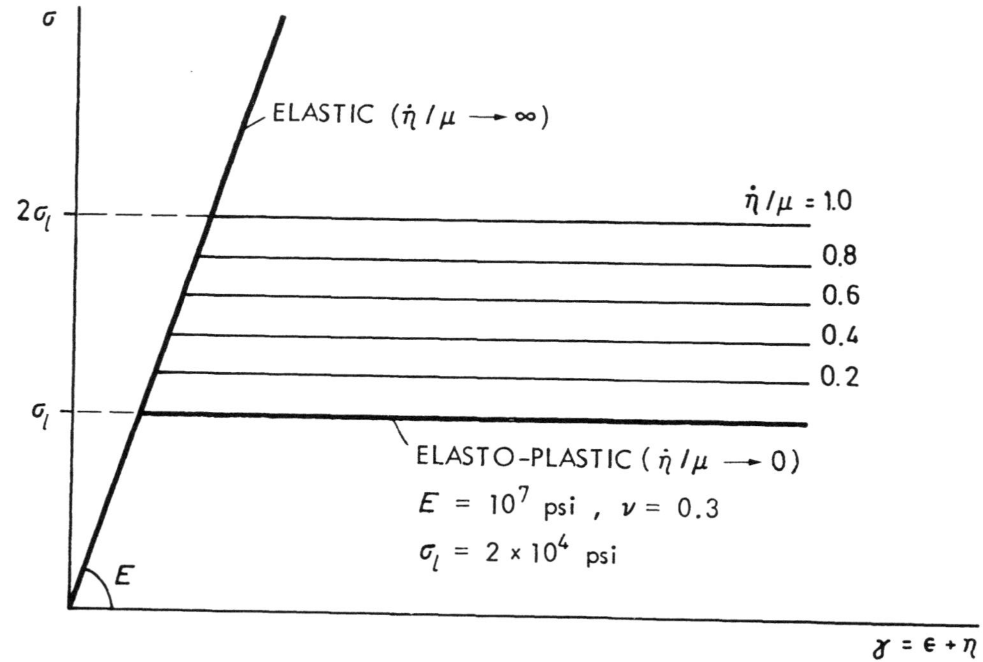

ELASTO-VISCOPLASTIC MATERIAL PROPERTIES

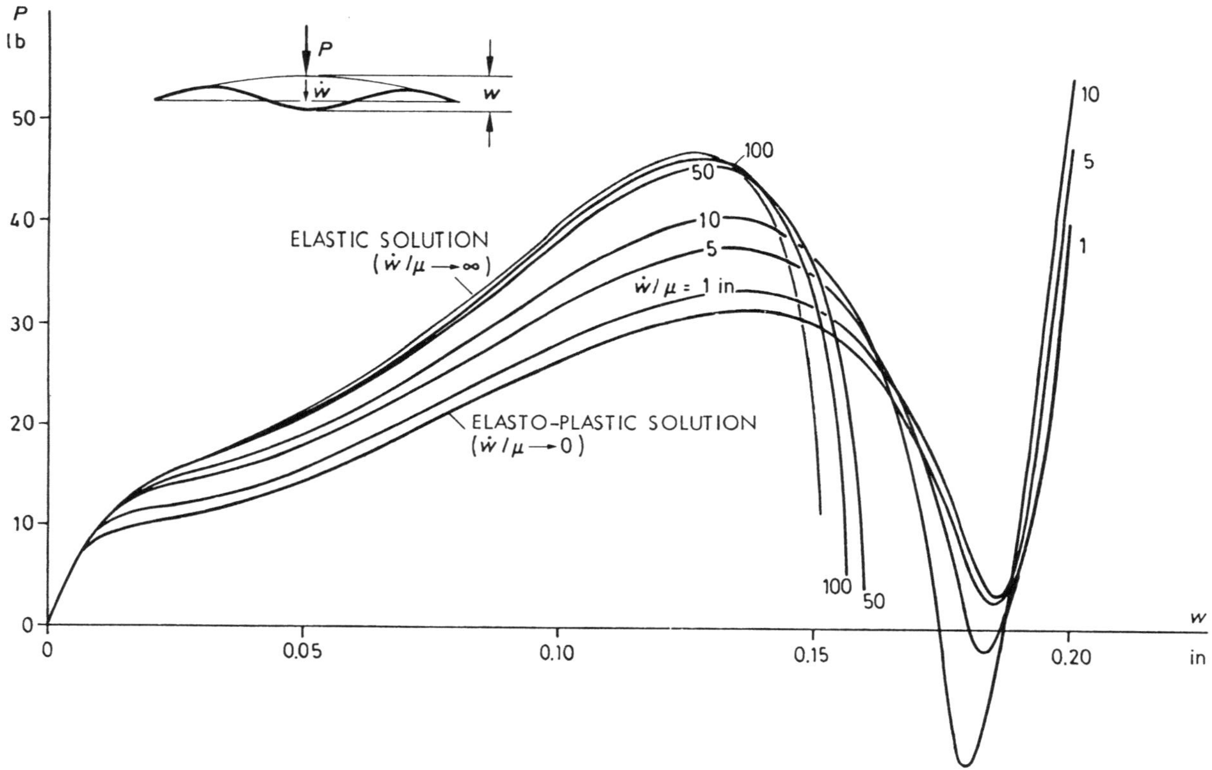

LOAD-DEFLECTION CURVES FOR DIFFERENT APPLIED VELOCITIES

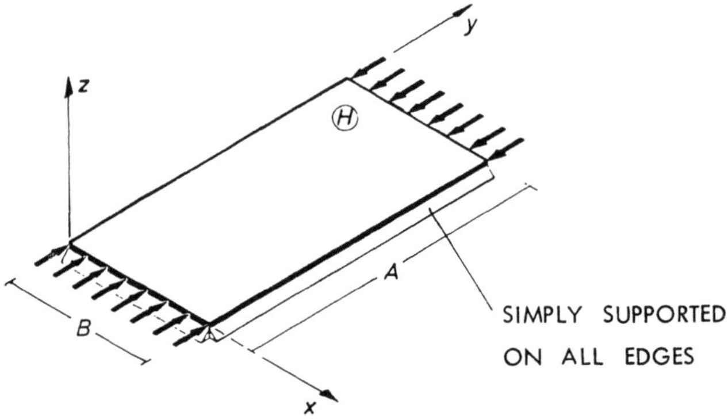

GEOMETRICAL DATA:

H = 1 mm

A = 80 mm

B = 40 mm

MATERIAL DATA

E = 7000 kp/mm^2

ν = 0.3

$$D = \frac{EH^3}{12(1-\nu^2)} = 641.0 \text{ kp mm}$$

ANALYTIC SOLUTION (TIMOSHENKO)

MODES: $W = \sin \frac{n\pi x}{B} \sin \frac{n\pi y}{A}$

WHERE

n	m	σ_{kr} [kp/mm^2]
1	2	15.80
1	3	18.52
1	1	24.69
1	4	24.69
1	5	33.22

BUCKLING OF A SIMPLY SUPPORTED RECTANGULAR PLATE: DATA AND ANALYTIC RESULTS

	ERROR %	43 d.o.f 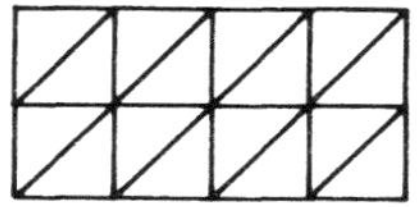			167 d.o.f. 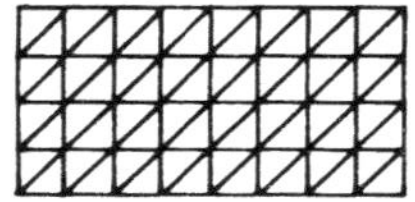			655 d.o.f. 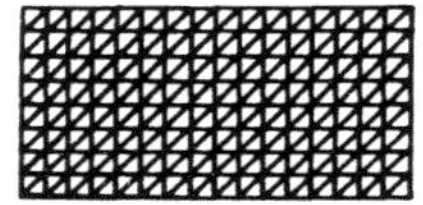		
		TRUNC	TRUMP	TRIB3	TRUNC	TRUMP	TRIB3	TRUNC	TRUMP	TRIB3
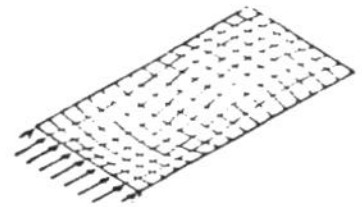 $\lambda_1 = 15.80$	f_1	− 3,4	− 5,9	+ 19,6	− 1,5	− 2,3	+ 4,6	− 0,4	− 0,1	+ 1,2
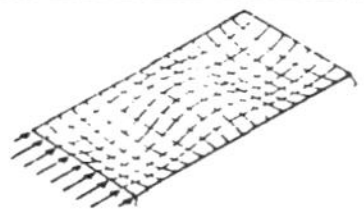 $\lambda_2 = 18.52$	f_2	− 3,5	+ 13,3	+ 36,7	− 1,5	+ 2,1	+ 8,1	− 0,4	+ 0,5	+ 2,1
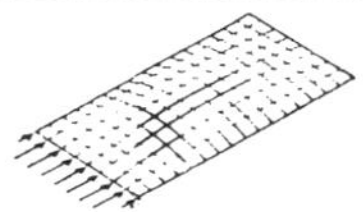 $\lambda_3 = 24.69$	f_3	− 2,2	− 3,4	+ 13,4	− 0,7	− 1,1	+ 3,1	− 0,1	− 0,2	+ 0,7
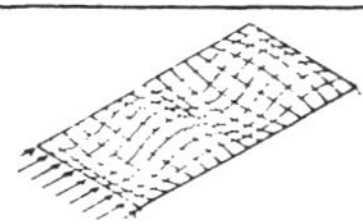 $\lambda_4 = 24.69$	f_4	+ 40,5	+ 7068.9	+ 57,8	− 1,1	+ 12,1	+ 13,5	− 0,6	+ 2,8	+ 3,0

CONVERGENCE TEST FOR TRUNC GEOMETRICAL STIFFNESS

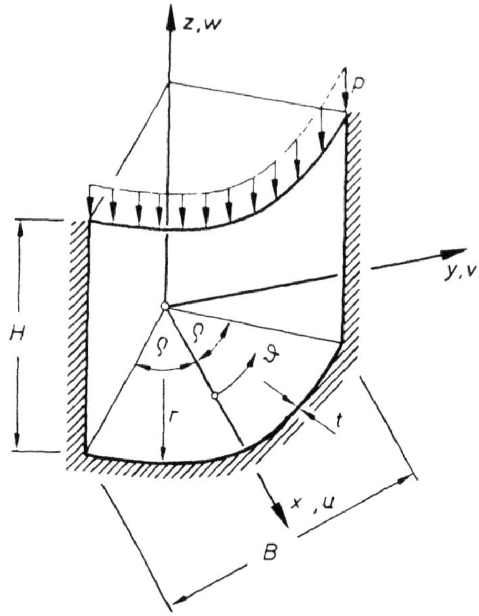

GEOMETRICAL DATA:

$B = 100$ cm

$H = B$

$\rho = 60°$

$r = \dfrac{B}{2\sin\rho} = 57.735$ cm

$t = r/500 = 0.11547$ cm

MATERIAL DATA:

$E = 2.1 * 10^6$ kp cm^{-2}

$\nu = 0.3$

LOADING: LINE-LOAD p PARALLEL TO z AXIS

LINEAR BUCKLING OF CYLINDRICAL PANEL UNDER LINE-LOAD
GEOMETRY AND ELASTIC PROPERTIES

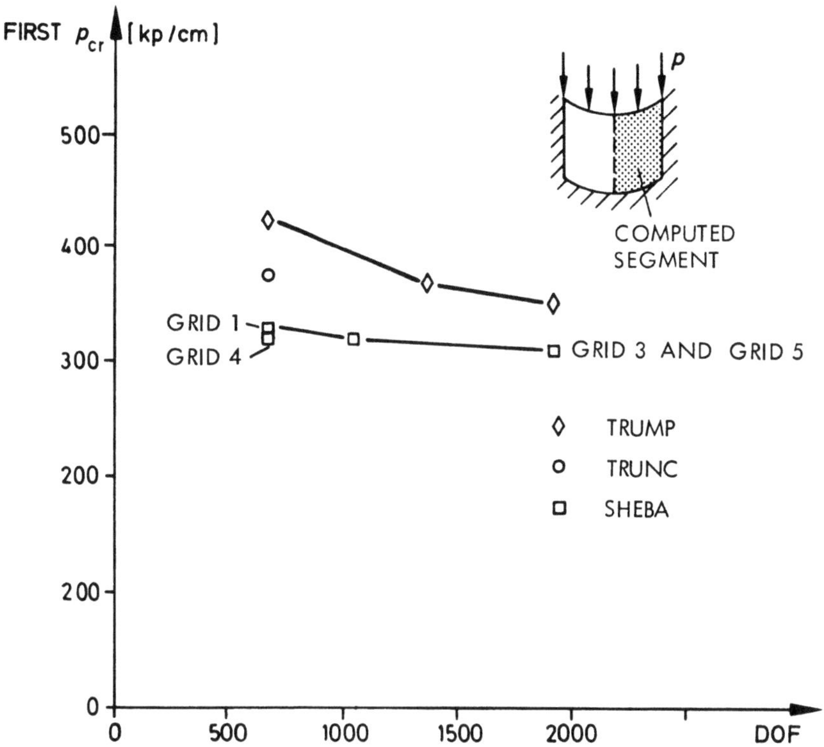

ELEMENT	DOF	FIRST p_{cr}	SECOND p_{cr}
TRUMP	658	422.7 (+ 37.6%)	423.4
"	1354	368.0 (+ 19.8%)	368.8
"	1954	350.0 (+ 14.0%)	352.3
TRUNC	658	363.6 (+18.4%)	370.8
SHEBA	616	307.2 (ref.sol.)	307.9

LINEAR BUCKLING OF A CYLINDRICAL SHELL UNDER LINE LOAD: CRITICAL LINE LOAD VALUES

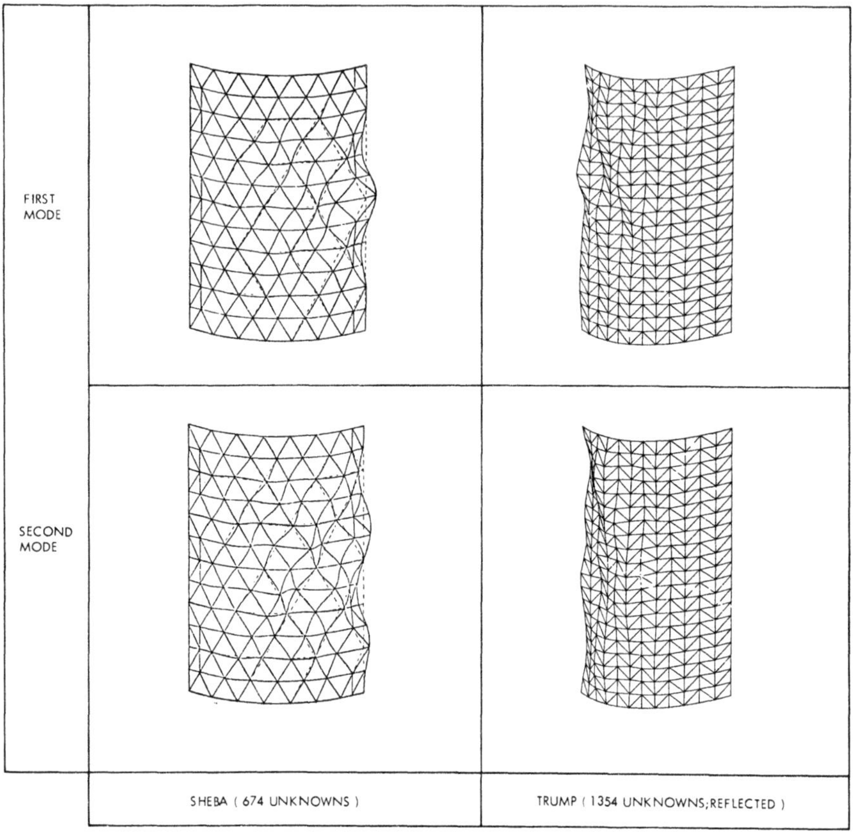

LINEAR BUCKLING OF CYLINDRICAL SHELL UNDER LINE LOAD

BUCKLING MODE SHAPES

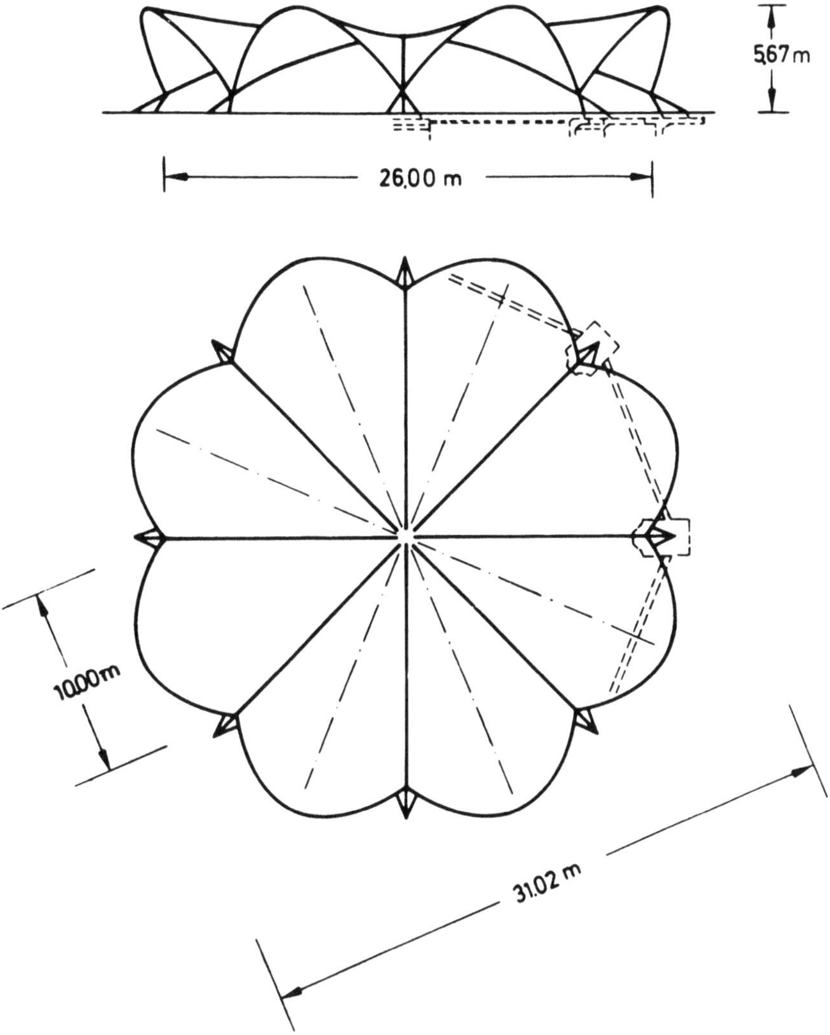

FRONT ELEVATION AND PLAN VIEW OF THE EXHIBITION HALL AT THE NATIONAL GARDEN SHOW IN STUTTGART (1977)

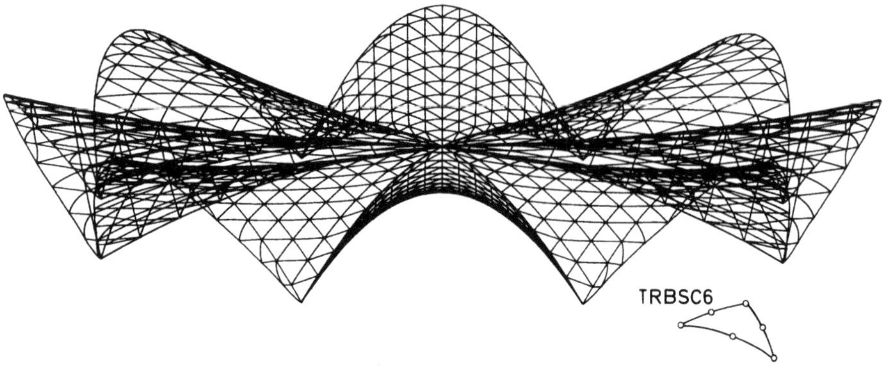

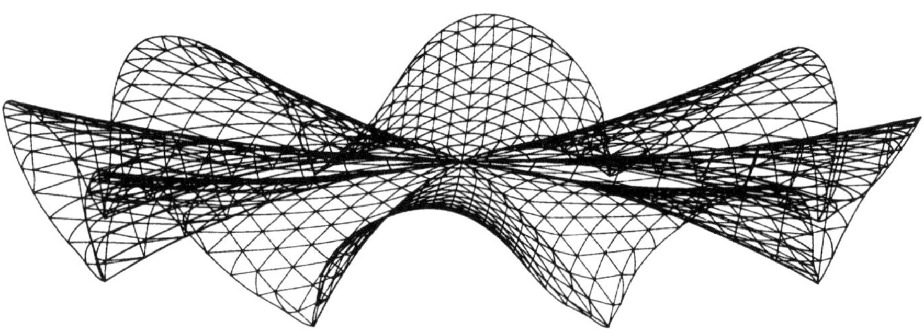

	TRBSC6 WITH TRANSVERSE SHEAR	TRUMP PURE BENDING
IDEALISATION ELEMENTS (1 SECTOR) DOF (1 SECTOR)	214 1729	828 2662
CRITICAL LOAD FACTOR	1.33195	1.47682

EXHIBITION HALL UNDER DEAD AND NONUNIFORM SNOW LOAD

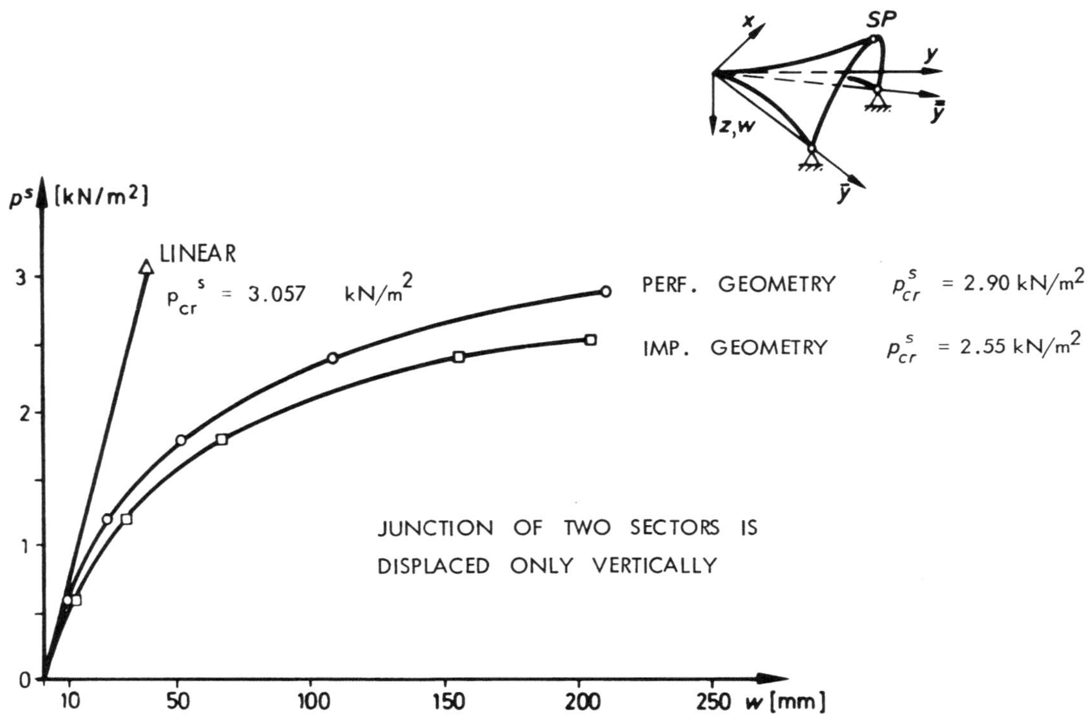

EXHIBITION HALL

VERTICAL DISPLACEMENT OF SUMMIT POINT (SP)

DEAD AND UNIFORM SNOW LOAD

Different Levels of Nonlinear Shell Theory in Finite Element Stability Analysis

E. STEIN, A. BERG, W. WAGNER

Institut für Baumechanik und Numerische Mechanik
Universität Hannover

Summary

On the basis of Kirchhoff-Love-hypothesis and small elastic strains a geometrically nonlinear theory of thin shells is derived in symbolic tensor notation. Furthermore, an incremental version for finite increments of the displacements is formulated and the corresponding incremental principle of virtual work. This is the starting point of FEM with curved triangular elements, each containing 63 kinematical DoFs. Different levels of geometrical nonlinearity are implemented in the developed computer programs for nonlinear, critical and postcritical calculations. As examples quadrilateral sections of circular cylindrical shells with different boundary conditions are investigated for normal pressure, normal point loads, axial loads and combined loads, considering additionally imperfections. Comparisons with published results are made as far as possible. The results show the validity of Donnell's approximation in a wide range.

1. Introduction

In the frame of Kirchhoff-Love-hypothesis (K-L-H) for thin elastic shells we start from the fundamental works on nonlinear theories by Koiter [1], Naghdi [2], [3], Simmonds, Danielson [4], Pietraszkiewicz [5].

In view of the complicated and voluminous representation of incremental equations in index notation, Berg [6] formulated a nonlinear theory, an incremental theory and the corresponding incremental principle of virtual work in a complete symbolic tensor notation. This is used throughout this paper. The symbolic notation is suitable to represent the complete nonlinear theory without any approximations. The decomposition of the nonlinear operators in the incremental kinematic equations into three essential parts can be directly continued into the formulation

Buckling of Shells, Ed.: E. Ramm
Proceedings of a State-of-the-Art Colloqium

of nonlinear algebraic equations of the discretized system. So, different levels of nonlinear theory, i.e. Donnell [7] and the moderate rotation concept of Pietraszkiewicz [5] can be realized numerically. Concerning the formulation of energy functionals for shells, the papers of Schmidt and Pietraszkiewicz [8] and Stumpf [9] are of interest in this context.

Special considerations of adequate theories for stability problems and postcritical deformations were made by Koiter [1o] Basar and Krätzig [11], Wittek [12], Stein [13] et.al.

The fully conforming SHEBA element used here was proposed by Argyris and Scharpf [14]. The treatment of nonlinear systems of algebraic equations is wellknown in undercritical deformation states, see Stein and Kessel [15], e.g. The effective calculation of postcritical deformation paths gave stimulation for many investigations within the last years. Essential contributions were made by Riks [16], Wempner [17], Wessels [18], Crisfield [19] and Brendel and Ramm [2o]. In our algorithms a modified Newton-Raphson process with an automatic step control, alternately by loads or displacements, is used.

2. Geometry of the shell

In Fig. 1, the undeformed and deformed shell with the essential notations is shown.

All further quantities are introduced in the text.

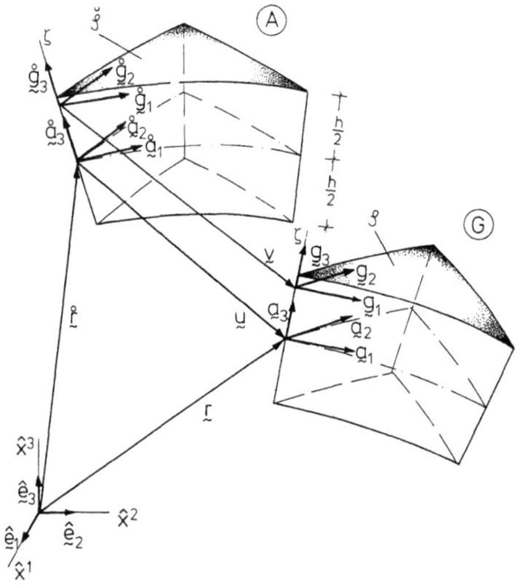

Fig. 1. Geometry of a thin shell in the reference and in the current configuration

3. Nonlinear theory of thin shells with Kirchhoff-Love constraints

3.1 Kinematics of shell space

The reference configuration $\overset{\circ}{\mathcal{S}}$ of the shell is given by normal coordinates with the straight coordinate ζ. Then, each point $\overset{\circ}{P} \in \overset{\circ}{\mathcal{S}}$ is adjoined uniquely to a point $\overset{\circ}{R}(\overset{\circ}{P}) \in \overset{\circ}{\mathcal{M}}$ in the form

$$\overset{\circ}{\underset{\sim}{p}} = \overset{\circ}{\underset{\sim}{r}} + \zeta \overset{\circ}{\underset{\sim}{a}}_3 \; . \tag{3.1}$$

As the vectors $\overset{\circ}{\underset{\sim}{r}}$ and $\overset{\circ}{\underset{\sim}{a}}_3$ are not dependent from ζ the base vectors $\overset{\circ}{g}_i$ in shell space $\overset{\circ}{\mathcal{S}}$ are given by

$$\overset{\circ}{\underset{\sim}{g}}_\alpha = \overset{\circ}{\underset{\sim}{p}}_{,\alpha} = \overset{\circ}{\underset{\sim}{a}}_\alpha + \zeta \overset{\circ}{\underset{\sim}{a}}_{3,\alpha} \tag{3.2}$$

$$\overset{\circ}{\underset{\sim}{g}}_3 = \overset{\circ}{\underset{\sim}{p}}_{,\zeta} = \overset{\circ}{\underset{\sim}{a}}_3 \tag{3.3}$$

The gradient of a threedimensional vector field

$$\underset{\sim}{u} = u^\alpha \overset{\circ}{\underset{\sim}{a}}_\alpha + u_3 \overset{\circ}{\underset{\sim}{a}}_3 \tag{3.4}$$

on the surface $\overset{\circ}{\mathcal{M}}$ is defined by

$$\Gamma\text{rad}\ \underset{\sim}{u} = \underset{\sim}{u}_{,\alpha} \otimes \overset{\circ}{\underset{\sim}{a}}{}^\alpha \; . \tag{3.5}$$

The 2. fundamental form of the surface $\overset{\circ}{\mathcal{M}}$ is adjoined to the tensor

$$\overset{\circ}{B} = \overset{\circ}{b}_{\alpha\beta}\, \overset{\circ}{\underset{\sim}{a}}{}^\alpha \otimes \overset{\circ}{\underset{\sim}{a}}{}^\beta = - \overset{\circ}{\underset{\sim}{a}}_{3,\beta} \otimes \overset{\circ}{\underset{\sim}{a}}{}^\beta = - \Gamma\text{rad}\ \overset{\circ}{\underset{\sim}{a}}_3 . \tag{3.6}$$

The threedimensional fundamental tensor of the surface $\overset{\circ}{\mathcal{M}}$ is

$$\overset{\circ}{\underset{=}{1}} = \overset{\circ}{\underset{\sim}{a}}_\alpha \otimes \overset{\circ}{\underset{\sim}{a}}{}^\alpha + \overset{\circ}{\underset{\sim}{a}}_3 \otimes \overset{\circ}{\underset{\sim}{a}}_3 \; . \tag{3.7}$$

With (3.6) and (3.7), (3.2) can be represented in the form

$$\overset{\circ}{\underset{\sim}{g}}_\alpha = \overset{\circ}{\underset{\sim}{a}}_\alpha - \zeta \overset{\circ}{B}\, \overset{\circ}{\underset{\sim}{a}}_\alpha = (\overset{\circ}{\underset{=}{1}} - \zeta \overset{\circ}{B}) \overset{\circ}{\underset{\sim}{a}}_\alpha \; . \tag{3.8}$$

Further, with

$$\overset{o}{\underset{\sim}{g}}_3 = \overset{o}{\underset{=}{1}}\, \overset{o}{\underset{\sim}{a}}_3 \tag{3.9}$$

and

$$\overset{o}{B}\, \overset{o}{\underset{\sim}{a}}_3 = \underset{\sim}{0} \ , \tag{3.10}$$

the base vectors $\overset{o}{\underset{\sim}{g}}_i$ in shell space $\overset{o}{\mathscr{S}}$ are given in the general form

$$\overset{o}{\underset{\sim}{g}}_i = \overset{o}{\underset{=}{g}}\, \overset{o}{\underset{\sim}{a}}_i \tag{3.11}$$

with the shifter tensor in the reference configuration

$$\overset{o}{\underset{=}{g}} = \overset{o}{\underset{=}{1}} - \zeta \overset{o}{B} = \overset{o}{\underset{\sim}{g}}_i \otimes \overset{o}{\underset{\sim}{a}}{}^i \ . \tag{3.12}$$

It holds

$$\overset{o}{\mu} = \det \overset{o}{\underset{=}{g}} = \frac{[\overset{o}{\underset{=}{g}}\,\overset{o}{\underset{\sim}{a}}_1, \overset{o}{\underset{=}{g}}\,\overset{o}{\underset{\sim}{a}}_2, \overset{o}{\underset{=}{g}}\,\overset{o}{\underset{\sim}{a}}_3]}{[\overset{o}{\underset{\sim}{a}}_1, \overset{o}{\underset{\sim}{a}}_2, \overset{o}{\underset{\sim}{a}}_3]} = \frac{[\overset{o}{\underset{\sim}{g}}_1, \overset{o}{\underset{\sim}{g}}_2, \overset{o}{\underset{\sim}{g}}_3]}{[\overset{o}{\underset{\sim}{a}}_1, \overset{o}{\underset{\sim}{a}}_2, \overset{o}{\underset{\sim}{a}}_3]} = \frac{\sqrt{\overset{o}{g}}}{\sqrt{\overset{o}{a}}} \ . \tag{3.13}$$

Assuming that $\sqrt{\overset{o}{g}}$ and $\sqrt{\overset{o}{a}}$ don't vanish, this is true also for $\mu = \det \overset{o}{\underset{=}{g}}$. Then the inverse shifter tensor $\overset{o}{\underset{=}{g}}{}^{-1}$ exists

$$\overset{o}{\underset{=}{g}}{}^{-1} = \overset{o}{\underset{\sim}{a}}_j \otimes \overset{o}{\underset{\sim}{g}}{}^j \ . \tag{3.14}$$

Further we get

$$\overset{o}{\underset{=}{g}}{}^T = \overset{o}{\underset{\sim}{a}}{}^i \otimes \overset{o}{\underset{\sim}{g}}_i \ , \quad \overset{o}{\underset{=}{g}}{}^{T-1} = \overset{o}{\underset{\sim}{g}}{}^j \otimes \overset{o}{\underset{\sim}{a}}_j \ . \tag{3.15}$$

Before a strain measure for the shell can be introduced, simplifying assumptions for the deformations are necessary. The goal to express the displacements of the shell $\mathscr{S}$ in the current configuration only by the middle surface $\mathscr{M}$ in the same configuration, is achived with the wellknown Kirchhoff-Love hypothesis (K-L-H) which contains the following assumptions:

1. All material points which are situated on a normal of the middle surface $\overset{o}{\mathscr{M}}$, remain on a normal of $\mathscr{M}$ in the current

configuration.

2. The distances of material points P form adjoined points R on the middle surface with respect to the current configuration are equal to distances of points $\overset{\circ}{P}$ and $\overset{\circ}{R}$ in the reference configuration.

As a consequence of K-L-H, the current configuration is also described by normal coordinates. Each point $P \in \mathcal{S}$ is again uniquely correlated with a point $R \in \mathcal{M}$ of the middle surface of the shell in the current configuration. The position vector of an arbitrary point $P \in \mathcal{S}$ is given, according to Fig. 1, by

$$\underset{\sim}{p} = \underset{\sim}{r} + \zeta \, \underset{\sim}{a}_3 \quad . \tag{3.16}$$

As in the reference configuration one gets

$$\underset{\sim}{g}_\alpha = \underset{\sim}{p}_{,\alpha} = \underset{\sim}{a}_\alpha + \zeta \, \underset{\sim}{a}_{3,\alpha} \tag{3.17}$$

$$\underset{\sim}{g}_3 = \underset{\sim}{p}_{,\zeta} = \underset{\sim}{a}_3 \quad . \tag{3.18}$$

The gradient of a vector field defined on the middle surface $\mathcal{M}$, namely

$$\bar{\underset{\sim}{u}} = \bar{u}^\alpha \, \underset{\sim}{a}_\alpha + \bar{u}_3 \, \underset{\sim}{a}_3 \; , \tag{3.19}$$

is introduced in the form

$$\gamma\text{rad} \, \bar{\underset{\sim}{u}} = \bar{\underset{\sim}{u}}_{,\alpha} \otimes \underset{\sim}{a}^\alpha \quad . \tag{3.20}$$

Then, corresponding to (3.6), a tensor

$$B = b_{\alpha\beta} \, \underset{\sim}{a}^\alpha \otimes \underset{\sim}{a}^\beta = - \underset{\sim}{a}_{3,\beta} \otimes \underset{\sim}{a}^\beta = - \gamma\text{rad} \, \underset{\sim}{a}_3 \tag{3.21}$$

is correlated with the 2. fundamental form of the middle surface. The threedimensional fundamental tensor of the middle surface $\mathcal{M}$ is

$$\underset{=}{1} = \underset{\sim}{a}_\alpha \otimes \underset{\sim}{a}^\alpha + \underset{\sim}{a}_3 \otimes \underset{\sim}{a}_3 \quad . \tag{3.22}$$

With (3.21) and (3.22), (3.17) can be written as

$$\underset{\sim}{g}_\alpha = \underset{\sim}{a}_\alpha - \zeta B \underset{\sim}{a}_\alpha = (\underset{=}{1} - \zeta B) \underset{\sim}{a}_\alpha \quad .$$

Using further

$$\underset{\sim}{g}_3 = \underset{=}{1}\ \underset{\sim}{a}_3 \tag{3.23}$$

and

$$B\ \underset{\sim}{a}_3 = \underset{\sim}{O}\ , \tag{3.24}$$

we get - analogously to the reference configuration - the general representation of base vectors $\underset{\sim}{g}_i$ in the shell space of the current configuration

$$\underset{\sim}{g}_i = \underset{=}{g}\ \underset{\sim}{a}_i \tag{3.25}$$

with the shifter tensor

$$\underset{=}{g} = \underset{=}{1} - \zeta B = \underset{\sim}{g}_i \otimes \underset{\sim}{a}^i \tag{3.26}$$

Corresponding to (3.13) we get

$$\mu = \det \underset{=}{g} = \frac{\sqrt{g}}{\sqrt{a}}\quad . \tag{3.27}$$

Assuming that $\sqrt{g}$ and $\sqrt{a}$ don't vanish, the inverse shifter tensor exists

$$\underset{=}{g}^{-1} = \underset{\sim}{a}_j \otimes \underset{\sim}{g}^j\quad . \tag{3.28}$$

We further get

$$\underset{=}{g}^T = \underset{\sim}{a}^i \otimes \underset{\sim}{g}_i\quad ,\quad \underset{=}{g}^{T-1} = \underset{\sim}{g}^j \otimes \underset{\sim}{a}_j \tag{3.29}$$

The material deformation gradient F is defined for convective coordinates as

$$F = \underset{\sim}{g}_i \otimes \overset{o}{\underset{\sim}{g}}{}^i\quad . \tag{3.3o}$$

Using (3.15) and (3.25), (3.3o) yields

$$F = (\underset{=}{g}\ \underset{\sim}{a}_i) \otimes (\overset{o}{\underset{=}{g}}{}^{T-1}\ \overset{o}{\underset{\sim}{a}}{}^i) = \underset{=}{g}\ (\underset{\sim}{a}_i \otimes \overset{o}{\underset{\sim}{a}}{}^i)\overset{o}{\underset{=}{g}}{}^{-1}\quad . \tag{3.31}$$

Corresponding to Pietraszkiewicz [5] we introduce the three-dimensional deformation gradient G on the middle surface $\mathcal{M}$ of the shell

$$G = F(\zeta = 0) = \underset{\sim}{a}_i \otimes \overset{o}{\underset{\sim}{a}}{}^i = \underset{\sim}{a}_\alpha \otimes \overset{o}{\underset{\sim}{a}}{}^\alpha + \underset{\sim}{a}_3 \otimes \overset{o}{\underset{\sim}{a}}_3 \ . \tag{3.32}$$

Introducing (3.32) into (3.31), F finally results in

$$F = \underline{\underline{g}}\ G\ \overset{o}{\underline{\underline{g}}}{}^{-1} . \tag{3.33}$$

The original and current forms of the tensors in (3.33) are illustrated in Fig. 2.

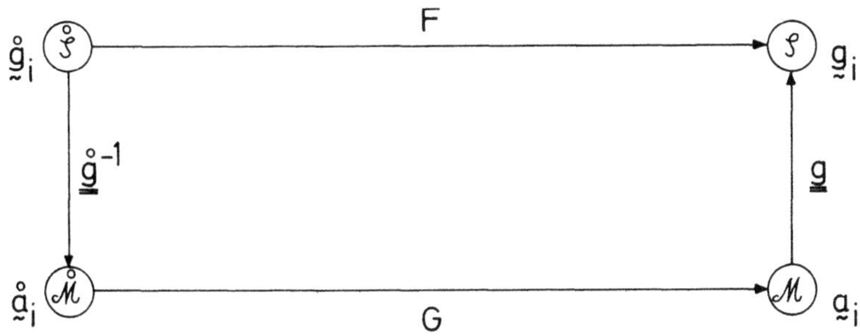

Fig. 2. The deformation gradient as a composed mapping

The Green strain tensor is

$$E = \frac{1}{2} [\, F^T F - \underline{\underline{1}}_\zeta \,] = \frac{1}{2}\ [\, g_{ij} - \overset{o}{g}_{ij} \,]\ \overset{o}{\underset{\sim}{g}}{}^i \otimes \overset{o}{\underset{\sim}{g}}{}^j \tag{3.34}$$

With (3.33) we get

$$E = \overset{o}{\underline{\underline{g}}}{}^{T-1}\ [\, \frac{1}{2}\ (G^T \underline{\underline{g}}^T \underline{\underline{g}} G - \overset{o}{\underline{\underline{g}}}{}^T\ \overset{o}{\underline{\underline{g}}})]\ \overset{o}{\underline{\underline{g}}}{}^{-1} \tag{3.35}$$

Introducing (3.12) and (3.26) yields

$$\begin{aligned} E &= \overset{o}{\underline{\underline{g}}}{}^{T-1} \{ \tfrac{1}{2} G^T (\underline{\underline{1}} - \zeta B^T)\ (\underline{\underline{1}} - \zeta B)\ G - (\overset{o}{\underline{\underline{1}}} - \zeta \overset{o}{B}{}^T)(\overset{o}{\underline{\underline{1}}} - \zeta \overset{o}{B}) \}\ \overset{o}{\underline{\underline{g}}}{}^{-1} \\ &= \overset{o}{\underline{\underline{g}}}{}^{T-1} \{ \frac{1}{2}\ [\, G^T G - \overset{o}{\underline{\underline{1}}} \,] - \zeta \frac{1}{2} [\, G^T (B^T + B) G - (\overset{o}{B}{}^T + \overset{o}{B})] \\ &\quad + \zeta^2\ \frac{1}{2}\ [\, G^T B^T B G - \overset{o}{B}{}^T B] \}\ \overset{o}{\underline{\underline{g}}}{}^{-1} \quad . \end{aligned} \tag{3.36}$$

Similar to Pietraszkiewicz [5], the Green strain tensor becomes with K-L-H

$$E = \overset{o}{\underline{\underline{g}}}{}^{T-1}\ (\underline{\underline{\gamma}} + \zeta\ \underline{\underline{\chi}} + \zeta^2\ \underline{\underline{\nu}})\ \overset{o}{\underline{\underline{g}}}{}^{-1} \ . \tag{3.37}$$

with the strains in the middle surface

$$\underline{\underline{\gamma}} = \frac{1}{2}\,(G^T G - \overset{\circ}{\underline{\underline{1}}}) \quad , \tag{3.38}$$

the elastic curvatures

$$\underline{\underline{\chi}} = -\,(G^T B G - \overset{\circ}{B}) \tag{3.39}$$

and with

$$\underline{\underline{\nu}} = \frac{1}{2}\,(G^T B^T B G - \overset{\circ}{B}{}^T \overset{\circ}{B}) \quad . \tag{3.4o}$$

With (3.6), (3.21) and (3.32) we get

$$\underline{\underline{\gamma}} = \frac{1}{2}\,(\underset{\sim}{a}_\alpha \cdot \underset{\sim}{a}_\beta - \overset{\circ}{\underset{\sim}{a}}_\alpha \cdot \overset{\circ}{\underset{\sim}{a}}_\beta)\; \overset{\circ}{\underset{\sim}{a}}{}^\alpha \otimes \overset{\circ}{\underset{\sim}{a}}{}^\beta \tag{3.41}$$

$$\underline{\underline{\chi}} = (\underset{\sim}{a}_\alpha \cdot \underset{\sim}{a}_{3,\beta} - \overset{\circ}{\underset{\sim}{a}}_\alpha \cdot \overset{\circ}{\underset{\sim}{a}}_{3,\beta})\; \overset{\circ}{\underset{\sim}{a}}{}^\alpha \otimes \overset{\circ}{\underset{\sim}{a}}{}^\beta \quad , \tag{3.42}$$

$$\underline{\underline{\nu}} = \frac{1}{2}\,(\underset{\sim}{a}_{3,\alpha} \cdot \underset{\sim}{a}_{3,\beta} - \overset{\circ}{\underset{\sim}{a}}_{3,\alpha} \cdot \overset{\circ}{\underset{\sim}{a}}_{3,\beta})\; \overset{\circ}{\underset{\sim}{a}}{}^\alpha \otimes \overset{\circ}{\underset{\sim}{a}}{}^\beta \quad . \tag{3.43}$$

and in index notation

$$\underline{\underline{\gamma}} = \frac{1}{2}\,(a_{\alpha\beta} - \overset{\circ}{a}_{\alpha\beta})\; \overset{\circ}{\underset{\sim}{a}}{}^\alpha \otimes \overset{\circ}{\underset{\sim}{a}}{}^\beta \quad , \tag{3.44}$$

$$\underline{\underline{\chi}} = -\,(b_{\alpha\beta} - \overset{\circ}{b}_{\alpha\beta})\; \overset{\circ}{\underset{\sim}{a}}{}^\alpha \otimes \overset{\circ}{\underset{\sim}{a}}{}^\beta \quad , \tag{3.45}$$

$$\underline{\underline{\nu}} = \frac{1}{2}\,(b^\lambda_\alpha\, b_{\lambda\beta} - \overset{\circ}{b}{}^\lambda_\alpha\, \overset{\circ}{b}_{\lambda\beta})\; \overset{\circ}{a}{}^\alpha \otimes \overset{\circ}{a}{}^\beta \quad . \tag{3.46}$$

So, in the frame of K-L-H, the Green strain tensor of every point in shell space $\mathcal{S}$ is representable by the first and second fundamental tensor of the middle surfaces $\mathcal{M}$ and $\overset{\circ}{\mathcal{M}}$.

For practical calculations, the strain measures are to be expressed by displacements. According to Fig. 1, the displacement field $\underset{\sim}{u}$ of the middle surface $\overset{\circ}{\mathcal{M}}$ is

$$\underset{\sim}{u} = \underset{\sim}{r} - \overset{\circ}{\underset{\sim}{r}} \quad . \tag{3.47}$$

Corresponding to (3.47), the difference vector $\underset{\sim}{\beta}$ of the normal unit vectors is

$$\underset{\sim}{\beta} = \underset{\sim}{a}_3 - \overset{\circ}{\underset{\sim}{a}}_3 \quad . \tag{3.48}$$

With K-L-H the vector field $\underset{\sim}{\beta}$ is representable by the displacement field $\underset{\sim}{u}$. The rotation vector $\underset{\sim}{\beta}$ is used for further derivations but doesn't represent additional indepent kinematical variables. For example, the displacement vector $\underset{\sim}{v}$ of an arbitrary point of the shell space, $\overset{\circ}{P} \in \overset{\circ}{\mathcal{S}}$, can be represented in the form, see Fig. 1,

$$\underset{\sim}{v} = \underset{\sim}{u} + \zeta \underset{\sim}{\beta} \tag{3.49}$$

From (3.41) - (3.43) it follows

$$\underset{=}{\gamma} = \frac{1}{2} (\underset{\sim}{u}_{,\alpha} \cdot \overset{\circ}{\underset{\sim}{a}}_\beta + \underset{\sim}{u}_{,\beta} \cdot \overset{\circ}{\underset{\sim}{a}}_\alpha + \underset{\sim}{u}_{,\alpha} \cdot \underset{\sim}{u}_{,\beta}) \overset{\circ}{\underset{\sim}{a}}{}^\alpha \otimes \overset{\circ}{\underset{\sim}{a}}{}^\beta \quad , \tag{3.50}$$

$$\underset{=}{\chi} = (\underset{\sim}{u}_{,\alpha} \cdot \underset{\sim}{a}_{3,\beta} + \underset{\sim}{\beta}_{,\beta} \cdot \overset{\circ}{\underset{\sim}{a}}_\alpha + \underset{\sim}{u}_{,\alpha} \cdot \underset{\sim}{\beta}_{,\beta}) \overset{\circ}{\underset{\sim}{a}}{}^\alpha \otimes \overset{\circ}{\underset{\sim}{a}}{}^\beta \quad , \tag{3.51}$$

$$\underset{=}{\nu} = \frac{1}{2} (\underset{\sim}{\beta}_{,\alpha} \cdot \overset{\circ}{\underset{\sim}{a}}_{3,\beta} + \underset{\sim}{\beta}_{,\beta} \cdot \overset{\circ}{\underset{\sim}{a}}_{3,\alpha} + \underset{\sim}{\beta}_{,\alpha} \cdot \underset{\sim}{\beta}_{,\beta}) \overset{\circ}{\underset{\sim}{a}}{}^\alpha \otimes \overset{\circ}{\underset{\sim}{a}}{}^\beta . \tag{3.52}$$

We define the displacement gradient

$$H = \Gamma\text{rad}\ \underset{\sim}{u} = \underset{\sim}{u}_{,\alpha} \otimes \overset{\circ}{\underset{\sim}{a}}{}^\alpha \tag{3.53}$$

and the twodimensional fundamental tensor $\overset{\circ}{\underset{=}{1}}_2$ on $\mathcal{M}$

$$\overset{\circ}{\underset{=}{1}}_2 = \overset{\circ}{\underset{\sim}{a}}_\alpha \otimes \overset{\circ}{\underset{\sim}{a}}{}^\alpha \quad . \tag{3.54}$$

Then we get the strains of the middle surface in symbolic tensor notation

$$\underset{=}{\gamma} = \frac{1}{2} (H^T \overset{\circ}{\underset{=}{1}}_2 + \overset{\circ}{\underset{=}{1}}_2 H + H^T H) \tag{3.55}$$

and the elastic curvatures of the middle surface

$$\begin{aligned} \underset{=}{\chi} &= H^T\ \Gamma\text{rad}\ \overset{\circ}{\underset{\sim}{a}}_3 + H^T\ \Gamma\text{rad}\underset{\sim}{\beta} + \overset{\circ}{\underset{=}{1}}_2\ \Gamma\text{rad}\underset{\sim}{\beta} \\ &= H^T\ \Gamma\text{rad}\ \underset{\sim}{a}_3 + \overset{\circ}{\underset{=}{1}}_2\ \Gamma\text{rad}\underset{\sim}{\beta} \quad . \end{aligned} \tag{3.56}$$

Corresponding to Koiter [1] and Pietraszkiewicz [5] we define

$$\underset{\sim}{u}_{,\alpha} = \phi^\beta{}_\alpha\ \overset{\circ}{\underset{\sim}{a}}_\beta + \phi_{3\alpha}\ \overset{\circ}{\underset{\sim}{a}}_3 \tag{3.57}$$

and

$$\underset{\sim}{\beta}_{,\alpha} = \beta^{\beta}_{\cdot\alpha} \overset{\circ}{\underset{\sim}{a}}_{\beta} + \beta_{3\alpha} \overset{\circ}{\underset{\sim}{a}}_{3} \quad . \tag{3.58}$$

The we get

$$\underset{=}{\gamma} = \frac{1}{2} (\phi_{\beta\alpha} + \phi_{\alpha\beta} + \phi^{\lambda}_{\cdot\alpha} \phi_{\lambda\beta} + \phi_{3\alpha} \phi_{3\beta}) \overset{\circ}{\underset{\sim}{a}}{}^{\alpha} \otimes \overset{\circ}{\underset{\sim}{a}}{}^{\beta} \tag{3.59}$$

$$\underset{=}{\chi} = (- \overset{\circ}{b}{}^{\lambda}_{\beta} \phi_{\lambda\alpha} + \beta_{\alpha\beta} + \phi^{\lambda}_{\cdot\alpha} \beta_{\lambda\beta} + \phi_{3\alpha} \beta_{3\beta}) \overset{\circ}{\underset{\sim}{a}}{}^{\alpha} \otimes \overset{\circ}{\underset{\sim}{a}}{}^{\beta} \ . \tag{3.60}$$

Using

$$\underset{\sim}{a}_{\alpha,\beta} \cdot \underset{\sim}{a}_{3} = - \underset{\sim}{a}_{\alpha} \cdot \underset{\sim}{a}_{3,\beta} \quad , \tag{3.61}$$

an alternative form of (3.42) is obtained

$$\begin{aligned} \underset{=}{\chi} &= (- \underset{\sim}{a}_{\alpha,\beta} \cdot \underset{\sim}{a}_{3} + \overset{\circ}{\underset{\sim}{a}}_{\alpha,\beta} \cdot \overset{\circ}{\underset{\sim}{a}}_{3}) \overset{\circ}{\underset{\sim}{a}}{}^{\alpha} \otimes \overset{\circ}{\underset{\sim}{a}}{}^{\beta} \\ &= (- \underset{\sim}{a}_{\alpha}|_{\beta} \cdot \underset{\sim}{a}_{3} + \overset{\circ}{\underset{\sim}{a}}_{\alpha}|_{\beta} \cdot \overset{\circ}{\underset{\sim}{a}}_{3}) \overset{\circ}{\underset{\sim}{a}}{}^{\alpha} \otimes \overset{\circ}{\underset{\sim}{a}}{}^{\beta} \ . \end{aligned} \tag{3.62}$$

With

$$\underset{\sim}{a}_{\alpha}|_{\beta} = \overset{\circ}{\underset{\sim}{a}}_{\alpha}|_{\beta} + \underset{\sim}{u}|_{\alpha\beta} \tag{3.63}$$

we get

$$\underset{=}{\chi} = (- \underset{\sim}{u}|_{\alpha\beta} \cdot \overset{\circ}{\underset{\sim}{a}}_{3} - \underset{\sim}{u}|_{\alpha\beta} \cdot \underset{\sim}{\beta} - \overset{\circ}{\underset{\sim}{a}}_{\alpha}|_{\beta} \cdot \underset{\sim}{\beta}) \overset{\circ}{\underset{\sim}{a}}{}^{\alpha} \otimes \overset{\circ}{\underset{\sim}{a}}{}^{\beta} . \tag{3.64}$$

For further considerations it is convenient to keep the normal unit vector of $\overset{\circ}{\mathcal{M}}$ in (3.63).
One gets

$$\underset{=}{\chi} = (-\underset{\sim}{u}|_{\alpha\beta} \cdot \underset{\sim}{a}_{3} - \overset{\circ}{\underset{\sim}{a}}_{\alpha}|_{\beta} \cdot \underset{\sim}{\beta}) \overset{\circ}{\underset{\sim}{a}}{}^{\alpha} \otimes \overset{\circ}{\underset{\sim}{a}}{}^{\beta} \ . \tag{3.65}$$

With

$$\overset{\circ}{\underset{\sim}{a}}{}^{\alpha} \otimes \overset{\circ}{\underset{\sim}{a}}{}^{\beta} \otimes \underset{\sim}{u}|_{\alpha\beta} = \overset{\circ}{\underset{\sim}{a}}{}^{\alpha} \otimes \Gamma\text{rad}^{T} \underset{\sim}{u}_{,\alpha} \tag{3.66}$$

and

$$\overset{\circ}{\underset{\sim}{a}}{}^{\alpha} \otimes \overset{\circ}{\underset{\sim}{a}}{}^{\beta} \otimes \overset{\circ}{\underset{\sim}{a}}_{\alpha}|_{\beta} = \overset{\circ}{\underset{\sim}{a}}{}^{\alpha} \otimes \Gamma\text{rad}^{T} \overset{\circ}{\underset{\sim}{a}}_{\alpha} \ , \tag{3.67}$$

the elastic curvatures become in symbolic tensor notation

$$\underset{=}{\chi} = - (\overset{\circ}{\underset{\sim}{a}}{}^{\alpha} \otimes \Gamma\text{rad}^{T} \underset{\sim}{u}_{,\alpha}) \underset{\sim}{a}_{3} - (\overset{\circ}{\underset{\sim}{a}}{}^{\alpha} \otimes \Gamma\text{rad}^{T} \overset{\circ}{\underset{\sim}{a}}_{\alpha}) \underset{\sim}{\beta} \ . \tag{3.68}$$

3.2 Equilibrium conditions in spatial representation

For statical equilibrium of a body $\mathcal{B}$ in the current configuration there is a massforce

$$\bar{\underset{\sim}{f}}_b = \int_{\mathcal{B}} \overset{*}{\rho}\; \overset{*}{\bar{\underset{\sim}{b}}}\, dv \qquad (3.69)$$

and a surface force

$$\bar{\underset{\sim}{f}}_c = \int_{\partial\mathcal{B}} \bar{\underset{\sim}{t}}\, da \quad . \qquad (3.70)$$

The global equilibrium condition is

$$\bar{\underset{\sim}{f}}_b + \bar{\underset{\sim}{f}}_c = \underset{\sim}{0} \quad . \qquad (3.71)$$

For a thin shell $\mathcal{S}$ the dependence of mass forces $\overset{*}{\bar{\underset{\sim}{b}}}$ from the thickness coordinate ζ is neglected. With an integration over the thickness h, according to

$$\bar{\underset{\sim}{f}}_b = \int_{\mathcal{S}} \overset{*}{\rho}\, \overset{*}{\bar{\underset{\sim}{b}}}\, dv = \int_{\mathcal{M}} \rho\; \bar{\underset{\sim}{b}}\, da \;, \qquad (3.72)$$

we get the mass density ρ with respect to the middle surface $\mathcal{M}$ as

$$\rho = \int_h \overset{*}{\rho}\; \mu\, d\zeta \quad . \qquad (3.73)$$

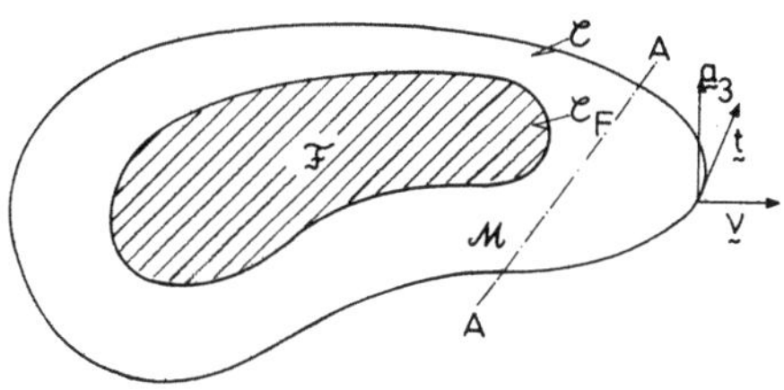

Fig. 3.a. middle surface $\mathcal{M}$ with edge curve $\mathcal{C}$ and corresponding subdomains.

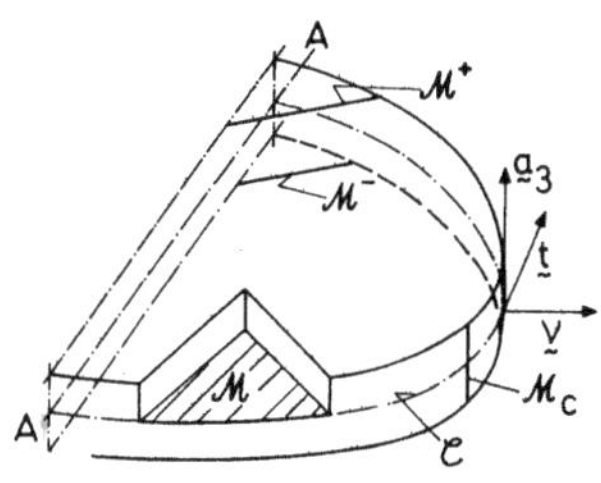

Fig. 3.b. Part of Fig. 3.a showing upper and lower surfaces $\mathcal{M}^+$, $\mathcal{M}^-$ and edge surface $\mathcal{M}_c$.

It is important for a shell theory that given surface forces are acting only at the part $\mathcal{M}_c$ of the edge of $\mathcal{S}$, see Fig. 3.b. This means no essential restrictions of loading facilities because given surface loads on $\mathcal{M}^+$ and $\mathcal{M}^-$ can be transformed into given loads $\bar{\underset{\sim}{b}}$ referred to the middle surface $\mathcal{M}$. Only the local stresses due to given forces of $\mathcal{M}^+$ or $\mathcal{M}^-$ are not considered as in beam and plate theory. Furthermore, moments arising from displacing the given surface loads from $\mathcal{M}^+$ and $\mathcal{M}^-$ to the middle surface $\mathcal{M}$ are neglected for thin shells. These excentricites can be considered in the equilibrium conditions for the moments in connection with given moments at the middle surface $\mathcal{M}$.

So, in the frame of the shell theory treated here, given surface forces $\bar{\underset{\sim}{f}}_c$ are covered by a surface integral over $\mathcal{M}_c$

$$\bar{\underset{\sim}{f}}_c = \int_{\mathcal{M}_c} \bar{\underset{\sim}{t}}\, da_c \; . \tag{3.74}$$

It is assumed that the edge load vector $\bar{\underset{\approx}{q}}$, referred to unit length of $\mathcal{C}$, exists. Then, $\bar{\underset{\sim}{f}}_c$ is representable in the form

$$\bar{\underset{\sim}{f}}_c = \int_{\mathcal{M}_c} \bar{\underset{\sim}{t}}\, da_c = \int_{\mathcal{C}} \bar{\underset{\approx}{q}}\, ds \; . \tag{3.75}$$

The edge load vector $\bar{\underset{\approx}{q}}$ can be calculated by introducing an edge load tensor

$$\bar{\underset{\approx}{Q}} = \underset{\approx}{q}^{\alpha} \otimes \underset{\sim}{a}_{\alpha} = \bar{Q}^{\alpha\beta}\, \underset{\sim}{a}_{\alpha} \otimes \underset{\sim}{a}_{\beta} + \bar{Q}^{3\beta}\, \underset{\sim}{a}_{3} \otimes \underset{\sim}{a}_{\beta} \tag{3.76}$$

and a unit normal vector $\underset{\sim}{\nu}$ of the edge curve $\mathcal{C}$ in the tangential plane of the middle surface $\mathcal{M}$

$$\bar{\underset{\approx}{q}} = \bar{\underset{\approx}{Q}}\underset{\sim}{\nu} = \bar{\underset{\approx}{Q}}\, \nu_{\alpha}\, \underset{\sim}{a}^{\alpha} = \nu_{\alpha}\, \bar{\underset{\approx}{q}}^{\alpha} \; . \tag{3.77}$$

Equ. (3.75) can be formulated then as

$$\bar{\underset{\sim}{f}}_c = \int_{\mathcal{C}} \bar{\underset{\approx}{q}}\, \nu_{\alpha}\; ds = \int_{\mathcal{C}} \bar{\underset{\approx}{Q}}\, \underset{\sim}{\nu}\, ds \; . \tag{3.78}$$

For deriving local equilibrium conditions we consider an arbitrary subdomain $\mathcal{P} \subset \mathcal{S}$ with the sub-middle surface $\mathcal{F} \subset \mathcal{M}$ which is cut out by the choice of a closed edge curve $\mathcal{C}_F$, see Fig. 3a. Corresponding to (3.72), the mass force is

$$\underset{\sim}{f}_b\ (\mathcal{P}) = \int_{\mathcal{F}} \rho\, \underset{\sim}{\bar{b}}\ da \quad . \tag{3.79}$$

The total stress resultant in the sectional surface $\mathcal{M}_{c_F}$ is representable by introducing the force vector $\underset{\sim}{q}$ per unit length of $\mathcal{C}_F$.

$$\underset{\sim}{f}_c\ (\mathcal{P}) = \int_{\mathcal{M}_{c_F}} \underset{\sim}{t}\ da_c = \int_{\mathcal{C}_F} \underset{\sim}{q}\ ds \quad . \tag{3.80}$$

We can assign the sectional force tensor Q to the vector $\underset{\sim}{q}$ according to

$$Q = \underset{\sim}{q}^{\alpha} \otimes \underset{\sim}{a}_{\alpha} = Q^{\alpha\beta}\ \underset{\sim}{a}_{\alpha} \otimes \underset{\sim}{a}_{\beta} + Q^{3\beta}\ \underset{\sim}{a}_3 \otimes \underset{\sim}{a}_{\beta} \tag{3.81}$$

and

$$\underset{\sim}{q} = Q\underset{\sim}{\nu} = Q\ \nu_{\alpha}\ \underset{\sim}{a}^{\alpha} = \nu_{\alpha}\ \underset{\sim}{q}^{\alpha} \quad . \tag{3.82}$$

Due to the relations between the area element of the edge surface and the curve element on the middle surface we get, see Naghdi [3]

$$\begin{aligned} \int_{\mathcal{M}_{c_F}} \underset{\sim}{t}\ da_c &= \int_{\mathcal{C}_F} \left[\left(\int_h \underset{\sim}{t}^1 \mu d\zeta\right)\sqrt{a}\ d\theta^2 - \left(\int_h \underset{\sim}{t}^2 \mu d\zeta\right)\sqrt{a}\ d\theta^1 \right] \\ &= \int_{\mathcal{C}_F} (\underset{\sim}{q}^1\ \nu_1 + \underset{\sim}{q}^2\ \nu_2)\, ds = \int_{\mathcal{C}_F} \underset{\sim}{q}^{\alpha}\ \nu_{\alpha}\ ds \ . \end{aligned} \tag{3.83}$$

So, the sectional force vectors (vectors of stress resultants) are defined by

$$\underset{\sim}{q}^{\alpha} := \int_h \underset{\sim}{t}^{\alpha}\ \mu d\zeta \qquad , \ \alpha = 1,2 \quad . \tag{3.84}$$

From the decomposition of the sectional forces

$$\underset{\sim}{q}^{\alpha} = Q^{\beta\alpha}\ \underset{\sim}{a}_{\beta} + Q^{3\alpha}\ \underset{\sim}{a}_3 \tag{3.85}$$

and the representation of the stress vectors in the form

$$\underset{\sim}{t}^{\alpha} = T^{i\alpha}\ \underset{\sim}{g}_i = T^{\beta\alpha}\ \underset{\sim}{g}_{\beta} + T^{3\alpha}\ \underset{\sim}{g}_3 \tag{3.86}$$

we get the components

$$Q^{\beta\alpha} = \int_h T^{\rho\alpha} \mu^{\beta}_{\rho} \mu \, d\zeta \quad , \tag{3.87}$$

$$Q^{3\alpha} = \int_h T^{3\alpha} \, \mu d\zeta \quad . \tag{3.88}$$

Applying Gaussian integral theorem to (3.8o) yields

$$\underset{\sim}{f}_c \, (\mathcal{P}) = \int_{\mathcal{C}_\mathcal{F}} \underset{\sim}{q} \, ds = \int_{\mathcal{C}_\mathcal{F}} \underset{\sim}{q}^{\alpha} \, \underset{\sim}{\nu}_{\alpha} ds = \int_{\mathcal{F}} \underset{\sim}{q}^{\alpha}\|_{\alpha} \, da$$

$$= \int_{\mathcal{C}_\mathcal{F}} Q\underset{\sim}{\nu} \, ds = \int_{\mathcal{F}} \delta iv \, Q \, da \, . \tag{3.89}$$

We define

$$\delta iv \, Q := Q_{,\alpha} \, \underset{\sim}{a}^{\alpha} \tag{3.9o}$$

with the partial derivatives of tensor Q

$$Q_{,\alpha} = (\underset{\sim}{q}^{\beta} \otimes \underset{\sim}{a}_{\beta})_{,\alpha} = \underset{\sim}{q}^{\beta}_{,\alpha} \otimes \underset{\sim}{a}_{\beta} + \underset{\sim}{q}^{\beta} \otimes \underset{\sim}{a}_{\beta,\alpha}$$

$$= \underset{\sim}{q}^{\gamma}_{,\alpha} \otimes \underset{\sim}{a}_{\gamma} + \underset{\sim}{q}^{\beta} \otimes (\Gamma_{\beta\alpha}^{\gamma} \, \underset{\sim}{a}_{\gamma} + b_{\beta\alpha} \, \underset{\sim}{a}_3)$$

$$= \underset{\sim}{q}^{\gamma}\|_{\alpha} \otimes \underset{\sim}{a}_{\gamma} + b_{\gamma\alpha} \, \underset{\sim}{q}^{\gamma} \otimes \underset{\sim}{a}_3 \tag{3.91}$$

and

$$\delta \, iv \, Q = \underset{\sim}{q}^{\alpha}\|_{\alpha} \quad . \tag{3.92}$$

The global force equilibrium condition for the subdomain $\mathcal{P} \subset \mathcal{S}$ can be written with (3.89) in the form

$$\int_{\mathcal{F}} (\rho \underset{\sim}{\bar{b}} + \delta iv \, Q) \, da = \underset{\sim}{O} \tag{3.93}$$

With arbitrary $\mathcal{P}$ and presuming the continuity of the integrand, we get the local force equilibrium condition

$$\delta iv \, Q + \rho \, \underset{\sim}{\bar{b}} = \underset{\sim}{q}^{\alpha}\|_{\alpha} + \rho \, \underset{\sim}{\bar{b}} = \underset{\sim}{O}. \tag{3.94}$$

The moment $\bar{d}_b$ of the mass forces $\bar{b}$ acting on a body $\mathcal{B}$ is calculated with respect to an arbitrary point

$$\underset{\sim}{\bar{d}}_b = \int_{\mathcal{B}} (\underset{\sim}{p} - \underset{\sim}{p}_o) \times \overset{*}{\rho} \, \underset{\sim}{\overset{*}{\bar{b}}} \, dv \quad , \tag{3.95}$$

and the moment $\underset{\sim}{\bar{d}}_c$ of surface forces $\underset{\sim}{\bar{t}}$ on $\partial \mathcal{B}$ is

$$\bar{\underset{\sim}{d}}_c = \int_{\partial\mathcal{B}} (\underset{\sim}{p} - \underset{\sim}{p}_o) \times \bar{\underset{\sim}{t}}\, da \quad . \tag{3.96}$$

The global moment equilibrium condition is given by

$$\bar{\underset{\sim}{d}}_b + \bar{\underset{\sim}{d}}_c = \underset{\sim}{0} \quad . \tag{3.97}$$

For thin shells we get - similar (3.72) - by neglect the dependence of mass forces from the coordinate ζ

$$\bar{\underset{\sim}{d}}_b = \int_{\mathcal{M}} [(\underset{\sim}{r} - \underset{\sim}{p}_o) \times \rho \bar{\underset{\sim}{b}} + \underset{\sim}{a}_3 \times \overset{(2)}{\rho} \bar{\underset{\sim}{b}}] da, \tag{3.98}$$

with

$$\overset{(2)}{\rho} = \int_h \overset{*}{\rho} \zeta \mu \, d\zeta \quad . \tag{3.99}$$

This integral vanishes for constant density $\overset{*}{\rho}$ only in the case of a plane middle surface $\mathcal{M}$. In the following $\overset{(2)}{\rho}$ is put equal to zero. This is no essential restriction, see also Naghdi [3]
Corresponding to (3.75) the moment of given surface forces is

$$\bar{\underset{\sim}{d}}_c = \int_{\mathcal{M}_c} [(\underset{\sim}{r} - \underset{\sim}{p}_o) \times \bar{\underset{\sim}{t}} + \zeta \underset{\sim}{a}_3 \times \bar{\underset{\sim}{t}}] da_c \quad . \tag{3.1oo}$$

It is assumed that a second, generalized edge load vector $\bar{\underset{\sim}{m}}$, referred to the arc length l , exists by means of which $\bar{\underset{\sim}{d}}_c$ is representable in the form

$$\bar{\underset{\sim}{d}}_c = \int_{\mathcal{C}} [(\underset{\sim}{r} - \underset{\sim}{p}_o) \times \bar{\underset{\sim}{q}} + \underset{\sim}{a}_3 \times \bar{\underset{\sim}{m}}] ds \quad . \tag{3.1o1}$$

The assigned generalized edge load tensor is introduced as

$$\bar{M} := \bar{\underset{\sim}{m}}^{\beta} \otimes \underset{\sim}{a}_{\beta} = \bar{M}^{\alpha\beta} \underset{\sim}{a}_{\alpha} \otimes \underset{\sim}{a}_{\beta} \quad . \tag{3.1o2}$$

It holds

$$\bar{\underset{\sim}{m}} = \hat{M}\underset{\sim}{\nu} = \bar{M}\nu_{\alpha} \underset{\sim}{a}^{\alpha} = \nu_{\alpha} \bar{\underset{\sim}{m}}^{\alpha} \quad . \tag{3.1o3}$$

Then we get for (3.1o1)

$$\begin{aligned} \bar{\underset{\sim}{d}}_c &= \int_{\mathcal{C}} [(\underset{\sim}{r} - \underset{\sim}{p}_o) \times \bar{\underset{\sim}{q}}^{\alpha} + \underset{\sim}{a}_3 \times \bar{\underset{\sim}{m}}^{\alpha} \nu_{\alpha}] ds \\ &= \int_{\mathcal{C}} [(\underset{\sim}{r} - \underset{\sim}{p}_o) \times \bar{Q} + \underset{\sim}{a}_3 \times \bar{M}] \underset{\sim}{\nu}\, ds \quad . \end{aligned} \tag{3.1o4}$$

Furthermore the edge moment tensor is introduced as

$$\bar{S} := \underset{\sim}{a}_3 \times \bar{M} = \underset{\sim}{a}_3 \times \bar{\underset{\sim}{m}}^{\alpha} \otimes \underset{\sim}{a}_{\alpha} , \tag{3.1o5}$$

with the representation

$$\bar{S} = \bar{\underset{\sim}{s}}^{\alpha} \otimes \underset{\sim}{a}_{\alpha} \quad , \quad \bar{\underset{\sim}{s}}^{\alpha} = \underset{\sim}{a}_3 \times \bar{\underset{\sim}{m}}^{\alpha} . \tag{3.1o6}$$

It follows

$$\bar{\underset{\sim}{s}} = \bar{S}\underset{\sim}{\nu} = \nu_{\alpha} \bar{\underset{\sim}{s}}^{\alpha} = \nu_{\alpha} \underset{\sim}{a}_3 \times \bar{\underset{\sim}{m}}^{\alpha} . \tag{3.1o7}$$

For (3.1o4) we get the alternative form

$$\begin{aligned} \bar{\underset{\sim}{d}}_c &= \int_{\mathcal{C}} [(\underset{\sim}{r} - \underset{\sim}{p}_o) \times \bar{\underset{\sim}{q}}^{\alpha} + \bar{\underset{\sim}{s}}^{\alpha}] \nu_{\alpha} \, ds \\ &= \int_{\mathcal{C}} [(\underset{\sim}{r} - \underset{\sim}{p}_o) \times \bar{Q} + \bar{S}] \underset{\sim}{\nu} \, ds . \end{aligned} \tag{3.1o8}$$

For deriving the local moment equilibrium conditions we consider - like before - an arbitrary subdomain $\mathcal{P} \subset \mathcal{S}$ with a part of the middle surface $\mathcal{F} \subset \mathcal{M}$. The moment of the mass forces is

$$\underset{\sim}{d}_b (\mathcal{P}) = \int_{\mathcal{F}} (\underset{\sim}{r} - \underset{\sim}{p}_o) \times \rho \, \bar{\underset{\sim}{b}} \, da , \tag{3.1o9}$$

and the moment of the sectional forces (contact stresses $\underset{\sim}{t}$) results in

$$\begin{aligned} \underset{\sim}{d}_c (\mathcal{P}) &= \int_{\mathcal{M}_{\mathcal{C}_F}} [(\underset{\sim}{r} - \underset{\sim}{p}_o) \times \underset{\sim}{t} + \zeta \underset{\sim}{a}_3 \times \underset{\sim}{t}] \, da_c \\ &= \int_{\mathcal{C}_F} [(\underset{\sim}{r} - \underset{\sim}{p}_o) \times \underset{\sim}{q} + \underset{\sim}{a}_3 \times \underset{\sim}{m}] \, ds . \end{aligned} \tag{3.11o}$$

We can assign a generalized sectional force tensor M to the generalized sectional force vector m as

$$M := \underset{\sim}{m}^{\beta} \otimes \underset{\sim}{a}_{\beta} = M^{\alpha\beta} \underset{\sim}{a}_{\alpha} \otimes \underset{\sim}{a}_{\beta} \tag{3.111}$$

with

$$\underset{\sim}{m} = M\underset{\sim}{\nu} = M\nu_{\alpha} \underset{\sim}{a}^{\alpha} = \nu_{\alpha} \underset{\sim}{m}^{\alpha} . \tag{3.112}$$

Then (3.11o) becomes

$$\begin{aligned} \underset{\sim}{d}_c (\mathcal{P}) &= \int_{\mathcal{C}_F} [(\underset{\sim}{r} - \underset{\sim}{p}_o) \times \underset{\sim}{q}^{\alpha} + \underset{\sim}{a}_3 \times \underset{\sim}{m}^{\alpha}] \nu_{\alpha} \, ds \\ &= \int_{\mathcal{C}_F} [(\underset{\sim}{r} - \underset{\sim}{p}_o) \times Q + \underset{\sim}{a}_3 \times M] \underset{\sim}{\nu} \, ds \end{aligned} \tag{3.113}$$

We define a sectional moment vector S in the form

$$S := \underset{\sim}{a}_3 \times M = \underset{\sim}{a}_3 \times \underset{\sim}{m}^{\alpha} \otimes \underset{\sim}{a}_{\alpha} \quad . \tag{3.114}$$

It has the representation

$$S = \underset{\sim}{s}^{\alpha} \otimes \underset{\sim}{a}_{\alpha} \quad , \qquad \underset{\sim}{s}^{\alpha} = \underset{\sim}{a}_3 \times \underset{\sim}{m}^{\alpha} \quad , \tag{3.115}$$

and it holds

$$\underset{\sim}{s} = S\underset{\sim}{\nu} = \nu_{\alpha}\underset{\sim}{s}^{\alpha} = \nu_{\alpha}\, \underset{\sim}{a}_3 \times \underset{\sim}{m}^{\alpha} \; . \tag{3.116}$$

For (3.113) we get the alternate form

$$\underset{\sim}{d}_c\,(\mathcal{P}) = \int_{\mathcal{C}_F} [\,(\underset{\sim}{r} - \underset{\sim}{p}_o) \times \underset{\sim}{q}^{\alpha} + \underset{\sim}{s}^{\alpha}\,]\, \nu_{\alpha}\, ds$$

$$= \int_{\mathcal{C}_F} [\,(\underset{\sim}{r} - p_o) \times Q + S\,]\, \underset{\sim}{\nu}\, ds \quad . \tag{3.117}$$

The generalized sectional force vectors are defined similar to (3.84) as

$$\underset{\sim}{m}^{\alpha} := \int_h \underset{\sim}{t}^{\alpha}\zeta\mu d\zeta \; , \tag{3.118}$$

and the components of the generalized sectional force vectors are

$$M^{\beta\alpha} = \int_h T^{\rho\alpha}\, \mu^{\beta}_{\rho}\zeta\mu d\zeta \quad . \tag{3.119}$$

The sectional moment vectors $\underset{\sim}{s}^{\alpha}$ are situated in tangential planes of the middle surface $\mathcal{M}$. They have the form

$$\underset{\sim}{s}^{\alpha} = \underset{\sim}{a}^3 \times \underset{\sim}{m}^{\alpha} = \underset{\sim}{a}_3 \times M^{\beta\alpha}\, \underset{\sim}{a}_{\beta} = M^{\beta\alpha}\, \varepsilon_{\beta\lambda}\underset{\sim}{a}^{\lambda} \; . \tag{3.120}$$

Applying the Gaussian integral theorem to (3.110) yields

$$\underset{\sim}{d}_c\,(\mathcal{P}) = \int_{\mathcal{F}} [\,(\underset{\sim}{r} - \underset{\sim}{p}_o) \times \underset{\sim}{q}^{\alpha} + \underset{\sim}{a}_3 \times \underset{\sim}{m}^{\alpha}\,]\|_{\alpha}\, da$$

$$= \int_{\mathcal{F}} [\,(\underset{\sim}{r} - \underset{\sim}{p}_o) \times \underset{\sim}{q}^{\alpha} + \underset{\sim}{s}^{\alpha}\,]\|_{\alpha}\, da$$

$$= \int_{\mathcal{F}} \delta iv [\,(\underset{\sim}{r} - \underset{\sim}{p}_o) \times Q + \underset{\sim}{a}_3 \times M\,]\, da$$

$$= \int_{\mathcal{F}} \delta iv [\,(\underset{\sim}{r} - \underset{\sim}{p}_o) \times Q + S\,]\, da \quad . \tag{3.121}$$

As $\mathcal{P}$ is arbitrary we get the local moment equilibrium condition - regarding the force equilibrium conditions - in vector notation

$$\underset{\sim}{a}_{\alpha} \times \underset{\sim}{q}^{\alpha} + (\underset{\sim}{a}_3 \times \underset{\sim}{m}^{\alpha})\|_{\alpha}$$

$$= \underset{\sim}{a}_{\alpha} \times \underset{\sim}{q}^{\alpha} + \underset{\sim}{s}^{\alpha}\|_{\alpha} = \underset{\sim}{0} \tag{3.122}$$

and in symbolic tensor notation

$$\gamma\text{rad } \underset{\sim}{r} \times Q + \gamma\text{rad } \underset{\sim}{a}_3 \times M + \underset{\sim}{a}_3 \times \delta\text{iv } M$$

$$= \gamma\text{rad } \underset{\sim}{r} \times Q + \delta\text{iv } S = \underset{\sim}{0} \quad . \tag{3.123}$$

In (3.123) the vector-valued outer product TxS of two tensors is used corresponding to de Boer [21] . We split up additively the sectional force tensor in the form

$$Q = Q_t + Q_n \tag{3.124}$$

with the tangential part

$$Q_t = Q^{\alpha\beta} \underset{\sim}{a}_{\alpha} \otimes \underset{\sim}{a}_{\beta} \tag{3.125}$$

and the normal part

$$Q_n = Q^{3\beta} \underset{\sim}{a}_3 \otimes \underset{\sim}{a}_{\beta} \quad . \tag{3.126}$$

From (3.123),it follows the symmetry condition

$$Q_t - \gamma\text{rad } \underset{\sim}{a}_3 \, M^T = Q_t^T - M \, \gamma\text{rad}^T a_3 \quad . \tag{3.127}$$

This leads to the definition of a symmetrical tensor

$$N = N^T = N^{\alpha\beta} \underset{\sim}{a}_{\alpha} \otimes a_{\beta} = \underset{\sim}{n}^{\beta} \otimes \underset{\sim}{a}_{\beta} \tag{3.128}$$

in the form

$$N := Q_t - \gamma\text{rad } \underset{\sim}{a}_3 \, M^T \quad . \tag{3.129}$$

Furthermore, it follows from (3.123)

$$Q_n = (\underset{\sim}{a}_3 \otimes \delta\text{iv } M) \, \underline{\underline{1}}_2 \tag{3.130}$$

with the twodimensional fundamental tensor of $\mathcal{M}$

$$\underline{\underline{1}}_2 = \underset{\sim}{a}_{\alpha} \otimes \underset{\sim}{a}^{\alpha} \quad . \tag{3.131}$$

With (3.129) and (3.13o) one can formulate

$$Q = N + \gamma\text{rad}\, \underset{\sim}{a}_3\, M^T + \underset{\sim}{a}_3 \otimes \delta\text{iv}\, M\, \underset{=}{1}_2 \qquad (3.132)$$

and gets as a condition for the equilibrium of forces and moments of a thin shell

$$\delta\text{iv}\,(N + \gamma\text{rad}\, \underset{\sim}{a}_3\, M^T + \underset{\sim}{a}_3 \otimes \delta\text{iv}\, M\, \underset{=}{1}_2) + \rho\, \underset{\sim}{\bar{b}} = \underset{\sim}{0}\ . \qquad (3.133)$$

3.3 <u>Equilibrium conditions in Lagrangean representation</u>

The mass force of a subdomain $\mathcal{P} \subset \mathcal{S}$ is calculated by integration over the corresponding subdomain $\mathring{\mathcal{P}} \subset \mathring{\mathcal{S}}$ of the reference configuration. One gets

$$\underset{\sim}{f}_b = \int_{\mathring{\mathcal{F}}} \mathring{\rho}\, \underset{\sim}{\bar{b}}\, d\mathring{a}\ . \qquad (3.134)$$

For the total sectional force can be represented as

$$\underset{\sim}{f}_c = \int_{\mathring{\mathcal{C}}_F} \underset{\sim}{q}_R\, d\mathring{s}\ . \qquad (3.135)$$

The sectional force vector $\underset{\sim}{q}_R$ can be assigned to a sectional force tensor Q_R

$$Q_R = \underset{\sim}{q}_R^{\alpha} \otimes \underset{\sim}{\mathring{a}}_{\alpha} = Q_R^{\alpha\beta}\, \underset{\sim}{a}_{\alpha} \otimes \underset{\sim}{\mathring{a}}_{\beta} + Q_R^{3\beta}\, \underset{\sim}{a}_3 \otimes \underset{\sim}{\mathring{a}}_{\beta}\ . \qquad (3.136)$$

It holds

$$\underset{\sim}{q}_R = Q_R\, \underset{\sim}{\mathring{\nu}} = Q_R \mathring{\nu}_{\alpha}\, \underset{\sim}{\mathring{a}}^{\alpha} = \mathring{\nu}_{\alpha}\, \underset{\sim}{q}_R^{\alpha}\ , \qquad (3.137)$$

and one gets

$$\underset{\sim}{f}_c = \int_{\mathring{\mathcal{C}}_F} \underset{\sim}{q}_R\, d\mathring{s} = \int_{\mathring{\mathcal{C}}_F} q_R^{\alpha}\, \mathring{\nu}_{\alpha}\, d\mathring{s} = \int_{\mathring{\mathcal{C}}_F} Q_R\, \underset{\sim}{\mathring{\nu}}\, d\mathring{s}\ . \qquad (3.138)$$

Due to

$$\underset{\sim}{\nu}\, ds = \det G\ G^{T-1}\, \underset{\sim}{\mathring{\nu}}\, d\mathring{s} \qquad (3.139)$$

one obtains

$$Q_R = \det G\ Q\ G^{T-1}\ , \qquad (3.140)$$

and we define

$$\tilde{Q} := G^{-1} Q_R = \det G \; G^{-1} Q \, G^{T-1} \quad . \qquad (3.141)$$

For this tensor we get the representation

$$\tilde{Q} = \tilde{q}^{\alpha} \otimes \overset{\circ}{a}_{\alpha} = \tilde{Q}^{\alpha\beta} \overset{\circ}{a}_{\alpha} \otimes \overset{\circ}{a}_{\beta} + \tilde{Q}^{3\beta} \overset{\circ}{a}_{3} \otimes \overset{\circ}{a}_{\beta} \quad . \qquad (3.142)$$

As the force equilibrium condition in Lagrangean representation we get

$$\Delta iv \, Q_R + \overset{\circ}{\rho} \bar{b} = q_R^{\alpha}|_{\alpha} + \overset{\circ}{\rho} \bar{b} = 0 \qquad (3.143)$$

or alternately

$$\Delta iv \, G \tilde{Q} + \overset{\circ}{\rho} \bar{b} = (G\tilde{q}^{\alpha})|_{\alpha} + \overset{\circ}{\rho} \bar{b} = 0 \quad . \qquad (3.144)$$

The moment of the massforces acting on a subdomain $\mathcal{P} \subset \mathcal{S}$ is calculated by integration over the subdomain $\overset{\circ}{\mathcal{P}} \subset \overset{\circ}{\mathcal{S}}$

$$d_b = \int_{\overset{\circ}{\mathcal{P}}} (r - p_o) \times \overset{\circ}{\rho} \, \bar{b} \, d\overset{\circ}{a} \quad . \qquad (3.145)$$

The moment of the sectional forces is given by

$$d_c = \int_{\overset{\circ}{\mathcal{C}}_F} [\, (r - p_o) \times q_R + a_3 \times m_R \,] \, d\overset{\circ}{s}. \qquad (3.146)$$

With

$$M_R = m_R^{\alpha} \otimes \overset{\circ}{a}_{\alpha} = M_R^{\alpha\beta} \, a_{\alpha} \otimes \overset{\circ}{a}_{\beta} \qquad (3.147)$$

and

$$m_R = M_R \overset{\circ}{\nu} = M_R \overset{\circ}{\nu}_{\alpha} \, \overset{\circ}{a}^{\alpha} = \overset{\circ}{\nu}_{\alpha} \, m_R^{\alpha} \qquad (3.148)$$

we get

$$d_c = \int_{\overset{\circ}{\mathcal{C}}_F} [\, (r - p_o) \times q_R^{\alpha} + a_3 \times m_R^{\alpha} \,] \overset{\circ}{\nu}_{\alpha} d\overset{\circ}{s}$$

$$= \int_{\overset{\circ}{\mathcal{C}}_F} [\, (r - p_o) \times Q_R + a_3 \times M_R \,] \overset{\circ}{\nu} \, d\overset{\circ}{s} \quad . \qquad (3.149)$$

It holds

$$M_R = \det G \; M \, G^{T-1} \, , \qquad (3.150)$$

and we define

$$\tilde{M} := G^{-1} M_R = \det G \; G^{-1} M \, G^{T-1} \quad . \tag{3.151}$$

$\tilde{M}$ has the representation

$$\tilde{M} = \tilde{m}^{\alpha} \otimes \overset{\circ}{a}_{\alpha} = \tilde{M}^{\alpha\beta} \, \overset{\circ}{a}_{\alpha} \otimes \overset{\circ}{a}_{\beta} \; . \tag{3.152}$$

The moment equilibrium condition in Lagrangean consideration becomes in vector notation

$$a_{\alpha} \times q_R^{\alpha} + (a_3 \times m_R^{\alpha})|_{\alpha} = 0 \tag{3.153}$$

or alternately

$$G\overset{\circ}{a}_{\alpha} \times G\,\tilde{q}^{\alpha} + (G\,\overset{\circ}{a}_3 \times G\,\tilde{m}^{\alpha})|_{\alpha} = 0 \; . \tag{3.154}$$

In symbolic tensor notation, (3.153) has the form

$$\Gamma\mathrm{rad}\; r \times Q_R + (\Gamma\mathrm{rad}\; a_3) \times M_R + a_3 \times \Delta\mathrm{iv}\; M_R = 0 \; , \tag{3.155}$$

and alternately (3.154) becomes

$$(G\,\overset{\circ}{1}_2) \times (G\,\tilde{Q}) + (\Gamma\mathrm{rad}\; G\overset{\circ}{a}_3) \times (G\,\tilde{M}) + G\overset{\circ}{a}_3 \times \Delta\mathrm{iv}\; G\tilde{M} = 0 \; . \tag{3.156}$$

We split up the tensor $\tilde{Q}$ additively into a tangential and a normal part as

$$\tilde{Q} = \tilde{Q}_t + \tilde{Q}_n \quad . \tag{3.157}$$

From (3.156) it follows that the following symmetrical tensor $\tilde{N}$ exists

$$\tilde{N} = \tilde{N}^T = \tilde{Q}_t - G^{-1}\,\Gamma\mathrm{rad}\; a_3\, \tilde{M}^T . \tag{3.158}$$

Furthermore we obtain from (3.156)

$$\tilde{Q}_n = \{\overset{\circ}{a}_3 \otimes G^{-1}\,\Delta\mathrm{iv}\;(GM)\}\,\overset{\circ}{1}_2 \quad . \tag{3.159}$$

Then we get

$$G\tilde{Q} = G\tilde{N} + \Gamma\mathrm{rad}\; a_3\, \tilde{M}^T + \{a_3 \otimes G^{-1}\,\Delta\mathrm{iv}\;(G\tilde{M})\}\,\overset{\circ}{1}_2 . \tag{3.160}$$

The force and moment equilibrium condition in Lagrangean consideration is obtained by introducing (3.160) into (3.144)

$$\Delta iv\ (G\tilde{N} + \Gamma rad\ \underset{\sim}{a}_3\ \tilde{M}^T + \{\underset{\sim}{a}_3 \otimes G^{-1}\ \Delta iv\ (GM)\}\ \overset{\circ}{\underset{=}{1}}_2) + \overset{\circ}{\rho}\ \overline{\underset{\sim}{b}}$$

$$= \Delta iv\ (G\tilde{N} + \Gamma rad\ \underset{\sim}{a}_3\ \tilde{M}^T$$

$$+ \{\underset{\sim}{a}_3 \otimes G^{-1}(\Gamma rad\ G[\tilde{M}])\}\ \overset{\circ}{\underset{=}{1}}_2$$

$$+ (\underset{\sim}{a}_3 \otimes \Delta iv\ \tilde{M})\ \overset{\circ}{\underset{=}{1}}_2) + \overset{\circ}{\rho}\ \overline{\underset{\sim}{b}} = \underset{\sim}{0}\ . \tag{3.161}$$

For clearness of (3.161), the unit normal vector $\underset{\sim}{a}_3 = G\ \overset{\circ}{\underset{\sim}{a}}_3$ is left in the current configuration.

3.4 The principle of virtual work in Lagrangean consideration

The principle of virtual work, suitable for practical calculations, is a weak formulation of the force and moment equilibrium conditions in the considered domain and at the boundary. We can formulate the statical boundary conditions by multiplying the statical field equations (3.144) with a virtual displacement, integrating over the middle surface $\overset{\circ}{\mathcal{M}}$ of the reference configuration, as

$$\int_{\overset{\circ}{\mathcal{M}}} \Delta iv\ G\tilde{Q} \cdot \delta\underset{\sim}{u}\ d\overset{\circ}{a} + \int_{\overset{\circ}{\mathcal{M}}} \overset{\circ}{\rho}\ \overline{\underset{\sim}{b}} \cdot \delta\underset{\sim}{u}\ d\overset{\circ}{a} = 0, \tag{3.162}$$

using partial integration

$$\int_{\overset{\circ}{\mathcal{M}}} \Delta iv\ G\tilde{Q} \cdot \delta\underset{\sim}{u}\ d\overset{\circ}{a} =$$

$$\int_{\overset{\circ}{\mathcal{M}}} \Delta iv\ (\tilde{Q}^T G^T \delta\underset{\sim}{u})\ d\overset{\circ}{a} - \int_{\overset{\circ}{\mathcal{M}}} G\tilde{Q} \cdot \Gamma rad\ \delta\underset{\sim}{u}\ \ d\overset{\circ}{a} \tag{3.163}$$

and applying the Gaussian divergence theorem in the form

$$\int_{\overset{\circ}{\mathcal{M}}} \Delta iv(\tilde{Q}^T G^T \delta\underset{\sim}{u})\ d\overset{\circ}{a} = \int_{\overset{\circ}{\mathcal{C}}} \tilde{Q}^T G^T \delta\underset{\sim}{u} \cdot \overset{\circ}{\underset{\sim}{\nu}}\ d\overset{\circ}{s}$$

$$= \int_{\overset{\circ}{\mathcal{C}}} \delta\underset{\sim}{u} \cdot G\tilde{Q}\ \underset{\sim}{\nu}\ d\overset{\circ}{s} = \int_{\overset{\circ}{\mathcal{C}}} \delta\underset{\sim}{u} \cdot G\tilde{\underset{\sim}{q}}\ d\overset{\circ}{s}\ . \tag{3.164}$$

On the part $\overset{\circ}{\mathcal{C}}_q$ of the boundary with prescribed edge forces we get the equilibrium condition

$$G\tilde{\underset{\sim}{q}} = \underset{\sim}{q}_R = \overline{\underset{\sim}{q}}_R = G\ \overline{\underset{\sim}{q}}\ . \tag{3.165}$$

So, from (3.162) follows a first form of the wanted principle

$$\int_{\mathcal{A}} G\tilde{Q} \cdot \Gamma\text{rad}\,\delta\underset{\sim}{u} = \int_{\mathcal{E}_q} \overline{\underset{\sim}{q}}_R \cdot \delta\underset{\sim}{u}\, d\overset{\circ}{s} + \int_{\mathcal{A}} \overset{\circ}{\rho}\overline{\underset{\sim}{b}} \cdot \delta\underset{\sim}{u}\, d\overset{\circ}{a}\,. \qquad (3.166)$$

In (3.168) the virtual work of the surface loads and of the edge loads $\overline{\underset{\sim}{q}}_R$ (including the congruents in the direction of the normal of the surface) are represented. Like in (3.161) we introduce

$$G\tilde{Q} \cdot \Gamma\text{rad}\,\delta\underset{\sim}{u} = (G\tilde{N} + \Gamma\text{rad}\,\underset{\sim}{a}_3\, \tilde{M}^T + \{\underset{\sim}{a}_3 \otimes G^{-1}\Delta\text{iv}(GM)\}\overset{\circ}{\underline{\underline{1}}}_2) \cdot \Gamma\text{rad}\delta\underset{\sim}{u}\,. \qquad (3.167)$$

We use the following transformation

$$\begin{aligned} G\tilde{Q} \cdot \Gamma\text{rad}\delta\underset{\sim}{u} &= G\tilde{N} \cdot \Gamma\text{rad}\delta\underset{\sim}{u} \\ &+ \Delta\text{iv}\,(\tilde{M}^T\, \Gamma\text{rad}\delta\underset{\sim}{u}\, \underset{\sim}{a}_3) \\ &+ \tilde{M}^T \cdot (\Gamma\text{radG})^{\overset{13}{T}}\, (G^{T-1}\, \Gamma\text{rad}^T\delta\underset{\sim}{u}\, \underset{\sim}{a}_3) \\ &- \tilde{M}^T \cdot (\Gamma\text{rad}\, \Gamma\text{rad}\delta\underset{\sim}{u})^{\overset{13}{T}}\, \underset{\sim}{a}_3\,. \end{aligned} \qquad (3.168)$$

The application of the Gaussian divergence theorem with consideration of the boundary condition for the moments on the part $\overset{\circ}{\mathcal{E}}_m$ of the edge where a moment loading is prescribed,

$$\overline{\underset{\sim}{m}}_R = \underset{\sim}{m}_R \quad \text{on } \overset{\circ}{\mathcal{E}}_m\,, \qquad (3.169)$$

we get for the second term of (3.169)

$$\begin{aligned} \int_{\overset{\circ}{\mathcal{A}}} \Delta\text{iv}(\tilde{M}^T\, \Gamma\text{rad}^T\delta\underset{\sim}{u}\, \underset{\sim}{a}_3)\, d\overset{\circ}{a} &= \int_{\overset{\circ}{\mathcal{E}}} (\tilde{M}^T\, \Gamma\text{rad}^T\delta\underset{\sim}{u}\, \underset{\sim}{a}_3) \cdot \overset{\circ}{\underset{\sim}{\nu}} d\overset{\circ}{s} \\ &= \int_{\overset{\circ}{\mathcal{E}}} (\Gamma\text{rad}^T\delta\underset{\sim}{u}\, \underset{\sim}{a}_3) \cdot \tilde{M}\overset{\circ}{\underset{\sim}{\nu}}\, d\overset{\circ}{s} \\ &= \int_{\overset{\circ}{\mathcal{E}}} \delta G^T\, \underset{\sim}{a}_3 \cdot \tilde{\underset{\sim}{m}}\, d\overset{\circ}{s} = -\int_{\overset{\circ}{\mathcal{E}}} G^T\, \delta\underset{\sim}{a}_3 \cdot \tilde{\underset{\sim}{m}}\, d\overset{\circ}{s} \\ &= -\int_{\overset{\circ}{\mathcal{E}}} \delta\underset{\sim}{a}_3 \cdot G\underset{\sim}{m}\, d\overset{\circ}{s} = -\int_{\overset{\circ}{\mathcal{E}}} \delta\underset{\sim}{a}_3 \cdot \underset{\sim}{m}_R\, d\overset{\circ}{s} = -\int_{\overset{\circ}{\mathcal{E}}} \overline{\underset{\sim}{m}}_R \cdot \delta\underset{\sim}{a}_3 d\overset{\circ}{s} \end{aligned} \qquad (3.170)$$

The principle of virtual work in Lagrangean consideration finally gets the form

$$\int_{\overset{o}{\mathcal{A}}} \{\tilde{N} \cdot G^T\ \Gamma\text{rad}\delta\underset{\sim}{u} + \tilde{M}^T \cdot [-(\Gamma\text{rad}\Gamma\text{rad}\delta\underset{\sim}{u})^{\overset{13}{T}}\ \underset{\sim}{a}_3 - (\Gamma\text{rad}G)^{\overset{13}{T}}\ \delta\underset{\sim}{a}_3]\}\ d\overset{o}{a}$$

$$= \int_{\overset{o}{\ell}_q} \bar{\underset{\sim}{q}}_R \cdot \delta\underset{\sim}{u}\ d\overset{o}{s} + \int_{\overset{o}{\ell}_m} \bar{\underset{\sim}{m}}_R \cdot \delta\underset{\sim}{a}_3\ d\overset{o}{s} + \int_{\overset{o}{\mathcal{A}}} \overset{o}{\rho}\bar{\underset{\sim}{b}} \cdot \delta\underset{\sim}{u}\ d\overset{o}{a} \quad . \tag{3.171}$$

Berg[6] shows that this form is equivalent with the usual notation

$$\int_{\overset{o}{\mathcal{A}}} \{N \cdot \delta\underset{=}{\gamma} + \tilde{M} \cdot \delta\underset{=}{\chi}\}\ d\overset{o}{a}$$

$$= \int_{\overset{o}{\ell}_q} \bar{\underset{\sim}{q}}_R \cdot \delta\underset{\sim}{u}\ d\overset{o}{s} + \int_{\overset{o}{\ell}_m} \bar{\underset{\sim}{m}}_R \cdot \delta\underset{\sim}{a}_3\ d\overset{o}{s} + \int_{\overset{o}{\mathcal{A}}} \overset{o}{\rho}\bar{\underset{\sim}{b}} \cdot \delta\underset{\sim}{u}\ d\overset{o}{a} \tag{3.172}$$

where $\underset{=}{\gamma}$ is the strain tensor and $\underset{=}{\chi}$ is the change of the curvature tensor of the middle surface of the shell, see chapter 3.1.

3.5 Constitutive equations for thin elastic shells

We assume a homogeneous and isotropic elastic material. Only small strains are admitted so that a linear relation between the 2. Piola-Kirchhoff stress tensor and Green strain tensor is valid for many engineering materials. Under these conditions it is usual and admissible to introduce the following constitutive shell equations for the symmetrical parts of the sectional force and moment tensors

$$\tilde{N} = \frac{Ch}{1-\nu^2}\ \mathbb{H}_2[\underset{=}{\gamma}] \ , \tag{3.173}$$

$$\tilde{M} = \frac{Ch^3}{12(1-\nu^2)}\ \mathbb{H}_2[\underset{=}{\chi}] \ . \tag{3.174}$$

C is Youngs modulus, and ν is Poisson's ratio in the reference configuration. The components of the fourth order tensor $\mathbb{H}_2$ are

$$H^{\alpha\beta\gamma\delta} = (1-\nu)\ \overset{o}{a}^{\alpha\beta}\ \overset{o}{a}^{\beta\delta} + \nu\ \overset{o}{a}^{\alpha\beta}\ \overset{o}{a}^{\gamma\delta} \tag{3.175}$$

The constitutive equations (3.173) and (3.174) are very suitable for practical calculations because of their simple structure. They are used by Naghdi [3], Koiter[1o], Pietraszkiewicz[5], Basar and Krätzig [11] and Stumpf [9].

4. Incremental theory of thin elastic shells

4.1 Notation for Lagrangean consideration

Besides the reference configuration A and a known current configuration G - which is in equilibrium with the given load - we introduce an arbitrary incremental equilibrium configuration D corresponding to Fig. 4.

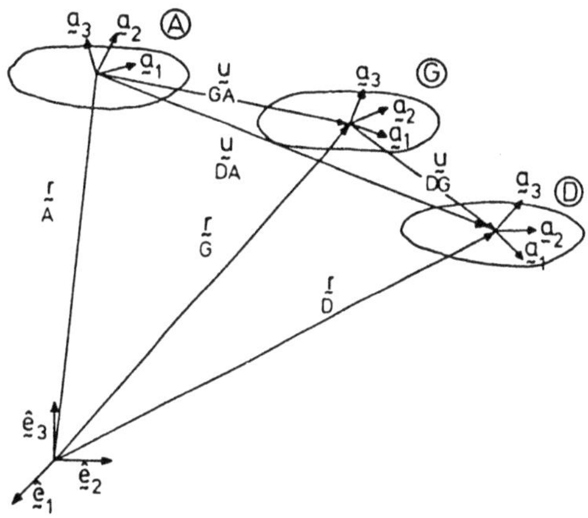

Fig. 4. Notations for the incremental deformation process

The quantities of the different configurations are marked by subindices A , G and D. Displacements from one configuration to another and incremental quantities, defined by differences, are marked by the letters of both configurations. In the case of gradients of incremental quantities, we must denote also the reference base. This can be omitted when we restrict to a total Lagrangean description.

4.2 Kinematics of shell space in incremental description

The displacement gradient of the current configuration has the form

$$\underset{GA}{H} = \Gamma\mathrm{rad}\, \underset{\tilde{GA}}{u} = \underset{\tilde{GA}}{u}{}_{,\alpha} \otimes \underset{\tilde{A}}{a}{}^{\alpha} \; . \tag{4.1}$$

Correspondingly, the gradient for the displacement increments is

$$H_{DA} = \Gamma\mathrm{rad}\,\tilde{u}_{DA} = \tilde{u}_{DA,\alpha} \otimes \tilde{a}_{A}^{\alpha} \quad . \tag{4.2}$$

It holds, see Fig. 4

$$\tilde{u}_{DA} = \tilde{u}_{DG} + \tilde{u}_{GA} \quad , \tag{4.3}$$

and we calculate

$$H_{DGA} = \Gamma\mathrm{rad}\,\tilde{u}_{DG} = \tilde{u}_{DG,\alpha} \otimes \tilde{a}_{A}^{\alpha} = H_{DA} - H_{GA} \quad . \tag{4.4}$$

The strains of the middle surface in the reference configuration are according to (3.55)

$$\underline{\underline{\gamma}}_{GA} = \frac{1}{2}\,(H^{T}_{GA}\,\underline{\underline{1}}^{2}_{A} + \underline{\underline{1}}^{2}_{A}\,H_{GA} + H^{T}_{GA}\,H_{GA}) \tag{4.5}$$

and in the incremental state

$$\underline{\underline{\gamma}}_{DA} = \frac{1}{2}\,(H^{T}_{DA}\,\underline{\underline{1}}^{2}_{A} + \underline{\underline{1}}^{2}_{A}\,H_{DA} + H^{T}_{DA}\,H_{DA})$$

$$= \frac{1}{2}\,\{(H^{T}_{DGA} + H^{T}_{GA})\,\underline{\underline{1}}^{2}_{A} + \underline{\underline{1}}^{2}_{A}\,(H_{DGA} + H_{GA}) + (H_{DGA} + H_{GA})^{T}(H_{DGA} + H_{GA})\} \quad . \tag{4.6}$$

As the incremental measure for the strains we get

$$\underline{\underline{\gamma}}_{DGA} = \underline{\underline{\gamma}}_{DA} - \underline{\underline{\gamma}}_{GA} = \frac{1}{2}\,(H^{T}_{DGA}\,\underline{\underline{1}}^{2}_{A} + \underline{\underline{1}}^{2}_{A}\,H_{DGA}$$

$$+ H^{T}_{GA}\,H_{DGA} + H_{DGA}\,H^{T}_{GA} + H^{T}_{DGA}\,H_{DGA}) \,. \tag{4.7}$$

So we can distinguish three terms

$$\underline{\underline{\gamma}}_{DGA} = \mathrm{rlin}\,\underline{\underline{\gamma}}_{DGA} + \mathrm{plin}\,\underline{\underline{\gamma}}_{DGA} + \mathrm{nlin}\,\underline{\underline{\gamma}}_{DGA} \tag{4.8}$$

with the usual purely linear term

$$\mathrm{rlin}\,\underline{\underline{\gamma}}_{DGA} = \frac{1}{2}\,(H^{T}_{DGA}\,\underline{\underline{1}}^{2}_{A} + \underline{\underline{1}}^{2}_{A}\,H_{DGA}) \quad , \tag{4.9}$$

a pseudolinear term with displacements of the configuration G

$$\mathrm{plin}\,\underline{\underline{\gamma}}_{DGA} = \frac{1}{2}\,(H^{T}_{GA}\,H_{DGA} + H^{T}_{DGA}\,H_{GA}) \tag{4.1o}$$

and a nonlinear term

$$\text{nlin}\ \underset{DGA}{\underline{\underline{\gamma}}} = \frac{1}{2}\ \underset{DGA}{H}^{T}\ \underset{DGA}{H}\ . \tag{4.11}$$

One can recognize that the first and the third term of $\underset{DGA}{\underline{\underline{\gamma}}}$ originate formally from strains $\underset{DA}{\underline{\underline{\gamma}}}$ where the total displacements $\underset{DA}{u}$ are replaced by the incremental displacements $\underset{D\tilde{G}}{u}$. The additional second term in (4.8) is characteristic for the total-Lagrangean representation. This second term consists of two parts which are obtained as follows: In the nonlinear part of the total strains $\underset{\overline{DA}}{\underline{\gamma}}$, we introduce $\underset{\tilde{GA}}{u}$ in the first and $\underset{\tilde{DA}}{u}$ in the second factor, and vice versa. The virtual strains are

$$\delta\ \underset{D\tilde{G}A}{\underline{\underline{\gamma}}} = \text{rlin}\ \delta\ \underset{DGA}{\underline{\underline{\gamma}}} + \text{plin}\ \delta\ \underset{D\tilde{G}A}{\underline{\underline{\gamma}}} + \text{nlin}\ \delta\ \underset{D\tilde{G}A}{\underline{\underline{\gamma}}} \tag{4.12}$$

with

$$\text{rlin}\ \delta\ \underset{DGA}{\underline{\underline{\gamma}}} = \frac{1}{2}\ (\delta\ \underset{DGA}{H}^{T}\ \underset{A}{\underline{\underline{1}}}_{2} + \underset{\bar{A}}{\underline{\underline{1}}}_{2}\ \delta\ \underset{DGA}{H})\ , \tag{4.13}$$

$$\text{plin}\ \delta\ \underset{D\tilde{G}A}{\underline{\underline{\gamma}}} = \frac{1}{2}\ (\ \underset{GA}{H}^{T}\ \underset{DGA}{\delta H} + \underset{DGA}{\delta H}^{T}\ \underset{GA}{H}\ )\ , \tag{4.14}$$

$$\text{nlin}\ \delta\ \underset{DGA}{\underline{\underline{\gamma}}} = \frac{1}{2}\ (\ \underset{DGA}{H}^{T}\ \underset{DGA}{H}\ )\ . \tag{4.15}$$

The elastic curvatures in the reference configurations are according to (3.68)

$$\underset{\underline{\underline{}}}{\chi} = -\ (\underset{\tilde{A}}{a}^{\alpha} \otimes \Gamma\text{rad}^{T}\ \underset{\tilde{GA}}{u}_{,\alpha})\underset{\tilde{G}}{a}_{3} - (\underset{\tilde{A}}{a}^{\alpha} \otimes \Gamma\text{rad}^{T}\ \underset{\tilde{A}}{a}_{\alpha})\ \underset{\tilde{GA}}{\beta}$$

$$= (\underset{\tilde{A}}{a}^{\alpha} \otimes \Gamma\text{rad}^{T}\ \underset{\tilde{A}}{a}_{\alpha})\underset{\tilde{A}}{a}_{3} - (\underset{\tilde{A}}{a}^{\alpha} \otimes \Gamma\text{rad}^{T}\ \underset{\tilde{G}}{a}_{\alpha})\underset{\tilde{G}}{a}_{3} \tag{4.16}$$

and in the incremental state

$$\underset{\overline{DA}}{\underline{\chi}} = (\underset{\tilde{A}}{a}^{\alpha} \otimes \Gamma\text{rad}^{T}\ \underset{\tilde{A}}{a}_{\alpha})\underset{\tilde{A}}{a}_{3} - (\underset{\tilde{A}}{a}^{\alpha} \otimes \Gamma\text{rad}^{T}\ \underset{\tilde{D}}{a}_{\alpha})\underset{\tilde{D}}{a}_{3}\ . \tag{4.17}$$

The incremental elastic curvatures are

$$\underset{DGA}{\underline{\underline{\chi}}} = \underset{\overline{DG}}{\underline{\chi}} - \underset{\overline{GA}}{\underline{\chi}} = -[\ (\underset{\tilde{A}}{a}^{\alpha} \otimes \Gamma\text{rad}^{T}\underset{D\tilde{G}}{u}_{,\alpha})\underset{\tilde{D}}{a}_{3} + (\underset{\tilde{A}}{a}^{\alpha} \otimes \Gamma\text{rad}^{T}\underset{\tilde{G}}{a}_{\alpha})\underset{D\tilde{G}}{\beta}\ ]\ . \tag{4.18}$$

Further decomposition of the base vectors $\underset{\tilde{D}}{a}_3$ and $\underset{\tilde{G}}{a}_\alpha$ yields

$$\underset{DGA}{\underline{\underline{\chi}}} = -\,[\,(\underset{\tilde{A}}{a}^\alpha \boxtimes \Gamma rad^T \underset{\tilde{DG}}{u}_{,\alpha})\underset{\tilde{A}}{a}_3 + (\underset{\tilde{A}}{a}^\alpha \boxtimes \Gamma rad^T \underset{\tilde{A}}{a}_\alpha)\underset{\tilde{DG}}{\beta}$$

$$+ (\underset{\tilde{A}}{a}^\alpha \; \Gamma rad^T \underset{\tilde{DG}}{u}_{,\alpha})\underset{\tilde{GA}}{\beta} + (\underset{\tilde{A}}{a}^\alpha \boxtimes \Gamma rad^T \underset{\tilde{GA}}{u}_{,\alpha})\underset{\tilde{DG}}{\beta}$$

$$+ (\underset{\tilde{A}}{a}^\alpha \Gamma rad^T \underset{\tilde{DG}}{u}_{,\alpha})\underset{\tilde{DG}}{\beta}\,]. \tag{4.19}$$

With the linear part lin $\underset{DG}{\beta}$ and the nonlinear part nlin $\underset{\tilde{DG}}{\beta}$ of $\underset{\tilde{DG}}{\beta}$ we get the elastic curvatures, corresponding to the strains, as

$$\text{rlin}\,\underset{DGA}{\underline{\underline{\chi}}} = -\,[\,(\underset{\tilde{A}}{a}^\alpha \boxtimes \Gamma rad^T \underset{\tilde{DG}}{u}_{,\alpha})\underset{\tilde{A}}{a}_3 + (\underset{\tilde{A}}{a} \boxtimes \Gamma rad^T \underset{\tilde{A}}{a}_\alpha)\,\text{lin}\underset{\tilde{DG}}{\beta}\,], \tag{4.20}$$

$$\text{plin}\,\underset{DGA}{\underline{\underline{\chi}}} = -\,[\,(\underset{\tilde{A}}{a}^\alpha \boxtimes \Gamma rad^T \underset{\tilde{DG}}{u}_{,\alpha})\underset{\tilde{GA}}{\beta} + (\underset{\tilde{A}}{a}^\alpha \boxtimes \Gamma rad^T \underset{\tilde{GA}}{u}_{,\alpha})\,\text{lin}\underset{\tilde{DG}}{\beta}\,], \tag{4.21}$$

$$\text{nlin}\,\underset{DGA}{\underline{\underline{\chi}}} = -\,[\,\;\underset{\tilde{A}}{a}^\alpha \boxtimes \Gamma rad^T \underset{\tilde{A}}{a}_\alpha)\,\text{nlin}\underset{\tilde{DG}}{\beta} + (\underset{\tilde{A}}{a} \boxtimes \Gamma rad^T \underset{\tilde{GA}}{u}_{,\alpha})\,\text{nlin}\underset{\tilde{DG}}{\beta}$$

$$+ (\underset{\tilde{A}}{a}^\alpha \boxtimes \Gamma rad^T \underset{\tilde{DG}}{u}_{,\alpha})\underset{\tilde{DG}}{\beta}\,]. \tag{4.22}$$

In total we get the sum

$$\underset{DGA}{\underline{\underline{\chi}}} = \text{rlin}\,\underset{DGA}{\underline{\underline{\chi}}} + \text{plin}\,\underset{DGA}{\underline{\underline{\chi}}} + \text{nlin}\,\underset{DGA}{\underline{\underline{\chi}}} \;. \tag{4.23}$$

The virtual elastic curvatures are

$$\delta\,\underset{D\bar{G}A}{\underline{\underline{\chi}}} = -\,[\,(\underset{\tilde{A}}{a}^\alpha \boxtimes \Gamma rad^T \delta \underset{\tilde{DG}}{u}_{,\alpha})\underset{\tilde{D}}{a}_3 + (\underset{\tilde{A}}{a}^\alpha \boxtimes \Gamma rad^T \underset{\tilde{D}}{a}_\alpha)\,\delta \underset{\tilde{D}}{a}_3\,]\,. \tag{4.24}$$

It holds

$$(\underset{\tilde{A}}{a}^\alpha \boxtimes \Gamma rad^T \underset{\tilde{D}}{a}_\alpha)\;\delta \underset{\tilde{D}}{a}_3 = -\underset{DAA}{L}^{\overset{13}{T}}(\Gamma rad^T \delta \underset{\tilde{DG}}{u}\;\underset{\tilde{D}}{a}_3)\;, \tag{4.25}$$

with

$$\underset{DAA}{L} = \underset{DA}{L}{}^{\mu}{}_{\alpha\beta}\;\underset{\tilde{A}}{a}_\mu \boxtimes \underset{\tilde{A}}{a}^\alpha \boxtimes \underset{\tilde{A}}{a}^\beta = ({}_D{}^{\mu}{}_{\alpha\beta} - {}_A{}^{\mu}{}_{\alpha\beta})\,\underset{\tilde{A}}{a}_\mu \boxtimes \underset{\tilde{A}}{a}^\alpha \boxtimes \underset{\tilde{A}}{a}^\beta\;. \tag{4.26}$$

So, we can replace (4.24) by

$$\delta \underset{DGA}{\underline{\underline{\chi}}} = - (\underset{\tilde{A}}{a^{\alpha}} \boxtimes \Gamma rad^{T} \underset{D\tilde{G},\alpha}{\delta u}) \underset{\tilde{D}}{a_3} + \overset{13}{\underset{DAA}{L}}{}^{T} (\Gamma rad^{T} \underset{D\tilde{G}}{\delta u} \underset{\tilde{D}}{a_3}) \ . \qquad (4.27)$$

4.3 Equilibrium conditions and principle of virtual work in incremental form

The force equilibrium conditions in the current configuration are

$$\Delta iv(\underset{GA}{G} \underset{GA}{Q}) + \underset{A}{\rho} \underset{G}{\bar{\underset{\sim}{b}}} = \underset{\sim}{O} \qquad (4.28)$$

and in the incremental state

$$\Delta iv(\underset{DA}{G} \underset{DA}{Q}) + \underset{A}{\rho} \underset{\tilde{D}}{b} = \underset{\sim}{O} \ . \qquad (4.29)$$

With the decomposition of the sectional force tensor

$$\underset{DA}{Q} = \underset{DGA}{Q} + \underset{GA}{Q} \qquad (4.30)$$

and the loading

$$\underset{\tilde{D}}{\bar{b}} = \underset{D\tilde{G}}{\bar{b}} + \underset{\tilde{G}}{\bar{b}} \qquad (4.31)$$

we get from (4.29)

$$\Delta iv\,[\,(\underset{DGA}{G} + \underset{GA}{G})(\underset{DGA}{Q} + \underset{GA}{Q})\,] + \underset{A}{\rho}\ (\underset{\tilde{D}G}{\bar{b}} + \underset{\tilde{G}}{\bar{b}}) = \underset{\sim}{O} \ . \qquad (4.32)$$

We subtract (4.26) from (4.22) and get for the incremental force equilibrium condition

$$\Delta iv\ (\underset{DA}{G} \underset{DGA}{Q} + \underset{DGA}{G} \underset{GA}{Q}) + \underset{A}{\rho} \underset{\tilde{D}G}{\bar{b}} = \underset{\sim}{O} \ . \qquad (4.33)$$

The moment equilibrium condition in the current configuration is, see (3.160)

$$\underset{GA}{G} \underset{GA}{Q} = \underset{GA}{G} \underset{GA}{N} + \Gamma rad \underset{\tilde{G}}{a_3} \underset{GA}{M}{}^{T}$$

$$+ (\underset{\tilde{G}}{a_3} \boxtimes \Delta iv \underset{GA}{M}) \underset{\underline{\underline{A}}}{1_2} + \underset{\tilde{G}}{a_3} \boxtimes \underset{GAA}{L} [\underset{GA}{M}] \underset{\underline{\underline{A}}}{1_2} \qquad (4.34)$$

and in the incremental state

$$\underset{DA}{G}\ \underset{DA}{Q} = \underset{DA}{G}\ \underset{DA}{N} + \Gamma\mathrm{rad}\ \underset{D}{\tilde{a}_3}\ \underset{DA}{M}^T$$

$$+ (\underset{D}{\tilde{a}_3} \boxtimes \Delta\mathrm{iv}\ \underset{DA}{M})\ \underset{A}{\underline{\underline{1}}}^2 + \underset{D}{\tilde{a}_3} \boxtimes \underset{DAA}{L}[\underset{DA}{M}]\ \underset{A}{\underline{\underline{1}}}^2 \tag{4.35}$$

We introduce (4.34) into (4.28) and (4.35) into (4.29) and subtract (4.34) from (4.35). The result is an unified force and moment equilibrium condition for the load increment which was derived by Berg [6] as

$$\Delta\mathrm{iv}\{\underset{DA}{G}\ \underset{DGA}{N} + \underset{DGA}{G}\ \underset{GA}{N} + \Gamma\mathrm{rad}\underset{D}{\tilde{a}_3}\ \underset{DGA}{M}^T + \Gamma\mathrm{rad}\underset{DG}{\tilde{\beta}}\ \underset{GA}{M}^T$$

$$+ (\underset{D}{\tilde{a}_3} \boxtimes \Delta\mathrm{iv}\ \underset{DGA}{M})\ \underset{A}{\underline{\underline{1}}}^2 + (\underset{DG}{\tilde{\beta}} \boxtimes \Delta\mathrm{iv}\ \underset{GA}{M})\ \underset{A}{\underline{\underline{1}}}^2$$

$$+ \underset{D}{\tilde{a}_3} \boxtimes \underset{DAA}{L}[\underset{DGA}{M}]\ \underset{A}{\underline{\underline{1}}}^2 + \underset{DG}{\tilde{\beta}} \boxtimes \underset{DAA}{L}[\underset{GA}{M}]\ \underset{A}{\underline{\underline{1}}}^2$$

$$+ \underset{G}{\tilde{a}_3} \boxtimes \underset{DGA}{L}[\underset{GA}{M}]\ \underset{A}{\underline{\underline{1}}}^2\} + \underset{A}{\rho}\ \underset{DG}{\tilde{\bar{b}}} = \tilde{0}\ . \tag{4.36}$$

By multiplication of (4.36) with an virtual incremental displacement $\delta\underset{DG}{\tilde{u}}$ and integration over $\mathring{\mathcal{M}}$ we get the incremental principle of virtual work

$$\int_{\mathring{\mathcal{M}}} (\underset{DGA}{N} \cdot \delta\underset{DGA}{\underline{\underline{\gamma}}} + \underset{GA}{N} \cdot \mathrm{nlin}\ \delta\underset{DGA}{\underline{\underline{\gamma}}}$$

$$+ \underset{DGA}{M} \cdot \delta\underset{DGA}{\underline{\underline{\chi}}} + \underset{GA}{M} \cdot \mathrm{nlin}\ \delta\underset{DGA}{\underline{\underline{\chi}}})\ d\mathring{a}$$

$$= \int_{\mathring{\mathcal{M}}} \underset{A}{\rho}\,\underset{DG}{\tilde{\bar{b}}} \cdot \delta\underset{DG}{\tilde{u}}\ d\mathring{a} + \int_{\mathring{\mathcal{C}}_q} \underset{DG}{\tilde{q}_R} \cdot \delta\underset{DG}{\tilde{u}}\ d\mathring{s}$$

$$+ \int_{\mathring{\mathcal{C}}_m} \underset{DG}{\tilde{m}_R} \cdot \delta\underset{DG}{\tilde{\beta}}\ d\mathring{s}\ . \tag{4.37}$$

An extended derivation of (4.31) is given by Berg [6] .

5. FE-formulation of the incremental principle of virtual work

In the following, the FE-formulation of the incremental principle of virtual work according to chapter 4 is carried out. By introducing the constitutive equations from chapter 3.5 and the incremental kinematical relations of chapter 4.2, the work principle is formulated in the incremental displacements $\underset{DG}{\underset{\sim}{u}}$. The approximation of these displacements in finite subdomains (finite elements) leads to the FE-displacement method.

5.1 FE-discretization of the incremental principle of virtual work

We introduce polynomial test functions of equal degree for the covariant components $\underset{DG}{u}_\alpha$ and $\underset{DG}{u}_3$ in the form

$$\underset{DG}{u}_j = \sum_{N=1}^{n} \Phi_N \, \hat{u}_{j_N} = \underline{\Phi} \, \underline{\hat{u}}_j \qquad j = 1,2,3 \qquad (5.1)$$

with the Ritz-parameter $\hat{u}_{j_N}$. With a proper choice of n and the introduction of the column matrix $\underline{v}_j$ of nodal displacements v_{j_N} by the help of the regular transformation matrix $\underline{G}_j$ we get the matrix representation

$$\underset{DG}{u}_j = \underline{\Phi} \; \underline{G}_j \; \underline{v}_j = \underline{\Omega}_j \; \underline{v}_j \; . \qquad j = 1,2,3 \qquad (5.2)$$

Arranging the three displacements in a column matrix $\underline{u}$ gives the condensed form

$$\underline{u} = \underline{\Omega} \; \underline{v} \qquad (5.3)$$

with the unit displacement functions in matrix $\underline{\Omega}$. The formal matrix representation of the kinematical equations is introduced as

$$\underline{\gamma} = (\text{rlin } \underline{D}_\gamma + \text{plin } \underline{D}_\gamma + \text{nlin } \underline{D}_\gamma)\underline{v} \qquad (5.4)$$

and

$$\underline{\chi} = (\text{rlin } \underline{D}_\chi + \text{plin } \underline{D}_\chi + \text{nlin } \underline{D}_\chi)\underline{v} \qquad (5.5)$$

with the operational matrices $\underline{D}$ which contain the transformations between the tensor components of the strains $\underset{DGA}{\gamma}_{\alpha\beta}$ and the elastic curvatures $\underset{DGA}{\chi}_{\alpha\beta}$ at one hand and the nodal displacements at the other.

With the collection

$$\underline{\varepsilon} = \begin{bmatrix} \underline{\gamma} \\ \hline \underline{\chi} \end{bmatrix} \tag{5.6}$$

(5.4) and (5.5) can be written as

$$\underline{\varepsilon} = (\text{rlin } \underline{D}_{\varepsilon} + \text{plin } \underline{D}_{\varepsilon} + \text{nlin } \underline{D}_{\varepsilon})\underline{v} \quad . \tag{5.7}$$

The constitutive equations have the matrix form

$$\underline{N} = \underline{E}_N \, \underline{\gamma} \tag{5.8}$$

$$\underline{M} = \underline{E}_M \, \underline{\chi} \quad , \tag{5.9}$$

and with the collection

$$\underline{\sigma} = \begin{bmatrix} \underline{N} \\ \hline \underline{M} \end{bmatrix} \tag{5.10}$$

we get the short form

$$\underline{\sigma} = \underline{E} \;\; \underline{\varepsilon} \quad . \tag{5.11}$$

The incremental virtual work of the interior forces and moments of an element $\mathring{\mathcal{M}}_e$ is

$$\underset{DG}{\delta W} = \int_{\mathring{\mathcal{M}}_e} \left(\underset{DGA}{\delta \underline{\varepsilon}^T} \; \underset{DGA}{\underline{\sigma}} + \delta \text{nlin } \underset{DGA}{\underline{\varepsilon}^T} \; \underset{GA}{\underline{\sigma}} \right) d\mathring{a} \tag{5.12}$$

and after introducing the constitutive equations

$$\underset{DG}{\delta W} = \int_{\mathring{\mathcal{M}}_e} \left(\underset{DGA}{\underline{\varepsilon}^T} \; \underline{E} \; \underset{DGA}{\underline{\varepsilon}} + \delta \text{nlin } \underset{DGA}{\underline{\varepsilon}^T} \; \underset{GA}{\underline{\sigma}} \right) d\mathring{a} \quad . \tag{5.13}$$

With (5.7) we get

$$\underset{DG}{\delta W} = \delta \underline{v}^T \, (\underline{k}_{\varepsilon} + \underline{k}_{\sigma} + \overset{*}{\underline{k}}_{\sigma}) \underline{v} = \delta \underline{v}^T \underline{k} \; \underline{v} \tag{5.14}$$

with

$$\underline{k}_{\varepsilon} = \int_{\mathring{\mathcal{M}}_e} \underline{D}_{\varepsilon}^T \; \underline{E} \; \underline{D}_{\varepsilon} \;\; d\mathring{a} \; , \tag{5.15}$$

$$\underline{k}_{\sigma} = \int_{\mathring{\mathcal{M}}_e} \underline{D}_{\sigma}^T \; \underline{E}_{\sigma} \; \underline{D}_{\sigma} \; d\mathring{a} \; , \tag{5.16}$$

$$\overset{*}{\underline{k}}_{\sigma} = \int_{\mathring{\mathcal{M}}_e} \overset{*}{\underline{D}}{}_{\sigma}^T \; \overset{*}{\underline{E}}_{\sigma} \; \overset{*}{\underline{D}}_{\sigma} \; d\mathring{a} \; . \tag{5.17}$$

The matrices (5.16) and (5.17) result from the second integrand in (5.13) which contains - differing from a geometrically nonlinear threedimensional theory - besides quadratic terms in the unknown nodal displacements (contained in (5.16)) also products of higher degree which result form the elastic curvatures (contained in (5.17)). We decompose the stiffness matrix as

$$\underline{k}_\varepsilon = \text{rlin}\ \underline{k}_\varepsilon + \text{plin}\ \underline{k}_\varepsilon + \text{nlin}\ \underline{k}_\varepsilon \ . \tag{5.18}$$

In (5.18),

$$\text{rlin}\ \underline{k}_\varepsilon = \int_{A_e} \text{rlin}\ \underline{D}_\varepsilon^T\ \underline{E}\ \text{rlin}\ \underline{D}_\varepsilon\ d\overset{\circ}{a} \tag{5.19}$$

is the usual linear stiffness matrix.
The matrix

$$\text{plin}\ \underline{k}_\varepsilon = \int_{A_e} (\text{rlin}\ \underline{D}_\varepsilon^T\ \underline{E}\ \text{plin}\ \underline{D}_\varepsilon + \text{plin}\ \underline{D}_\varepsilon^T\ \underline{E}\ \text{rlin}\ \underline{D}_\varepsilon + \text{plin}\ \underline{D}_\varepsilon^T\ \underline{E}\ \text{plin}\ \underline{D}_\varepsilon)\, d\overset{\circ}{a} \tag{5.2o}$$

is linear and quadratic in the known displacements - and nonlinear of higher degree in the case of consideration of nonlinear elastic curvatures. A special treatment needs the matrix nlin $\underline{k}_\varepsilon$ in which the unknown nodal displacements are contained. This matrix represents the geometrical nonlinearities which appear in a one step solution. Additionally, we can recognize terms in nlin $\underline{k}_\varepsilon$ in which unknown quantities are coupled linear, of higher degree or with known displacements. With the virtual work of the incremental loads

$$\delta \underset{DG}{A_a} = \delta\, \underline{v}^T\, \underline{\bar{p}} \ , \tag{5.21}$$

the principle becomes

$$\delta \underset{DG}{W} = \delta \underset{DG}{A_a} \ . \tag{5.22}$$

After assembling and introducing the geometrical boundary conditions, we get the global stiffness matrix $\underline{K}$, the global nodal displacement vector $\underline{V}$ and the global right-hand side $\underline{\bar{P}}$. The nonlinear system of algebraic equations is

$$\underline{K}\ \underline{V} = \underline{\bar{P}} \ . \tag{5.23}$$

As on the element level we decompose formally

$$\underline{K} = \underline{K}_{\varepsilon} + \underline{K}_{\sigma} + \overset{*}{\underline{K}}_{\sigma} \tag{5.24}$$

with

$$\underline{K}_{\varepsilon} = \text{rlin } \underline{K}_{\varepsilon} + \text{plin } \underline{K}_{\varepsilon} + \text{nlin } \underline{K}_{\varepsilon} \quad . \tag{5.25}$$

We define

$$\text{lin } \underline{K} = \text{rlin } \underline{K}_{\varepsilon} + \text{plin } \underline{K}_{\varepsilon} + \underline{K}_{\sigma} \tag{5.26}$$

and

$$\underline{K}(\underline{V}) = \text{nlin } \underline{K}_{\varepsilon} + \overset{*}{\underline{K}}_{\sigma} \ . \tag{5.27}$$

So we get the linearized system of equations for a load increment

$$\text{lin } \underline{K}\ \underline{V} = \underline{\overline{P}} - \underline{K}\ (V)\underline{V} \ . \tag{5.28}$$

The system matrix lin $\underline{K}$ consists - see (5.26)- of a constant stiffness matrix rlin $\underline{K}_{\varepsilon}$ for all load steps, a stiffness matrix plin $\underline{K}_{\varepsilon}$ which depends from known displacements and a matrix $\underline{K}_{\sigma}$ which depends from known sectional forces and moments. $\underline{K}_{\sigma}$ is known as the geometrical stiffness matrix.

5.2 Choice of the finite element

The middle surface $\mathring{\mathcal{M}}$ is discretized by triangular elements with the area parameters $\theta^{\alpha}, \alpha = 1,2$ in the parameter domain. According to the used parameter representation of $\mathring{\mathcal{M}}$, three parameter couples define the corner nodes of a curved subdomain $\mathring{\mathcal{M}}_e^{(i)} \subset \mathring{\mathcal{M}}$ as a finite element, see Fig.5.

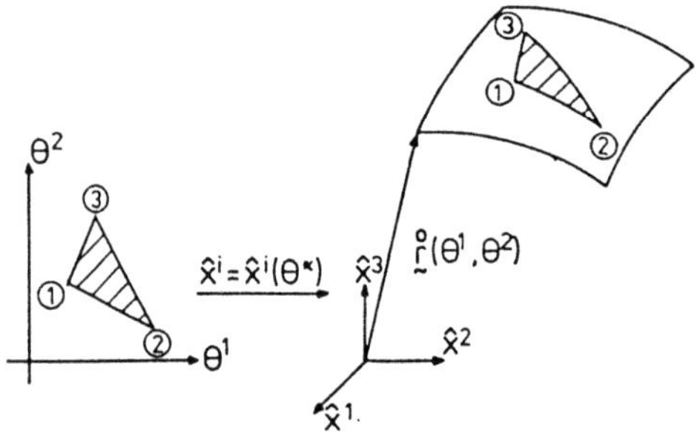

Fig.5 Triangular finite element in the parameter plane and in the middle surface of the shell.

We choose polynomial shape functions of 5.th degree, represented in the so-called area coordinates (λ_1, λ_2, λ_3), which is the twodimensional case of barycentric coordinates for simplexes. So, all required numerical calculations including differentiations und integrations can be performed schematically by the computer program. The sequence of the polynomial terms is chosen as follows.

$$\begin{aligned}\underline{\Phi}=\{ & \lambda_1^5 ; \\ & \lambda_1^4 \lambda_2, \lambda_1^4 \lambda_3; \\ & \lambda_1^3 \lambda_2^2, \lambda_1^3 \lambda_2 \lambda_3, \lambda_1^3 \lambda_3^2; \\ & \lambda_1^2 \lambda_2^3, \lambda_1^2 \lambda_2^2 \lambda_3, \lambda_1^2 \lambda_2 \lambda_3^2, \lambda_1^2 \lambda_2^3; \\ & \lambda_1 \lambda_2^4, \lambda_1 \lambda_2^3 \lambda_3, \lambda_1 \lambda_2^2 \lambda_3^2 \ \lambda_1 \lambda_2 \lambda_3^3, \lambda_1 \lambda_3^4; \\ & \lambda_2^5, \lambda_2^4 \lambda_3, \lambda_2^3 \lambda_3^2, \lambda_2^2 \lambda_3^3, \lambda_2 \lambda_3^4, \lambda_3^5 \} . \end{aligned} \qquad (5.29)$$

This scheme is based an a formation rule which is useful for an automatic calculation. The polynomial shape functions (5.29) are also used for the approximation of all required geometrical quantities in the element domain. A corresponding generally curved triangular shell element is described by Argyris and Scharpf in [14] with the name SHEBA-element. It fulfils all essential conditions of the FE-displacement-method, as geometrical conformity, completeness and rigid body movement conditions. The 21 usual nodal values for each primarily approximated quantity are shown in Fig.6.

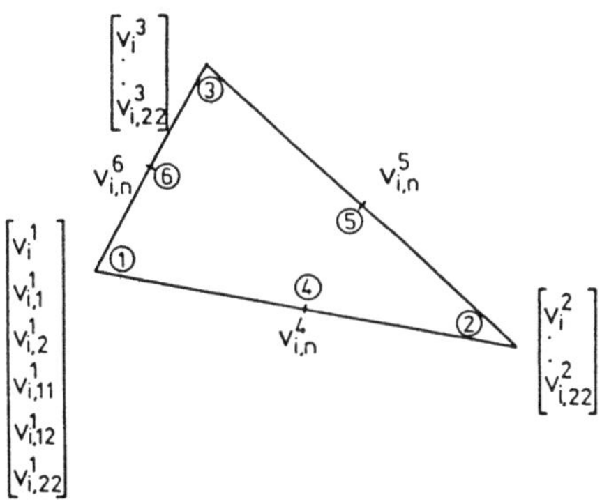

Fig.6 Local nodal numbers und nodal displacements.

5.3 Consideration of different levels of nonlinearity

In our computer program, described by Berg in detail [6], different levels of geometrical nonlinearity can be chosen. In the highest level we apply a theory of moderate rotations for $\underline{\underline{\gamma}}$ and $\underline{\underline{\chi}}$, corresponding to Pietraszkiewicz [5] . Here the strains of the middle surface are approximated in the form

$$\tilde{\underline{\underline{\gamma}}} = \frac{1}{2} \left(H^T \overset{\circ}{\underline{\underline{1}}}_2 + \overset{\circ}{\underline{\underline{1}}}_2 H + \tilde{H}^T H \right)$$

$$= \frac{1}{2} \left[\underset{\sim}{u}_{,\alpha} \cdot \overset{\circ}{\underset{\sim}{a}}_\beta + \underset{\sim}{u}_{,\beta} \cdot \overset{\circ}{\underset{\sim}{a}}_\alpha + (\underset{\sim}{u}_{,\alpha})_n \cdot (\underset{\sim}{u}_{,\beta})_n \right] \overset{\circ}{\underset{\sim}{a}}{}^\alpha \otimes \overset{\circ}{\underset{\sim}{a}}{}^\beta \qquad (5.30)$$

with

$$\tilde{H} = (\underset{\sim}{u}_{,\alpha})_n \otimes \overset{\circ}{\underset{\sim}{a}}{}^\alpha \qquad (5.31)$$

$$(\underset{\sim}{u}_{,\alpha})_n = \phi_{3\alpha} \overset{\circ}{\underset{\sim}{a}}_3 = (u_{3,\alpha} + \overset{\circ}{b}{}^\lambda_\alpha u_\lambda) \overset{\circ}{\underset{\sim}{a}}_3 \, . \qquad (5.32)$$

For (3.59) we get in this case

$$\tilde{\underline{\underline{\gamma}}} = \frac{1}{2} (\phi_{\beta\alpha} + \phi_{\alpha\beta} + \phi_{3\alpha} \phi_{3\beta}) \overset{\circ}{\underset{\sim}{a}}{}^\alpha \otimes \overset{\circ}{\underset{\sim}{a}}{}^\beta \, . \qquad (5.33)$$

In index notation we get for (5.33)

$$\tilde{\gamma}_{\alpha\beta} = \frac{1}{2} (u_\alpha|_\beta + u_\beta|_\alpha) - \overset{\circ}{b}_{\alpha\beta} u_3$$

$$+ \frac{1}{2} (u_{3,\alpha} + \overset{\circ}{b}{}^\lambda_\alpha u_\lambda)(u_{3,\beta} + \overset{\circ}{b}{}^\mu_\beta u_\mu) \, . \qquad (5.34)$$

The lowest level of geometrical nonlinearity comes out if we omit in (5.34) the terms which contain the curvatures of the reference configuration, i.e. a shallow shell theory. Then we get for the strains of the middle surface the wellknown generalized theory of Donnell [7] with the approximated $\phi_{3\alpha}$

$$\approx\!\gamma_{\alpha\beta} = \frac{1}{2} (u_\alpha|_\beta + u_\beta|_\alpha) - \overset{\circ}{b}_{\alpha\beta} u_3 + \frac{1}{2} u_{3,\alpha} u_{3,\beta} \, . \qquad (5.35)$$

$$\tilde{\phi}_{3\alpha} = u_{3,\alpha} \, . \qquad (5.35a)$$

The additional strains $\zeta \underline{\underline{\chi}}$ depending from the elastic curvatures are approximated linearly in the frame of moderate rotation theory. So we get for (3.56) the approximation

$$\tilde{\underline{\underline{\chi}}} = H^T \Gamma\text{rad}\, \overset{\circ}{\underset{\sim}{a}}_3 + \overset{\circ}{\underline{\underline{1}}}_2 \, \Gamma\text{rad}\, (\text{lin} \underset{\sim}{\beta}) \qquad (5.36)$$

with

$$\text{lin}\, \underset{\sim}{\beta} = - \phi_{3\lambda} \overset{\circ}{\underset{\sim}{a}}{}^\lambda = -(u_{3,\lambda} + \overset{\circ}{b}{}^\rho_\lambda u_\rho) \overset{\circ}{a}{}^\lambda \, . \qquad (5.37)$$

In index notation we get the linearized but not further approximated expression

$$\tilde{\chi}_{\alpha\beta} = -(u_3|_{\alpha\beta} + \overset{o}{b}{}^{\lambda}_{\alpha}|_{\beta} u_\lambda + \overset{o}{b}{}^{\lambda}_{\alpha} u_\lambda|_\beta + \overset{o}{b}{}^{\lambda}_{\beta} u_\lambda|_\alpha - \overset{o}{b}{}^{\lambda}_{\alpha}\overset{o}{b}_{\lambda\beta} u_3) . \tag{5.38}$$

5.4 Description of the incremental-iterative algorithm

For each load step j, the linear stiffness matrix $\underline{K}^j$, according to (5.28), is not changed during the following iteration process and is denoted by $\underline{K}$. The load steps are controlled by one parameter so that the shape of loading remains constant. Therefore the load-depending right-hand-side has to be generated only once. For an initial step the right-hand-side vector is marked with $\underline{\overline{P}}_a$. We calculate the initial displacement vector $\underline{V}_a$ with the linear system

$$\underline{K}\ \underline{V}_a = \underline{\overline{P}}_a \quad . \tag{5.39}$$

The work of the nodal loads is

$$2\ A_a = \underline{V}_a^T\ \underline{\overline{P}}_a = \underline{V}_a^T\ \underline{K}\ \underline{V}_a , \tag{5.40}$$

and the quotient λ_a of a constant comparative work A_v and A_a is

$$\lambda_a = \frac{A_v}{A_a} \quad . \tag{5.41}$$

At first we consider a system with decreasing stiffness for growing load. For one single load F the displacement v of the loadpoint in the direction of F is shown in Fig. 7.

The incremental nodal displacements $\underline{V}_a^j$ increase monotonously for constant load increments $\underline{\overline{P}}_a$, and λ_a^j decreases according to (5.41).

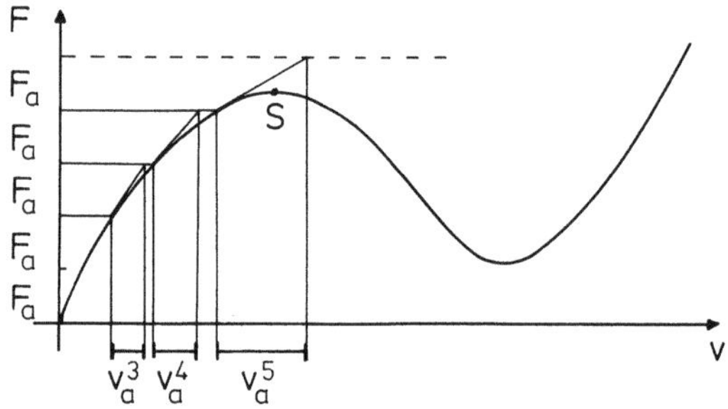

Fig. 7. Displacement-force diagram for decreasing stiffness

In advance the iteration process for a fixed load increment $\underline{\bar{P}}_a = \underline{\bar{P}}$ is described. In this case (5.39) can be written as

$$\underline{K}\ \underline{V}^o = \underline{\bar{P}} \quad .$$

Introducing $\underline{V}^o$ into the nonlinear part of (5.28) yields the error forces $\underline{R}_u^1$. For the i-th iteration, the error forces are

$$\underline{R}_u^{i+1} = \underline{K}\ (\underline{V}^i)\ \underline{V}^i \qquad i = 0,\ 1,\ 2\ \ldots\ , \tag{5.42}$$

$$\underline{R}_u^o = \underline{0} \quad .$$

The linear system of equations for the i-th iteration is

$$\underline{K}\ \underline{V}^i = \underline{\bar{P}} - \underline{R}_u^i \quad . \tag{5.43}$$

The iteration is finished with $\underline{V} = \underline{V}^n$ by an error mark. Then a new load step follows. This method is marked by an F (force-controlled). It is only stable if the solution vector $\underline{V}$ does not differ too much from the initial vector $\underline{V}_a$.
Method F becomes impracticable in the neighbourhood of a vertex point S of the F(v)-diagram.
A stable iteration process for flat tangents is possible by an automatic reducing of the initial load. At first $\underline{V}_a$, A_a and λ_a are calculated according to (5.39 - 5.41). Then the above described iteration is modified in the form

$$\underline{\bar{P}}^o = \lambda_a\ \underline{\bar{P}}_a \tag{5.44}$$

and an corresponding equation to (5.43)

$$\underline{K}\ \underline{V}^i = \underline{\bar{P}}^o - \underline{R}_u^i \tag{5.45}$$

with error forces (5.42).

This method, called FV (force-displacement-controlled) fails in the vicinity of the vertex point S.

By a little modification of FV, we can overcome the horizontal tangent. The right-hand-side in (5.45) is changed in every iteration by the postulate

$$\underline{V}^{iT}\ \underline{\bar{P}}^o = 2\ A_o = \text{const.}\ . \tag{5.46}$$

It holds

$$\underline{V}^i = \underline{V}_p^i + \underline{V}_u^i \tag{5.47}$$

with $\underline{V}_p^i$ from

$$\underline{K}\ \underline{V}_p^i = \underline{\bar{P}}^i \tag{5.48}$$

and $\underline{V}_u^i$ from

$$\underline{K}\ \underline{V}_u^i = -\ \underline{R}_u^i\ . \tag{5.49}$$

We name the changing load part of the right-hand-side in the i-th iteration

$$\underline{\bar{P}}^i = \lambda^i\ \underline{\bar{P}}^o \tag{5.50}$$

and the corresponding displacements

$$\underline{V}_p^i = \lambda^i\ \underline{V}^o\ . \tag{5.51}$$

The postulate (5.46) yields

$$(\lambda^i\ \underline{V}^o + \underline{V}_u^i)^T\ \underline{\bar{P}}^o =$$

$$\lambda^i\ \underline{V}^{oT}\ \underline{\bar{P}}^o + \underline{V}_u^{iT}\ \underline{\bar{P}}^o = 2\ A_o = \text{const.}\ . \tag{5.52}$$

with

$$2\ A_u^i = \underline{V}_u^{iT}\ \underline{\overline{P}}^o \tag{5.53}$$

we get

$$\lambda^i = \frac{A_o - A_u^i}{A^o} \tag{5.54}$$

and as iteration rule

$$\underline{K}\ \underline{V}^i = \lambda^i\ \underline{\overline{P}}^o - \underline{R}_u^i . \tag{5.55}$$

This method has the symbol V (way-controlled).

6. Examples

We consider circular cylindrical shell segments with different boundary conditions, thicknesses and loads.

6.1. Clamped edges; normal pressure

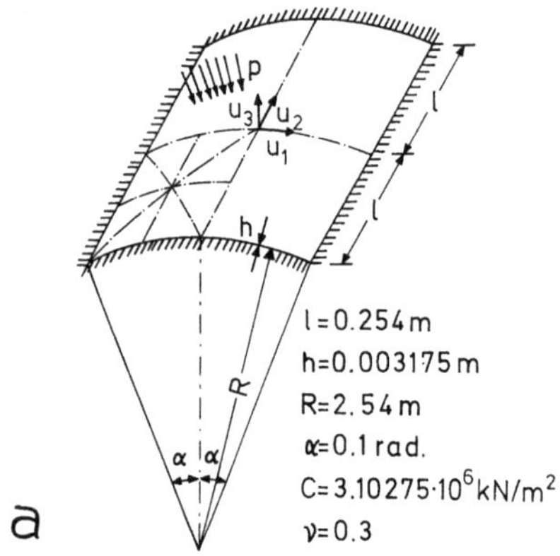

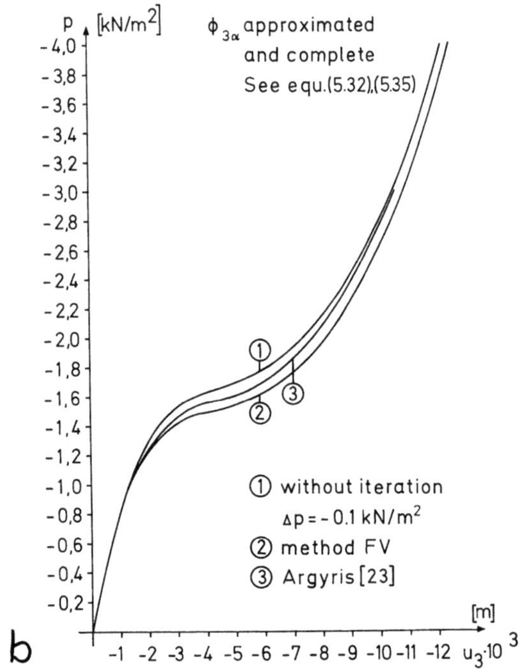

Fig. 8a. Circular cylindrical shell segment with all edges clamped, under normal pressure. 8 triangular elements in the shell quarter.

Fig. 8b. Normal displacement of the midpoint of the shell. Comparisons of different calculations.

This example was used for testing the described algorithms and to compare them with published results, see Fig. 8a,b. Curve ① was calculated with method F without iteration for uncontrolled load steps $\Delta p = -0,1\ KN/m^2$. The more accurate curve ② contains iterations of the equilibrium for each load step. The published curve ③ was calculated also with the SHEBA-element.

6.2. Two opposite edges hinged, two edges free; central point load and axial load.

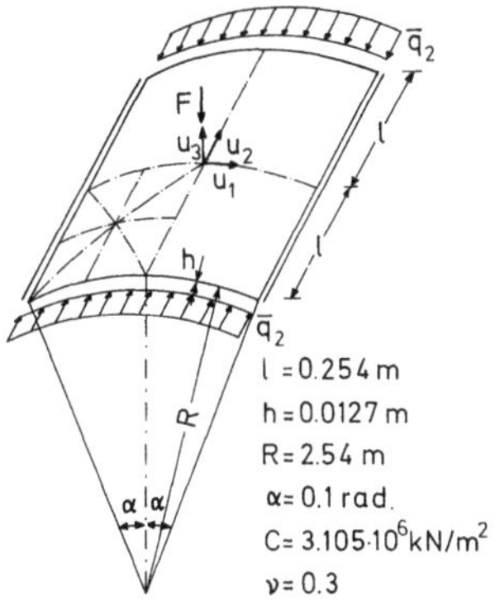

Fig. 9a. Circular cylindrical shell segment with two opposite edges hinged and two edges free.

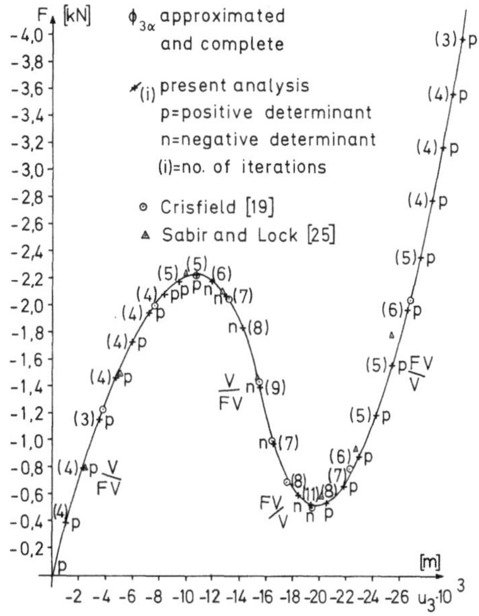

Fig.9b. Normal displacement of the midpoint of the shell under normal midpoint load. Comparisons of different results.

All displacement components (i.e. also the axial ones) are zero in the nodal points of the hinged edges. Fig. 9b shows a good agreement of our and Crisfield's results for a central point load. Our calculations with Ramm's refined NISA-program [26] (using 191 kinematic unknowns) also lead to the same results in the first decimal place. There is no difference in the first two decimal places of the displacements between Donnell's theory and the more accurate moderate rotation theory. The reason is that the circumferential displacements u_1 remain comparatively small. Only those influence (multiplied with the reference curvature) the strains of the middle surface. The different parts of the curve are marked with F, FV or V to show different control algorithms for load or displacement steps, see chapter 5.4.

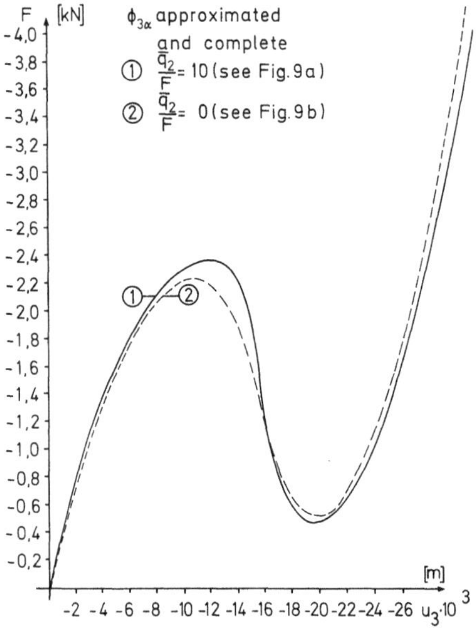

In Fig. 9c an affine growing central normal point load with an axial pressure are considered. Caused by the zero axial displacements at the hinged edges, the shell moves upwards for axial pressure in the case $\nu \neq 0$. Therefore the radial displacement u_3 is smaller than such in Fig. 9b. As the axial load alone remains below the critical one the results of Fig. 9c and Fig. 9b are approaching in the postcritical branch.

Fig. 9c. Normal displacement of the midpoint of the shell under a normal midpoint load and an axial pressure (curve ①).

In Fig. 9d the same example as in Fig. 9b - but with halved shell thickness - is considered and compared with Crisfield's results. Depending from the kind of step control one gets another branch in the postcritical domain for negatively growing loads and positive tangents. It is important to realize that the statical equilibrium in the postcritical branches U-L and B-E is unstable. This follows from the fact that for monogenetic loads the second variation of the total potential energy is $\delta \underline{V}^T \underline{K} \delta \underline{V}$ where $\underline{K}$ is the tangential incremental stiffness matrix. This quadratic form - and so the matrix $\underline{K}$ - is not positive definite in the descending branches, expressed by det $\underline{K} < 0$ in the numerical process. Therefore these branches - the same holds for Fig. 9b - are artificial because the system behaves dynamically, and the unstable statical equilibrium configurations depend on the imposed constraints of the step control of the incremental iterative algorithm.

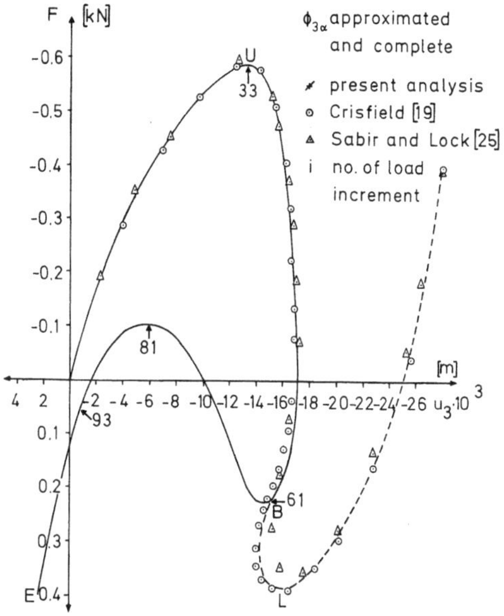

Fig. 9d. Normal displacement of the midpoint of the shell with half thickness under normal midpoint load.

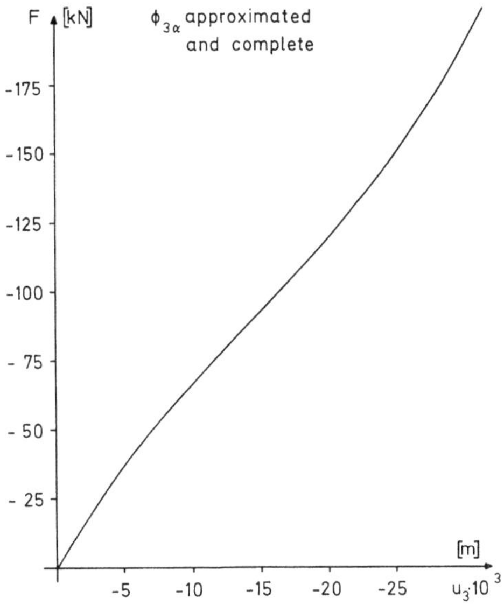

Fig. 9f. Normal displacement of the midpoint of the shell with four-fold thickness under normal midpoint load.

The plots in Fig. 9e show that the free edges snap through between the load increments 33 and 6o.

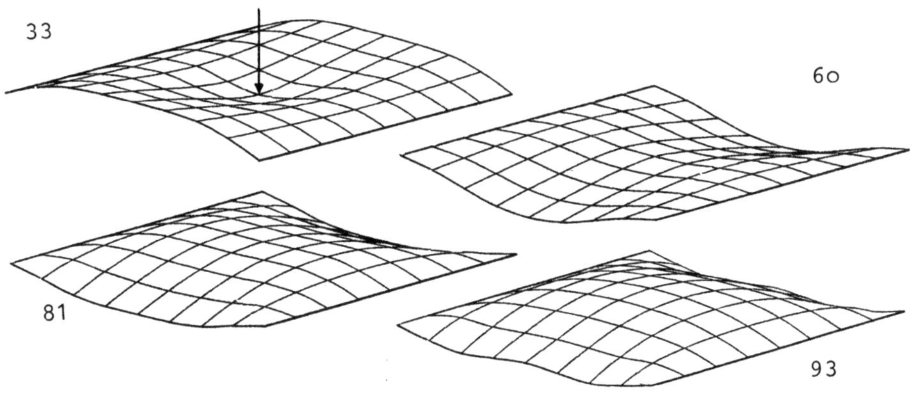

Fig. 9e. Plots of normal displacements at different load stages for Fig. 9d.

Fig. 9f shows the same system like in Fig. 9b and 9d - but with the four-fold shell thickness. One recognizes an essential change of the mechanical behaviour as there is no snapping through.

6.3. Two opposite edges hinged with free axial displacements, two edges free; axial pressure and geometrical disturbances

With these boundary conditions a homogeneous state of stress is achieved for constant axial pressure. The shell does not move upwards due to Poisson's ratio. Geometrical disturbances are introduced by generating once a geometrical stiffness matrix for an a priori given displacement pattern. This disturbance causes stresses, and therefore it is not called an imperfection which means an a priori, i.e. stress-free change of the shell geometry.

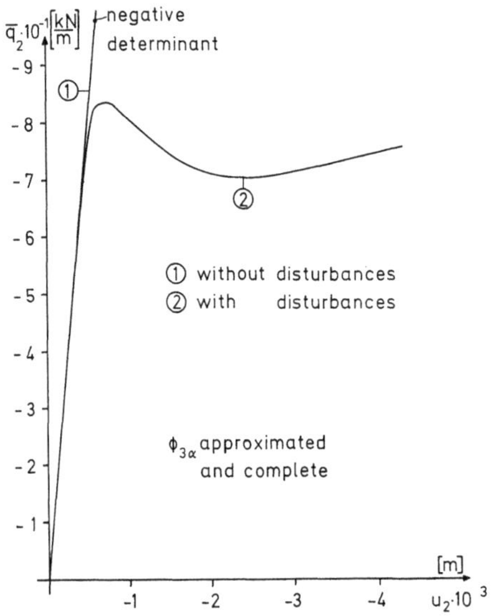

Fig. 10a. Axial displacement of the free edge midpoint for the shell of Fig. 9a with $\nu = 0$ under axial pressure with and without geometrical disturbances.

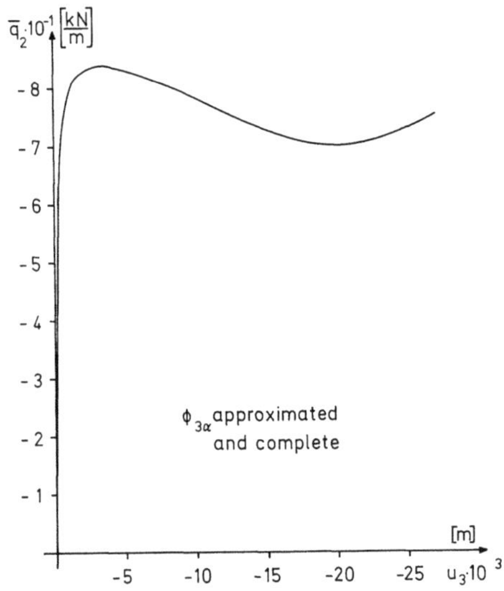

Fig. 10b. Normal displacement of the midpoint corresponding to Fig. 10a with geometrical disturbances.

In Fig. 10a, 10b axial and normal displacements are shown for a shell under axial pressure and a chosen geometrical disturbance of the midpoint of $u_3^0 = h/10$. The disturbance pattern is represented by shape functions of the adjacent elements.

References

1. Koiter, W.T.: On the nonlinear theory of thin elastic shells. Proc. Koninkl. Ned. Akad. Wetenshap., Ser. B. 69 (1966) 1-54.

2. Naghdi, P.M.: Foundations of elastic shell theory, in Progress in solid mechanics. Vol. 4, Amsterdam: North-Holland Publ. Co. 1963 1-9o.

3. Naghdi, P.M.: The theory of shells and plates, in Handbuch der Physik Vol. VI a/2, Berlin-Heidelberg-New York: Springer 1972 425-64o.

4. Simmonds, J.G. and D.A. Danielson: Nonlinear shell theory with finite rotation vector. Proc. Koninkl. Ned. Akad. Wetenshap., Ser. B 73 (197o) 46o-478.

5. Pietraszkiewicz, W.: Introduction to the non-linear theory of shells. Mitteilungen aus dem Institut für Mechanik, Nr. 1o (1977).

6. Berg, A.: Beiträge zur geometrisch nichtlinearen Theorie und inkrementellen Finite-Element-Berechnung dünner elastischer Schalen. Diss. Universität Hannover 1982, Forschungs- und Seminarberichte aus dem Bereich der Mechanik der Universität Hannover, Nr. 82/3 (in press).

7. Donnell, L.H.: A new theory for the buckling of thin cylinders under axial compression and bending. Trans. ASME 56 (1934) 795-8o6.

8. Schmidt, R. and W. Pietraszkiewicz: Variational principles in the geometrical non-linear theory of shells undergoing moderate rotations. Ing.-Arch. 5o (1981) 187-2o1.

9. Stumpf, H.: On the linear and nonlinear stability analysis in the theory of thin elastic shells. Ing.-Arch. 51 (1981) 195-213.

1o. Koiter, W.T.: On the stability of elastic equilibrium, ch.5, Shells with finite deflections. Thesis Delft. H.J. Paris, Amsterdam (1945).

11. Basar, Y. and W.B. Krätzig: Struktur konsistenter Grundgleichungen für das Beul- und Nachbeulverhalten allgemeiner Flächentragwerke. Der Stahlbau 46 (1977) 138-146.

12. Wittek, U.: Beitrag zum Tragverhalten der Strukturen bei endlichen Verformungen unter besonderer Beachtung des Nachbeulmechanismus dünner Flächentragwerke. Technisch-wissenschaftliche Mitteilungen des Instituts für Konstruktiven Ingenieurbau Bochum, Nr. 8o-1 (198o).

13. Stein, E.: Variational functionals in the geometrical nonlinear theory of thin shells and finite-element discretizations with applications to stability problems, in Theory of shells, W.T. Koiter and G.K. Mikhailov (Eds.), Amsterdam: North-Holland Publ. Co. 198o 5o9-535.

14. Argyris, J.H. and D. W. Scharpf: The SHEBA family of shell elements for the matrix displacement method. Aeron. J. Royal Aeron. Soc. 72 (1968) 873-883, 73 (1969) 423-426.

15. Stein, E. and M.H. Kessel: Numerische Methoden und deren Konvergenz zur statischen Berechnung geometrisch nichtlinearer Stabwerke im unter- und überkritischen Bereich. Ing.-Arch. 46 (1977) 323-335.

16. Riks, E.: The application of Newton's method to the problem of elastic stability. J. Appl. Mech. 39 (1972) 1o6o-1o66.

17. Wempner, G.A.: Discrete approximations related to nonlinear solids, Int. J. Solids Struct. 7 (1971) 1581-1599.

18. Wessels, M.: Das statische und dynamische Durchschlagproblem der imperfekten flachen Kugelschale bei elastischer rotationssymmetrischer Verformung. Diss. Technische Universität Hannover (1977), Mitt. Inst. Statik Hannover 23 (1977).

19. Crisfield, M.A.: A fast incremental/iterative solution procedure that handles "snap-through". Comp. & Struct. 13 (1981) 55-62.

2o. Brendel, B. and E. Ramm: Nichtlineare Stabilitätsuntersuchungen mit der Methode der Finiten Elemente. Ing.-Arch. 51 (1982) 337-362.

21. de Boer, R. and H. Prediger: Tensorrechnung - Grundlagen für Ingenieurwissenschaften. Forschungsberichte aus dem Fachbereich Bauwesen der Universität Essen Gesamthochschule, Nr. 5 (1978).

22. de Boer, R. and H. Prediger: Tensorrechnung in der Mechanik - Ausgewählte Kapitel. Forschungsberichte aus dem Fachbereich Bauwesen der Universität Essen Gesamthochschule, Nr. 8 (1979).

23. Argyris, J.H.: Entwicklung von Finiten Element Methoden für die Berechnung von Schalen allgemeiner Form. Vortrag anläßlich des Schlußkolloquiums des DFG-Schwerpunktes "Flächentragwerke im konstruktiven Ingenieurbau", 198o.

24. Horrigmoe, G.: Finite element instability analysis of free-form shells. Report Nr. 77-2 der Universität von Trondheim, Norwegen, 1977.

25. Sabir, A.B. and A.C. Lock : The application of finite elements to the large-deflection geometrically nonlinear behaviour of cylindrical shells. In Variational Methods in Engineering C.A. Brebbia and H. Tottenham (Eds.), Southampton: University Press 1972 7/54 - 7/65.

26. Brendel, B., L. Häfner, E. Ramm und J.M. Sättele: Programmdokumentation und Eingabebeschreibung zum Programmsystem NISA. Bericht Inst. f. Baustatik Universität Stuttgart 1977.

Analysis of Elastic-Plastic Buckling and Imperfection-Sensitivity of Shells of Revolution

W. Wunderlich, H.J. Rensch, H. Obrecht

Institut für Konstruktiven Ingenieurbau
Ruhr-Universität Bochum
Bochum, West-Germany

Summary

The nonlinear analysis of elastic-plastic shells of revolution is performed by combining an efficient one-dimensional transfer-matrix approach on the element level with the usual finite element assemblage of the system. It is based on a Fourier-representation of the loading, the shell variables and initial imperfections in the circumferential direction. This semi-analytical procedure is formulated for geometrically and physically nonlinear problems. It is applied to the detailed investigation of the buckling behavior and imperfection sensitivity of various dished end closures under external pressure.

1. Introduction

A realistic assessment of the actual load-carrying behavior of general shells of revolution necessitates detailed numerical analyses in which all nonlinear influences and initial conditions are accounted for, and an appropriate material description is chosen. Since such nonlinear calculations are, by nature, extensive, solution procedures are desirable which do not require excessive numerical effort. This aspect is particularly important when the response of the shell is nonaxisymmetric, due to the loading or initial imperfections, in addition to being nonlinear.

In most previous buckling analyses of elastic-plastic shells of revolution (see e.g. [1-4]) deflections were assumed to remain axisymmetric up to the bifurcation or limit loads. Nonaxisymmetric components appeared only in the bifurcation modes. Although most valuable in illuminating fundamental aspects of plastic shell buckling, e.g. the strong degrading influence of axisymmetric imperfections [4] , they neglected the fact that in

Buckling of Shells, Ed.: E. Ramm
Proceedings of a State-of-the-Art Colloqium

measurements on laboratory and industrially produced shells the vast majority of initial imperfections showed a decidedly nonaxisymmetric distribution [5,6]. Whether or not in the plastic range these nonaxisymmetric components lead to a decrease in load-carrying capacity beyond that of the axisymmetric ones, is a matter of considerable importance. However, the analysis of these problems by usual approximation techniques is expensive, and no systematic studies are available.

Two-dimensional discretization procedures employing finite elements or finite differences are versatile and universally applicable to general shell problems. However, storage requirements and computer times are considerable and may limit their use in practical situations. As is known, the linear elastic solution of nonaxisymmetric shell-of-revolution-problems by expansion of the loads and unknowns into Fourier series with respect to the circumferential coordinate has certain advantages compared to a fully two-dimensional analysis. It leads to uncoupled equations for each Fourier term and permits the use of one-dimensional ring elements. Their stiffness characteristics can then be determined via trial functions or by accurate numerical integration of the differential equations along the meridional coordinate [7,8]. Despite the wide use of Fourier decomposition in linear analyses, only few extensions to the nonlinear range have been presented. The method has been applied to geometrically nonlinear elastic problems in [9-12], to small deflection elastic-plastic problems in [13] and to large deflection elastic-plastic problems in [14] .

In the present paper this approach, which has proved to be efficient in linear problems, is applied to the geometrically and physically (elastic-plastic) nonlinear analysis of shells of revolution [15-19] , with the aim of preserving its advantages, and of developing a cost-effective numerical procedure. Starting from a Fourier decomposition of the loads and shell quantities, and treating all nonlinear terms as pseudo-loads, we again arrive at uncoupled differential equations for all Fourier harmonics. Integration of these sets of one-dimensional equations over a finite interval along the meridional coordinate line yields the local stiffness matrices of ring elements. Their assembly to a global finite element stiffness matrix is done in the usual fashion, and, due to the uncoupling between the equations for each Fourier term the numerical effort required in the iterative solution of the full nonlinear equations is reduced significantly. A drawback of the procedure is that the iterations diverge at limit points, and that no solution can be obtained in regions where the paths of load versus corresponding deflections decrease. Nevertheless, in its range of validity it permits the solution of nonaxisymmetric problems which might become prohibitive when using other discretization schemes.

In the following chapters the procedure as it has been incorporated into the computer program ROT B is first outlined in some detail. Next, numerical results of the analysis of the buckling behavior and sensitivity to axisymmetric as well as nonaxisymmetric imperfections of dished end closures under external pressure are given.

2. Formulation of the shell equations

2.1 General shell equations

In the linear analysis of thin shells of revolution it has proved advantageous to cast the partial differential equations of equilibrium, strain-displacement, and the constitutive equations into such a form that the quantities to be described on a circumferential boundary appear as independent variables. Then only first derivatives with respect to the meridional coordinate appear. If the dependence on the circumferential coordinate is described by a Fourier expansion the problem reduces to a set of one-dimensional differential equations, and is thus amenable to efficient and straightforward solution procedures. Here this approach is applied to geometrically nonlinear, non-axisymmetric deformations and elastic-plastic material behavior. It will be seen to have tangible advantages in comparison with two-dimensional, displacement based finite element analyses. Most important in this respect is the efficiency and accurarcy which can be realized in the numerical solution of one-dimensional problems. Choosing the independent variables as indicated above the basic equations for shells of revolution in incremental form are first expressed as a set of eight partial and several additional ordinary differential equations. After elimination of the latter the remaining eight equations may be written in matrix notation as

$$\frac{\partial}{\partial s}\,\underline{z}(s,\theta) = \underline{A}(s,\frac{\partial}{\partial\theta},\overset{o}{\underline{z}},E_t)\,\underline{z}(s,\theta) + \bar{\underline{z}}(s,\theta) \qquad (1)$$

In (1) s is the meridional coordinate of an orthogonal convected coordinate system, $\underline{z}^T = [\,v_1 \;\; v_2 \;\; v_3 \;\; \phi_2 \mid s_{21} \;\; n_{22} \;\; s_3 \;\; m_{22}\,]$ is the vector of the increments of the eight displacement and stress quantities that may be prescribed on a boundary, i.e. the three displacement components v_i, the meridional angle of rotation ϕ_2, the in-plane shear force s_{21}, the meridional stress resultants n_{22} and m_{22}, and the transverse shear force s_3. The above form is arrived at by eliminating the circumferential stress resultants n_{11} and m_{11} as well as the twisting moment m_{12} from the full set of field equations [7,8,17,18]. The 'load vector' $\bar{\underline{z}}$ contains the external load components $\bar{p}_i$ (i = 1,2,3) and, if present, 'initial strain' terms due to temperature, creep or other influences. The components of the matrix $\underline{A}$ depend explicitly on the geometric parameters of the shell, the current quantities of the above state vector $\overset{o}{\underline{z}}$, differentials with respect to the circumferential coordinate θ, and on terms resulting from the elastic-plastic constitutive equations. The latter are essentially the current values of the tangent moduli of the points through the thickness, represented schematically by E_t. They will be described in more detail below. Concerning the notation used in (1) and the following expressions it is noted that current quantities, evaluated at a given configuration from which increments under given loading rates $\bar{p}_i$ are sought, are denoted by a superscript (^{o}) whereas increments of these quantities are not marked specifically.

Equations (1) represent a set of eight partial differential equations which, together with the appropriate boundary conditions constitute a two-dimensional, incremental boundary-value problem. Either directly, or via an associated variational principle, they could be the basis of a suitable two-dimensional discretization procedure. In this way shells of revolution with loads and/or geometric and material parameters varying irregularly in both the meridional and circumferential directions could be analyzed.

It should be mentioned that the matrix form of equation (1) is not dependent on the particular shell theory used. Provided the field equations can be written as sets of first order equations the differences involved in the various approximations of the nonlinear midsurface stretching and bending strains, as well as the corresponding equilibrium equations, only lead to modifications in the components of the vector $\underline{z}$ and matrix $\underline{A}$ in (1). In the analysis of elastic-plastic shells the assumptions of small strains, small curvatures, and moderate rotations are usually made. These are also justified in most cases of plastic shell buckling as has been verified by numerical calculations and by experiments. They indicate that the deflections associated with the buckling load are generally of the order of the shell thickness.
In the shell equations used in the numerical calculations below the nonlinear term appearing in addition to the usual linear incremental stretching strain tensor is given by $\frac{1}{2}\left[(\overset{o}{\phi}_\alpha - \bar{\phi}_\alpha)\phi_\beta + (\overset{o}{\phi}_\beta - \bar{\phi}_\beta)\,\phi_\alpha\right]$ $(\alpha,\ \beta = 1,2)$, where $\bar{\phi}_\alpha$ denotes the angle of rotation of the initial imperfection. This implies that in the in-plane and transverse equilibrium equations the usual expression for the transverse shear force is augmented by the term $(\overset{o}{\phi}_\alpha - \bar{\phi}_\alpha)\cdot n^{\alpha\beta} + \overset{o}{n}{}^{\alpha\beta}\,\phi_\alpha$. The transverse shear force on the boundary is similarly modified, while the rest of the field equations remain linear. Thus, the nonlinear terms lead to products of current (or initial) and incremental values of the angles of rotation, and of current (or initial) and incremental values of the membrane forces and the angles of rotation (or of the derivatives of either).

2.2 Fourier decomposition

As already mentioned we choose an approach which amounts to a reduction of (1) to a set of ordinary differential equations by using Fourier expansions in the circumferential direction. It is felt that the advantages of such a procedure outweigh many of its drawbacks in the area of applications envisioned, i.e. the determination of bifurcation or limit loads of perfect and imperfect shells of revolution.

In order to reduce the computational effort as much as possible it is assumed that the loading and geometrical imperfections have a common plane of symmetry at $\Theta = 0$. Then the truncated Fourier representation of the incremental state vector reads

$$\underline{z}(s,\theta) = \overset{(0)}{\underline{z}}(s) + \sum_{n=1}^{N} \underline{\Psi}_n(\theta)\ \overset{(n)}{\underline{z}}(s) \tag{2}$$

where $\overset{(n)}{\underline{z}}(s)$ represents the n-th Fourier harmonic of $\underline{z}$, and the components of the diagonal matrices $\underline{\Psi}_n(\Theta)$ are sine and cosine functions of the argument $n\Theta/R$. R is a reference shell radius. Similar expansions may also be evaluated for the 'load vector' $\bar{\underline{z}}$ and for the known current distributions of the quantities appearing in $\overset{o}{\underline{z}}$, as well as for the tangent moduli.

Substitution of these series expansions into (1) does not, however, as is the case for linear elastic shells, lead to N+1 uncoupled ordinary differential equations for the coefficient vectors $\overset{(n)}{\underline{z}}$ $(n = 0,N)$. The difficulty is due to the nonlinear terms appearing in $\underline{A}$ of (1). As noted above they are products of incremental and current quantities involving the normal stress resultants, angles of rotation (as well as first derivatives of both) and the current values of the plane stress moduli.

For known Fourier representations these product terms may be expressed as sums of trigonometric series using e.g.

$$2\cos m\theta \cos n\theta = \cos(m-n)\theta + \cos(m+n)\theta \tag{3}$$

Differentation with respect to Θ can then be carried out explicitly and equation (1) reduces to N + 1 sets of coupled ordinary differential equations for the coefficient vectors $\overset{(n)}{\underline{z}}$. They can be written concisely in the form of a single matrix equation

$$\frac{d\underline{Z}}{ds} = (\underline{A}_L + \underline{A}_G + \underline{A}_{PL})\ \underline{Z} + \bar{\underline{Z}} \tag{4}$$

or as

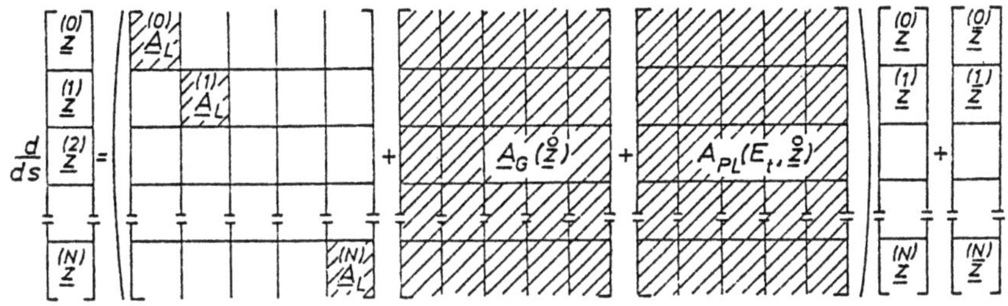

(4a)

As illustrated in (4a) the column vector $\underline{Z}$ in (4) contains all $N+1$ Fourier harmonics $\overset{(n)}{\underline{z}}$. Matrix $\underline{A}_L$ has diagonal form, and its submatrices $\overset{(n)}{\underline{A}}_L$ depend only on geometric parameters and on the Fourier index n. It governs the linear behavior of the shell. Matrices $\underline{A}_G$ and $\underline{A}_{PL}$ result from the geometrically nonlinear terms and from the plastic parts of the tangent moduli. All $8 \times (N+1)$ equations in (4) are coupled unlike in the linear elastic case where $\underline{A}_G$ and $\underline{A}_{PL}$ do not appear.

Direct integration of (4) is possible. However, depending on the number of Fourier terms necessary to accurately describe existing circumferential variations, the size of the $\underline{A}$-matrices may be considerable. Such a solution procedure may thus turn out to be as involved and costly as a two-dimensional discretization of the partial differential equations (1). These disadvantages are minimized if all coupled terms in (4) are treated as pseudo-'loads'. In this way the original problem (1) is reduced to linear ordinary differential equations of first order of the form

$$\underline{Z}' = \underline{A}_L \, \underline{Z} + \bar{\underline{Z}} + \hat{\underline{Z}} \qquad (5)$$

where the prime denotes differentiation with respect to s, and the additional pseudo-'load' increment $\hat{\underline{Z}}$ is given by

$$\hat{\underline{Z}} = \underline{Z}_G + \underline{Z}_{PL} = (\underline{A}_G + \underline{A}_{PL})\underline{Z}. \qquad (5a)$$

Since $\hat{\underline{Z}}$ depends on the unknown vector $\underline{Z}$ equation (5) must be solved consecutively by iteration.

Basing the solution procedure on equation (5), rather than (4), has the advantage that, due to the diagonal form of $\underline{A}_L$, the equations for the various harmonics $\overset{(n)}{\underline{z}}$ are uncoupled and that the homogeneous solution of (5) is identical to that of the linear elastic case. Thus one has obtained N + 1 separate sets of eight first order differential equations for each Fourier index n

$$\frac{d\overset{(n)}{z}}{ds} = \underline{A}_L \overset{(n)}{\underline{z}} + \overset{(n)}{\underline{\bar{z}}} + \overset{(n)}{\underline{\hat{z}}} \qquad (n = 0, 1, \ldots N) \qquad (6)$$

where e.g. $\overset{(n)}{\underline{z}}$ represents the respective subcolumn vector of $\underline{Z}$ in (5). The numerical procedure fully exploits the fact that the homogeneous and particular solutions of (6) can be obtained separately for each value of n, and that matrix $\underline{A}_L$ remains unchanged during the loading history. This, and the fact that the iteration procedure only involves the repeated evaluation of the vectors $\overset{(n)}{\underline{\hat{z}}}$ rather than matrices analogous to $\underline{A}_G$ and $\underline{A}_{PL}$ in (4) reduces the numerical effort considerably. However, difficulties in connection with the rate of convergence may arise at certain points of the load-deflection path.

2.3 Material description

The question of the appropriate choice of a constitutive law in buckling calculations is still largely unresolved [20,21]. Physically acceptable, simple flow theories with smooth yield surfaces have sometimes been found to give bifurcation loads which are in marked disagreement with experimental observation. In contrast, remarkably good agreement has been obtained in certain cases with the use of simple deformation theories despite the fact that physical arguments can be raised against them in principle. Attempts have been made recently to reconcile these differences by appealing to more complex flow theories of plasticity which either abandon the assumption of smoothness of the yield surface [20,21] or that of normality of the plastic strain increment[22,23]. The results, however, are still inconclusive. Therefore, and since the emphasis here is more on numerical considerations, this important point is not pursued in detail. Rather, J_2-flow theory is used throughout, and in some cases the results are compared to those of J_2-deformation theory, which may be viewed as a sufficiently simple limiting case of more complex theories involving the formation of yield surface corners.

In both cases the constitutive equations relating the increments of the stress resultants $n^{\alpha\beta}$, $m^{\alpha\beta}$ ($\alpha = 1,2$) to those of the mid-surface stretching and bending strains $\alpha_{\alpha\beta}$, $\beta_{\alpha\beta}$ may be written as [1,20,18]:

$$n^{\alpha\beta} = \overset{(1)}{H}{}^{\alpha\beta\gamma\delta}\alpha_{\gamma\delta} + \overset{(2)}{H}{}^{\alpha\beta\gamma\delta}\beta_{\gamma\delta} \tag{7a}$$

$$m^{\alpha\beta} = \overset{(2)}{H}{}^{\alpha\beta\gamma\delta}\alpha_{\gamma\delta} + \overset{(3)}{H}{}^{\alpha\beta\gamma\delta}\beta_{\gamma\delta} \tag{7b}$$

where Greek indices take on the values 1,2. The tensors $\overset{(i)}{H}{}^{\alpha\beta\gamma\delta}$ ($i = 1,2,3$) are defined as integrals over the shell thickness of the current in-plane tangent moduli. At each point through the shell thickness they depend on the values of the stress deviators and on the parameters describing the state of hardening. The latter are obtained from a uniaxial stress-strain relation, here taken to be a piecewise power law. The constitutive relation in the form (7a,b) is needed in the derivation of (4, 4a) and in the formulation of the bifurcation problem.

For the solution of the incremental problem (1) by the 'initial stress approach' underlying (5,6) equations (7a,b) are reformulated. Starting from the general incremental plane-stress constitutive relation

$$\sigma^{\alpha\beta} = E^{\alpha\beta\lambda\rho}\,(\gamma_{\lambda\rho} - \gamma_{\lambda\rho}^{(pl)}) \tag{8}$$

where $\sigma^{\alpha\beta}$ denotes the in-plane components of the increments of the second Piola-Kirchhoff stress, $\gamma_{\alpha\beta}$ are the increments of the Lagrangian strain, and $E^{\alpha\beta\lambda\rho}$ is the plane-stress tensor of elastic moduli (assumed constant everywhere) one obtains

$$n^{\alpha\beta} = \int_{-t/2}^{t/2} \sigma^{\alpha\beta}\,dz = t\,E^{\alpha\beta\lambda\rho}\alpha_{\lambda\rho} - n^{\alpha\beta}_{(pl)} \tag{9a}$$

$$m^{\alpha\beta} = \int_{-t/2}^{t/2} \sigma^{\alpha\beta}\,zdz = \frac{t^3}{12}\,E^{\alpha\beta\lambda\rho}\beta_{\lambda\rho} - m^{\alpha\beta}_{(pl)} \tag{9b}$$

In (9a, b), z is the normal distance from the middle surface, t is the shell thickness and the pseudo-stress resultants $n^{\alpha\beta}_{(pl)}$, $m^{\alpha\beta}_{(pl)}$ are given by

$$n^{\alpha\beta}_{(pl)} = E^{\alpha\beta\lambda\rho} \int_{-t/2}^{t/2} \gamma^{(pl)}_{\lambda\rho} \, dz \quad ; \qquad m^{\alpha\beta}_{(pl)} = E^{\alpha\beta\lambda\rho} \int_{-t/2}^{t/2} \gamma^{(pl)}_{\lambda\rho} \, z dz. \qquad (9c)$$

The plastic part of the strain increment $\gamma^{(pl)}_{\lambda\rho}$ can, in turn, be expressed in terms of the total strain increment $\gamma_{\alpha\beta}$. Thus, like the moduli $\overset{(i)}{H}{}^{\alpha\beta\gamma\delta}$ in (7a, b) the quantities $n^{\alpha\beta}_{(pl)}$, $m^{\alpha\beta}_{(pl)}$ are obtained by numerical integration through the thickness, and the integrands depend nonlinearly on the current, nonlinear, distribution of stresses and hardening parameters.

Both quantities in (9c) are related to the components of the Fourier harmonics $\overset{(n)}{\hat{\underline{z}}}$ in (6) which act as pseudo-'load' vectors. In the actual calculation they are determined in the following manner. First, the Fourier representation of the total strain increments is used to determine the values of $\gamma_{\alpha\beta}$ at discrete stations around the circumference. From them, and the known values of the current quantities appearing in the integrands of (9c), the plastic strain increments $\gamma^{(pl)}_{\alpha\beta}$ at these stations are evaluated, and integration is performed numerically. Finally, a Fourier expansion of $n^{\alpha\beta}_{(pl)}$ and $m^{\alpha\beta}_{(pl)}$ is determined, the coefficients of which are then related to $\overset{(n)}{\hat{\underline{z}}}$. This is done at each iteration step throughout the load history.

3. Numerical solution procedure

As already indicated, the numerical solution of (6), as performed in the computer program ROT B, is based on a combination of approaches. On the local level a transfer matrix and an associated stiffness matrix for an element are derived by accurate numerical integration [7,17,18], whereas on the structural level the standard finite element procedure is followed.

3.1 Element evaluation

Regarding the ordinary differential equations for each Fourier harmonic (6) as an initial value problem with respect to s, and performing the integration over a finite segment of the shell meridian, a transfer matrix $\overset{(n)}{\underline{U}}(s)$ for a ring element can be derived explicitly from $\overset{(n)}{\underline{A}}_L$ which relates the respective values of $\overset{(n)}{\underline{z}}$ at the end of the interval of integration to its corresponding initial values $\overset{(n)}{\underline{z}}_o = \overset{(n)}{\underline{z}}(s = 0)$ in the following manner

$$\overset{(n)}{\underline{z}}(s) = \overset{(n)}{\underline{U}}(s)\,\overset{(n)}{\underline{z}}_o + \overset{(n)}{\bar{\underline{z}}}{}^p + \overset{(n)}{\hat{\underline{z}}}{}^p \qquad (10)$$

Matrix $\overset{(n)}{\underline{U}}(s)$ is an exponential function of $\overset{(n)}{\underline{A}}_L$ in (6) and s, and $\overset{(n)}{\bar{\underline{z}}}{}^p$, $\overset{(n)}{\hat{\underline{z}}}{}^p$ are the particular solutions of (6) which depend on $\overset{(n)}{\underline{U}}$ and the sum of the 'load' vectors $\overset{(n)}{\bar{\underline{z}}}$ and $\overset{(n)}{\hat{\underline{z}}}$. In the actual analysis $\overset{(n)}{\underline{U}}$ is developed into a power series with respect to s. It can then be evaluated to any desired degree of accuracy by taking into account as many terms as necessary. In this way a priori error bounds can be given. This is, in general, not possible for standard finite element methods, and constitutes one of the main advantages of the approach. Moreover, it is found to be more efficient compared to other first order integration schemes, e.g. the Runge-Kutta method [7].

As is known, the accuracy of (10) depends strongly on the length of the interval of integration. For this reason, and in order to obtain more versatility in the analysis of arbitrary shell configurations, solution (10) has been incorporated into a displacement finite element procedure which is used to calculate the response of the entire structure. To do this the transfer matrix $\overset{(n)}{\underline{U}}$, valid over a finite curved ring segment, is first transformed by elementary operations to a standard element stiffness matrix $\overset{(n)}{\underline{k}}_L$ [7,17,18]. Equation (10) then becomes

$$\overset{(n)}{\underline{k}}_L\,\overset{(n)}{\underline{v}} = \overset{(n)}{\bar{\underline{p}}} + \overset{(n)}{\hat{\underline{p}}} \qquad (11)$$

where $\overset{(n)}{\underline{v}}$ is the vector of the n-th Fourier harmonic of the incremental displacement quantities at the edges of the element, $\overset{(n)}{\bar{\underline{p}}}$ is the respective vector of the nodal load increments and $\overset{(n)}{\hat{\underline{p}}}$ is that of the pseudo-load derived from $\overset{(n)}{\hat{\underline{z}}}{}^p$ in (10).

3.2 System evaluation

Superposition of the stiffness matrices and load vectors of each element and for each Fourier index n is done in the same way as in a standard finite element procedure to give

$$\overset{(n)}{\underline{K}}_L \, \overset{(n)}{\underline{V}} = \overset{(n)}{\underline{\bar{P}}} + \overset{(n)}{\underline{\hat{P}}} \qquad (n = 0, 1, \ldots N) \tag{12}$$

for the element assembly of the total structure. Recall that $\overset{(n)}{\underline{V}}$ and $\overset{(n)}{\underline{\bar{P}}}$ are incremental quantities. Since the global, elastic, stiffness matrix $\overset{(n)}{\underline{K}}_L$ does not change during loading its assemblage and triangularization needs only be performed once, thus simplifying the repeated iterative solution of (12) considerably.

Writing all N + 1 linear equations (12) in the form of a single matrix equation, analogous to what was done in (4), one obtains

$$\underline{K}_L \, \underline{V} = \underline{\bar{P}} + \underline{\hat{P}} \tag{13}$$

where now $\underline{K}_L$ is the stiffness matrix of the entire structure. A similar equation would have been obtained in a two dimensional discretization procedure employing e.g. finite elements. To briefly illustrate the differences between the two approaches, the numerical approximations involved and the forms of the resulting stiffness matrices $\underline{K}_L$ are shown schematically in figures 1a und 1b.

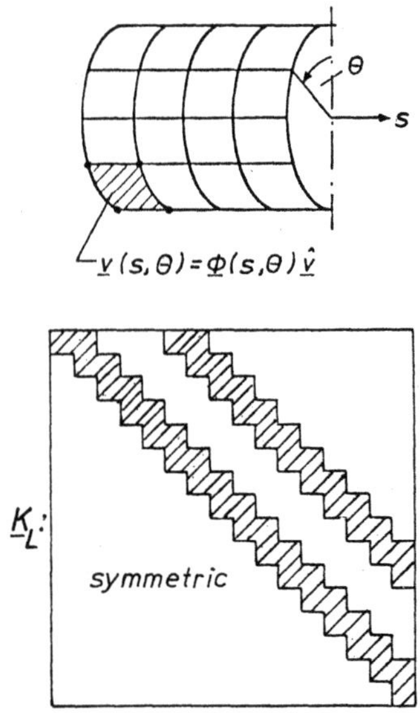

Fig.1a. Two-dimensional discretization

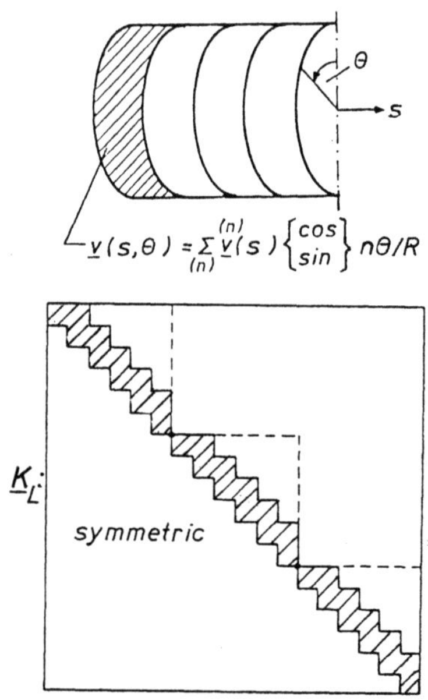

Fig. 1b. One-dimensional discretization

In a standard finite element approach the distribution of the displacement variables is approximated by interpolating functions $\underline{\Phi}$ (s,θ). The vector of unknowns $\underline{V}$ in (13) then consists of the degrees of freedom $\underline{\hat{v}}$ at the nodal points of the two-dimensional mesh, and $\underline{K}_L$ is symmetric and banded as indicated schematically in the lower part of figure 1a. The present approach based on Fourier expansions in the circumferential direction is illustrated in figure 1b. It leads to curved ring elements, and the unknown components of $\underline{V}$ are the Fourier coefficients $\overset{(n)}{\underline{V}}$ of the displacement variables at the nodal lines. Again, $\underline{K}_L$ is symmetric and banded. The main difference between the two stiffness matrices is that in the one-dimensional case the equations corresponding to the various Fourier indices n are uncoupled (see (12) and figure 1b). Moreover, the bandwidth is quite small, whereas in the two-dimensional case it depends on the complexity of the elements and may be considerable. Thus, the solution of (13) in the form (12) requires much less storage allocation and computation time.

It is noted that, compared to the direct stiffness method the outlined combination of a direct integration approach on the element level with a finite element procedure on the structural level is based on fewer approximations. It also yields more accurate stress resultants because their evaluation requires only multiplications of matrices which do not involve derivatives of assumed displacement fields.

3.3 Nonlinear iteration

Using a constant, in this case the elastic, stiffness matrix throughout, and accounting for all nonlinear influences via pseudo-loads has the numerical advantages detailed above, but necessitates the iterative adjustment of the vector $\overset{(n)}{\underline{\hat{P}}}$ in (12) in each loading step. As is known, this type of iteration may experience slow convergence, or divergence, in certain cases. As shown in figure 2 for a 'softening' structure the number of iterations and the relative magnitude of the additional vector $\underline{\hat{P}}$ (where $\underline{\hat{P}}_{(3)}$ in figure 2 indicates the value of $\underline{\hat{P}}$ in the third iteration step) may become quite large if the current tangent stiffness is considerably smaller than the initial stiffness indicated

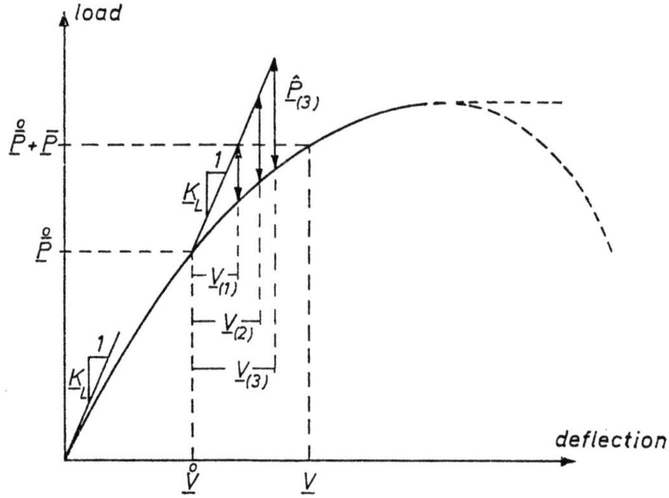

Fig. 2. 'Softening' structure

by $\underline{K}_L$. The iteration diverges at or near a load maximum and in regions where the tangent stiffness becomes negative. Similar convergence problems may be encountered in structures exhibiting significant geometric 'stiffening', i.e. continuously increasing slopes of their load-deflection paths. However, since we shall be dealing with elastic-plastic structures with decreasing material moduli and with generally decreasing geometric stiffnesses, the latter case will not become important.

Our main interest lies in the determination of the maximum support load of a structure. For this purpose, the present incremental and iterative procedure, in combination with an algorithm for convergence acceleration due to Aitken [24] has been found to be stable and effective. It permits the analysis of elastic-plastic shells of revolution of essentially arbitrary shape and under general loading.

3.4 Bifurcation analysis

Depending on the value of certain geometric parameters, the distribution of initial imperfections, and the type of loading a shell may either reach a load maximum, leading to instantaneous collapse, or to bifurcation. As can be shown on the basis of Hill's general theory of bifurcation and uniqueness of incremental boundary value problems, bifurcation in the plastic range always occurs under increasing load [25,20,21]. Thus, the behavior of the shell in the immediate vicinity of the bifurcation point is stable. Normally, however, the additional load-carrying capacity available on the bifurcated path is eroded rapidly due to the destabilizing interaction of nonlinear geometric effects and the continually decreasing material moduli. Thus, the maximum support load of the bifurcated structure is found to be only slightly higher than the bifurcation load [1,20,21].

Bifurcation from a nonlinear axisymmetric prebuckling state, obtained through the successive solution of (12), first becomes possible when one or more of the following homogeneous boundary value problems

$$\frac{d\overset{(m)}{\underline{z}}}{ds} = \overset{(m)}{\underline{A}_T}\,\overset{(m)}{\underline{z}} \qquad (m = 1,2,3\ldots) \tag{14}$$

have nontrivial solutions. Matrix $\overset{(m)}{\underline{A}_T}$ is the sum of matrices analogous to $\underline{A}_L$, $\underline{A}_G$, $\underline{A}_{PL}$ introduced in (4). In the derivation of (14) from (1), or (4), the condition that no elastic unloading can take place at bifurcation (Shanley condition) [20,21,25] has been used, and the bifurcation solution has been expanded into a Fourier series. Due to the assumed axisymmetry of the prebuckling state and the linearity of the eigenvalue problem the ordinary differential equations (14) are automatically uncoupled. Assuming the eigenmode to be distinct, which is often the case in plastic buckling, equations (14) are satisfied for only one value of the circumferential wave number m. The associated eigenvector $\overset{(m)}{\underline{z}}$ gives the meridional distribution of the eigenmode at the nodal lines. As explained in connection with (1) the components of matrix $\overset{(m)}{\underline{A}_T}$ depend explicitly on m, as well as on the current values of $\overset{o}{\underline{z}}$ and of the tangent moduli $\overset{(i)}{H}{}^{\alpha\beta\gamma\delta}$ introduced in (7a, b). These quantities thus influence the value of the bifurcation load directly.

In the manner described previously a transfer matrix and a related stiffness matrix for each element can be derived from (14). After assembly the eigenvalue problem for the entire shell takes the form

$$\overset{(m)}{\underline{K}_T}\,\overset{(m)}{\underline{V}} = 0. \tag{15}$$

Equation (15) is the homogeneous analog of the incremental relation (12), with the difference that now all nonlinear geometric and material influences are contained in the current tangent stiffness matrix $\overset{(m)}{\underline{K}_T}$, and that $\overset{(m)}{\underline{V}}$ is the discrete eigenmode associated with the critical wave number m. It is noted that, although linear, equation (15) is not a usual linear eigenvalue problem since it does not contain the critical loading parameter λ_{cr} explicitly.

Non-trivial solutions of (15) are sought systematically by continually approximating $\overset{(m)}{\underline{K}_T}$ in such a way that (15) takes the familiar form of linear eigenvalue problems (see e.g. [26,18])

$$\left[\overset{(m)}{\underline{K}_L} + \frac{\lambda^*}{\lambda_i}\left(\overset{(m)}{\underline{K}_T}(\lambda_i) - \overset{(m)}{\underline{K}_L}\right)\right]\overset{(m)}{\underline{V}} = 0 \,. \tag{16}$$

In (16) $\overset{(m)}{\underline{K}_L}$ is identical to the respective matrix in (12), λ_i are the values of the loading parameter λ at various load steps, and λ^* denotes estimates of the critical bifurcation value λ_{cr}. The approximations leading to (16) essentially involve linear extrapolations of the quantities appearing in $\overset{(m)}{\underline{K}_T}$ of (15) from a known state on the nonlinear prebuckling path. Equations (16) can be solved by standard methods. Convergence is obtained when the solutions of (15) and (16) coincide. This is the case when $\lambda^*/\lambda_i \rightarrow 1$, i.e. when $\lambda^* \rightarrow \lambda_i \rightarrow \lambda_{cr}$. This procedure yields the lowest eigenvalue of (15), the corresponding eigenmode, and the number of circumferential waves.

4. Imperfection sensitivity

In the presence of imperfections the maximum support load of an elastic-plastic shell is smaller than that of the respective perfect shell, and generally falls below the bifurcation load. This imperfection sensitivity may be quite severe even in cases where no such sensitivity exists if the same structure remained elastic during buckling [1,20,21]. Almost all previous plastic buckling analyses of shells have been concerned with bifurcation, or limit, loads of shells with no or with axisymmetric imperfections [1-4]. The results of these calculations indicate that, in general, the load-carrying capacity is reduced, sometimes strongly, and that there is a dependence on the imperfection shape. Since, as already noted, the majority of actual shells contain nonaxisymmetric initial imperfections, it is important to know whether or not the nonsymmetric components have a strongly detrimental influence. To shed some light on this question numerical calculations of dished end closures have been performed using the computer program ROT B described above. The results presented in the following section illustrate the influence of various geometric and material parameters on their buckling behavior and on their imperfection sensitivity.

5. Buckling behavior and imperfection-sensitivity of dished end closures under external pressure

Until comparatively recently interest has been centered on the buckling behavior of individual shell types, e.g. circular cylinders, cones, spherical caps, hyperbolic paraboloids. Improved computational procedures and increasing demands on safety and economy have, however, shifted much of the attention to the more complex buckling behavior of combinations of shells. A better, although still incomplete, understanding of the factors determining the buckling behavior of these shells has been gained through a growing number of numerical and experimental investigations (see e.g.[2,3,27-32]). Thus, it is now widely recognized that, in general, the critical and post-critical load-carrying behavior of built-up and/or curved thin-walled structures is governed by a rather complex interaction of nonlinear geometric and material influences, and that the details of this interaction are quite difficult to anticipate. Therefore, appealingly simple approximations, such as the assumption of a linear membrane prebuckling state, or the exclusive reliance on, possibly simplified, bifurcation calculations to estimate load-carrying capacities, should be viewed with great caution. In important cases they may well lead to severe misrepresentations, and thus uncertain predictions, of the actual structural behavior.

The value and importance of detailed analyses becomes apparent in the investigation of the buckling behavior of dished end closures (see figure 3).

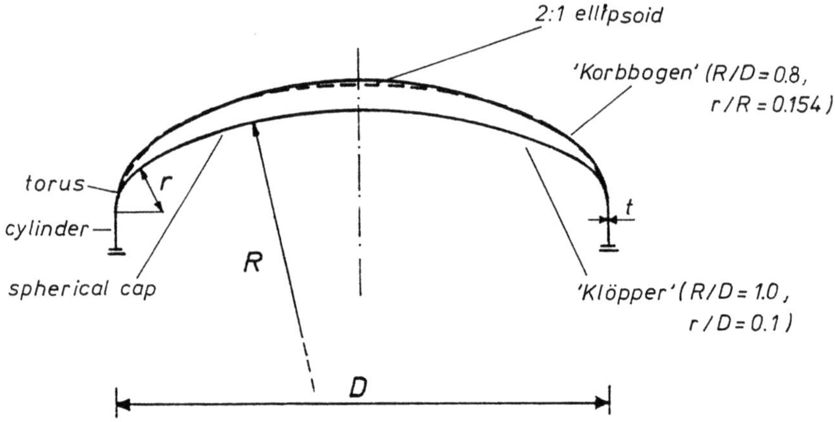

Fig. 3. Dished end closures

They are somewhat unique in that, when subjected to sufficiently high internal pressure, buckling is observed in the toroidal segment despite the occurrence of predominantly tensile stresses in most of the structure. Extensive numerical and experimental investigations of these shells [2,3, 29-32] have led to a reasonably complete understanding of their behavior under internal pressure.

Much fewer details are known at present in the case of externally applied pressure. Valuable results have been reported in [29] and other work referenced therein. Nevertheless, systematic numerical studies of the influence of the dominant geometric and material parameters on their buckling behavior are lacking. Moreover, except for the asymptotic analysis of complete spheroidal shells presented in [33], virtually no information concerning the imperfection-sensitivity of this type of shell is available.

In this chapter numerical results of the elastic-plastic buckling behavior and the sensitivity to axisymmetric and nonaxisymmetric initial imperfections of three types of dished ends are presented. Two of them are torispheres, i.e. the 'Klöpper' and 'Korbbogen' types commonly used in Germany, while the third is a 2:1 ellipsoid. Their respective cross-sections and parameter ratios are given in figure 3. It shows that the 'Klöpper'-end is quite shallow, and that the geometries of the 'Korbbogen' and ellipsoid differ very little. Despite this close resemblance, however, their buckling behavior shows significant differences, as will be seen.

For completeness it is noted that the numerical results given in table I of [31] indicate that, depending on the D/t ratios and the yield stresses investigated there, the internal buckling pressures of the 'Korbbogen' may be from 65 % to 110 % larger than those of the 'Klöpper' ends. In turn, for the 2:1 ellipsoids 25 % to 35 % higher internal pressures have been computed than for the 'Korbbogen'. Thus, from the point of view of bifurcation pressures, the 2:1 ellipsoid is, in a sense, an optimally shaped dished end. However, it has also been shown [34] that, generally speaking, optimized structures are very prone to exhibit a strong imperfection sensitivity, which then erodes much of the increase in maximum support load realized by the respective perfect structure. This point will be raised when discussing the imperfection-sensitivity results.

In the following, first some basic aspects of the load-carrying and buckling behavior of 'Klöpper' ends will be treated. Next, differences in comparison with that of the 'Korbbogen' and the 2:1 ellipsoid are discussed, and finally their relative imperfection-sensitivities and load-carrying capacities will be assessed.

5.1 'Klöpper' end closure

A typical plot of the prebuckling axisymmetric distribution of the normal displacement component v_3 is given in figure 4. As would be expected, in the spherical and cylindrical sections the deflections are directed inward, accompanied by compression in these segments. In contrast, the toroidal part moves outward, despite the acting external pressure, as shown in figure 4. The thus induced circumferential tension causes it to act as a stiffening component. This influence, the particulars of which depend on D/t and on the yield stress σ_o of the material, also reveals itself in the rather complex stress distribution shown in figure 5. There it is seen that while the stresses are distributed quite uniformly in most of the adjoining sections, they vary considerably, and may change sign repeatedly, in the torus and in a tran-

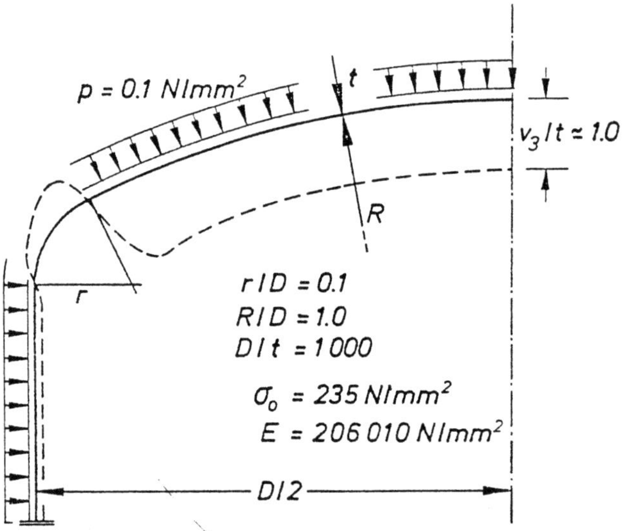

Fig. 4. Normal displacements

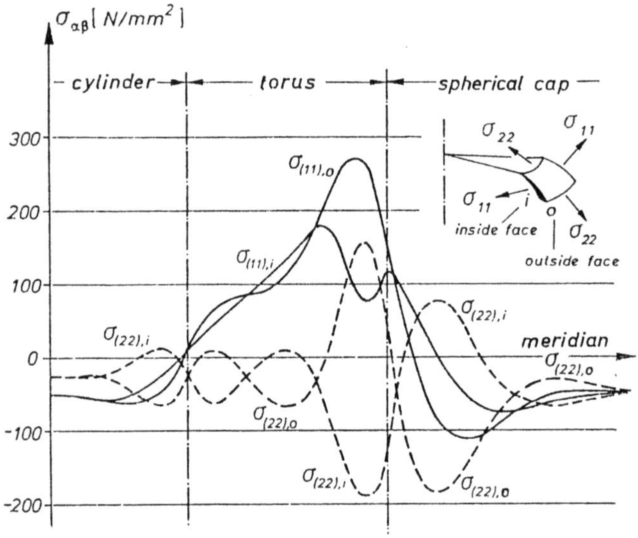

Fig. 5. Stress distribution

sition region of the spherical cap. Outside this region essentially a state of membrane compression prevails in both the meridional and circumferential directions, whereas inside it the shell is subjected to significant bending. This leads to the marked differences between the stresses on the inside (i) and outside (o) faces. Nevertheless, the mean circumferential stresses are tensile whereas those in the meridional direction are still compressive.

For a shell with the same parameter values as in figure 4 and elastic-ideally plastic material behavior, the extent and growth of the plastic zone is plotted in Fig. 6. It shows

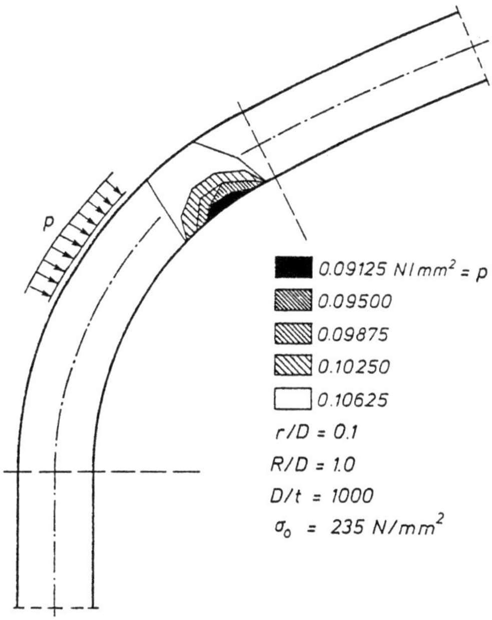

Fig. 6. Extent of plastic zones

that plastic strains first appear on the inside and that up to the buckling pressure (which is reached shortly after the configuration shown) the plastic zone is confined to a comparatively narrow region at the upper end of the toroidal segment. It does not extend into the spherical cap as do the bending stresses shown in Fig. 5. Thus, collapse occurs before the plastic zone has reached appreciable dimensions. This is typical of torispheres under external pressure. Nevertheless, its influence on the value of the maximum support load and on the buckling mode may be significant. Depending on the values of the main parameters (D/t, σ_o) the latter may be axisymmetrical or nonaxisymmetrical. In the previous example buckling occurs axisymmetrically at a limit point. The respective load vs. crown-deflection curve plotted in figure 7 remains nearly linear until yielding starts on the inside.

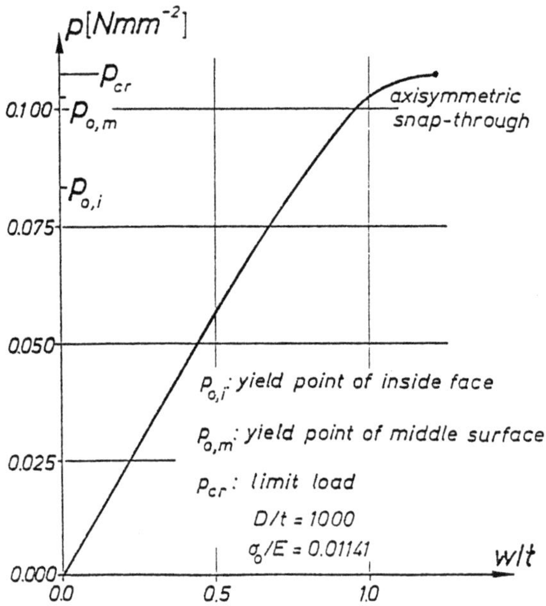

Fig. 7. Pressure vs. crown deflections

A rapid decrease in overall stiffness is observed after the plastic zone has reached the middle surface, and collapse occurs soon thereafter. From this and the appearance of the plastic zone in figure 6 one should not, however, draw the conclusion that collapse is due simply to the formation of a 'plastic hinge'. Rather, it is caused by the interaction of geometrical and material influences which contribute equally to the rapid loss of structural stiffness above $p_{o,m}$. Finally it is noted that the deflection at the limit load is only somewhat larger than the shell thickness.

As mentioned in connection with figures 4 and 5 the toroidal part is mainly subjected to tension, so that nonaxisymmetric buckles, when they appear, are concentrated in either the cylinder or the spherical segment. A typical meridional distribution of such a 'cap-buckling mode' is plotted in figure 8. It shows that the largest amplitude occurs quite close to the torus transition

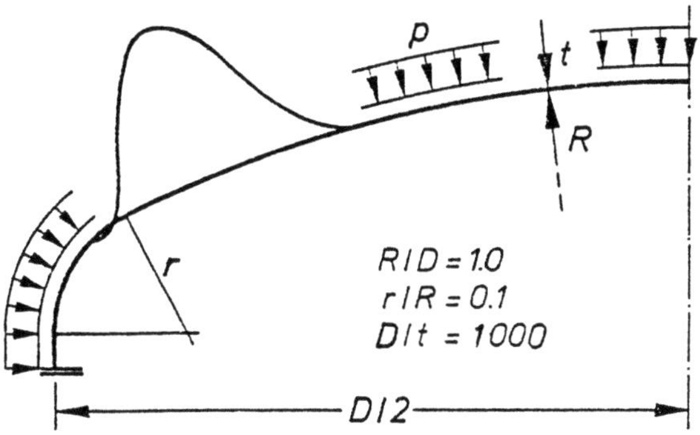

Fig. 8. Non-axisymmetric buckling mode (m = 11)

but that a relatively large portion of the sphere is also affected. Note that this mode shape has a periodic circumferential variation (here with 11 waves), and that only the upper portion is shown.

The elastic bifurcation results summarized in figure 9 are given for completeness. They illustrate the influence of the length of the cylinder L on the location of the buckling mode and on the circumferential wave number m. For the parameters considered, it is seen that the bifurcation pressures decrease sharply with increasing cylinder lengths. Moreover, except for small values of L/D, the buckling mode is largely confined to the cylinder, and m varies from 6 to 15 as indicated by the numbers in parentheses. The dashed line in figure 9 represents the known Batdorf results [35] for a closed circular cylinder with classical simple support conditions.

Non-axisymmetric buckling of the relatively stiff end closures is possible only if the cylinder length is smaller than about 20 percent of its diameter. The number of circumferential waves is then 11, and the buckling mode has the meridional distribution given in figure 8. Generalizing from the results of figure 9 to other geometries, all following numerical results were obtained for end closures with a short cylindrical attachment (L/D = 0.05), the bottom end of which is free to move perpendicular to the axis of rotation (see the inset of figure 10).

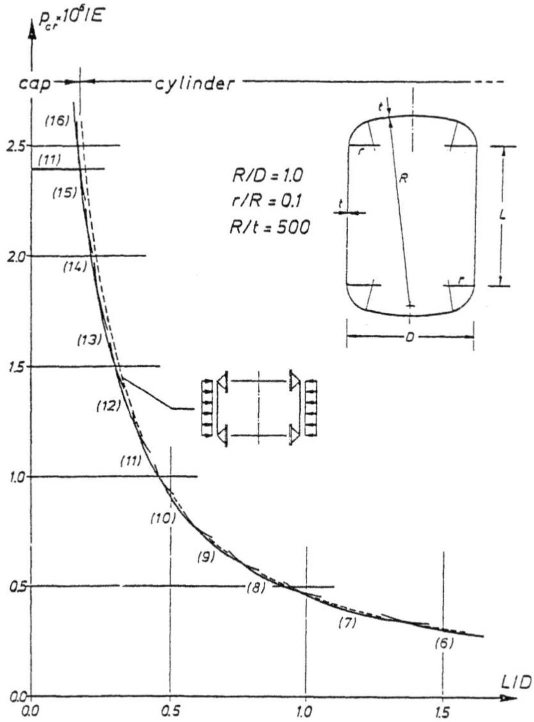

Fig. 9. Elastic bifurcation pressures of cylindrical vessels with 'Klöpper' ends

For this configuration the elastic bifurcation pressures are plotted in figure 10 for R/t values ranging from 400 to 1500. The corresponding wave numbers, again given in parentheses, are seen to increase towards higher ratios of R/t, i.e. for thinner shells. Also given are the respective limit pressures at which axisymmetric snap-through would occur. Over the whole range of R/t considered they lie between 60 and 100 percent above the bifurcation pressures. In the elastic range non-axisymmetric bifurcation is thus seen to be the dominant buckling mechanism. Somewhat surprisingly the lower curve of figure 10 may be represented to within an accuracy of about 2.5 % by the simple expression

$$p_{cr} = \alpha\, p_{cr}^{s} = \alpha \frac{2E}{\sqrt{3(1-\nu^2)}} \left(\frac{t}{R}\right)^2 \qquad (17)$$

where $\alpha = 0.5$ and p_{cr}^{s} is the classical critical pressure of a complete spherical shell with the same radius as the spherical

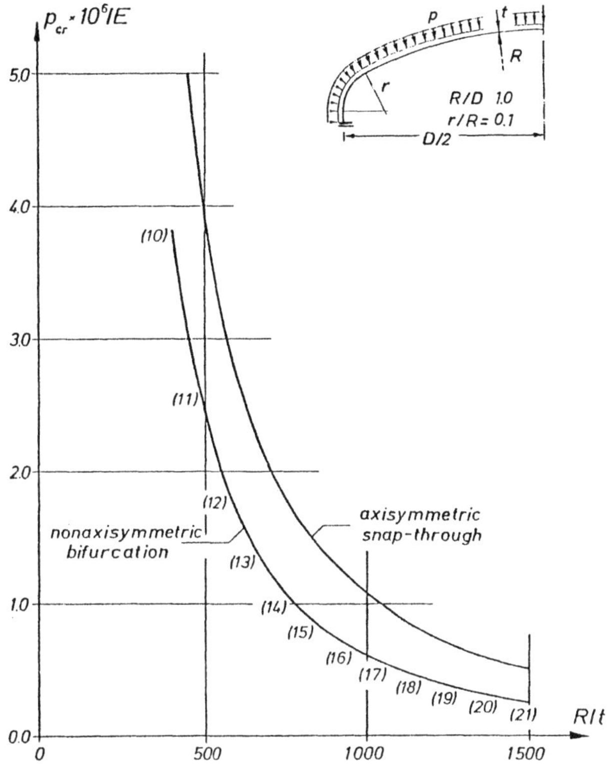

Fig. 10. Elastic bifurcation and limit pressures

cap of the 'Klöpper' end.

The occurrence of plastic strains has a significant influence on the buckling pressures and modes even though the plastic zones are rather small. This can be seen from figure 11, where normalized buckling pressures, evaluated for J_2-flow theory and elastic-ideally plastic material behavior, are plotted versus R/t for three values of the yield stress σ_o. The lower one of the elastic curves in figure 10 appears here as an upper limit ($\sigma_o/E \rightarrow \infty$). But while previously buckling was always due to non-axisymmetric bifurcation, it may now also occur at a limit point if the yield stress is sufficiently small or the relative shell thickness t/R is sufficiently large. The respective limit pressures lie on the dashed sections of the lower two curves in figure 11.

The overall effect of plasticity is to lower the buckling pressures below their elastic values. It is more pronounced for

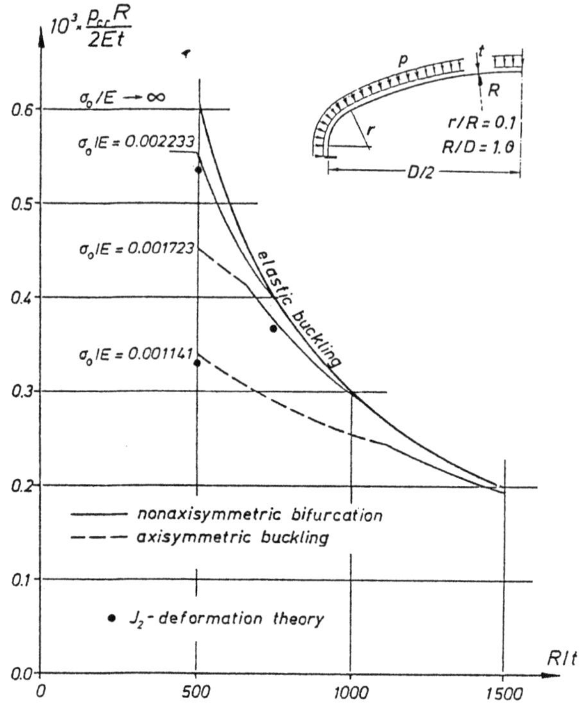

Fig. 11. Elastic-plastic buckling pressures

smaller yield stresses, where the decrease may be up to about 45 percent. Figure 11 also shows that, except for the lowest curve, the elastic-plastic and the elastic curves merge at a certain R/t value. Thus, for a given yield stress (e.g. $\sigma_0/E = 0.002233$) no plastic buckling will occur for large enough values of R/t (e.g. R/t > 750). For three sets of σ_0 and R/t results were also computed using J_2-deformation theory. They are represented in figure 11 by solid dots. Two of them correspond to bifurcation points, whereas the lowest one is associated with a limit point. It is seen that in the present case the material description does not have a strong influence. The respective buckling pressures differ very little, but the results for J_2-deformation theory are below those for J_2-flow theory, as is usually the case. One explanation for the slightness of the difference may lie in the fact that the limit, or bifurcation, pressures are only somewhat larger than the pressure at which yielding starts (see figure 7), and in the limited extent of the plastic zones.

The influence of material hardening may be seen from figure 12. The results given there were obtained using a piece-wise power law with the power n as the hardening parameter, and the lowest yield stress ($\sigma_o = 0.00114E$) of figure 11. Setting n equal to 1, or to a large number ($n \to \infty$) the elastic and the elastic-ideally plastic cases of figure 11 are recovered.

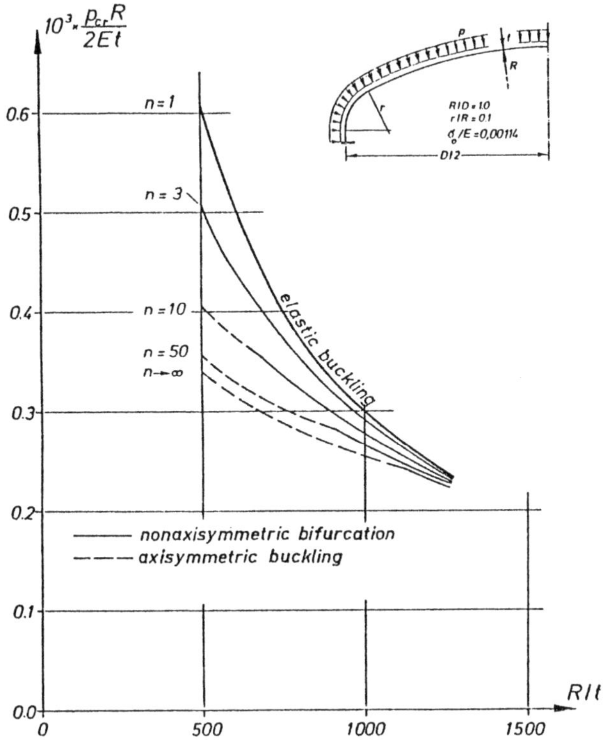

Fig. 12. Influence of material hardening

In figure 12 they represent the upper and lower limits. Curves for intermediate values of n are also given. They show the expected result that a material with high hardening can raise the critical pressure considerably, especially if the shell is relatively thick, in which case the influence of plasticity is most pronounced. Thus, for e.g. $R/t = 500$ a modest amount of hardening ($n = 10$) results in an appreciable increase (of approximately 20 percent) of the buckling pressure. For a high hardening material ($n = 3$) an even larger increase (about 50 percent) is possible. This trend is similar to that found for higher yield stresses (see fig. 11). Similarly, small values of n (i.e. high hardening) and/

or large values of R/t lead to nonaxisymmetric bifurcation whereas in the opposite case axisymmetric snap-through predominates. It is interesting to note that in fig. 12 the elastic and plastic curves do not merge. Instead, the plastic buckling pressures are always below the elastic ones, with the differences becoming quite small, however, for the larger R/t values.

In the last part of this chapter an assessment of the imperfection-sensitivity of 'Klöpper' ends under external pressure will be given. It is based on initial deviations from the perfect geometry which are in the shape of the nonaxisymmetric bifurcation mode (see e.g. figure 8), and on an appropriate choice of parameters (see figures 11 and 12). This is done to account for the fact that in most practical situations imperfections are nonaxisymmetric, rather than axisymmetric, in a simple manner. A second reason is to simplify the calculations, and also to obtain a close correspondence with, and a continuous transition to, the bifurcation results in the case of vanishing imperfection amplitudes. Conceivably this choice is nonconservative and more detrimental imperfection shapes exist. Nevertheless, in the absence of sufficiently representative measurements the results thus obtained give a suitable estimate of the relative decrease in load-carrying capacity that is to be expected if initial imperfections are present.

For a particular set of parameters the meridional shape of the bifurcation mode and its circumferential wave number were obtained from the numerical solution of the eigenvalue problem (15). After normalizing the maximum value of the normal displacement component v_3 of the eigenmode to a prescribed fraction δ of the shell thickness t, the nonlinear equations (12) were solved incrementally until a pressure maximum was reached. In figure 13 the elastic limit pressures p_{max}, normalized by the bifurcation pressures p_{cr} of the corresponding perfect shell (see figure 10), are plotted as functions of δ for two values of R/t. A relatively strong imperfection sensitivity, particularly for smaller amplitudes, is found. Thus, imperfection amplitudes of only about 35 percent of the shell thickness ($\delta = 0.35$) lead to reductions in the limit pressure of approximately 40 percent. The decrease for larger δ values is less pronounced. Both cases tend to a reduc-

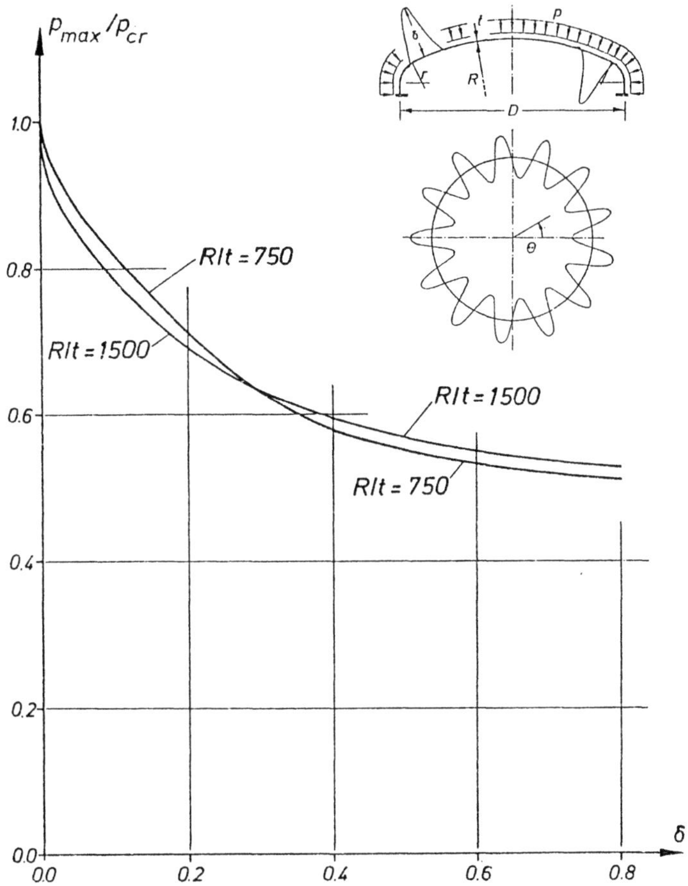

Fig. 13. Imperfection sensitivity in the elastic range

tion factor of about 0.5 for amplitudes which are roughly equal to the thickness.

Remarkably, both curves in figure 13 nearly coincide, although for R/t = 750 the circumferential wave number is 15, whereas for R/t = 1500 it equals 21 (see figure 10). Nevertheless, thinner shells will have a smaller load-carrying capacity not only because their bifurcation pressures are lower, but also because in practical situations their δ-values may be presumed to be larger. For comparison it is noted that the imperfection sensitivity of a 'Klöpper' end is less severe than that of a circular cylinder under axial compression (see e.g. [4]) or of a spherical cap [19], both of which are notorious in this respect. Yet, it is much stronger than for cylinders under external pressure [16].

The influence of plasticity on the imperfection-sensitivity of 'Klöpper' ends is shown in figure 14 for $R/t = 750$ and $\sigma_0 = 0.001723E$. There the maximum pressure p_{max} is normalized by the respective elastic-ideally plastic bifurcation value p_{cr} (see figure 11). With this choice of parameters the ratio of the initial yield-pressure to the bifurcation pressure is $p_{o,i}/p_{cr} = 0.81$.

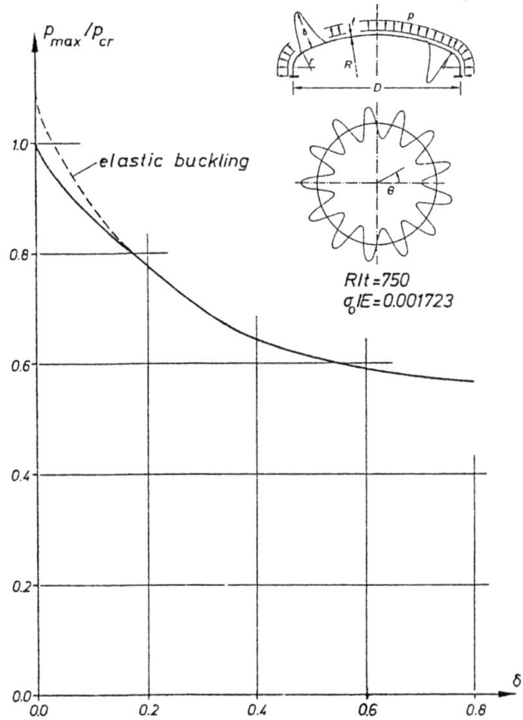

Fig. 14. Imperfection-sensitivity in the elastic-plastic range

The influence of plasticity is noticeable only for values of δ less than 0.17, that is for nearly perfect shells. Consequently, if imperfection amplitudes are small the shell is stressed into the plastic range before collapse, and its maximum pressure is below the corresponding elastic value. For δ becoming larger, however, the elastic and plastic curves again merge. As a result, for $\delta > 0.17$ the destabilizing geometric nonlinearities lead to collapse before the yield stress has been reached anywhere in the shell, even though the perfect shell bifurcates plastically. In this case then, as in others [1,4,20,21], for realistic imperfection levels the elastic results hold. This should not however, be generalized without caution because in any given case the range of validity of the elastic results depends strongly on the particular geometric and material parameters. Finally it is mentioned that the above curve is based on J_2-flow theory. Due to the small differences in bifurcation pressures (see figure 11) no comparison with J_2-deformation theory has been made.

5.2 'Korbbogen' end closure

Since both the 'Klöpper' and 'Korbbogen' end-closures have torispherical shapes their load-carrying behaviors are essentially alike, except for differences in degree. Thus, the displacement and stress distributions, the location of the plastic zone, the load-deflection curve and the shape of the bifurcation mode, as well as the influence of the material parameters given in the previous figures for the 'Klöpper' head can be considered to be typical of a 'Korbbogen' as well. A repeated discussion of these aspects is therefore not necessary. Instead, attention will be focused on the differences which exist with respect to the magnitudes of the bifurcation pressures and the imperfection sensitivities. For simplicity, and because the elastic results hold for realistic imperfection levels (see figure 14), consideration is here restricted to the elastic case.

As noted above, for comparable sets of parameters the critical internal pressure of a 'Korbbogen' may be significantly higher than that of a 'Klöpper' head. Under external pressure this relative increase is less pronounced, but may still amount to approximately 15 precent. If the conjecture that, loosely speaking, optimality (expressed in terms of higher buckling loads) generates increased imperfection-sensitivity, then the respective curves for the 'Korbbogen' would be expected to lie below those of the 'Klöpper' head. This is indeed the case as can be seen from figure 15 where the results for the two end closures are plotted. While in each case there is no significant difference between the results for each of the two values of R/t, the 'Korbbogen' is clearly more imperfection-sensitive. Thus, for the same value of δ the maximum pressure of the 'Korbbogen' is reduced by a larger factor than for a 'Klöpper' head. The difference increases for larger δ values and amounts to 15 % at $\delta \simeq 0.7$. This implies that at

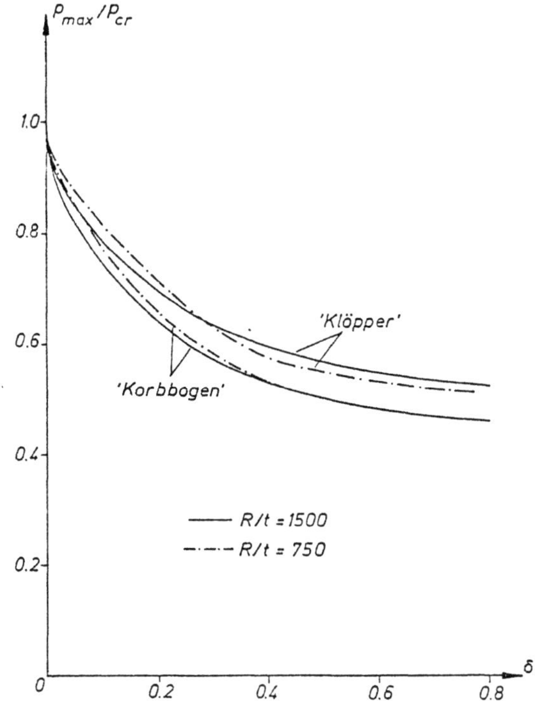

Fig. 15. Imperfection-sensitivity in the elastic range

imperfections of this, relatively small, magnitude most of the additional load-carrying capacity of the perfect 'Korbbogen' has been eroded. This point will again be raised at the end of the next chapter.

5.3 2:1 Ellipsoid

The load-carrying and buckling behavior of the 2:1 ellipsoid differs in several important respects from that of the two torispheres discussed above, even though at first sight its shape differs very little from that of the 'Korbbogen' (see figure 3). One major difference is that, to a very good approximation, its prebuckling state up to the buckling pressure is one of linear membrane compression over most of the shell surface. This was assumed in the asymptotic study of the buckling behavior of complete spheroidal shells [33], and was corroborated by the present numerical results. The bending moments are very localized in the high curvature regions. Contrary to the torispherical shells, however, their influence is negligible because the bifurcation mode of the ellipsoid is nearly always axisymmetric. In addition, the maximum displacements occur at the pole, and their values at the equator are zero. A typical plot is given in figure 16. As noted in [33]

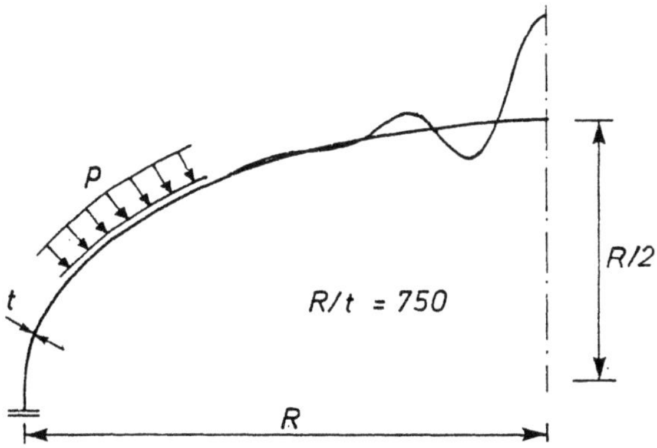

Fig. 16. Axisymmetric buckling mode

nonaxisymmetric modes with low values of the circumferential wave number are possible for very thin oblate spheroids, but their meridional distribution differs little from that of figure 16, and the respective buckling pressures almost coincide with the axisymmetric ones.

In the numerical calculations only axisymmetric eigenmodes similar to that in figure 16 were found. Using imperfections in this shape the results plotted in figure 17 were obtained. It shows the imperfection-sensitivity of the 2:1 ellipsoid relative to that of the 'Klöpper' head and the 'Korbbogen', as well as asymptotic results (shown dashed) taken from [33]. It should be noted that the asymptotic analysis in [33] revealed that the initial postbuckling behavior of oblate spheroids is nonsymmetric and thus depends on the sign of the buckling mode. In the present case this means that an inward dimple at the pole results in a decrease of the buckling load, whereas the same need not be true for an identically shaped outward dimple. This point was not investigated in detail but all numerical results shown in figure 17 were obtained by taking imperfections with the opposite sign of the eigenmode as shown in figure 16.

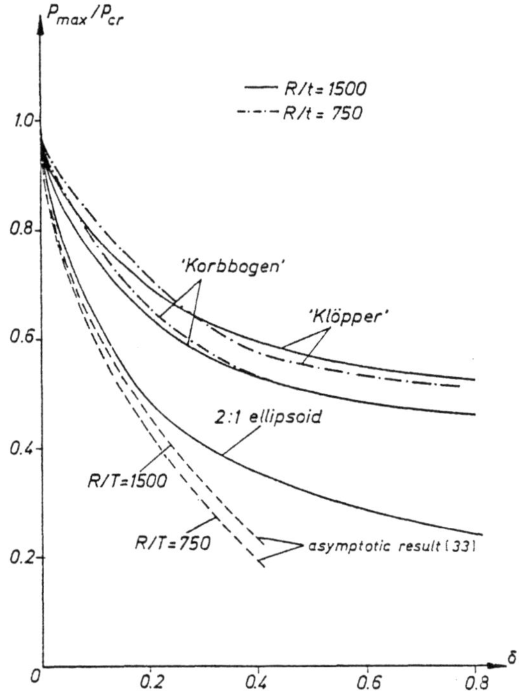

Fig. 17. Imperfection-sensitivity in the elastic range

The results for the 2:1 ellipsoid in figure 17 are identical for both R/t = 750 and 1500 (R is now the radius of curvature at the pole). They support the previously made observations concerning the imperfection-sensitivity of optimally shaped configurations. On the one hand, the bifurcation pressure of the ellipsoid is larger by about 80 percent than that of the 'Korbbogen' and by about 100 percent than that of the 'Klöpper' head. On the other, its imperfection-sensitivity is significantly stronger than that of either of the two torispheres. Thus, identical, especially larger, values of δ again lead to larger reduction factors. It is also seen that the dashed asymptotic curves overestimate the effect of imperfections.

In comparing the various curves in figures 15 and 17 it should be remembered that they are based on quite different imperfection shapes. Nevertheless, even though more, or less, detrimental imperfection shapes are conceivable, the results in figure 17 give useful estimates of their relative influence and may be used to evaluate upper bounds of limit loads.

Finally it is noted that the values of p_{cr} normalizing the maximum pressures differ considerably as mentioned above. To permit a more direct comparison of the dependence of the actual load-carrying capacities on the imperfection levels the results for R/t = 750 of figure 17 have been replotted in figure 18 using the critical pressure p_{cr}^{s} of the perfect sphere (see (17)) as a reference value. The resulting curves show that the various end

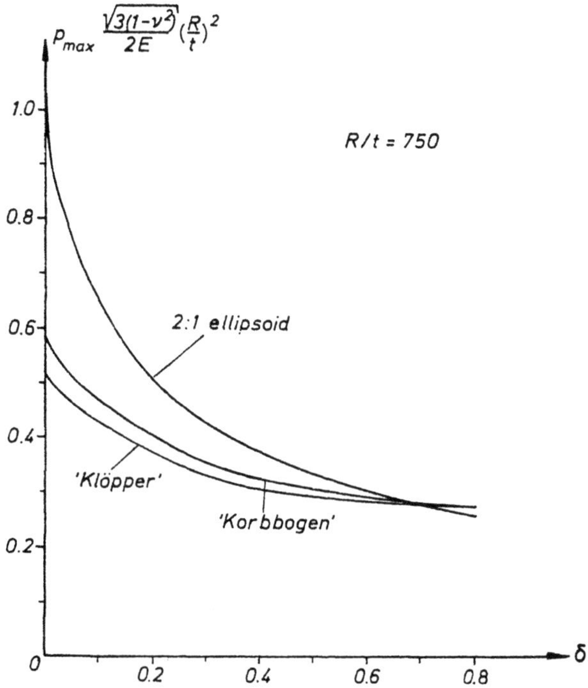

Fig. 18. Imperfection-sensitivity and maximum support pressures

closures are able to support pressures of very different magnitudes (one even exceeds that of a complete sphere) as long as they are perfect, or nearly so. For increasing imperfection levels, however, these differences diminish and they disappear at amplitudes of about 70 percent of the thickness of the shells. Thus, no real advantage can be taken of the large relative load-carrying capacitiy of the optimally shaped 2:1 ellipsoid if the imperfections are not guaranteed to be very small.

References

1. Hutchinson, J.W.: On the postbuckling behavior of imperfection-sensitive structures in the plastic range. J. Appl. Mech. (1972) 155-162.

2. Bushnell, D.: Plastic Buckling. LMSC Report-D673763, Lockheed Palo Alto Research Laboratory, April (1979).

3. Bushnell, D.: Buckling of shells - Pitfall for designers. AIAA J. 19 (1981) 1183-1226.

4. Gellin, S.: Effect of an axisymmetric imperfection on the plastic buckling of an axially compressed cylindrical shell. J.Appl. Mech. 46 (1979) 125-131.

5. Singer, J.: Buckling experiments on shells- A review of recent developments. TAE Report No. 403, Dept. Aeronautical Engineering, Technion, Haifa, April (1980).

6. Arbocz, J.: Past, present and future of shell stability analysis. Report LR-320, Dept. Aerospace Engineering, Delft University of Technology, May (1981).

7. Wunderlich, W.: On the analysis of shells of revolution by transfer matrices (in German). Ing. Archiv 36 (1967) 262-279.

8. Wunderlich, W.: Shell analysis of cooling towers (in German). KIB-Berichte, vol. 1, W. Zerna (ed.), Vulkan-Verlag, Essen (1968).

9. Stricklin, J.A. et al.: Nonlinear analysis of shells of revolution by the matrix displacement method. AIAA J. 6 (1968) 2303-2312.

10. Ball, R.E.: A computer program for the geometrically nonlinear static and dynamic analysis of arbitrarily loaded shells of revolution. NASA CR-1987 (1972).

11. Svalbonas, V.: Numerical analysis of stiffened shells of revolution - Vol. I: Theory manual. NASA CR-2273 (1973).

12. Esslinger, M. and Geier, B.: Large deflection stress analysis of shells of revolution under axisymmetric and non-axisymmetric loading (in German). Stahlbau 50 (1981) 263-270.

13. Tao, K. and Takezono, S.: Elasto-viscoplastic analysis of axisymmetrical shells under asymmetrical loading. Proc. 5th SMIRT Conference, Berlin (1979), Paper M4/8.

14. Klein, S.: The nonlinear dynamic analysis of shells of revolution with asymmetric properties by the finite element method. J. Pressure Vessel Technology (1975) 163-171.

15. Wunderlich, W.: On the nonlinear analysis of shells of revolution (in German). Proc. Int. Cong. on "Applications of Mathematics in Engineering", Weimar, July (1981).

16. Rensch, H.J. and Wunderlich, W.: A semi-analytical finite element procedure for non-linear elasto-plastic analysis of arbitrarily loaded shells of revolution. Proc. 6th SMIRT Conference, Paris (1981), Paper M4/1.

17. Wunderlich, W.: Mixed models for plates and shells. Proc. Int. Symp. on "Hybrid and Mixed Finite Element Methods", Atlanta, April (1981). S.N. Atluri, R.H. Gallagher, O.C. Zienkiewicz (eds.). J. Wiley, N.Y. (1982).

18. Rensch, H.J.: Elastic-plastic buckling and imperfection sensitivity of torispherical shells (in German). Dissertation. To be published in Techn. Wiss. Mitt., Inst. für Konstruktiven Ingenieurbau, Ruhr-Universität Bochum (1982).

19. Rensch, H.J.: Elastic-plastic buckling behavior of spherical caps with nonsymmetric initial imperfections. ZAMM 62 (1982) T 7032.

20. Hutchinson, J.W.: Plastic buckling. In: "Advances in Applied Mechanics - Vol. 14". C.S. Yih (ed.). Academic Press, N.Y. (1974) 67-144.

21. Needleman, A. and Tvergaard, V.: Aspects of plastic postbuckling behavior. In: "Mechanics of Solids - The R. Hill 60th Anniversary Volume". H.G. Hopkins and M.J. Sewell (eds.) Pergamon Press, Oxford (1982) 453-498.

22. Bruhns, O.: Bifurcation loads of inelastic shells (in German). ZAMM 57 (1977) 165-174.

23. Zander, G.: On the determination of bifurcation loads of thin-walled circular cylinders under combined axial and torsional loads (in German). Dissertation. Institut für Mechanik, Ruhr-Universität Bochum (1981).

24. Irons, B. and Tuck, R.: A version of the Aitken accelerator for computer iteration. Int. J. Num. Meth. Eng. 1 (1969) 275-277.

25. Hill, R.: Bifurcation and uniqueness in nonlinear mechanics of continua. In: "Problems of Continuum Mechanics". SIAM, Philadelphia (1961) 155-164.

26. Brendel, B. and Ramm, E.: Nonlinear stability investigations by the finite element method (in German). Ingenieur Archiv 51 (1982) 337-362.

27. Baltus, R. and Massonet, Ch.: Use of computer programs BOSOR 4 and 5 in the stability analysis of two civil engineering structures. Proc. 2nd Int. Coll. on "Stability of Steel Structures", Liège (1977) 609-618.

28. Vandepitte, D.: Model investigation of the collapse of a steel water tower. Ibid. 599-607.

29. Galletly, G.D.: On elastic and elastic-plastic asymmetric buckling of pressurized combinations of thin shells. Stahlbau (1979) 340-346.

30. Galletly, G.D.: Plastic buckling of torispherical and ellipsoidal shells subjected to internal pressure. Proc. Instn. Mech. Engrs. 195 (1981) 329-345.

31. Galletly, G.D.: Elastic-plastic buckling of dished ends subjected to internal pressure. Proc. 6th SMIRT Conference, Paris (1981) Paper G4/6.

32. Wunderlich, W.: Discussion of [30]. Proc. Instn. Mech. Engrs. 195 (1982) S39-S40.

33. Danielson, D.A.: Buckling and initial postbuckling behavior of spheroidal shells under pressure. AIAA J. 7 (1969) 936-944.

34. Thompson, J.M.T.: Optimization as a generator of structural instability. Int. J. Mech. Sci. 14 (1972) 627, and 15 (1973) 855.

35. Batdorf, S.B.: A simplified method of elastic stability analysis for thin cylindrical shells. NACA Rep. 874 (1947).

Postbuckling Instability Analysis of Shells Using the Mixed Method

B.-H. KRÖPLIN

Institut für Statik
TU Braunschweig
W. Germany

Summary

The paper deals with a method to calculate stable pre- and postbuckling behaviour of thin shells. First some features of the used mixed shell element, with displacements and stress resultants as nodal unknowns, are outlined and discussed. Second a quasi viscous strategy is presented, which allows to pass unstable regions and limit points. The approach uses variable ficticious damping. With the load applied all at once, stable equilibrium states are approached without step by step incrementation. Examples are given for the buckling of cylindrical shells.

1. Introduction

From experimental investigations of shell buckling, static unstable regions are a well known phenomenon. The structure passes through a sequence of rapid changing buckling modes, while the load drops often considerably. The sequence of buckling modes dependsoften on very small structural or dynamic perturbations, whereas the stable states in the postbuckling range are less history dependent. For thin cylinders the behaviour has been studied extensively in /1/. Over all the behaviour is influenced by imperfections in geometry, boundary conditions and material properties.
The frequent limit and bifurcation points in the region of rapid changing buckling modes make complete path tracing by numerical methods difficult. Roughly speaking, the problems consist of finding a limit or bifurcation point by accompanying measures, determining the lowest eigenvalue and the eigenmode in the critical point and superimposing the eigenmode and the displacement field such, that the iteration to the next equilibrium point is numerically stable /2/.This strategy does not

Buckling of Shells, Ed.: E. Ramm
Proceedings of a State-of-the-Art Colloqium

account for dynamic effects.
With respect to the efficiency, it is sufficient for many applications in engineering, to investigate only a few points of the post-buckling range. However this is normally beyond the numerical stability limits of the common incremental approach. Thus, as a first step, in this paper a strategy is lined out, which leads to a numerical stable iteration process for arbitrary large load steps if a stable state of equilibrium can be found. The method interprets the buckling as a quasi viscous process and uses variable, ficticious damping in order to stabilize the iteration. It can be used, first, to jump in one step to a stable equilibrium state in the post buckling range and investigate the post-buckling region by large unloading steps, and second as creep strategy with approximately tracing of the solution path. It has been applied with success for elastic and elastoplastic beam and plate problems / 3 / . Here emphasis is laid on the numerical stability properties in the situation of shell buckling.
The quasi viscous strategy is applied with a curved thin shell element in mixed formulation, which was derived in / 4 / .
Thus, first some specialities of the mixed formulation are discussed with respect to the nonlinear application.

2. Remarks on the used mixed shell element

2.1 General

The sufficient accurate description of shell behaviour with only displacements and rotations being the nodal variables, is facing problems: On the one hand, low order curved elements don't include the rigid body motions completely / 5 / and don't satisfy the compatibility along the element edges, while on the other hand high order elements / 6 / require nodal unknowns of higher order e.g. curvatures, with complicated element matrices. As alternative the mixed formulations express the element energy with displacements and stresses or stress resultants as independent (nodal) variables. The approach has certain advantages, as simplicity of the nonlinear element matrices, low order interpolation functions for the unknowns, compatibility in plate bending without extra measures and no separate stress calculation with loss of accuracy. Main drawbacks are the large number of nodal unknowns, the semidefiniteness of the element matrices and the continuity condition of the stresses over the element assemblage. The latter restricts the method to structures without stress discontinuities, as long as no extra measures are employed, and narrows the application range for practical purpose.

Equation 1 gives the element energy, where π_i represents

$$\pi = \pi_i + \pi_e \quad , \quad \pi_i = \frac{1}{2}\int_A \underline{s}^t \, \underline{\varepsilon} \, dA \tag{1}$$

the internal energy of an element, while π_e means external energy of the load terms. $\underline{s}$ are the stress components

$$\underline{s}^t = (n^{\alpha\beta} , m^{\alpha\beta}) \quad , \quad \alpha,\beta = 1,2 \tag{2}$$

and $\underline{\varepsilon}$ the strains and curvatures with respect of the middle surface of the shell,

$$\underline{\varepsilon}^t = (\alpha_{\alpha\beta} , \beta_{\alpha\beta}) \; . \tag{3}$$

Introducing the constitutive relationship

$$\underline{s} = \underline{E}\,\underline{\varepsilon} \; , \; \underline{\varepsilon} = \underline{F}\,\underline{s} \; , \; \underline{E}^{-1} = \underline{F} \tag{4}$$

equation 1 yields the element energy dependent only on $\underline{\varepsilon}$

$$\pi_i = \int_A \frac{1}{2} \underline{\varepsilon}^t \, \underline{E}\,\underline{\varepsilon} \; dA \tag{5}$$

or the complementary form depending on strains and stress resultants,

$$\pi_i = \int_A \underline{s}^t \underline{\varepsilon} - \frac{1}{2} \underline{s}^t \, \underline{F}\,\underline{s} \; dA \; . \tag{6}$$

It should be noted, that the strain energy of (5) is of quadratic order in $\underline{\varepsilon}$, while the mixed form (6) is of linear order in $\underline{\varepsilon}$. Hence, using Greens strain with terms up to second order,

$$\underline{\varepsilon} = \underline{\varepsilon}(u_i, u_i^2), \quad i = 1,2,3 \tag{7}$$

(6) leads to a energy of second order in u_i while (5) leads to an expression of 4. order. Thus the derivation of nonlinear element in matrices for the nonlinear case is simpler with the mixed principle.

2.2 Secant and Tangent element matrices

Assuming shear rigidity for thin member bending problems, (3) may be expressed in terms of the three displacements u_i and two rotations φ_α of the middle surface of the shell. According to the accuracy requirements of the actual problem, different shell theories may be employed for the linear and the nonlinear strains and the stress resultants. Here the Green strain (8) is used in the middle surface

$$\alpha_{\alpha\beta} = \alpha^{L}_{\alpha\beta}(u_i) + \alpha^{N}_{\alpha\beta}(u_i^2) \tag{8}$$

The linear and the nonlinear strain expression read as follows

$$\alpha^{L}_{\alpha\beta} = \frac{1}{2}(u_{\alpha|\beta} + u_{\beta|\alpha} - 2b_{\alpha\beta}u_3) \tag{9}$$

and

$$\alpha^{NL}_{\alpha\beta} = \frac{1}{2}\Big[(u^{\varrho}|_{\alpha} - b^{\varrho}_{\alpha}u_3)(u_{\varrho}|_{\beta} - b_{\varrho\beta}u_3) + (b^{\varrho}_{\alpha}u_{\varrho} + u_{3,\alpha})(b^{\lambda}_{\beta}u_{\lambda} + u_{3,\beta})\Big]. \tag{10}$$

The curvature depends only linear on the rotations and the displacements, thus

$$\beta_{\alpha\beta} = \beta^{L}_{\alpha\beta} = \frac{1}{2}(\varphi_{\alpha}|_{\beta} + \varphi_{\beta}|_{\alpha} - b^{\varrho}_{\beta}u_{\varrho}|_{\alpha} + 2b^{\varrho}_{\alpha}b_{\varrho\beta}u_3). \tag{11}$$

The above theory covers moderate large deflections.

In equation 12 matrix notation is adopted in order to reduce the complexity of the strain-displacement relationship,

$$\underline{\varepsilon} = (\underline{D}^{L} + \frac{1}{2}\underline{U}_A^{t}\underline{D}^{N})\underline{u}. \tag{12}$$

$\underline{D}^L$ means the linear and $\underline{D}^N$ the nonlinear operator matrix /4/. $\underline{U}_A^T$ is a matrix containing displacements.

Introducing (12) in (6) the secant matrix C^S is obtained by the first variation only, see fig. 1

$$\delta\pi = \int_A \delta\underline{z}^t\, \underline{C}^s\, \underline{z}\; dA - \int_A \delta\underline{z}^t\, \bar{\underline{p}}\; dA = 0\;, \tag{13}$$

where

$$\underline{z}^t = (\,\underline{u}^t\,,\,\underline{s}^t\,)\,. \tag{14}$$

$\underline{C}^S$ consists of a linear and a nonlinear part $\underline{C}^{SL}$ and $\underline{C}^{SN}$ and reads

$$\underline{C}^S = \underbrace{\left[\begin{array}{c|c} 0 & (\underline{D}^L)^t \\ \hline \underline{D}^L & \underline{F} \end{array}\right]}_{\underline{C}^{SL}} + \underbrace{\left[\begin{array}{c|c} 0 & (\underline{D}^N)^t\,\underline{U}_A \\ \hline \frac{1}{2}\underline{U}_A^t\,\underline{D}^N & 0 \end{array}\right]}_{\underline{C}^{SN}} \tag{15}$$

The nonlinear part $\underline{C}^{SN}$ is unsymmetric, but can be made symmetric as shown in /7/ by the identity

$$(\underline{D}^N)^t\,\underline{U}_A\,\underline{s} = \frac{1}{2}(\underline{D}^N)^t\,\underline{S}_A\,\underline{u} + \frac{1}{2}(\underline{D}^N)^t\,\underline{U}_A\,\underline{s}\;, \tag{16}$$

which holds for the matrices $\underline{U}$ and $\underline{S}$. Hence $\underline{C}^{SN}$ can be written as symmetric matrix with

$$\underline{C}^{SN} = \frac{1}{2}\left[\begin{array}{c|c} (\underline{D}^N)^t\,\underline{S}_A & (\underline{D}^N)^t\,\underline{U}_A \\ \hline \underline{U}_A^t\,\underline{D}^N & 0 \end{array}\right] \tag{17}$$

In contrast to the displacement method equation 17 provides the option to use a secant stiffness in nonlinear calculations and to calculate the internal stress and strain reactions $\underline{p}^i$ directly by the relationship, see fig. 1

$$\underline{p}^i = \underline{C}_A^S\,\underline{z}_A\,. \tag{18}$$

Following the same procedure as described for the secant element matrix, the tangent element matrix can be obtained from the

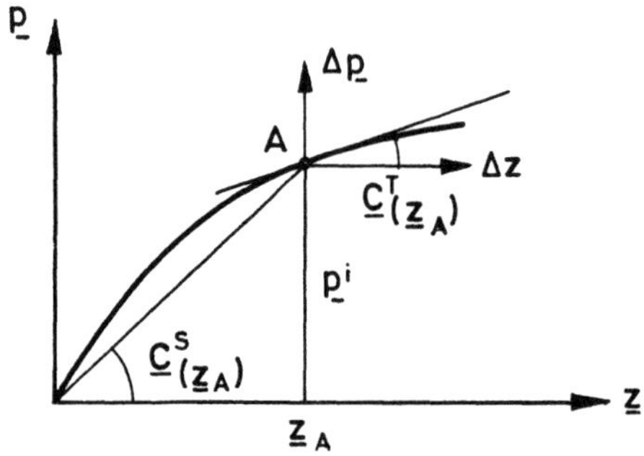

Fig. 1 : Secant and tangent matrix

second variation of π , as

$$\delta^2 \pi = \int_A \delta \Delta \underline{z}^t \, \underline{C}^T \, \Delta \underline{z} \, dA - \int_A \delta \Delta \underline{z}^t \Delta \bar{\underline{p}} \, dA \qquad (19)$$

$\underline{C}^T$ consist of a linear and a nonlinear part, $\underline{C}^{TL}$ and $\underline{C}^{TN}$. The linear parts of the secant and the tangent matrix are identical, while the nonlinear part $\underline{C}^{TN}$ is twice the secant matrix $\underline{C}^{SN}$, given in equation 17,

$$\underline{C}^{TN} = \left[\begin{array}{c|c} (D^N)^t \, \underline{S}_A & (D^N)^t \, \underline{U}_A \\ \hline \underline{U}_A^t \, D^N & 0 \end{array} \right] \qquad (20)$$

For comparison, deriving the tangent matrix for the displacement approach would mean to insert (12) in (5) and perform (19). The result are four nonlinear matrices instead of one /7/ with considerable complexity. Further it should be noted, that in displacement analysis the secant matrix cannot be obtained by simple operations and the internal reactions are normally calculated by the virtual work principle.

2.3 Independent variables

Beside the numerical interpolation and discretisation the accuracy of the results depends on the proper choice of the integration variables /5/. Basicly all stress resultants, $n^{\alpha\beta}$, $m^{\alpha\beta}$, the displacements u_i and the rotations φ_α are possible variables in a shear rigid thin walled shell element. Using all variables (as a differential equation of first order) in a shell theorie, which satisfies the requirements of a first approximation /8/, results in oszillating solutions of slow convergence, although all convergence requirements are satisfied. The convergence properties can be improved by using a second order differential equations system for the bending part. Thus the rotations are eliminated /4/ and the remaining variables are $\underline{z}^t = (u_i , n^{\alpha\beta}, m^{\alpha\beta})$. The axial forces $n^{\alpha\beta}$ are not eliminated because they lead to better convergence with respect to the rigid body motions.The reduced principle contains only first order derivatives with respect to the curved shell coordinates.

2.4 Interpolation functions

The convergence of the reduced mixed functional requires only first order interpolation functions. Hence all variables are approximated with linear shape functions in rectangular four mode elements, see fig. 2.

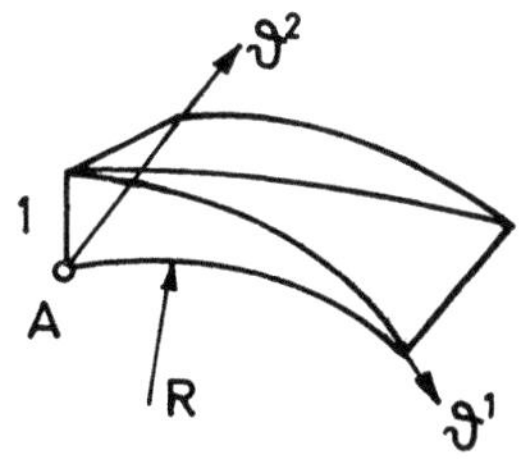

Fig. 2: Shape function of point A and cylindrical shell element

Recalling equation 6 the obtained set of algebraic equations may be written as

$$\delta \pi = \int_A \delta \underline{z}^t (\underline{C}^s \underline{z} - \underline{\bar{p}}) \, dA = 0 \tag{21}$$

with $\underline{z}^t = (u_i , n^{\alpha\beta} , m^{\alpha\beta})$ for every node.

Linear case

Equation 21 consists of three equilibrium equations u_i and six compatibility equations, with the linear part given in (22).

$$\delta \underline{u}: \quad \underline{D}^t \, \underline{s} = \underline{\bar{p}}$$

$$\delta \underline{s}: \underline{D} \, \underline{u} - \underline{F} \, \underline{s} = \underline{\bar{\varepsilon}} \, , \quad \underline{s}^t = (n^{\alpha\beta}, m^{\alpha\beta}). \tag{22}$$

Regarding the compatibility, the linear shape functions introduce constraints to the element. For example for a cylindrical arch the compatibility equation in ϑ^1 direction reads

$$u_{1,1} + \frac{1}{R} u_3 - F n^{11} = 0 \tag{23}$$

or $\quad n^{11} = E\,(u_{1,1} + \frac{1}{R} u_3)$.

which has to vanish in case of a rigid body motion $\underline{\varepsilon} = 0$ .Since $u_{1,1}$ is a constant function in the element and u_3 a linear one,both functions cancel only for a constant u_3 , which is in contradiction to the curved shell coordinates. Hence the best approximation is $u_{1,1} = - u_3^m$, see fig. 3 so that

$$n^{11} = E \frac{\xi a}{2R} \Delta u_3 \quad , \tag{24}$$

which leads to oszillating results of n^{11} or u_3 . An easy way to block the oszillations is, to prescribe the axial forces at the boundaries, which is permitted, or to use small elements among the regular mesh. Examples can be found in /5/.

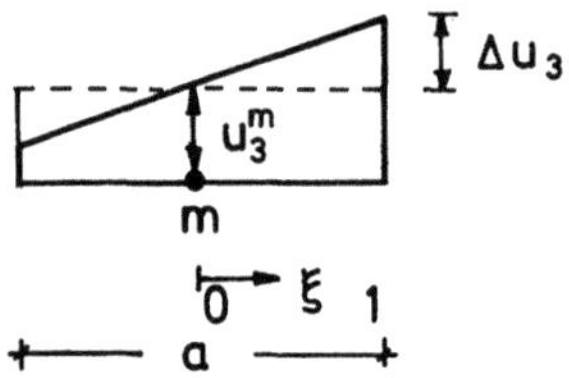

Fig. 3: Shape function for n^{11}

Independent on the oszillations the mixed approach approximates the mean value of the element. This is gained by the independent unknowns $n^{\alpha\beta}$, but payed with three nodal degrees of freedom. Without $n^{\alpha\beta}$, in displacement formulation polynominals of higher order would be necessary to model the rigid motions in curved elements.

Nonlinear case

The Green strain of the middle surface (8) and (9) contains derivatives of first order for u_α and u_3. Since u_α and u_3 appear of linear and quadratic order, u_1, u_3^2, have to match.

In mixed analysis they do, since u_α and u_3 are both linear interpolated. In displacement analysis the interpolation order of u_α has to be adapted to the order of the u_3 functions. Otherwise considerably ficticious stresses occur in the element /9/.

2.5 Summing up

Concluding this section, the mixed approach provides a tool for the derivation of compatible elements with low order shape functions for linear and nonlinear shell analysis. The shape functions for the tangent and normal deflection are of the same order and cause no ficticious strain in the nonlinear case. The nonlinear element matrices are simple structured and fast to calculate. Due to the second order of the energy expression a secant matrix can be derived and directly used for the calculation of the internal reactions. The nonlinear part of the secant matrix is half of the nonlinear part of the tangent matrix.

3. Incremental and finite creep iteration

Nonlinear shell calculations are normally performed with small load steps $\Delta\underline{p}$, with the structural matrix being the tangent matrix $\underline{C}^T$ and the unknowns being the increments of the nodal variables, see fig. 4. The incrementation is necessary to trace the solution path of path dependent problems. In path independent, elastic problems the incrementation is used in order to achieve numerical stability. The tangent structural matrix gives a trial solution in the surrounding of the current state, which can be improved by iterations in every increment. The iterations are performed, until the unbalanced force $\Delta\underline{p}^U$ vanishes,

$$\Delta\underline{p}^u = \underline{p} - \underline{p}^i . \qquad (25)$$

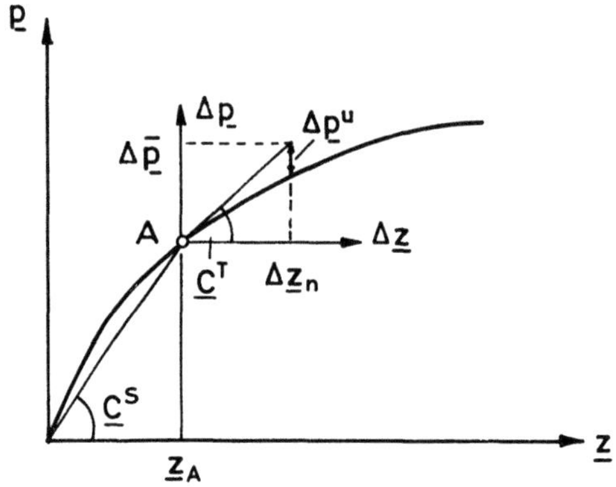

Fig. 4: Incremental relationship

Around points of static instability the tangent matrix gives a poor trial solution and the iteration in the incremental step diverges, when no other measures are taken, e.g. constant arc length method, or others /10,11/ . These methods constrain the iteration such, that a stable state of equilibrium is reached at the end of the increment.

In this paragraph a method is lined out, with which stable states of equilibrium are found with an unconditionally stable iteration. The iteration is performed until the components of the unbalanced forces $\underline{p}^{\dot{u}}$, see fig. 5 satisfy the convergence tolerance $\tilde{\varepsilon}$.

Ficticious nodal damping forces stabilize the structure during the iteration and vanish as a stable state of equilibrium is approached. The nonlinear calculation is interpreted as a creep process with the displacements depending on a virtual time step. The strategy is similar to dynamic relaxation /12/, with the difference, that here an implicit formulation and no dynamic terms are used. Instead of a load increment the full load can be applied at once. The iteration steps are governed by the effective matrix $\underline{K}$ which consists of the tangent structural matrix and a diagonal damping matrix. The iteration can be performed in finite or in incremental manner.

see fig. 5.

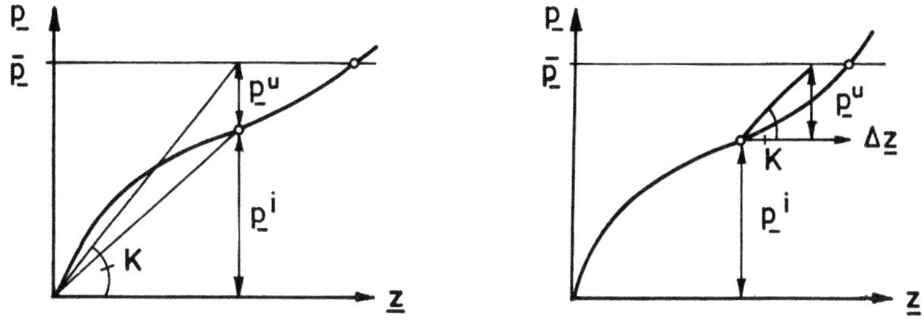

$$\underline{K}\,\underline{z} = \bar{\underline{p}} \qquad\qquad \underline{K}\,\Delta\underline{z} = \underline{p}^u$$

Fig. 5: Quasi viscous iteration in finite and incremental manner

The creep iteration is based on the viscous equation (26)

$$\underline{D}\,\dot{\underline{z}} + \underline{C}^s\underline{z} = \bar{\underline{p}}\,, \tag{26}$$

where $\underline{z}$ means the vector of nodal unknowns, $\underline{C}^s$ is the structural secant matrix, $\underline{D}$ means an artificial damping matrix and $\bar{\underline{p}}$ the applied load. A $(\dot{\ })$ describes the derivatives with respect to the time t . Equation 26 can be discretized in time domain with first order interpolation functions, see fig. 6.

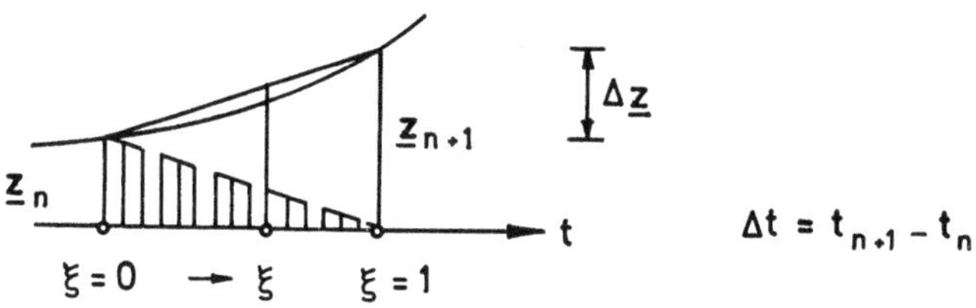

Fig. 6: Interpolation function in time domain

Using the interpolation functions (27) and their derivatives (28)

$$\underline{z}_{(t)} = (\ \xi\ ,\ 1-\xi\ )(\ \underline{z}_{n+1}\,,\ \underline{z}_n\ )^t \tag{27}$$

$$\underline{z}_{(t)} = \frac{1}{\Delta t}(\ 1\,,\ \ -1\ \ )(\ \underline{z}_{n+1}\,,\ \underline{z}_n\ )^t \tag{28}$$

for $\xi = 1$ the backward difference formular

$$\left(\frac{1}{\Delta t} \underline{D} + \underline{C}^S \right) \underline{z}_{n+1} = \bar{\underline{p}} + \frac{1}{\Delta t} \underline{D} \, \underline{z}_n \tag{29}$$

is obtained. $\bar{\underline{p}}$ was introduced as time independent load. Equation (29) is the base for the iteration in finite form.
The incremental equations are obtained by indroducing the incremental relation

$$\Delta \underline{z} = \underline{z}_{n+1} - \underline{z}_n \tag{30}$$

to (29). Thus the base for the incremental iteration reads

$$\left(\frac{1}{\Delta t} \underline{D} + \underline{C}^S \right) \Delta \underline{z} = \bar{\underline{p}} - \underline{C}^S \underline{z}_n \ . \tag{31}$$

For nonlinear structural analysis, the right hand side of (31) is interpreted as unbalanced force $\underline{p}^u$ recalling equation (25) and (18), where in mixed analysis $\underline{p}^i$ is calculated directly from the secant matrix.
The left hand side of (31) is regarded as effective matrix $\underline{K}$, which rules the size and the direction of $\underline{z}$ in the space of the nodal unknowns. Hence $\underline{K}$ can be used for manipulating the iteation path towards a stable equilibrium point, e.g. path tracing and may be chosen in a wide range without endangering the numerical stability. Since most structural program codes are based on the tangent matrix $\underline{C}^T$, in the examples the effective matrix $\underline{K}$ was chosen as

$$\underline{K} = \underline{D}^* + \underline{C}^T \quad , \quad \underline{D}^* = \frac{1}{\Delta t} \underline{D} \ . \tag{32}$$

Thus the difference to the incremental approach lies in the use of an effective matrix $\underline{K}$ instead of the tangent matrix $\underline{C}^T$ and the full unbalanced force $\underline{p}^u$ instead of the incremental one $\Delta \underline{p}^u$.

3.1 Numerical stability

The incremental iteration

$$\Delta \underline{z}_{n+1} = \left(\underline{D}^* + \underline{C}_n^T \right)^{-1} \underline{p}_n^u \tag{33}$$

is numerically stable and convergent for

$$\Delta \underline{z}_{n+1} = \lambda \Delta \underline{z}_n \ , \ |\lambda| < 1. \qquad (34)$$

Equation(34)holds for every component i of $\mathbf{z}$. Introducing (34) in (33) yields for the ith component

$$(D^*_{ij} + C^T_{ij})\lambda_j \Delta z_{j,n} = p^u_{i,n} \ , \qquad (35)$$

where

$$D^*_{ij} \lambda_j \ \Delta z_{j,n} = \Delta p^d_{i,n} \qquad (36)$$

represents the incremental damping force, while

$$C^T_{ij} \ \lambda_j \ \Delta z_{j,n} = \Delta p^i_{i,n} \qquad (37)$$

gives the incremental internal reaction of the ith component. Assuming $\underline{D}^*$ as lumped diagonal matrix, $\underline{D}^*$ can be written from equation (35) as

$$D^*_{ii} = \frac{p^u_{i,n} - p^i_{i,n}}{\lambda_j \Delta z_{i,n}} \qquad (38)$$

Any D^*_{ii} greater than D^*_{ii} of equation (38) leads to a stable iteration with (33). Since $p^i_{i,n}$ is greater zero in the surrounding of a stable state of equilibrium, the damping matrix can be written as

$$D^*_{ii} = \text{abs}\left(\frac{p^u_{i,n}}{\Delta z_{i,n}}\right) \quad \text{with} \quad \Delta z_{i,n} > \tilde{\varepsilon} \ . \qquad (39)$$

In (39) only positive values of the damping are admitted. $\Delta z_{i,n} > \varepsilon$ avoids locking for small $\Delta z_{i,n}$. It allows the damping to vanish near an equilibrium state.

4. Examples

The examples are carried out in order to show the performance of the quasi viscous iteration around static unstable regions. They are based on the described mixed functional with the displacements u_i, the axial forces $n^{\alpha\beta}$ and the bending moments $m^{\alpha\beta}$ as nodal unknowns. All unknowns are interpolated with linear shape functions. The Flügge shell theory is used. The quasi viscous strategy in its incremental form (31) is applied. Only the deflection u_3 is damped.

The first example deals with a shallow cylindrical shell panel. It shows the influence of small changes in geometry and boundary conditions and combines investigations with small and large steps.

The second example is a cylinder under axial compression. Here the buckling behaviour is calculated with large steps in the pre- and postbuckling range and the stability and uniqueness of the solution is commented.

In the third example the quasi viscous strategies are applied to a cylindrical shell panel under external pressure.

The examples have been carried out with a coarse finite element mesh in order to investigate the numerical performance with little numerical effort. In spite of qualitatively and quantitatively reasonable results the effect of mesh refinement needs to be further investigated.

4.1 Cylindrical panel

The shallow cylindrical panel of fig. 7 is loaded by the end shortening $\bar{u}$. The edges are simply supported, free in longitudinal and fixed in transverse direction. Half of the panel is discretised with 36 elements.

As a plate strip with imperfections of sine-shape the buckling behaviour has been studied in /3/. There the plate buckles first affin to the imperfection. Then the middle bends down and a snap to the third buckling mode occurs. On the unloading path the structure switches over to the second buckling mode and returns later to the first mode, see fig. 8.

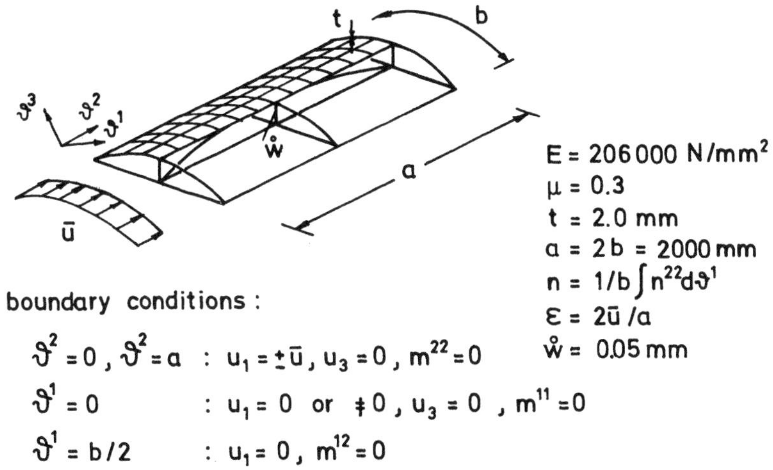

Fig. 7: Analyzed shell panel

Here, the influence of initial cylindrical curvatures and boundary conditions on the buckling behaviour is compared to the

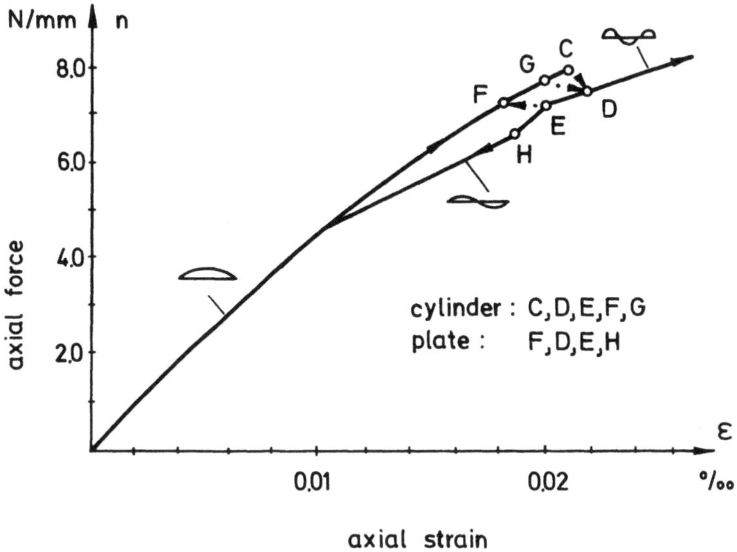

Fig. 8: Buckling behaviour of cylinder ($r = 25 \cdot 10^5$ mm) and plate with sine-imperfection ($\mathring{w} = 0.05$ mm)

plate. First, a perfect cylindrical shell with the curvature adapted to the maximum imperfection in the middle of the plate strip of /3/, $\mathring{w}$ = 0.05 mm $\hat{=}$ r = 25 10^5 mm, is loaded with a large end-shortening (ε = 0.021 ‰) in one step. The creep iteration converges to point D, see fig. 8.

A third mode affin buckling mode is obtained. The axial force coincides with the axial force of the plate strip. However further investigation of the region before point D by unloading and loading leads to another snap through behaviour. In unloading the second mode affin shape of the plate is not found, the cylinder snaps back to the prebuckling mode, points E, F

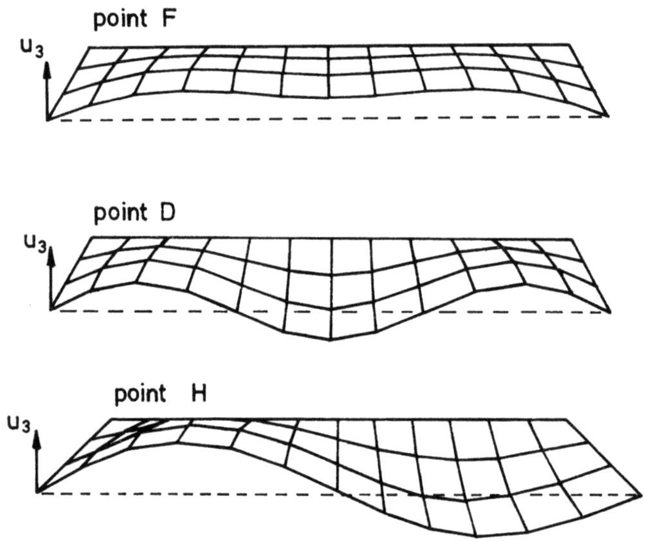

Fig. 9 : Buckling shapes around the snap region

see fig. 8 and 9. For loading from point F the cylindrical shell displays a higher stable peak load, points G, C than the imperfect plate. The investigation of the snap region can be started from every point of the post-buckling curve. The

first point can be calculated with the load applied in one step. Fig. 9 shows some of the calculated buckling shapes.
For investigation of the influence of another curvature (r = 10600 mm) and other boundary conditions (transverse displacements along the long edges free and fixed), one creep iteration is performed to the deep postbuckling range (ε = 0.1 ‰). The cylindrical panels are analyzed with and without a local geometric imperfection in the center, see fig. 11. The results are given in fig. 10. With fixed edges (point A), no postbuckling

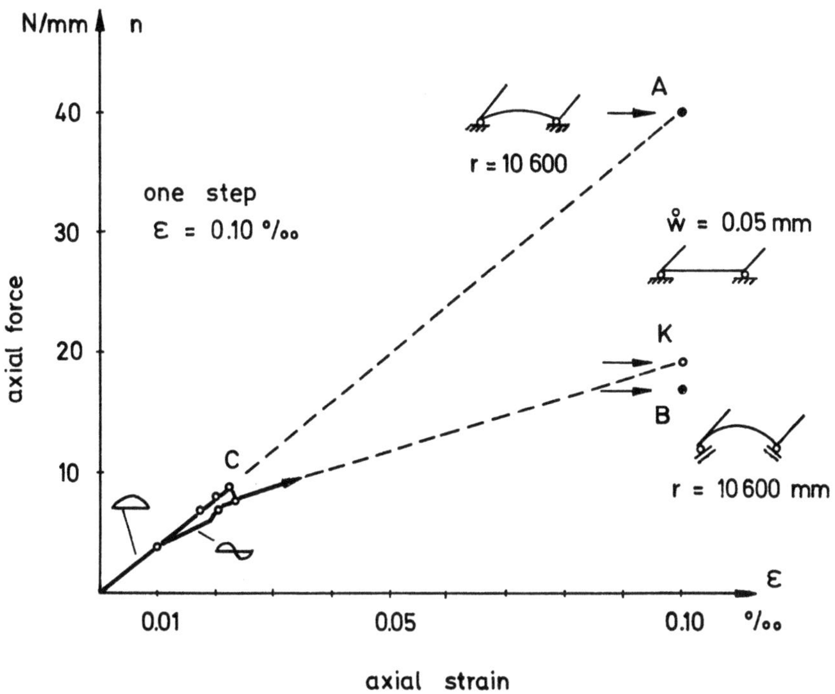

Fig. 10: Comparison of plate and shell buckling behaviour

occurs with and without the local imperfection. With free edges in plane and without imperfection point A is reached too, while with the local imperfection point B is obtained. The buckling shapes differ considerably, see fig. 11. For comparison, the third mode was found to be stable for the plate with sine-imperfection (point K). The axial force is little higher than that one of the cylinder with free edges.

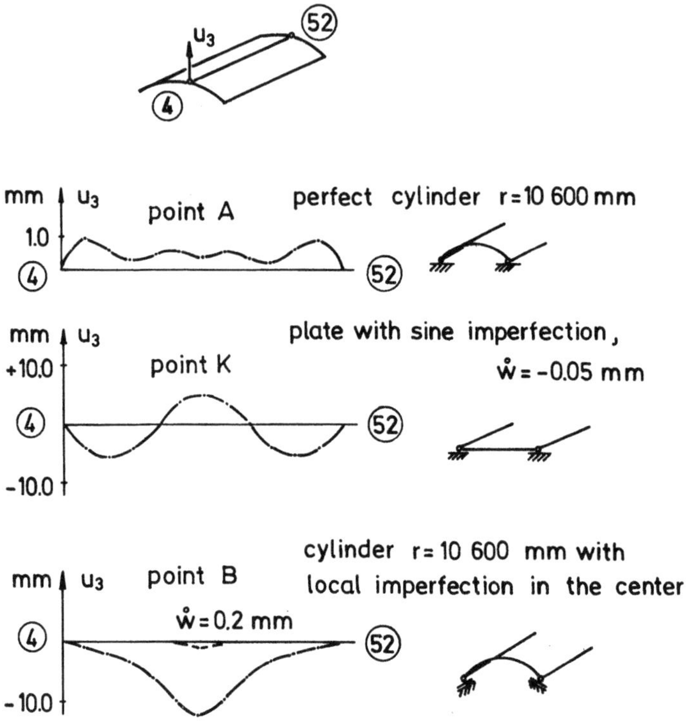

Fig. 11: Buckling patterns ε = 0.1 ‰

4.2 Cylindrical shell under axial compression

The cylindrical shell section of fig. 12 with r/t = 100 is discretised with 16 elements and loaded by an uniform end shortening. The boundary conditions and the size of the section are chosen such, that the experimental buckling mode in the deep postbuckling range can be represented. The experiments /1/ show a rhombic shape with large buckles in the middle and small buckles near the edges, see fig. 14.

First, a large end-shortening is applied (ε = 40 ‰) together with a local imperfection. The iteration leads to the stable equilibrium state point A in fig. 13. The obtained buckling shape is in good agreement to the rhombic experimental result with the large and the small buckles, see fig. 14. Clearly the

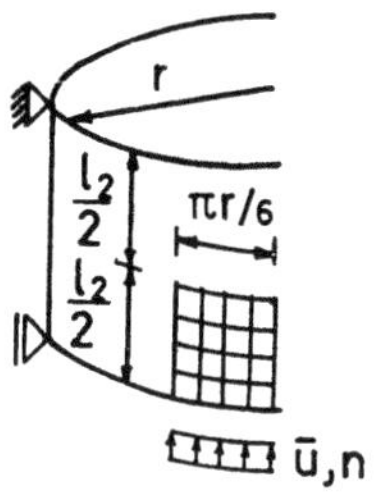

$r = 100$ mm
$l_2 = 100$ mm
$t = 1.0$ mm
$E = 2.06 \cdot 10^5$ N/mm^2
$\mu = 0.3$
$n_{cr} = 1246$ N/mm
$n = \frac{1}{l_1} \cdot \int n^{22} d\vartheta^1$

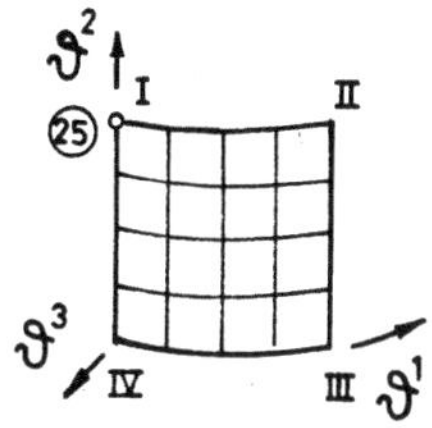

boundary conditions :

$\vartheta^1 = 0, \vartheta^1 = \frac{\pi r}{6}$: $u_1 = 0, m^{12} = 0$

$\vartheta^2 = 0$: $u_2 = \bar{u}, u_3 = 0, m^{22} = 0$

$\vartheta^2 = l/2$: $u_2 = 0, m^{12} = 0$

Fig. 12: Analyzed shell section

axial forces, see the plot in fig. 14, concentrate in the stiffer sections between the buckles. The same buckling pattern and about the same axial force is calculated with a sine-imperfection of ten times greater amplitude.

In order to control the path dependency of the solution points the cylinder is unloaded with small and with large steps. In all cases the same unloading path was found. In the deep post-buckling range the calculated pattern is independent on the imperfection, while the reentry point to the prebuckling range depends on the imperfection.
Since no measures are employed in the iteration, which ensure, that always the lowest possible equilibrium state is reached, the iteration may converge to any possible stable state. The number of existing possible states depends beside the physical behaviour on the discretisation and the used functional. In the calculation equilibrium states beside the unloading path were found when the post buckling range is entered with small

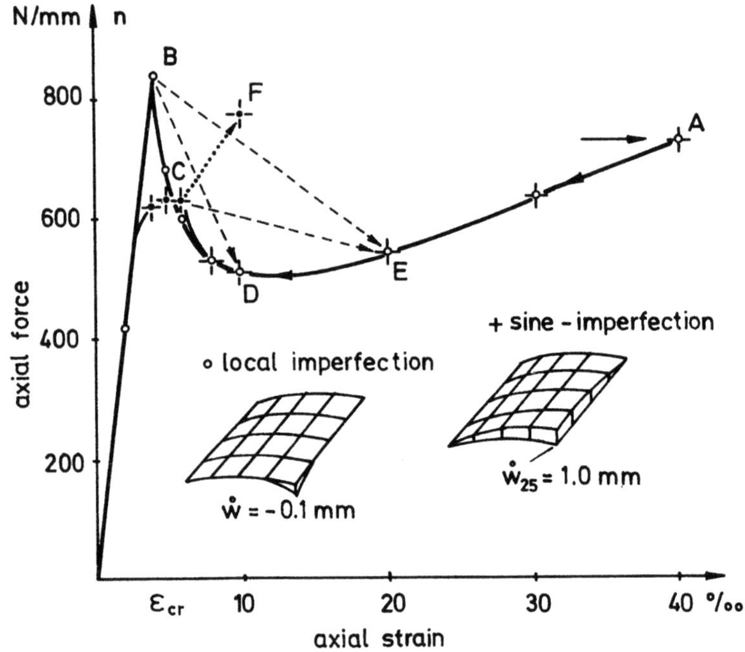

Fig. 13: Postbuckling of an axially compressed cylinder r/t = 100

load steps, e.g. point C to F in fig. 13. With a larger step size, e.g. point C to E, or point B to E, the low post buckling curve is approached, both, from the pattern related to the local imperfection and from the pattern related to the sine imperfection, see fig. 15.

The essential difference between the pattern of point C and point E is, that in E a small buckle near point III exists.This buckle does not appear, if a small step from C to F, fig. 14 is applied. The small buckle leads to the reduction of the load in the postbuckling range.

The larger the applied step, the more influences the damping the iteration path. It operates as perturbation with the size depending on the unbalanced force p^u. Hence sufficient large steps lead in the example to the low postbuckling curve. As another effect of the perturbation the pattern may turn in the iteration from the right to the left and vice versa. The 4 x 4 mesh is indicated in fig. 14 and 15 by the thick solid lines.

4.3 Cylinder under external forces

The cylinder section of fig. 17 is subject to conservative uniform external forces.

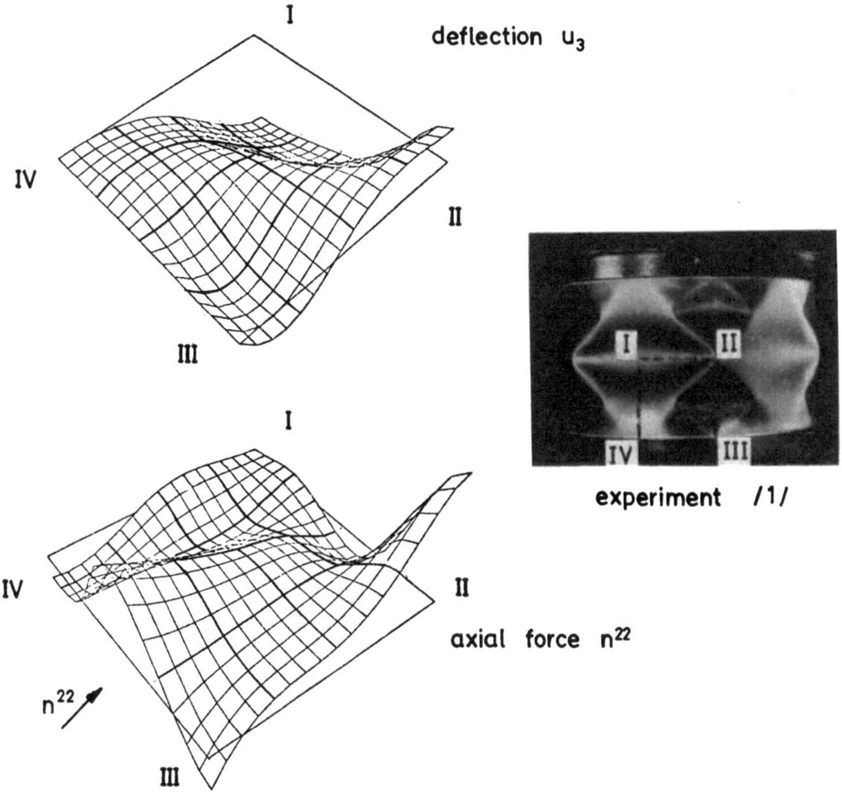

Fig. 14: Rhombic postbuckling pattern in point A

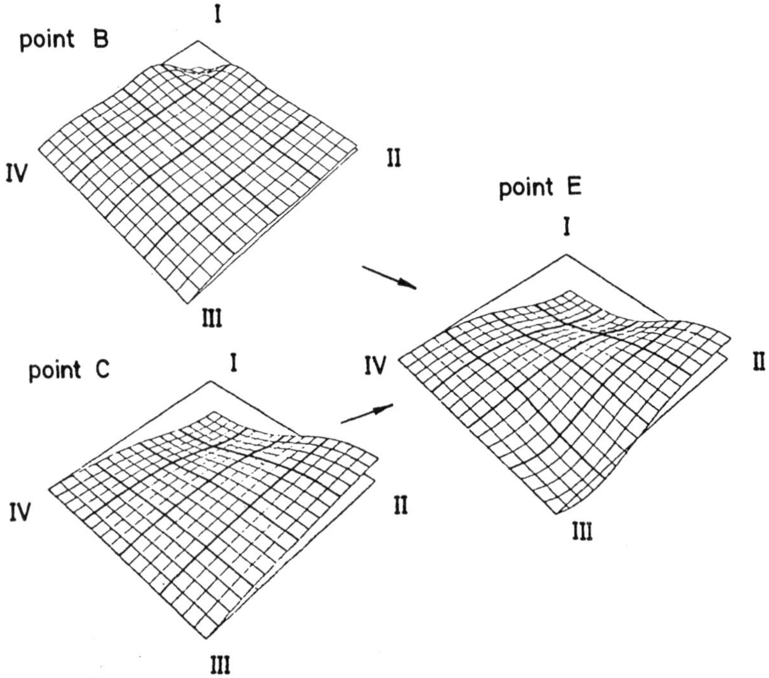

Fig. 15: Postbuckling pattern, see fig. 14

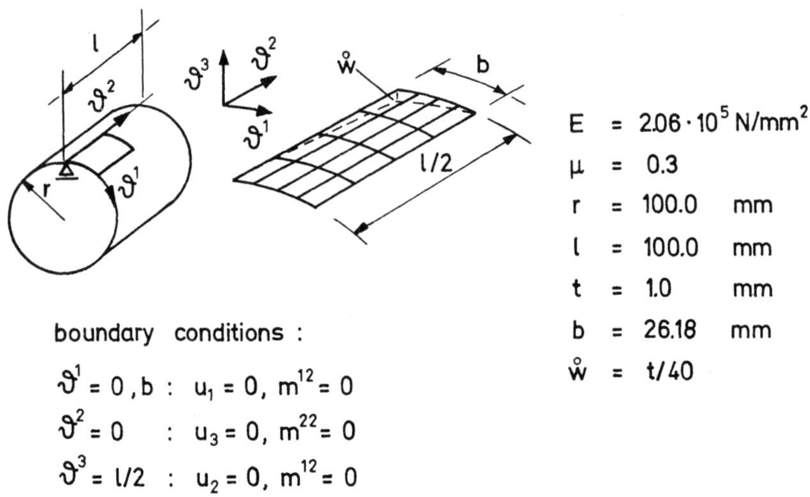

Fig. 16: System and boundary conditions

A 4 by 4 element section is analyzed. The size is chosen related to the experimental postbuckling pattern of /1/, see fig. 17. A small imperfection, see fig. 16 is imposed. The load is applied in one step of $\bar{p}$ = 4.9 N/mm^2. The postbuckling curves of fig. 17 were obtained by unloading from this point in several steps. In fig. 17 the behaviour under external radial forces and radial forces plus loaded top and bottom are compared. Without the axial force a stiffer postbuckling behaviour is calculated. In both cases no obvious unstable region has been passed. However, an attempt to calculate the postbuckling for $\bar{p}$ = 4.9 N/mm^2 without damping in one increment failed. The iteration diverged. It should be noted, that the example differs from the experiments in the unrestrained edge in axial direction.

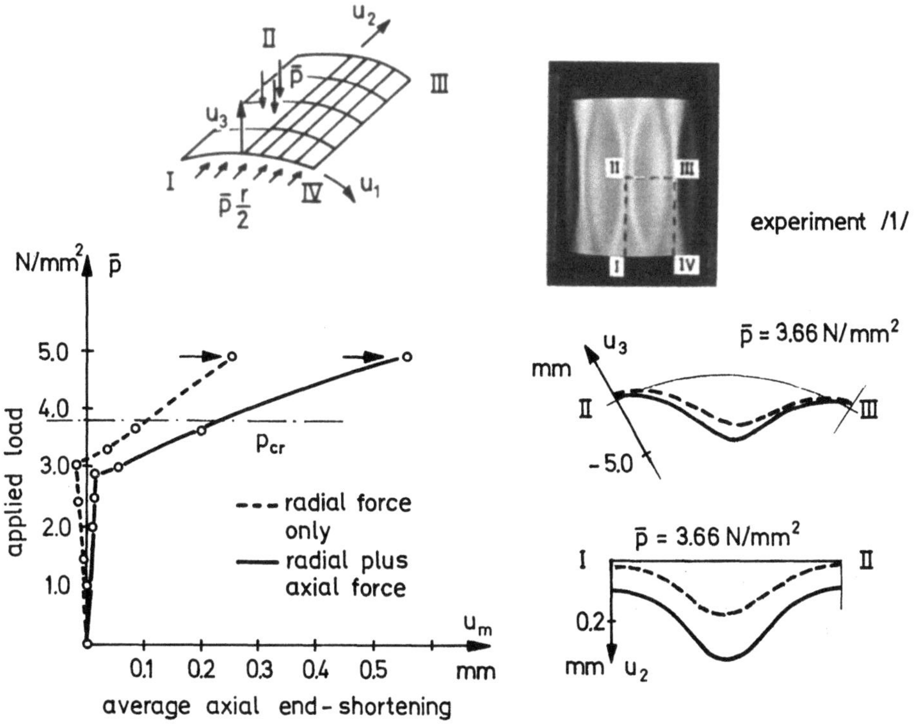

Fig. 17: Cylinder under external forces

Conclusion

The presented creep iteration with artificial damping can be used in order to find stable states of equilibrium in the pre- and postbuckling range. Static unstable regions are surmounted without numerical instability. However it cannot be ensured that always the lowest possible solution path is reached. The effectivity of the strategy is hard to compare, since most of the examples could not be calculated with the common incremental strategy. In all examples less than 50 iterations were necessary, in order to arrive at the final equilibrium point in one step. The number of the iterations depends on the number of unknowns and on the iteration path, which is influenced by the imperfection and the distribution of the damping.
The used mixed functional leads to nonlinear element matrices of simple structure and good convergenze of the nonlinear solution. However, the effect of mesh refinement should be studied further, in order to test the reliability in the complex post-buckling range.

Acknowledgement

The author acknowledges the support provided by the Heisenberg-Stipendium of the Deutsche Forschungsgemeinschaft and the work of Mr. J. Hillmann on the numerical examples.

References

1. Esslinger, M.; Geier, B.: Postbuckling behavior of Structures. CISM Courses and lectures No. 236 Springer Verlag 1975.

2. Argyris, J.H.; Hindenlang, U.; Menz, W.; Hilpert, O.; Malejannakis, G.A.: Geometrisch nichtlineare Stabilitätsanalyse eines dünnwandigen Schalentragwerks aus Glasfaserbeton. VDI Forschungsheft Nr. 604 (1981).

3. Kröplin, B.-H.; Dinkler, D.: A Creep Type Strategy for Tracing the Load Path in Elastoplastic Post Buckling Response. To appear in proceedings of FENOMECH' 81.

4. Harbord, R.: Berechnungen von Schalen mit endlichen Verschiebungen - Gemischte finite Elemente - . Bericht Nr. 72-7 aus dem Institut für Statik, TU Braunschweig.

5. Harbord, R.; Kröplin, B.; Schröder, R.: Schalenelemente in gemischter Darstellung: Theorie - Kritik - Beispiele. Ing.- Archiv 47 (1978) 207-222.

6. Dawe, D.H.: High-Order Triangular Finite Element for Shell Analysis. Int. J. Solids Struct. 11 (1975) 1097-1110.

7. Harbord,R.: Lectures on Finite Elements. TU Braunschweig (1975).

8. Krätzig,W.B.: Optimale Schalengrundgleichungen und deren Leistungsfähigkeit. Z. Angewandte Math. Mech. 54 (1974) 265-276.

9. Remseth,S.N.: Nonlinear Static and Dynamic Analysis of Space Structures. Report No. 78-2, Institutt für Statikk, Norges Tekniske Høgskole, University of Trondheim.

10. Bergan, P.G.: Solution Algorithms for Nonlinear Structural Problems. Prodeedings Int. Conf. on Eng. Applications of the Finite Element Method. A.S. Computas, Høvik, 9-11 May 1979.

11. Ramm, E.: Strategies for Tracing the Nonlinear Response Near Limit Points. Nonlinear Finite Element Analysis in Structural Mechanics, eds. Wunderlich, Stein, Bathe, Springer Verlag (1981) 63-89.

12. Key, S.W.; Stone, C.M.; Krieg, R.D.: Dynamic Relaxation Applied to the Quasi Static, Large Deformation, Inelastic Response of Axisymmetric Solids. Nonlinear Finite Element Analysis in Structural Mechanics, eds. Wunderlich, Stein, Bathe, Springer Verlag (1981) 585-620.

The Displacement Finite Element Method in Nonlinear Buckling Analysis of Shells

E. RAMM

H. STEGMÜLLER

Institut für Baustatik
Universität Stuttgart, West-Germany

Abstract

The paper compiles the current status of the finite element method in linear and nonlinear buckling analysis of shells. The classical concept via shell theory, the degeneration method, continuum mechanics based and corotational formulations used in the displacement approach and the corresponding incremental stiffness expression are briefly described. Some comments on the problem of non-uniqueness and stability of the solution and their practical evaluation are given. A classification of displacement dependent pressure loads is presented discussing the symmetry of the problem. The main characteristics of the different classes of shell elements are outlined. Besides flat and curved elements derived from shell theory the survey concentrates on degenerated elements. A detailed review on the main solution strategies in nonlinear shell analyses is presented. Among these are quasi-Newton methods combined with line search and iteration techniques in the displacement and load space. Finally selected numerical examples are described applying isoparametric degenerated elements to bifurcation buckling and nonlinear collapse analyses of shells.

1. Introduction

Buckling of thin shells is one of the most complex phenomenon in structural mechanics [1], [2]. The analytical description of the structural response requires a high standard in the solution algorithms and - at least as important - a profound knowledge of fundamental characteristics of the structure. Only by that the anticipated response can be modelled and the results of the analysis can be judged. It was pointed out by Bushnell in a recent survey paper [3] that a "feel" for shell buckling is an essential condition. Excellent accomplishments in numerical techniques are necessary but not sufficient.

The present paper focuses on the application of the finite element method to static buckling problems of shells including bifurcation, nonlinear collapse

Buckling of Shells, Ed.: E. Ramm
Proceedings of a State-of-the-Art Colloqium

and postbuckling behaviour. The topic also has been discussed in various articles by Gallagher, see for example [4], and in several papers of [5]. Even if the present study stresses primarily the displacement method many subjects can be seen in a more general context. The paper describes formulations used in large deformation analysis of shells, the non-uniqueness and stability problem, the influence of displacement dependent pressure loads, nonlinear displacement shell elements, the inclusion of the material nonlinearity and solution strategies in nonlinear analyses. Methods for arbitrary shell geometries are emphasized. The considerable literature on special algorithms for axisymmetric shells is not included. The paper discusses primarily conventional incremental methods. Special techniques, e.g. the perturbation method or dynamic relaxation, are not studied.

2. Nonlinear Formulation

2.1 Concepts

Nonlinear buckling investigations of thin shells can be mostly based upon small strain, large displacement analyses with moderate to large rotations. Basically two concepts have been persued deriving thin shell finite elements (Figure 1).

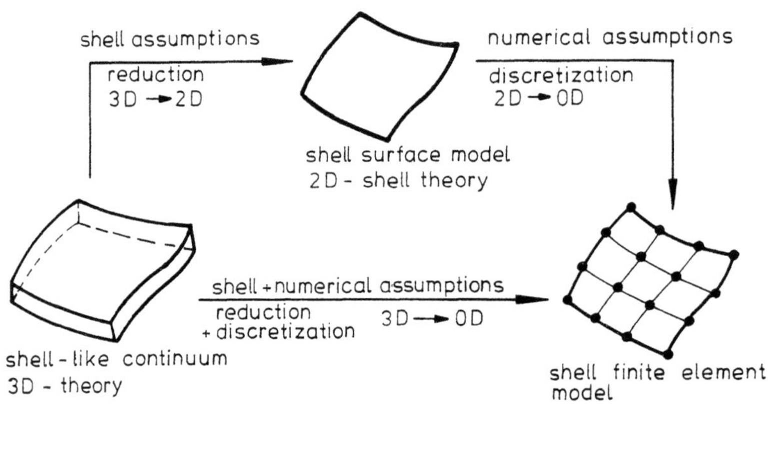

Fig. 1: Classical versus degeneration concept

* The classical concept: The finite element discretization is introduced into the shell surface model. This two-dimensional idealization is described by a nonlinear shell theory usually derived by introducing thin shell assumptions into the field equations of the three-dimensional continuum. In most cases the Kirchhoff-Love hypothesis is presumed so that the C^1-continuity requirement has to be satisfied. Deep shell or shallow shell theories are applied using either curvilinear or cartesian displacement components, see for example [6 - 8]. Often the shell is initially approximated by flat elements.

* The degeneration concept: This formulation directly discretizes the 3D-field equations in terms of midsurface variables applying simultaneously corresponding shell assumptions: In general it is assumed that the normal of the shell surface remains straight after deformation allowing for shear deformations. Thus displacements and rotations are independent variables so that only C^o-continuity is required. The term "degeneration" may be misleading since a shell theory also represents a degeneration from 3D to 2D.

2.2 Formulations

In geometrically nonlinear analysis either a total Lagrangian or an updated Lagrangian formulation is used depending on the reference state chosen: either the initial configuration or the configuration at the beginning of the next imcrement [9] (Figure 2). In shell analyses no preference for the one or the other formulation can be observed. In general an incremental approach is required applying the load stepwise because the problem is highly nonlinear or path dependent. Besides this the entire load - deflection path is frequently desired.

Since the main geometrically nonlinear effect in the analysis of thin shells is caused by finite rotations a simplified version of an updated approach is often applied, in particular for elements derived by the classical concept: the corotational formulation, often called a moving coordinate or rigid convected formulation, see for example [8], [10 - 16]. Rotations and pure deformations are approximately decomposed in an additive manner.

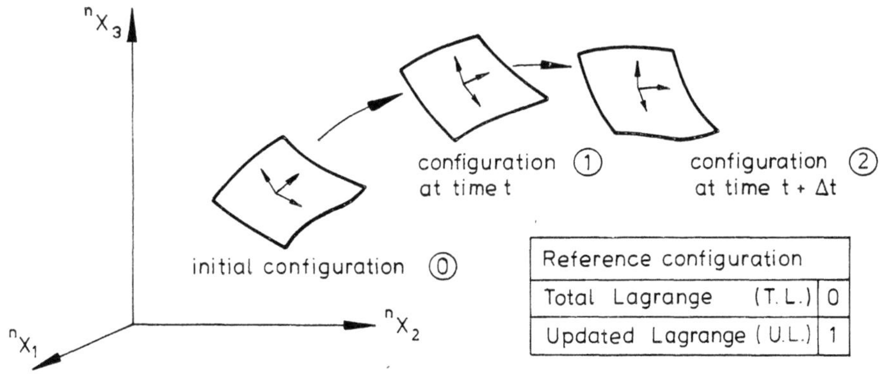

Fig. 2: Motion of a shell finite element

The essential feature of the method is that all element variables are referred to a local coordinate system $^{n}\bar{x}_i$ which is attached to each element and follows the rigid body movements of the element (Fig. 3). In the simplest version of the formulation the element matrices may be derived from

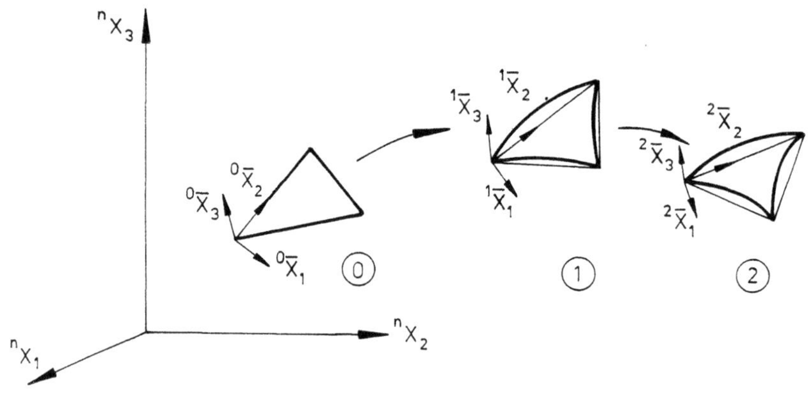

Fig. 3: Corotational formulation (initially flat element)

the usual small displacement analysis, including only the pure deformation modes, c. f. the natural stress and strain concept introduced by

Argyris [10]. All geometrically nonlinear effects enter through pure rigid body kinematics using proper transformations between the moving element attached coordinate system and the global cartesian frame. The formulation allows for arbitrarily large rotations but neglects the relative rotations within the element. The method may be refined if local coordinate systems are introduced at all integration points [15].

2.3 Stiffness Expression

Displacement finite element models can be derived from the principle of virtual work. The variational expression is either formulated in terms of resultant variables according to shell theory [6], [14] or on the basis of the 3D-continuum as indicated for the degeneration concept, see for example [17 - 19]. For the latter one the incremental virtual work expression and its algebraic counterpart is given in Figure 4 for a deformed configuration ①. Here ${}_n\mathbf{e}$ and ${}_n\eta$ are the linear and nonlinear terms of the

$$\int_{^nV} {}_n\boldsymbol{\varepsilon}^T\, {}^1_n\mathbf{C}\, \delta_n\boldsymbol{\varepsilon}\, dv + \int_{^nV} {}^1_n\mathbf{s}^T\, \delta_n\boldsymbol{\eta}\, dv = {}^2\delta W_{ext} - \int_{^nV} {}^1_n\mathbf{s}^T\, \delta_n\mathbf{e}\, dv \qquad (1)$$

linearized

$$\int_{^nV} {}_n\mathbf{e}^T\, {}^1_n\mathbf{C}\, \delta_n\mathbf{e}\, dv$$

discretized

$$\left({}^1_n\mathbf{K}_i + {}^1_n\mathbf{K}_g\right)\mathbf{u} = {}^2\mathbf{P} - {}^1_n\mathbf{F} \qquad (2)$$

$${}^1_n\mathbf{K} = {}^1_n\mathbf{K}_i + {}^1_n\mathbf{K}_g, \qquad {}^2\mathbf{R} = {}^2\mathbf{P} - {}^1_n\mathbf{F}$$

T.L.	n = 0	${}^1_0\mathbf{K} = {}^1_0\hat{\mathbf{K}}_i + {}^1_0\mathbf{K}_u + {}^1_0\mathbf{K}_g$	(2a)
U.L.	n = 1	${}^1_1\mathbf{K} = {}^1_1\mathbf{K}_i + {}^1_1\mathbf{K}_g$	(2b)

Fig. 4: Incremental equilibrium equations

incremental strain ${}_n\boldsymbol{\varepsilon}$, ${}^1_n\mathbf{s}$ is the total stress and ${}^1_n\mathbf{C}$ the incremental material tensor of configuration ①, ${}^2\delta\, W_{ext}$ expresses the external virtual work. n indicates the reference state (n = 0, 1). Solving the nonlinear equations by an iteration technique requires a linearization of the first expression eq. (1) (${}_{n}\boldsymbol{\varepsilon} \rightarrow {}_{n}\mathbf{e}$). The discretization leads to a stiffness equation, eq. (2), with the tangent stiffness matrix ${}^{1}_{n}\mathbf{K}$ which may be decomposed into the large deflection stiffness matrix ${}^{1}_{n}\mathbf{K}_{i}$ and the geometric stiffness or initial stress matrix ${}^{1}_{n}\mathbf{K}_{g}$. In the total Lagrangian formulation the initial displacement part ${}^{1}_{o}\mathbf{K}_{u}$ can be, in turn, separated from ${}^{1}_{o}\mathbf{K}_{i}$, eq. (2 a). In elastic analyses ${}^{1}_{1}\mathbf{K}_{i}$ (U. L.) or ${}^{1}_{o}\hat{\mathbf{K}}_{i}$ (T. L.) plays the role of the elastic stiffness matrix ${}^{1}_{n}\mathbf{K}_{e}$. $\mathbf{u}$ is the vector of the incremental displacements, the right hand side expresses the out-of-balance forces ${}^{2}\mathbf{R}$ of configuration (2) at the beginning of the increment, external forces ${}^{2}\mathbf{P}$ in (2) minus internal forces ${}^{1}\mathbf{F}$ of configuration (1). If the external load depends on the actual displacement field an additional load stiffness matrix develops (see Chapter 4).

3. Non-Uniqueness and Stability

Bifurcation points are locations on the load - deflection path with a non-unique solution of the fundamental state. Only an additional investigation of the secondary path, for example Koiter's initial postbuckling analysis, can decide whether or not the bifurcation point is in stable equilibrium. Limit points indicate an unstable configuration. They are unique in the sense that they are merely a pathological case on the load - deflection path with vanishing load increase. Despite a difference meaning both points satisfy either the static or the kinetic criterion defined in Fig. 5 for the

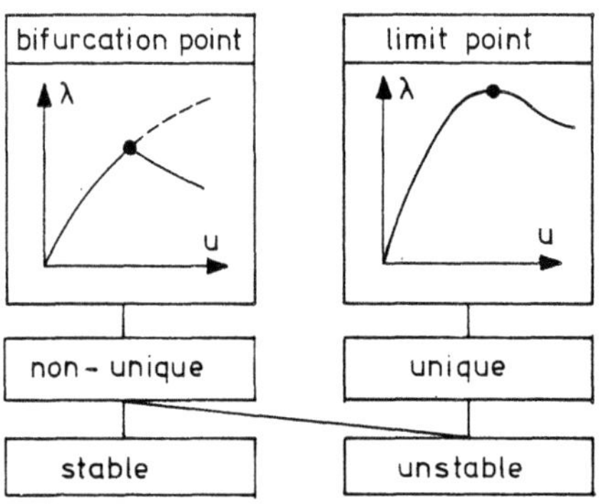

criterion	
static	${}^{c}_{n}\mathbf{K}\,{}^{c}\mathbf{v} = \mathbf{0}$
kinetic	$({}^{c}_{n}\mathbf{K} - \omega^{2}\mathbf{M})\,{}^{c}\mathbf{v} = \mathbf{0}$

Fig. 5: Critical points and criteria

finite element displacement method. Both criteria are equivalent for conservative problems. For non-conservative problems the kinetic criterion is obligatory even for static loading in order to detect flutter situations. The homogeneous equation (3) (Fig. 5) has only a nontrivial solution if the determinant of the stiffness matrix vanishes.

$$^{c}\boldsymbol{\Delta} = \det |\, ^{c}\mathbf{K} \,| = 0 \qquad (5)$$

The determinant can easily be obtained during factorization of the tangent stiffness matrix since it is the product of all diagonal terms of the upper triangular matrix.

Provided that the prebuckling response is linear with $^{m}\mathbf{P} = {}^{m}\lambda\, {}^{r}\mathbf{P}$, $^{m}\lambda$ being a load factor and $^{r}\mathbf{P}$ a reference load vector, the bifurcation point ($^{m}\lambda \equiv {}^{c}\lambda$, c = critical) is usually obtained by a linear eigenvalue analysis. In a displacement finite element analysis the following should be noted:

* There is no unique definition of a linear buckling analysis. This depends on the choice of formulation and on the assumption which part of the stiffness matrix is linearly magnified by the factor $^{m}\lambda$, for a detailed discussion see [20], [21]. For example the role of the initial displacement matrix may be interpreted in a different way:

T. L. $$[{}^{o}_{o}\mathbf{K}_{e} + {}^{c}\lambda\, ({}^{r}_{o}\mathbf{K}_{u} + {}^{r}_{o}\mathbf{K}_{g})]\; {}^{c}\mathbf{v} = 0 \quad \text{or} \qquad (6\text{ a})$$

$$[{}^{o}_{o}\mathbf{K}_{e} + {}^{r}_{o}\mathbf{K}_{u} + {}^{c}\lambda\, {}^{r}_{o}\mathbf{K}_{g}]\; {}^{c}\mathbf{v} = 0 \qquad (6\text{ b})$$

U. L. $$[{}^{r}_{1}\mathbf{K}_{e} + {}^{c}\lambda\, {}^{r}_{1}\mathbf{K}_{g}]\; {}^{c}\mathbf{v} = 0 \qquad (6\text{ c})$$

* In analytical investigations usually a "classical" linear buckling analysis is performed, that is an undeformed membrane fundamental state is assumed. In a general finite element analysis all influences like bending and deformations are included. The slightly nonlinear behaviour often occurring before buckling is suppressed in initial linear buckling analysis.

If a nonlinear prebuckling response is present the following procedures may be adopted to detect critical points (Fig. 6):

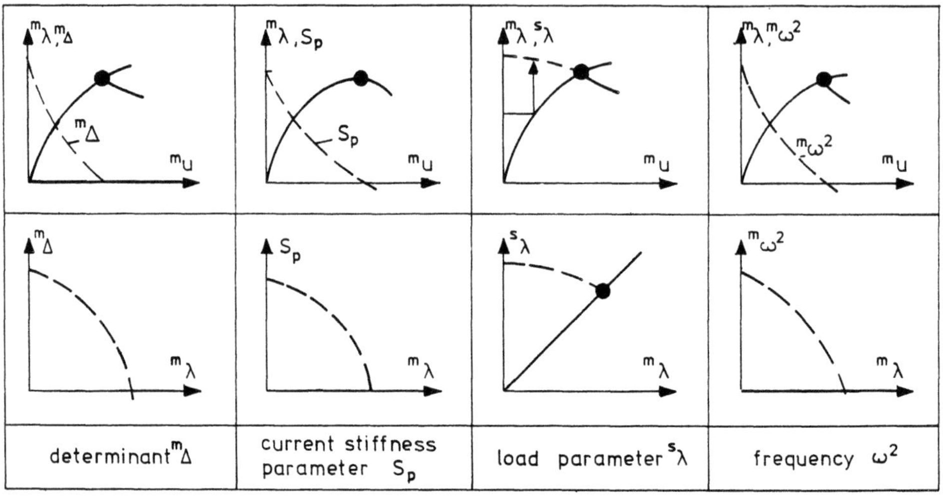

Fig. 6: Supplementary techniques

* The determinant according to eq. (5) can be monitored during the loading process. Note that a positive value is a necessary but not sufficient condition for stable equilibrium. Unfortunately the determinant does not express the sensitivity of the response with respect to load changes. Shallow intersections of the determinant function with the abscissa and convergence problems near critical points often do not allow to identify the neutral equilibrium position.

* The current stiffness parameter S_p [23] which is essentially the inverse of the incremental energy is a much better parameter than the determinant. Unfortunately it does not indicate a bifurcation point on the prebuckling path.

* Supplementary eigenvalue analyses for the load parameter defined in eq. (6) but performed in a deformed configuration [20 - 22] allow to judge the current status of the structure. Because eigenvalue functions intersect the load-deflection path at critical points bifurcation or limit points can already be anticipated at an earlier stage. Since this procedure is expensive it should be considered to determine the lowest eigenvalue $^S\lambda$ approximately.

* Eigenvalue analyses for the natural frequency at different load levels according to eq. (4), Figure 5, lead to a load-frequency diagram. The load for a vanishing frequency defines the critical static load [20], [21].

4. Displacement Dependent Loads

In the literature the existence of displacement dependent loads is often automatically identified with a non-conservative problem resulting in unsymmetric load stiffness matrices. In many cases this is not true. One of the most important load cases in buckling analysis of shells is the pressure load which always acts normal to the deformed structure. This case is discussed in detail in a recent thesis by Schweizerhof [24]. Following this work two different kinds of pressure loads have to be classified (Fig. 7):

origin of load stiffness terms			pressure load	
			body attached	field in space
domain	load on surface	not const.	N	S
		const.	S	= S
boundary	arbitrary b.c.		N	N
	specific b.c.		S	S

Fig. 7: Symmetry (S) and non-symmetry (N) of load stiffness terms

In addition to the varying direction the magnitude of the load may or may not depend on the displacement field. In the first case the load represents a pressure field in space and the shell deforms within this prescribed load field, for example a water imbedded pipe. Then the part of an arbitrarily distributed load in the domain which contributes to the stiffness leads to symmetrical matrices. Only the boundary terms may or may not result in non-symmetrical matrices depending on the kind of boundary conditions. For example a free but loaded boundary leads to unsymmetry whereas a boundary with two prescribed displacement components does not. In this latter case the problem is conservative in the sense that the linearized load stiffness matrix shows symmetry. If the load is body attached and not constant on the surface non-symmetric terms arise in the domain. For the boundary terms the same holds as in the previous case. It should be mentioned that both kinds of loads coincide when the pressure on the surface is constant. Then only the boundary conditions determine whether or not the problem is conservative. Note that if a pressure field in space is falsely classified as body attached it may lead to a non-conservative problem and be misinterpreted.

In finite element analyses the boundary terms play the role of compatibility conditions between adjacent elements. Therefore the element load stiffness matrices of displacement dependent loads are a priori non-symmetric. Nevertheless the assembled stiffness matrix is symmetric if the problem is conservative, that is the non-symmetric terms cancel each other. For a detailed discussion see [24].

The buckling problem of shells under pressure loads is also discussed in [25 - 27]. From the practical point of view the influence of non-symmetric terms on the static buckling load seems to be a minor one. In nonlinear analyses the parts can be included on the right hand side of the stiffness expression, eq. (2), in an iterative manner. The terms have to be introduced in a kinetic stability analysis, eq. (4), if flutter becomes important.

5. Nonlinear Elements

As mentioned in Chapter 2 thin shell elements are either derived from a shell theory or directly from the three-dimensional theory via degeneration. Figure 8 shows the element classes which can be established. Some characteristics and problems associated with each class will be summarized; see also extensive reviews of the literature in [28 - 30].

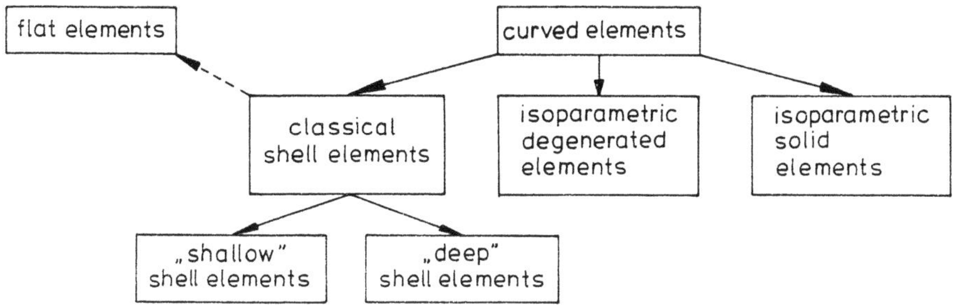

Fig. 8: Shell elements

Flat elements:

* The simplest approximation of a curved shell; often known elements for plane stress and plate bending are superimposed; for an arbitrary geometry only triangular elements can be used.

* Stretching and bending is coupled only at the nodes via coordinate transformation; bending moments at the kinks not present in the actual shell decrease with mesh refinement.

* Different interpolation schemes for stretching and bending lead to inter-element incompatibility; coplanar adjacent elements may require extra treatment to avoid singularities.

* The influence of the relatively crude finite element model on the nonlinear response of imperfection sensitive shells is not yet clarified.

There is an extensive literature on the application of flat elements in linear and nonlinear shell analysis, see the review in [6], [29]. Simple facet triangular elements are described in [16], [31], [32] based on a corotational updated formulation. The quadrilateral element composed of four triangular elements can be also included in this class [33]. Due to their simplicity and efficiency flat elements seem to have a good chance in the future despite some drawbacks mentioned above.

Curved elements via shell theory:

These elements are either based on shallow shell [34 - 36] or on deep shell theory [37], [38]. Shallow elements are also used to model deep shell problems [8].

* The representation of deep shells by shallow shell elements leads to a geometric error similar to that encountered in a shell discretized by flat elements.

* Incomplete representation of rigid body modes already known for linear shell elements [4], [6] effects the solution even more in nonlinear analysis, in particular if large rotations are present; rigid body modes may lead to strains.

* The inter-element compatibility requirement is a major problem; it is impossible to achieve complete compatibility if shallow shell elements are used in deep shell analysis. Enforcement of the compatibility leads to a loss of rigid body modes and vice versa.

The situation may be improved applying the following techniques:

* To satisfy all requirements of assumed displacement elements high order shape functions have to be used, see for example the quintic SHEBA element [38]; at least membrane and flexural displacement fields should be interpolated by functions of the same order, for example by cubic polynomials [14], [36], [37]; to avoid membrane action under pure bending ("membrane locking") even higher order polynomials for the in-plane displacements have been proposed [8].

* For the representation of rigid body modes, in particular in large rotation analysis, one should resort to a corotational formulation [8], [11], [14]; then the shape functions have to be defined locally for the relative displacements which result in pure deformation modes only.

* Additional constraints can be introduced to enforce inter-element compatibility [4], [37] following a hybrid displacement or generalized potential energy approach.

At present it seems that the development of nonlinear pure displacement models stagnates in favour of less complex elements. Due to less requirements either mixed/hybrid models or degenerated elements are preferred.

Isoparametric degenerated elements:

A detailed review of the development of degenerated elements in linear and nonlinear shell analysis up to the present time is given in [18] and in several papers of [5] and will not be repeated here. The elements include transverse shear deformation, require only C^o-continuity and originally have the same interpolation scheme for all three displacements and the two rotations: bilinear, biquadratic or bicubic either a serendipity or Lagrangian type. The following characteristics should be mentioned:

* The elements are compatible and include the rigid body modes.
* Large rotations are easily incorporated by proper use of trigonometric functions [17] or by use of relative displacements on the shell top surface [18], [19].
* Some element versions produce an excessively stiff solution (Fig. 9):
  - If the shell becomes thin the transverse shear is overestimated; this phenomenon is referred to in the literature as "shear locking" [17 - 19], [39] and has been explained in several papers in the meantime.
  - If a curved element model is used instead of a facet idealization low order elements tend to produce membrane effects under pure bending [39], [40]; the phenomenon may be termed "membrane locking".

Fully integrated element versions react in a different way on the locking phenomena (Fig. 9). This question depends also on the problem analysed, the boundary conditions, the idealization etc.

* Lagrange elements are less susceptible to shear - locking than serendipity elements.

* The bilinear 4-node element locks in thin shell analysis.

* The bicubic 16-node element has the lowest sensitivity to either kind of locking; it is the most reliable but obviously the most expensive element [17], [40].

		fully integrated elements tendency to		reduced integration obligatory if
		shear locking	membr. locking when curved	
Serendipity	S12 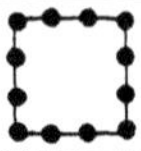	large	low	thin
Serendipity	S8 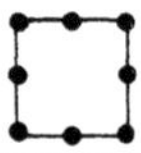	large	large	curved thin
Lagrange	S16 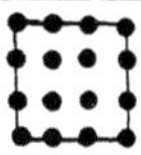	low	low	not necessary
Lagrange	S9 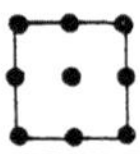	low	large	curved
S/L	S4 	extreme	extreme	always

Fig. 9: Locking and reduced integration

Several schemes have been proposed in order to avoid locking. The most commonly used procedures are:

* The reduced integration scheme: Either a uniform reduced integration of the entire stiffness matrix or a selective scheme is used in which only a part of the matrix is under-integrated; in the latter case it is important for the low order linear and quadratic element versions to reduce the integration order not only in the transverse shear stiffness but also in the membrane stiffness to avoid membrane locking [39 - 41]; although the main purpose of the reduced integration is to remove excessive stiffness, the structural matrix may result in undesired internal mechanisms or zero energy spurious modes (hour-glass modes) [18], [19].

* In the discrete-Kirchhoff-theory (DKT) additional constraints based on the Kirchhoff-hypothesis are introduced allowing also to reduce the number of d.o.f. on the element level; these constraints are applied either pointwise or in an integral way; a typical application of this procedure is the SEMILOOF concept in which the rotational d.o.f. are interpolated one order lower than the displacements using the Loof sampling points [15], [42], [43].

Several schemes to avoid locking as well as spurious modes are compiled in Figure 10 for the most effective bilinear 4-node element version. They may be applied to higher order elements in a similar way. A shallow shell degenerated element is described in [45].

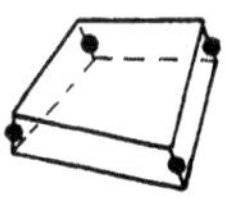 S4	different sampling points for transverse shear (o) (x o x / o o / x o x)	Mc Neal [44]
	uniform reduced integration + hourglass control	Belytschko et. al [13]
	selective integration „heterosis" scheme	Hughes / Liu [18] Kanok-Nukulchai et. al. [19]
	low order SEMILOOF (discrete Kirchhoff assumptions) SLICK	Nagtegaal / Slater [43]
	discrete penalty least square method solid element + strain constraint	Brockman [30]
	mixed finite element models (equivalent displacement models)	Noor et. al. [39]

Fig. 10: 4-node degenerated elements

In the last years the majority of shell elements developed for nonlinear analyses is based on the degeneration concept [5]. This is a strong indicator for the success of these elements. On the other side it seems that the two avenues in shell element development approach each other; compare for example the 4-node elements (Fig. 10) with the "classical" elements derived in [46].

Solid elements:

Continuum elements ignoring shell assumptions can be used to model shells as long as every individual element has a reasonable length to thickness ratio. In thin shell analysis this is not very economic. In addition the procedure may lead to ill-conditioned algebraic systems because the transverse displacements on both surfaces are nearly the same. Several continuum elements across the thickness have been used to model the complex three-dimensional stress state at zones where two shells intersect.

6. Material Nonlinearity

In elastic buckling analysis of shells no special scheme has to be introduced to incorporate the material law. The numerical integration in surface direction also covers inelastic material behaviour. In thickness direction basically three different models can be employed (Fig. 11):

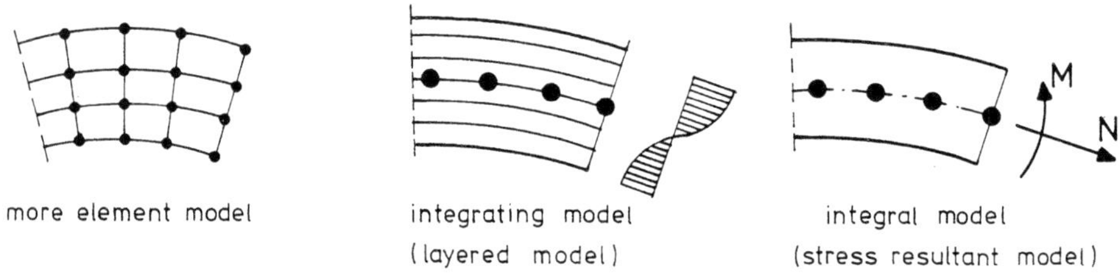

Fig. 11: Material models in thickness direction

* If solid elements are used more than one element will be placed across the thickness allowing each individual element to have its actual material state. The procedure is simple but expensive because of an excessive number of d. o. f.

* If shell elements are used the most common approach is a layered model. The procedure is applicable to classical as well as degenerated shell elements. It also allows to monitor the progression of material yielding or cracking perpendicular to the

shell surface. Various integration schemes are employed using different numbers of layers, for example Gauss integration [18], Simpson rule [40], Newton-Cotes quadrature formula [42] or a piecewise constant stress profile over each layer [47]. Although the multi-layer approach is the most versatile scheme it may not be the most economic one.

* In the integral model the material law is defined in stress resultants rather than in stresses. It requires that the formulation is based on stress resultants. A detailed discussion of integral material laws in plasticity is given in [48] where essentially two different yield surfaces proposed by Crisfield [49] and Bienieck [50] are compared, see also [15].

Numerical algorithms for integrating rate constitutive equations (subincrementation, radial return method etc.) are discussed in [18].

7. Solution Strategies

Efficient algorithms for the solution of the nonlinear equations are one essential part in nonlinear buckling and collapse analysis of shells. Combined incremental and iterative solution procedures based on the pure or modified Newton-Raphson (NR) method are fairly standard. Since these procedures are still too time consuming and not successful in the vicinity of critical points the search for more efficient, less sensitive and automatic schemes is quite natural. A detailed review on the subject is given in [51]. The main procedures which have been proposed are (Fig. 12):

* Quasi-Newton methods [51 - 53] compromise both kinds of NR methods: Avoiding factorization in each step but having a faster convergence rate than the modified NR method. The essential feature of quasi-Newton methods is the update of the inverse of the coefficient matrix to provide a secant approximation from one iterate to the next. In praxis this update can even be simplified if the method is implemented in a vector form [51], [53]. It is important that the correction procedure does not destroy the banded nature of the stiffness matrix. The update can start in every phase

of the solution either at the beginning of the first or any other increment or even after the first iteration in a step. The most successful quasi-Newton version is the BFGS update [51 - 53]. Crisfield [51] also describes various modifications which he calls secant-Newton methods.

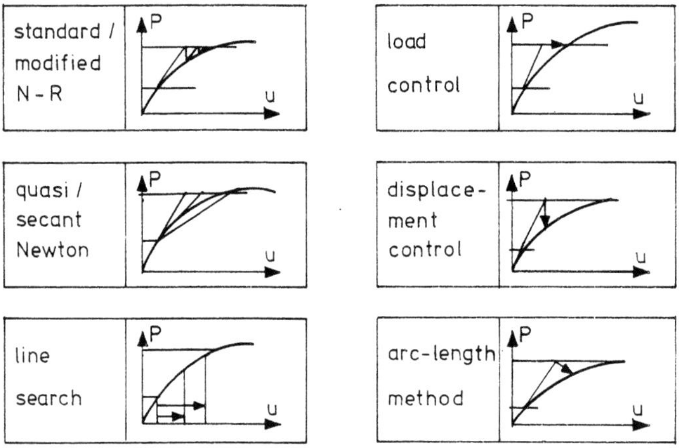

Fig. 12: Solution strategies

* Line - search [51 - 54]: The quasi-Newton methods often come along with line search methods where the incremental displacement vector is updated by a scalar factor η. The condition for an optimal step-length parameter η is that the product of the incremental displacement vector times the out-of-balance forces which, in turn, depend on η, vanishes. In praxis this can be an elaborate iterative operation. Therefore it is often abbreviated by successive inter- or extrapolation schemes [52], [54].

* Iteration methods in displacement and load space [55 - 58]: Since the regular load controlled Newton methods fail in the vicinity of critical points the load parameter has been introduced as an additional variable. This requires, in turn, an additional constraint equation controlling the iteration path in

the displacement and load space. Methods of this group are the displacement control procedure where one displacement parameter controls the solution or the arc-length-methods originally introduced by Riks and Wempner. For a detailed discussion with additional references see [55]. In arc-length-methods the solution is constrained to lie either in a "plane" normal to the tangent to the equilibrium path at the beginning of the increment or on a "sphere" with a radius equal to the length of the tangent, i. e. the arc-length. The arc-length originally held constant from one increment to the other should be adapted to the solution characteristic (see automatic step adjustment). The original arc-length method has been modified [55], [56] and is a versatile iteration technique in the entire load range. It allows to trace "snap-through" as well as "snap-back" behaviour. In principle an extension of the method to other constraint equations is possible, see for example [58].

Automatic step adjustment: Even if it is absolutely necessary in a nonlinear analysis of shells to check the solution at intermediate stages to prevent physically absurd results an automatic step adjustment is important. The adaptation of the constraint parameter discussed above to the number of iterations [55], [56], to a desired value of the current stiffness parameter [57] or another energy limit seems to be a reasonable approach. Bergan [59] has proposed a method in which the unbalanced forces and the external force vector are adjusted in each iteration so that both vectors are orthogonal. Until now there is not too much experience with the technique in many d. o. f. systems [58]. In any case the increment size must be reduced if convergence is not achieved within a specified maximum number of iterations. The restart can be easily automated. It should be noted that path-dependent constitutive laws require a limitation of the step-length.

Summarizing this part it can be stated that each of the above mentioned techniques may drastically improve the solution quality. It seems that a combination of several schemes, for example the arc-length approach together with quasi-Newton methods controlled by the current stiffness parameter is the most advantageous way. Nevertheless, there will not be the unique scheme applicable to all kinds of nonlinear buckling or collapse analyses [51], [54].

8. Remarks to Practical Application

The finite element method is especially suited to model imperfections of any kind. For example any geometrical deviation of a perfect shell can be included in the finite element idealization of the imperfect shell which is then taken as a new independent structure. But it should be realized that the definition of the relevant and realistic imperfection is still a major question in shell stability analysis. For example in a nonlinear stability analysis of an axially loaded cylindrical shell the geometrical imperfection can be based upon the buckling mode at the bifurcation load, on any post-buckling mode, on a combination of several modes or on realistic deviations of the ideal geometry occurring during the manufacturing process. The kind of imperfection chosen can lead, in turn, to another idealization since a different sector angle is required in either model.

It was mentioned in the last chapter that limit points do not require a special treatment if the iteration scheme used operates in the displacement and load space. To follow a secondary path after a bifurcation either a small perturbation in geometry or load is introduced (imperfect approach) or a fraction of the eigenmode is superimposed on the displacement field at the critical load (perfect approach).

A numerical stability solution is often checked against an analytical result. It is important to realize that the assumptions made in either kind of analyses may be different. For instance a rigerous buckling analysis of the axially loaded cylinder by a finite element approach including all pre-buckling effects does not render exactly the "classical" buckling load [3].

Even if trivial some comments to the finite element idealization should be made:

* It cannot be expected that the performance of a finite element model in a shell stability analysis is better than in a linear static analysis. On the contrary the anticipated buckling or failure mode has to be considered in advance.

* The limitation of the finite element model to only one part of the structure because of apparent symmetry of the problem often excludes other buckling modes; for example if in analyses of shells of revolution only a sector of the structure is idealized according to a buckling half-wave, jumping from one mode to another is usually suppressed in the model.

* The proper representation of the boundary conditions is essential in stability analyses of shells. This is especially important if parts of the structure are not included in the finite element model because buckling does not occur in this region; for example if free edges are assumed they may cause inextensional or nearly inextensional modes.

* Numerical divergence and critical structural response should not be mixed up even if they are often interrelated. The more information on the structural behaviour can be obtained the better the judgement of the numerical solution (see Chapter 3).

* If path-dependent material laws are applied the step-length should be small enough.

Even if the finite element method allows a sophisticated idealization the final structural model should be as small as possible. In this context improved initial stability analyses including more information on the buckling problem compared to usual linear stability analyses are to be mentioned [21].

9. Numerical Examples

The isoparametric degenerated elements [17], [40] with a Lagrangian interpolation scheme described in Figure 9 have been applied in the following buckling analyses. Cubic, quadratic and linear element versions are incorporated in the nonlinear finite element program NISA 80. All examples are run on a CDC 6600 or Cyber 174 computer. Eigenvalue analyses follow the subspace iteration technique. In nonlinear problems the total Lagrangian formulation and a modified arc-length iteration procedure (Riks - Wempner) [55] are used. The integration of the elastic plastic material properties according to classical plasticity is based on Simpson's rule with 6 layers in thickness direction. Locally defined boundary conditions are formulated by proper transformations [60].

9.1 Axially Loaded Cylindrical Shell

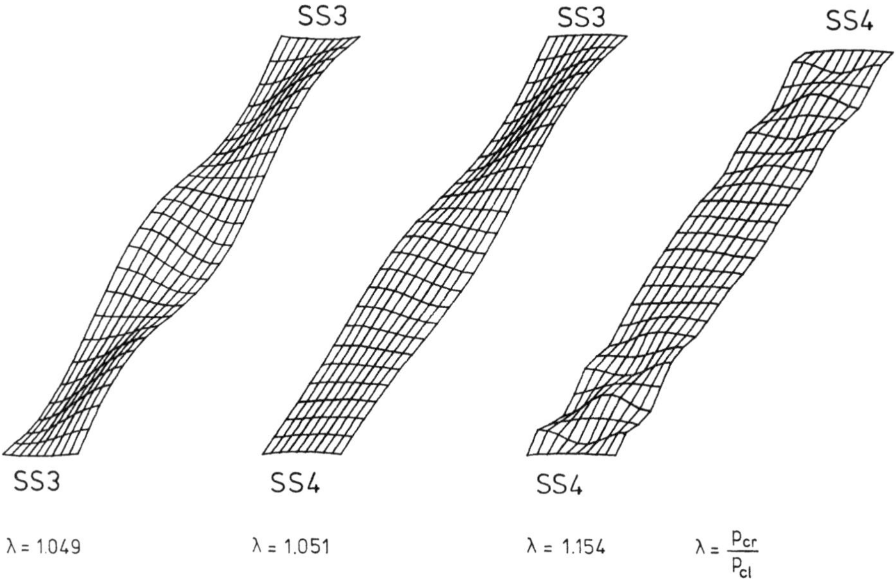

Fig. 13: Linear buckling loads and modes of cylindrical shell

In a preliminary study the linear buckling load of a model cylinder with a radius of 243.8 mm, a length of 400 mm and a thickness of 0.34 mm under axial load was determined. The material was elastic with a Poisson's ratio 0.3. In the linear initial buckling analysis eq. (6 a) was used. A sector of one half buckling wave which corresponds to 14 waves in the circumferential direction was idealized by 3 x 8 fully integrated bicubic elements with a regular mesh. At the longitudinal edges symmetry conditions (S) were assumed. Three different sets of boundary conditions at the curved ends of the cylinder were studied defining either "classical" boundary conditions (SS 3) or a warping constraint condition (SS 4) on one or both sides. The buckling loads normalized to the classical value $\lambda_{cr} = p_{cr}/p_{cl}$ and the corresponding eigenmodes are given in Figure 13. The example shows the influence of warping constraints on the buckling modes in the finite element model. The results which are in the near of the analytical solution may be improved if a finer mesh is used. Applying a warping constraint condition on one side there is only a slight difference in the eigenvalue and the buckling mode changes from 3 to 2 half waves in the longitudinal direction. The difference is more pronounced if warping constraints are applied on both sides. The eigenvalue increases by about 10 % and the buckling mode shows smaller waves in the near of both ends. In circumferential direction the number of half waves changes from 1 to 2. The results should not be generalized for other shell geometries but they show the sensitivity of buckling analyses if finite elements are used.

9.2 Cylindrical Panel under Axial Load

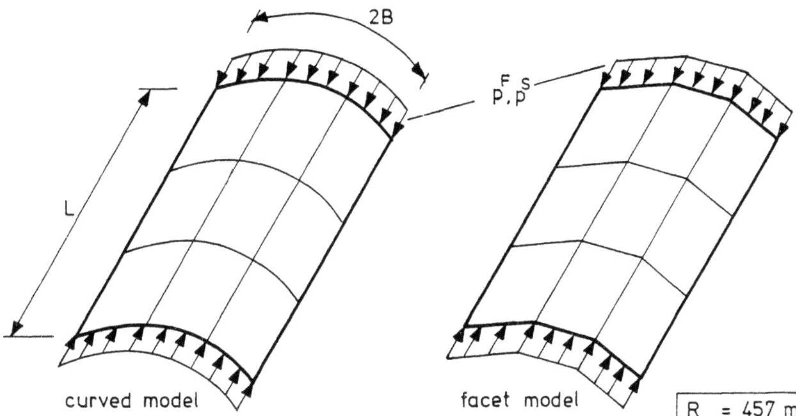

		free (F) — free	symm. (S) — symm.
n_L	S9	6	9
	S16	4	6

R = 457 mm
L = 686 mm
2B = 654 mm
t = 1.6 mm
E = 70400 $\frac{N}{mm^2}$
μ = 0.33

Fig. 14: Cylindrical panel under axial load

The purpose of the following investigation was to study the influence of the facet type of elements on the buckling load. The curved and facet idealization of the cylindrical panel is given in Figure 14 together with the geometrical and material properties. The aluminium panels with simply supported boundary conditions (SS 3) at the curved edges and free longitudinal edges (F) have been studied analytically and experimentally in [61]. In addition to the shell with free edges the present investigation is extended to a shell with symmetry conditions (S) at the longitudinal edges. By this a complete cylindrical shell under axial load is simulated but restricting the buckling mode to the chosen boundary condition. In other words not the exact buckling load and waves can be anticipated since the sector does not correspond to the buckling waves in circumferential direction according to classical theory. The number of elements in axial direction n_L is also indicated in Figure 14. Either the fully integrated quadratic S 9 or the cubic S 16 element is used. Since the curved quadratic element S 9 results in a too stiff solution because of membrane locking this element has only been applied in the facet mode l.

Linear buckling analyses:

First the curved cubic element S 16 has been applied with an increasing number of elements in the circumferential direction. For the shell with free edges (F) the solution converges to a value of 12.9 % of the classical buckling load p_{cl} indicating that the shell cannot develop the stiffening effect in hoop direction. In [61] three different results are given: the experimental buckling load 12.2 %, an analytical solution 14.9 % and a numerical result 14.1 %. The shell with symmetry conditions (S) shows a slightly lower convergence rate and finally renders to a linear buckling load which is 2 % above p_{cl}. The two different buckling modes in the circumferential direction are shown in Fig. 15. In axial direction either one half wave or two waves are indicated.

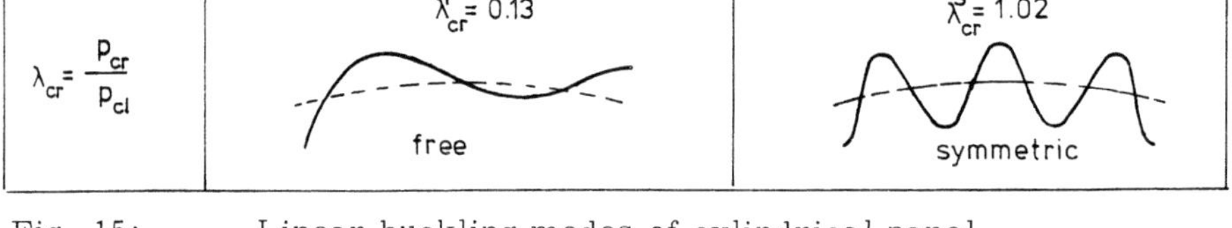

Fig. 15: Linear buckling modes of cylindrical panel

Next the influence of the facet idealization was studied varying the number of elements (facets) in the circumferential direction. The linear buckling loads are shown in Figure 16 for either boundary condition using both

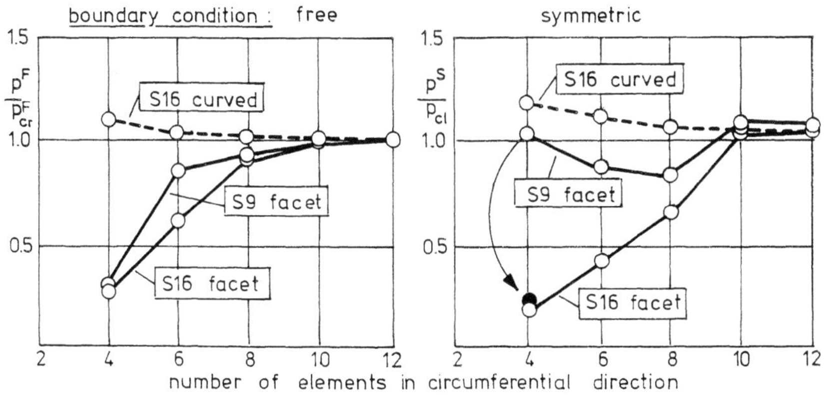

Fig. 16: Curved versus facet model

element kinds, the quadratic S 9 as well as the cubic S 16 version. The buckling loads p for different meshes are normalized to either the buckling load p^F_{cr} of the free panel or the classical buckling load p_{cl} for the symmetry conditions, resp. In both cases a remarkable influence of the facet modelling can be recognized when a relatively coarse mesh is used. In this case the buckling behaviour of flat plates seems to be dominant lowering the stability load. All solutions converge to the correct value. The strange result of the S 9 element in the right diagram shows that one element cannot reproduce the complex buckling mode of the folded plate. If two S 9 elements per facet are used the solution drops to a much lower value as marked by the arrow.

Nonlinear solution:

In this part of the study the influence of the flat elements on the exact stability load is investigated. The facet model of the panel with symmetry boundary conditions (S) has been chosen to continue the analysis into the nonlinear range. The only difference to the above discussed panels is that now the sector is restricted to one half wave of the buckling mode with an angle of 12. 87° using three plane quadratic elements S 9 in the circumferential direction. First the slightly nonlinear prebuckling path has been determined. The exact bifurcation point could be located by a supplementary eigenvalue analysis (see Chapter 3) at 90 % of the classical buckling load, compared to 88 % found by a parallel study with two cubic elements in the circumferential direction. This is the typical reduction known already from rigorous analytical studies, see the discussion in [3]. The eigenmode with a maximum amplitude of 1 % of the thickness has been superimposed on the displacement field at the bifurcation point. This allows the solution to follow the secondary path. In Figure 17 load -

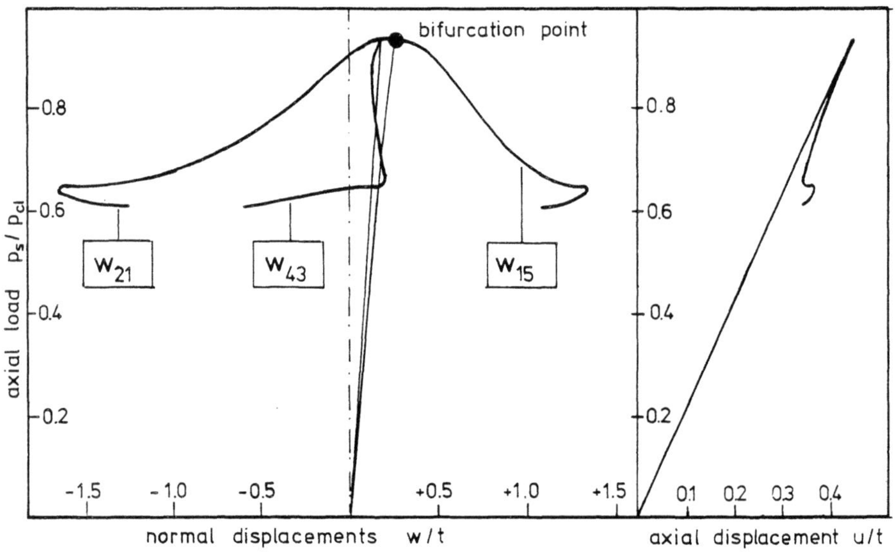

Fig. 17: Load - deflection diagram of cylindrical panel

deflection curves for three different lateral displacements, marked in Figure 18, and for the axial shortening are plotted. It can be seen that the displacement mode continuously varies during unloading. Finally it ends with a horizontal tangent indicating another bifurcation point (mode jumping). The final displacement pattern is given in Figure 18.

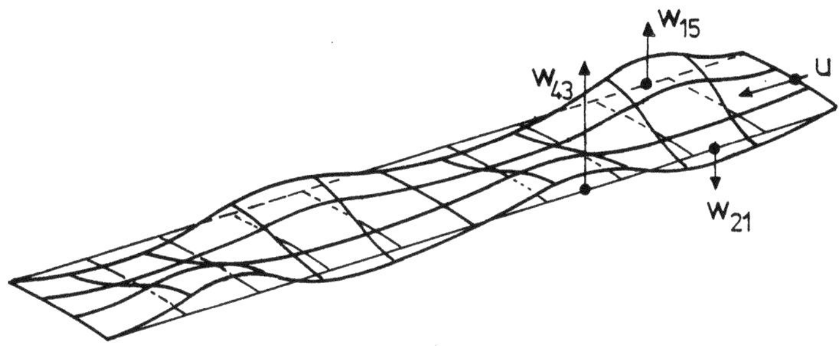

Fig. 18: Displacement pattern of cylindrical panel

Summarizing the results of this example it can be concluded that a facet idealization leads only to other buckling loads for a very coarse mesh. If a reasonable number of elements is used, necessary also to idealize a curved model, the influence on linear and nonlinear buckling loads is negligible. Of course this statement is valid only for the degenerated elements applied in this study.

9.3 Nonlinear Collapse Analysis of Stringer Stiffened Cylindrical Shells under Axial Load

The analysis of this example (Fig. 18) follows an experimental study of axially compressed stringer-stiffened cylindrical shells investigated in [62]. The geometrical and material properties given in Figure 19 correspond to a typical offshore component with a relatively small number of stringers compared to aeronautical structures. The example analysed was the specimen UC 8 in [62]. The shell has a radius of 291.6 mm, a wall thickness of 0.81 mm and a height of 323.7 mm. In the analysis only a panel with one stiffener and symmetry boundary conditions along the longitudinal edges was idealized, corresponding to the buckling mode which is one half wave between the stiffeners in circumferential direction. The failure mechanism - local buckling of shell and stringer while the junction between both remains straight - is typical for a broad-panelled cylinder.

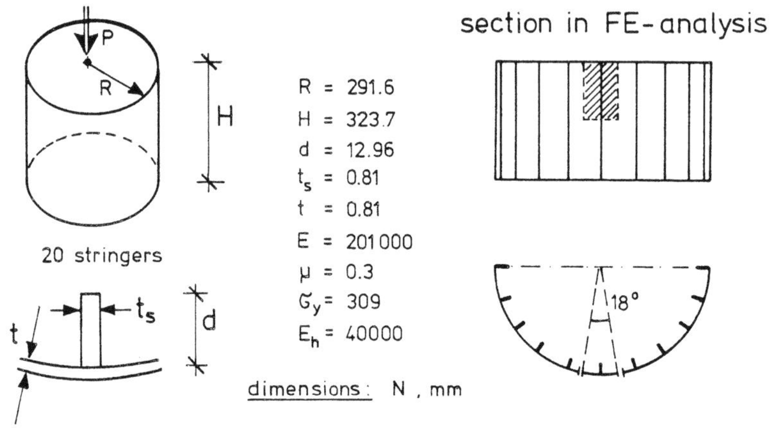

Fig. 19: Stringer stiffened cylindrical shell under axial load

For the imperfect shell the linear buckling mode, shown in Figure 20, was superimposed on the perfect geometry with a maximum initial imperfection amplitude equal to the wall thickness. The loads were applied by uniform axial shortening suppressing any warping of the loaded edge.

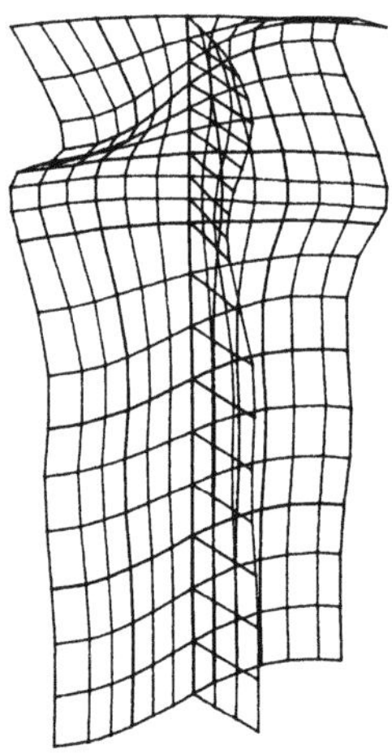

Fig. 20: Linear buckling mode

It should be noted here that in the case of constant edge loads there is a remarkable influence on the results. Two different material models were used in the analysis. First the computation was carried out with a linearly elastic ideally plastic material model without strain-hardening. In the second run strain-hardening was introduced to the material law. The results of the analyses compared to those of the experiment are shown in Figure 21. The average axial stress $\sigma = P/A$ is normalized to the yield

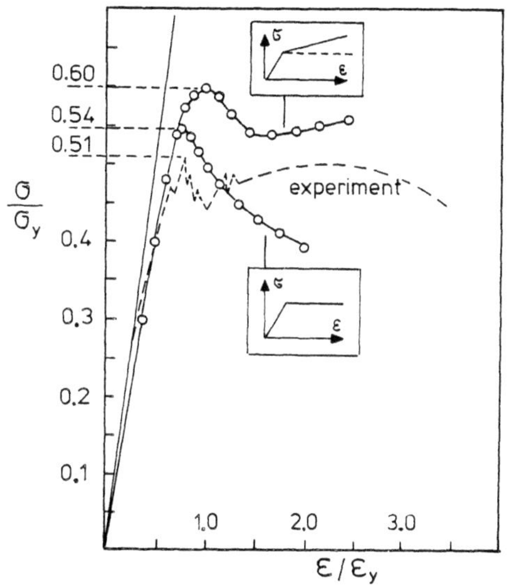

Fig. 21: Load - deflection diagram of stringer stiffened cylinder

stress σ_y and the end shortening ε is normalized to ε_y which corresponds to the linear elastic axial shortening due to the load $P_y = \sigma_y \cdot A$. The difference in the maximum load carrying capacity may be explained by two facts: The analysis did not consider residual stresses and no local imperfections in addition to the global mode described above were applied. Both effects reduce the ultimate load of the shell. This example shows that the finite element method is a powerful tool which allows to represent complex geometries and to include different nonlinearities with an acceptable effort.

10. Conclusions

This review allows the following conclusions:

* In the last years a remarkable progress has been made in nonlinear finite element techniques applied to buckling problems of shells; in particular this is true in the field of shell elements and solution strategies.

* Despite the enlarged range of applicability of the method it cannot be overlooked that nonlinear buckling and collapse analyses are still very expensive. Numerical methods often require a lot of parametric studies to get a sound knowledge of a buckling problem.

* Therefore, the current trend is directed towards more efficient but still reliable elements and algorithms. A typical example is the bilinear 4-node degenerated element (Fig. 10) and the reduced basis technique [57].

It seems that the effort is going too much into the development of new methods. It is at least as important to enlarge the physical insight of the complex buckling phenomenon of shells allowing a simpler and better modelling of the problem.

Acknowledgement

The study was sponsored, in part, by the Deutsche Forschungsgemeinschaft. The authors gratefully acknowledge this support. They also like to thank A. Diack and B. Brendel for performing some of the numerical calculations.

References

[1] Fung, Y. C.; Sechler, E.E. (eds.): Thin Shell Structures / Theory, Experiment and Design. Prentice-Hall, Englewood Cliffs, 1974.

[2] Harding, J. E.; Dowling, P. J.; Agelidis, N.(eds.): Buckling of Shells in Offshore Structures. Granada, 1982.

[3] Bushnell, D.: Buckling of Shells - Pitfall for Designers. AIAA-J. 19 (1981) 9, 1183-1226.

[4] Gallagher, R. H.: Finite Element Representations for Thin Shell Instability Analysis. In: Buckling of Structures (editor B. Budiansky), Springer-Verlag, New York, 1976, 40-52.

[5] Hughes, T. J. R.; Pifko, A.; Jay, A. (eds.): Nonlinear Finite Element Analysis of Plates and Shells. Proc. ASME - Winter Annual Meeting, Washington, 1981, AMD - Vol. 48.

[6] Ashwell, D. G.; Gallagher, R. H. (eds.): Finite Elements for Thin Shells and Curved Members. J. Wiley & Sons, 1976.

[7] Idelsohn, S.: On the Use of Deep, Shallow or Flat Shell Finite Elements for the Analysis of Thin Shell Structures. Comp. Meth. Appl. Mech. Eng. 26 (1981) 321-330.

[8] Jones, R. E.; Vos, R. G.: Development and Evaluation of Two Non-Linear Shell Elements. Int. J. Num. Meth. Eng. 16 (1980) 65-80.

[9] Bathe, K. J.; Ramm, E.; Wilson, E. L.: Finite Element Formulations for Large Deformation Dynamic Analysis. Int. J. Num. Meth. Eng. 9 (1975) 353-386.

[10] Argyris, J. H.; Dunne, P. C.; Scharpf, D. W.: On Large Displacement - Small Strain Analysis of Structures with Rotational Degrees of Freedom. Comp. Meth. Appl. Mech. Eng. 14 (1978) 401-451; 15 (1978) 99-135.

[11] Wempner, G.: Finite Elements Finite Rotations and Small Strains of Flexible Shells. Int. J. Solids Struct. 5 (1969) 117-153.

[12] Belytschko, T.; Glaum, L. W.: Applications of Higher Order Corotational Stretch Theories to Nonlinear Finite Element Analysis. Comp. & Struct. 10 (1979) 175-182.

[13] Belytschko, T.; Tsay, C.S.: Explicit Algorithms for Nonlinear Dynamics of Shells. In: [5], 1981, 209-232.

[14] Matsui, T. ; Matsuoka, O. : A New Finite Element Scheme for Instability Analysis of Thin Shells. Int. J. Num. Meth. Eng. 10 (1976) 145-170.

[15] Martins, R. A. F. ; Owen, D. R. J. : Elasto Plastic and Geometrically Nonlinear Thin Shell Analysis by the SEMILOOF Element. Comp. & Struct. 13 (1981) 505-513.

[16] Bathe, K. J. ; Ho, L. W. : A Simple and Effective Element for Analysis of General Shell Structures. Com. & Struct. 13 (1981) 673-681.

[17] Ramm, E. : A Plate/Shell Element for Large Deflections and Rotations. In: US-Germany Symp. on "Formulations and Computational Algorithms in Finite Element Analysis", MIT, 1976, MIT-Press 1976.

[18] Hughes, T. J. R. ; Liu, W. V. : Nonlinear Finite Element Analysis of Shells. Part I: Three-Dimensional Shells. Comp. Meth. Appl. Mech. Eng. 26 (1981) 331-362.

[19] Kanok-Nukulchai, W. ; Taylor, R. L. ; Hughes, T. J. R. : A Large Deformation Formulation for Shell Analysis by the Finite Element Method. Comp. & Struct. 13 (1981) 19-27.

[20] Brendel, B. : Zur geometrisch nichtlinearen Elastostabilität. Dissertation, Universität Stuttgart, 1979.

[21] Brendel, B. ; Ramm, E. : Nichtlineare Stabilitätsuntersuchungen mit der Methode der finiten Elemente. Ing. Archiv 51 (1982) 337-362.

[22] Brendel, B. ; Ramm, E. : Linear and Nonlinear Stability Analysis of Cylindrical Shells. Comp. & Struct. 12 (1980) 549-558.

[23] Bergan, P. G. ; Søreide, T. : Solution of Large Displacement and Instability Problems Using the Current Stiffness Parameter. Proc. Int. Conf. on "Finite Elements in Nonlinear Mechanics", pp. 647-670, Geilo, Norway (1977); Tapir (1978).

[24] Schweizerhof, K. : Nichtlineare Berechnung von Tragwerken unter verformungsabhängiger Belastung mit finiten Elementen. Dissertation, Universität Stuttgart, 1982.

[25] Loganathan, K. ; Chang, S. C. ; Gallagher, R. H. ; Abel, J. F. : Finite Element Representation and Pressure Stiffness in Shell Stability Analysis. Int. J. Num. Meth. Eng. 14 (1979) 1413-1425.

[26] Floegl, H. ; Mang, H. : Zum Einfluß der Verschiebungsabhängigkeit ungleichförmigen hydrostatischen Druckes auf das Ausbeulen dünner Schalen allgemeiner Form. Ing. Archiv 50 (1981) 15-30.

[27] Mang, H. A. ; Gallagher, R. H. : Finite Element Analysis of Thin Shells of General Form for Displacement Dependent Loads. In: [5], 1981, 65-82.

[28] Gallagher, R. H. : Shell Elements. Proc. First World Congress on Finite Element Methods in Structural Mechanics, Bournemouth, England, 1975.

[29] Horrigmoe, G. : Finite Element Instability Analysis of Free-Form Shells. Report No. 77-2, 1977, The Norwegian Inst. of Technology, The University of Trondheim, Norway.

[30] Brockman, R. A. : A Penalty Function Approach for the Nonlinear Finite Element Analysis of Thin Shells. Dissertation, University of Dayton, Dayton, Ohio, 1979.

[31] Argyris, J. H.; Dunne, P. C.; Malejannakis, G. A.; Schelkle, E.: A Simple Triangular Facet Shell Element with Applications to Linear and Nonlinear Equilibrium and Elastic Stability Problems. Comp. Meth. Appl. Mech. Eng. 10 (1977) 371-403; 11 (1977) 97-131.

[32] Argyris, J. H. ; Balmer, H. ; Kleiber, M. ; Hindelang, U. : Natural Description of Large Inelastic Deformations for Shells of Arbitrary Shape - Application of TRUMP Element. Comp. Meth. Appl. Mech. Eng. 22 (1980) 361-390.

[33] Sharifi, P. : Nonlinear Buckling Analysis of Composite Shells. AIAA-J. 13 (1975) 729-734.

[34] Brebbia, C. ; Connor, J. : Geometrically Nonlinear Finite Element Analysis. Proc. ASCE, J. Eng. Mech. Div. 95 (1969) 463-483.

[35] Bergan, P. G. ; Clough, R. W. : Large Deflection Analysis of Plates and Shallow Shells Using the Finite Element Method. I. J. Num. Meth. Eng. 5 (1973) 543-556.

[36] Batoz, J. L. ; Chattopadhyay, A. ; Dhatt, G. : Finite Element Large Deflection Analysis of Shallow Shells. I. J. Num. Meth. Eng. 10 (1976) 39 - 58.

[37] Thomas, G. R. ; Gallagher, R. H. :
A Triangular Thin Shell Finite Element
Linear Analysis NASA CR 2482
Nonlinear Analysis NASA CR 2483 (1975),
see also Chapter 9 in: [6].

[38] Argyris, J. H.; Scharpf, D. W.: The SHEBA Family of Shell Elements for the Matrix Displacement Method. Part III: Large Displacements. The Aeron. J. of the Roy. Aeron. Soc. 73 (1969) 423-426.

[39] Noor, A. K.; Andersen, C. M.: Mixed Models and Reduced/ Selective Integration Displacement Models for Nonlinear Shell Analysis. In: [5], 119-146.

[40] Ramm, E.; Sättele, J. M.: Elasto-Plastic Large Deformation Shell Analysis Using Degenerated Elements. In: [5], 265-282.

[41] Parisch, H.: A Critical Survey of the Nine-Node Degenerated Shell Element with Special Emphasis on Thin Shell Application and Reduced Integration. Comp. Meth. Appl. Mech. Eng. 20 (1979) 323-350.

[42] Javaherian, H.; Dowling, P. J.; Lyons, L. P. R.: Nonlinear Finite Element Analysis of Shell Structures Using the SEMI-LOOF Element. Comp. & Struct. 12 (1980) 147-159.

[43] Nagtegaal, J. C.; Slater, J. G.: A Simple Noncompatible Thin Shell Element Based on Discrete Kirchhoff Theory. In: [5], 167-192.

[44] Mac Neal, R. H.: A Simple Quadrilateral Shell Element. Comp. & Struct. 8 (1978) 175-183.

[45] Pica, A.; Wood, R. D.: Postbuckling Behaviour of Plates and Shells Using a Mindlin Shallow Shell Formulation. Comp. & Struct. 12 (1980) 759-768.

[46] Wempner, G.; Talaslidis, D.; Hwang, C.-M.: The Simplest Approximation of Shells via Finite Quadrilateral Elements. To be published J. Appl. Mech.

[47] Parisch, H.: Nonlinear Analysis of Shells Using Isoparametric Elements. In: [5], 47-63.

[48] Eidsheim, O. M.; Larsen, P. K.: A Study of Some Generalized Constitutive Models for Elasto-Plastic Shells. Proc. Europe-U.S. Workshop on "Nonlinear Finite Element Analysis in Structural Mechanics", Bochum, 1980, Springer-Verl., 1981.

[49] Crisfield, M. A.: On an Approximate Yield Criterion for Thin Steel Shells. TRRL Report LR 658, Transport and Road Research Laboratory, Crowthorne, Berkshire, 1974.

[50] Bienieck, M. P.; Funaro, J. R.: Elasto-Plastic Behaviour of Plates and Shells. Techn. Rep. DNA 3954 T, Defense Nuclear Agency, Weidlinger Ass., New York, 1976.

[51] Crisfield, M. A.: Incremental/Iterative Solution Procedures for Non-Linear Structural Analysis. In: "Numerical Methods for Non-Linear Problems", Pineridge, Swansea, 1980, 261-290.

[52] Bathe, K.-J.; Cimento, A. P.: Some Practical Procedures for the Solution of Nonlinear Finite Element Equations. Comp. Meth. Appl. Mech. Eng. 22 (1980) 59-85.

[53] Geradin, M.; Idelsohn, S.; Hogge, M.: Computational Strategies for the Solution of Large Nonlinear Problems via Quasi-Newton Methods. Comp. & Struct. 13 (1981) 73-81.

[54] Crisfield, M. A.: Accelerated Solution Techniques and Concrete Cracking. Conference FENOMECH 2, Stuttgart, 1981.

[55] Ramm, E.: Strategies for Tracing the Nonlinear Response Near Limit Points. Proc. Europe - U.S. Workshop on "Nonlinear Finite Element Analysis in Structural Mechanics", Bochum, 1980, Springer-Verlag, 1981.

[56] Crisfield, M. A.: A Fast Incremental/Iterative Solution Procedure that Handles "Snap - Through". Comp. & Struct. 13 (1981) 55-62.

[57] Noor, A. K.; Peters, J. M.: Tracing Post-Limit-Point Paths with Reduced Basis Technique. Comp. Meth. Appl. Mech. Eng. 28 (1981) 217-240.

[58] Powell, G.; Simons, J.: Improved Iteration Strategy for Nonlinear Structures. Int. J. Num. Meth. Eng. 17 (1981) 1455-1467.

[59] Bergan, P. G.: Solution Algorithms for Nonlinear Structural Problems. Comp. & Struct. 12 (1980) 497-509.

[60] Häfner, L.; Ramm, E.; Sättele, J. M.; Stegmüller, H.: Programmdokumentation - Programmsystem NISA 80. Report, Institut für Baustatik, Universität Stuttgart, 1980.

[61] Yang, T. H.; Guralnick, S. A.: Buckling of Axially Loaded Open Shells. Proc. ASCE, J. Eng. Mech. Div. 102 (1976) 199-211.

[62] Walker, A. C.; Sridharan, S.: Analysis of the Behaviour of Axially Compressed Stringer-Stiffened Shells. Proc. Instn. Civ. Engrs., Part 2, 69 (1980) 447-472.

Part II :

Cylindrical Shells

BUCKLING OF ORTHOGONALLY STIFFENED CYLINDRICAL SHELLS USED IN OFFSHORE ENGINEERING

P.J.DOWLING, J.E.HARDING, N.AGELIDIS and W.FAHY

Department of Civil Engineering,
Imperial College, London

Summary

This paper presents the results of tests on a series of small and large scale stringer stiffened cylinder models. Emphasis is placed on the fabrication techniques, methods of imperfection measurement and subsequent testing. A discussion of the resulting data and a comparison with available design rules have also been included.

Introduction

One of the fields in which stiffened cylindrical shells have found extensive use in recent years is offshore engineering. Of course, stiffened cylinders are not new in engineering construction: aerospace engineers, for example, have been using them with great success for decades. However, there are several factors which make offshore cylinders unique and limit the extrapolation of experience from other fields.

Firstly, offshore construction employs structural grade steel, whose stress-strain characteristics differ markedly from those of the alloys used in the aerospace industry. Secondly, conventional structural engineering fabrication techniques such as flame-cutting and welding are used, resulting in residual stresses and imperfections quite different from those produced by other fabrication methods. These factors, combined with the loading and environmental conditions present in offshore locations result in cylinder cross-sections of stockier proportions than those used elsewhere.

Although ring-stiffened cylinders have been used more

Buckling of Shells, Ed.: E. Ramm
Proceedings of a State-of-the-Art Colloqium

frequently in offshore construction to date, situations in which large axial loads, bending moments, shear forces and concentrated loads are present may make the orthogonally stiffened cylinders an economical alternative. It is interesting to note that the first tension leg platform, at present being constructed for the North Sea, uses ring stiffened construction for its main column members, except in areas of local peak stressing. A study of the use of stringer stiffeners under the type of loading relevant to this structure may show them to be generally more economic. The need for a more complete understanding of the behaviour of stiffened cylinders has prompted work at many centres around the world.

Any investigation into stiffened cylinder behaviour is incomplete without the corroboration of experimental work. The factors which invalidate the extrapolation of results from other fields, namely, material characteristics, residual stresses and imperfections, and the effects they have on cylinder behaviour need to be determined experimentally. A corollary to this is that any experimentation in which these parameters are not accurately quantified is of limited value. This paper describes testing techniques developed at Imperial College. Fabrication methods have been developed for the production of both small and large scale models. Procedures for monitoring and processing initial imperfections and testing data are also presented.

Small scale vs large scale models

Obviously, if experimental work were not constrained by economic factors, most testing would be carried out on large scale models. These allow the use of full scale fabrication methods and materials to produce residual stresses, imperfections and boundary conditions representative of the prototype structure. In addition, large models allow a higher degree of instrumentation and thereby a closer subsequent check on analytical procedures.

Small models, of course, are much cheaper; the small

cylinders described in this paper were approximately 1/8th of the cost of the large model. In small scale testing it is not always possible to represent precisely all details of the full size structure. The aim is, rather, to test an accurately described model, in order to investigate the effects of certain parameters and enable comparisons to be made with theoretical and design rule predictions. Although many parameters may not be representative of full scale construction, they must, nevertheless, be known in order that the causes of observed behaviour can be determined.

The two types of models are clearly complementary; reliance on one is expensive, on the other, incomplete.

Background to tests

This paper describes tests carried out at Imperial College on one large scale and nine small scale stringer-stiffened models. Details of all the models are given in Table 1.

The large and six of the small cylinders were part of an extensive testing programme, funded by the Department of Energy and carried out at several centres around the U.K. The geometric parameters of the models were selected by the D.En. to be representative of current structures. The six small models consisted of three pairs of nominally identical cylinders. One of each pair was tested under concentric axial load and the other under eccentric loading. The large model was loaded concentrically.

The three remaining small cylinders were part of a programme of twelve tests to investigate the interaction of combined axial and pressure loading, funded by Det norske Veritas and carried out jointly at Imperial College and Surrey University. Detailed initial imperfection measurements on all twelve cylinders and the testing under concentric axial load of three of them were carried out at I.C. The remaining nine cylinders were tested under various combinations of axial and pressure loading at Surrey.

Table 1. Model Details

CYLINDER	1	2	3	4	5	6	A	B	C	IC4 Centre panel	IC4 Outer panels
Radius (R) mm	160	160	160	160	160	160	160	160	160	599.2	600
Thickness (t) mm	0.84	0.84	0.84	0.84	0.84	0.84	0.84	0.84	0.84	3.53	4.0
Length (L) mm	65	65	180	65	65	180	64	128	192	666	666
Stiffener depth (d) mm	6.72	6.72	13.44	6.72	6.72	13.44	6.72	6.72	6.72	48	48
No. of stiffners	40	20	40	20	40	40	40	40	40	20	20
R/t	190	190	190	190	190	190	190	190	190	170	150
L/R	0.41	0.41	1.125	0.41	0.41	1.125	0.40	0.80	1.20	1.11	1.11
d/t	8.0	8.0	16.0	8.0	8.0	16.0	8.0	8.0	8.0	13.6	12.0
Panel b/t	30	60	30	30	60	30	30	30	30	53.3	47.1
Load eccentricity	0	0	0	.25R	.40R	.40R	0	0	0	0	0
Compressive yield N/mm²	344	344	344	344	344	344	344	344	344	289	344
Young's Modulus kN/mm²	202	202	202	202	202	202	202	202	202	205	213
Failure load kN	357	321	430	277	190	270	350	371	371	4190	
Failure stress N/mm²	334	336	332	383* 372+	348* 327+	378* 356+	327	347	352	251	

* Extreme values assuming single bending theory from load bearing position

\+ Extreme values assuming single bending theory from load cell readings

The six small cylinders in the D.En. programme and the three cylinders in the DnV work will be referred to as the first and second series and will be numbered 1 to 6 and A to C respectively throughout this paper. The large cylinder will be referred to by its original identification, IC4.

Small scale models

All the small models were manufactured to similar scales, from the same material, and using the same fabrication methods. Apart from the geometric parameters, the major difference between the models was the end condition detail used in the two series.

The models were all of nominally 320 mm diameter with 20 or 40 internal stringer stiffeners and were fabricated entirely from 0.84 mm steel sheet.

It was decided to use fully clamped ends for all the models due to the difficulty in achieving simply supported end conditions in practice. Also it was considered important to have an accurate estimate of the residual stress levels produced by the manufacturing process; however, the measurement of residual stresses is extremely difficult in models of this size and, in any case, it is unlikely that they would be representative of full size construction. It was decided, therefore, to stress-relieve all the models and then attach the end rings using a method which would not induce residual stresses. Consequently, it can be assumed that the residual stress levels were approximately zero for all the small models.

Fabrication

The first step in the manufacture of the models was to TiG weld the stiffeners to a flat sheet, the length of which was equal to the circumference plus a small allowance for shrinkage. This was subsequently rolled around a copper forming mandrel and the closing T-butt weld was laid. An outer copper former was then clamped to the assembly and, in this tightly-restrained condition, the model was stress-relieved in an inert atmosphere heat treatment oven. Once the assembly was dismantled, the ends of the cylinder were lapped parallel and flat in preparation for the attachment of the end rings.

In the first series the clamping of the ends was achieved by casting the model, using a sand-araldite mixture, in the annulus between two heavy steel rings as shown in Fig. 1. The shell projected slightly beyond the end rings, so that during the test the load was applied directly to the shell, with the end rings only providing torsional restraint. When these models were located in the test rig this small projection was lapped against the loading platens and an extremely even contact between the surfaces was achieved.

However, this end ring arrangement was unsuitable for the second series, where waterproofing requirements for the hyperbaric tests resulted in the choice of the end ring detail shown in Fig. 2. The ends of the cylinder were fitted against the small, accurately machined step in the bottom of the groove in the end rings, and a sand-araldite mixture was once again used to cast the cylinder in position.

Material properties

Each model was fabricated from a single sheet of material from which test coupons were also cut and subjected to heat treatment at the same time as the models. Tensile tests were carried out on specimens cut from these coupons. Different approaches exist for the calculation of compressive yield stress. A value of 344 N/mm^2 has been used in this paper, corresponding approximately to lower tensile yield + 10 N/mm^2 or lower tensile static yield + 10%.

Initial imperfections

The six models of the D.En. programme and all twelve models in the DnV series were scanned in considerable detail to determine the imperfections in the initial shapes of the cylinder surfaces. In all cases the grid of readings taken for each model consisted of 160 readings around the circumference and up to 26 readings along the cylinder length. The models in the second series were also scanned to determine the lateral imperfections of all the stiffeners.

Shell surface imperfections

The shell surface imperfections were obtained using a measuring frame devised for the first series and subsequently modified for the second series. The modified version is shown in Fig. 3, located in the testing rig in which it was also used to measure the radial deflections during loading of the models. In the modified version the frame enclosed the model and supported a ring carrying forty transducers, equally spaced around the circumference and in contact with the cylinder. The ring could be raised or lowered by means of the threaded columns which supported it.

The frame could also be rotated about the vertical axis and positioned using a locating pin. By successively rotating the frame through a quarter of the angle between adjacent transducers, the number of readings around the circumference was increased to 160.

It is estimated that the results obtained using this equipment were accurate to within 0.05 mm.

The results were analysed by a program which used a least squares fit to determine the perfect cylinder which best fitted the imperfection data and calculated the deviation of all the points from this perfect cylinder. It also calculated the coefficients of the Fourier series:

$$w_0 = \sum_{M=1,2,} \sum_{N=0,1,} (A_{MN} \cos N\vartheta + B_{MN} \sin N\vartheta) \sin\frac{M\pi x}{L}$$

which approximates the imperfect surface. It was decided to use a half-sine series expansion even though neither a sine series nor a cosine series is capable of representing the imperfect shape completely. The sine series results in zero imperfections along the upper and lower circumferences, whereas the cosine series produces zero slopes at these positions.

Stiffener imperfections

The equipment shown in Fig. 4 was built to measure the lateral imperfections of the stiffener tips for the models in the second series. It consisted of a spring loaded carriage which could roll along a machined beam of trapezoidal cross-section. Attached to the sliding carriage was a transducer, bearing against one end of a pivoting right-angle mechanism whose other end was in contact with the side of the stiffener. By sliding the carriage along the beam to predetermined positions, readings of the stiffener imperfections could be obtained.

The readings were then analysed to obtain the best fit straight line through them and the imperfections of the stiffeners from these imaginary lines. The equipment is estimated to be accurate to within 0.03 mm.

Imperfection results

The important features from the imperfection surveys are summarized in Table 2. The full results are presented in Refs. 1 and 2 for the first and second series respectively. The largest measured values of various imperfections specified in the DnV rules [3] are given, together with the DnV tolerances in brackets.

The manufacture of the second series of models is seen to be extremely accurate, satisfying all the DnV tolerances, while the first was somewhat less so. The major component of the imperfections of the first series was usually an ovality in the cylinder cross-sections, which was much less apparent in the second series. This difference was probably due to the end conditions used: the second type forced the models to conform to a circular shape. All the models were extremely straight

Table 2 Initial imperfections.

CYLINDER	Perfect Rad.	Max. outward reading, $\Delta r+$	Max. inward reading, $\Delta r-$	Out-of-straightness, δ_g	Deviation from circle ε	Stiffener out-of-straigthness w_s
2	159.91	.56(.80)	-.40(.80)	.12(.39)	.44(.44)	
3	159.93	.94(.80)	-.67(.80)	.19(.39)	.46(.58)	
4	159.90	.41(.80)	-.52(.80)	.17(.39)	.29(.44)	
5	159.89	.74(.80)	-.65(.80)	.23(.39)	.52(.44)	
6	159.98	.80(.80)	-.65(.80)	.15(.39)	.58(.58)	
A	159.99	.10(.80)	-.22(.80)	.10(.39)	.21(.44)	.08(.10)
B	159.90	.17(.80)	-.41(.80)	.20(.39)	.26(.53)	.12(.19)
C	159.91	.24(.80)	-.20(.80)	.17(.39)	.31(.59)	.21(.29)
IC4	599.2	2.15(3.00)	-2.90(3.00)	1.20(1.50)	1.80(3.40)	.80(1.00)

Notation is as used in DnV rules

Bracketed figures are the equivalent DnV tolerances

along their lengths, and this may have been an important factor in the high failure loads obtained.

Test rig

The models were tested in a circular frame 2.5 m high and 1.0 m in diameter (Fig. 5). The frame consisted of four 125 mm thick plates acting as a multiple sandwich for the model, load cells and screw jack. The screw jack was located between the two bottom plates and load was applied by reaction against the upper plate through six 50 mm diameter tension rods. The load was transferred directly through three load cells which provided the only means of support to the plate which carried the model and transducer frame.

Although self-reacting, the rig was also anchored to the laboratory strong floor by means of a 200 ton capacity cross-frame which considerably increased its stiffness and allowed the post-collapse path to be followed with greater control.

Instrumentation

All but one of the models were strain gauged. The main aims in the first series were to check the uniformity of loading and to determine the areas of first yield and initiation of failure.

In the second series, where cylinder length was the only variable, attention was focussed on the distributions of strains along the lengths of the models. For economy, only the top half of the cylinders were strain gauged. This was considered justified by the assumption of symmetry about the midheight circumference, which, at least until the elastic limit, was probably not severely violated.

The axial shortening of the models was measured by three displacement transducers located within the model, symmetrically placed about its centre.

Description of tests

The tests were carried out under displacement control, although the loads were continuously monitored. Initially, an elastic test was carried out for each model to check the line of action of the load and the measuring and loading systems. The failure test was then carried out, applying displacement increments up to and considerably beyond failure.

All the models exhibited a linear load-deflection response, typically until approximately 80% of the failure load. The strain gauge and radial deflection readings showed that the end restraints induced prebuckling deflections of the expected barrelling form. The response then departed smoothly from linear behaviour, and a reasonably gradual failure and unloading was obtained in most cases. The failure was most sudden and the unloading most rapid for the broad panelled, twenty stiffener models, nos. 2 and 5. The load-deflection plots for cylinder B and cylinder 2, typical of the narrow panelled and broad panelled models respectively are given in Fig. 6.

In all cases, the models failed at loads producing average stresses (in the axial tests) or maximum stresses (in the eccentric tests) near yield. The failure loads for all the tests are presented in Table 1.

In the first series of tests the largest strains were observed at the outer surface of the shell near the end rings, produced by the clamping of the barrelling deflections. Yield at these locations was followed by yielding of the stiffener outstands at midheight. In all cases the failure modes, although consisting of a pattern of buckles spreading around all or most of the model circumferences, were made up of buckles, each of which was contained within a single panel.

For the narrow panelled cylinders (nos. 1, 3, 4 and 6), the failure mode was a ring of alternating inward and outward ridge-type buckles around the cylinder midheight. Models 3 and

6 were longer and had more slender stiffeners, and for these cylinders the panel buckling was accompanied by sideways tripping of the stiffeners. Failure of the stiffeners probably resulted from their higher slenderness and the lower degree of lateral restraint provided by the end boundary conditions due to the longer stiffener span. The longer model length and rapid stiffener unloading produced a more sudden load fall-off for these cylinders than for the shorter models, 1 and 3.

The broad panelled cylinders, 2 and 5, on the other hand, failed in a mechanism near the ends of the models. The short model lengths and wide stiffener spacing enabled this mechanism to form.

A comparison of the axially loaded and eccentrically loaded cylinders shows that similar behaviour and failure modes were observed for the corresponding cylinders in each load type. In the narrow panelled models the stress at failure in the most heavily loaded sections was about 10% higher than the average stress in the corresponding axially loaded models, indicating that a redistribution of stress had taken place. Initial buckling occurred at one side of the cylinder, and the drop in load-carrying capacity of the buckled region resulted in an elastic unloading of the remainder.

Very little stress redistribution occurred in the broad panelled eccentrically loaded cylinder (no. 5), most probably due to the more slender nature of the panels. The stress in the critical region at failure was virtually the same as the average stress for cylinder no 2, the axially loaded counterpart.

The different end conditions used in the second series of models produced different failure mode shapes, although the ultimate loads were once again very near squash. It appears that the stiffer restraint against radial expansion provided by the portion of the end rings in contact with the shell produced greater curvatures near the ends and resulted in the end condition failure modes obtained. In all three cylinders the failure mode was a ring of outward buckles, restrained to some extent by the stiffeners, which spread around the entire circumference very near the bottom end ring.

In the case of the longest cylinder, an overall mode consisting of buckles, each spanning approximately five panels, occurred virtually independently and simultaneously with the end condition mode. Only one of these buckles was discernible by eye during the test; it is clearly visible in the right portion of Fig. 7.

However, the entire pattern can be seen in a plot of the radial deflection data for this cylinder in Fig. 8. The figure gives the radial deflections just before and just after failure along two circumferential sections of the cylinder. The lower diagram shows the buckling pattern near the lower end ring; it is not as uniform as in the shorter cylinders because the peaks in the ring of buckles were not all contained in a single set of readings. The upper plot, the radial deflections further up the cylinder, shows an entirely different mode shape: the eight circumferential waves can clearly be seen.

Although no other model of either series buckled in such an overall mode, similar mode shapes could be detected in the prebuckling deflections of several cylinders. These almost coincident, and perhaps interacting, mode shapes could easily have escaped detection had detailed radial deflection readings not been taken during the tests.

As an example of the results obtained from the strain gauging of the models, for a full account of which the reader is directed to Ref. 4, four diagrams (Figs. 9 - 12) have been included here showing the hoop strains and hoop stresses along the lengths of the two longer cylinders, B and C. The plots of the hoop strains are quite regular in form, with maximum tensile strains near the centre of the models and much smaller strains near the radially restrained ends.

The hoop stress plots are given in Figs. 11 and 12. In

both cases relatively large compressive stresses were observed near the ends, the effect of the radial restraints. Near the centre of the models the hoop stresses were approximately zero for cylinder B, while in the longer cylinder C they were tensile. Unfortunately, some of the gauges on the short cylinder did not function and a full set of readings is not available. However, by interpolating some of the missing readings, it can be confidently estimated that the hoop stresses in the centre of that model were compressive. These observations are all consistent with the expected effect of the radial restraint: as the model length increases, the restraint to barrelling near the centre decreases and the hoop stresses present there would become more tensile.

Large scale model

A three bay model with twenty stringer stiffeners was tested as part of the D.En programme. The three bays produced a more realistic boundary restraint for the centre bay, the panel under investigation. The centre panel was made from 3.5 mm Grade 43 steel plate, whereas the outer panels were of 4.0 mm Grade 50 steel. The geometric and material properties of this model are given in Table 1.

A full account of the residual strains, initial imperfections and strain gauge data is contained in Ref. 2. This discussion will deal with some of the more interesting features only.

Fabrication

Fig. 13 shows the fabrication details of the model. The first stage in the fabrication sequence was the butt welding of the flat plate panels. Then the combined plate was rolled, butt welded along the longitudinal seam and re-rolled to enhance its circularity. Next, the end flanges were attached. The ring stiffeners were made up in sections of one quarter of the circumference and attached to the cylinder using intermittent fillet welds. Finally, the stringers were welded, again using relatively light intermittent welds. The central bay stiffeners were added first, and then those in the outer bays; in each case diametrically opposite stiffeners were attached alternately. All the welding of the model was carried out using the MiG welding process.

Residual Strains

Residual strain readings were taken after each major stage of fabrication using an electrical extensometer of the type developed at Cambridge University [5]. The measurements were taken across 100 mm gauge lengths defined by small punch marks in corresponding locations on both sides of the plate, and were averaged to obtain midplate values.

A large increase in the circumferential strain readings occurred during rolling. This effect was primarily due to the reduction in the linear distance between gauge marks which occurs as a flat plate is bent to an arc, and produced readings of about 1000 $\mu\varepsilon$. The remaining fabrication produced a further average change in the hoop strain of about 80 $\mu\varepsilon$, although some individual readings increased by as much as 350 $\mu\varepsilon$.

The longitudinal strain readings, on the other hand, were little affected by the rolling of the cylinder. At that stage the longitudinal strains averaged about 85 $\mu\varepsilon$ and were probably due to a Poisson effect from both the butt welding of the flat plates and the rolling. As would be expected, the greater part of the longitudinal residuals was accumulated during the attachment of the stringers. At completion of the model, the longitudinal strain was fairly uniform throughout the shell, with an average value of about 350 $\mu\varepsilon$ or 21% of the yield strain.

Three longitudinal stiffeners were also gauged for residual strains before and after being welded to the cylinder. The maximum compressive strain occurred at the tip of the stiffener at midheight, except for the stiffener along the longitudinal butt weld; this area was moderately heat treated after completion in an attempt to reduce the local imperfections and may have resulted in the different residual strain distribution in this stiffener.

Initial imperfections

Initial imperfection readings of the shell surface were obtained by rotating the cylinder against a vertical bank of transducers. A straight reference was used to provide a datum and the distance between the datum line and the axis of rotation was accurately measured. The grid of readings obtained for the central panel was analysed as before to determine the best fit cylinder, the imperfections and the Fourier coefficients of the imperfect shape.

Fig. 14 is a pictorial representation of the cylinder surface, in which most of the important features are discernible. The largest imperfections occurred along the obvious inward kink produced by the longitudinal butt weld. Also clearly visible is the major imperfection mode of four full waves around the circumference, which was at least partly due to the rings having been manufactured and attached in quarter sections.

In the longitudinal direction the predominant mode was a single half-wave along the length (i.e. a barrelling imperfection). Superimposed upon this, however, were small inward kinks and adjacent outward bulges near the rings, the combined result of the circumferential butt welds and the attachment of the rings. These can be seen in a plot of the imperfections along a representative vertical section of the shell (Fig. 15).

Measurements were also taken of the imperfections of the stringers, using a dial gauge travelling along a straight-edge which was positioned against the stiffener.

The maximum values of various imperfection measurements and the DnV tolerances on them are given in Table 2. All the tolerances are satisfied, but not by such a large margin as to make the cylinder very different to full scale construction.

Test rig

The model was tested under concentric compressive loading in the test rig shown in Fig. 17. Twenty four jacks, each of 20 T rating, were equally spaced beneath the cylinder's circumference and were all fed from a common pressure supply to ensure a uniform concentric loading. Because the jacks were directly beneath the cylinder shell, a relatively light bottom platen was sufficient to distribute the load. At the top, however, the reaction against the cross-frame was through a central spherical bearing and the top platen had to be substantial to minimize bending distortions. The cylinder was bedded onto the platens using a sand and araldite mixture to ensure as uniform a load application as possible.

Instrumentation

Also visible in Fig. 17 is the rotating transducer frame used to measure the out-of-plane deflections during the test. The frame, which carried two vertical banks of transducers, rotated on an accurately machined steel ring bedded onto the top platen, and could be located at twenty predetermined positions using a spring loaded plunger. Using this method, a repeatability of readings to within 0.1 mm was achieved.

A large number of strain gauges were used to investigate the distribution of strains in the shell,stringers and rings.

Test description

Fig. 18 is the autographic record of the testing machine - a continuous plot of the load against the deflection of the loading platen of the rig. After an initial "bedding in", the load-deflection behaviour was linear. Because the model had already been loaded to 160 tons in an elastic test, and some plastic straining had occurred, on reloading a small change in slope was observed when this load was exceeded.

The first part of the test, up to 320 tons, was carried out under load control. After each load increment, a small amount of creep occurred, which is represented by the short horizontal sections of the graph. At 320 tons the machine was changed to displacement control, and from then on the model relaxed at the end of each increment; the vertical sections of the graph represent this shedding of load.

Visual inspection of the model first revealed the imminence of failure at a load of 410 tons: a small buckle had appeared near the top of a panel almost diametrically opposite the longitudinal weld. A slightly larger load fall-off can be seen on the autographic record. However, the model remained stable and testing continued until the load reached 420 tons or 0.87 times the squash load. New buckles gradually formed about 90^{0} around from the first one and then linked up to form a "chequerboard" pattern of local panel buckles extending around approximately one third of the circumference. The load dropped to 340 tons. The unloading part of the curve was obtained by applying further deflection increments, which increased both the magnitude and extent of the buckles.

It seems that the cylinder behaviour was strongly influenced by the initial imperfections of the model. Fig. 19 presents the radial deflections of the cylinder at 390 tons around two circumferential sections: the upper diagram for the midheight circumference and the lower diagram for a section about one quarter of the way up the centre panel. At midheight an overall deflection mode containing eight waves can be clearly seen. A comparison of this mode with the initial imperfections reveals a marked similarity in the locations of peaks and troughs. Lower down the cylinder, where the proximity to the rings restrained the deflections of the stringers, this overall mode was broken up by the superimposed local panel mode. The stiffener deflections were almost invariably smaller than those of the plate panels they enclosed.

If the deflections are plotted along vertical sections, and this is done for the panel in which the first buckle formed in Fig. 16, the importance of the small bulges adjacent to the rings can be seen. Plate-initiated panel buckling will usually occur as an inward buckle, even though the final overall pattern will also include outwardly deflecting panels. In this model, the inward buckle developed gradually in the concave region adjacent to the outward bulge near the top ring. In fact, the subsequent buckles, which formed independently in a region about one quarter of the circumference around from the first buckle, were located in a similar position near the bottom ring.

The final failure shape, in which the buckles were located fairly centrally in the panel, may, therefore, be misleading in that it obscures the location at which failure was initiated.

The strain gauge data also confirmed the importance of the area near the rings. Figs. 20 and 21 present the midplate longitudinal and hoop strains along a typical vertical section of the model. The hoop strains can be directly related to the shape of the radial deflections, with the largest strains in the regions of the outwardly deflecting bulges. The initially uniform longitudinal strains also tend towards this sort of distribution as the inner shell surface in these areas began to yield, resulting in a reduced stiffness and larger measured midplate strains.

Five of the stiffeners were strain gauged along their lengths, and the readings showed the somewhat surprising fact that at 400 tons all the stiffeners had yielded through virtually their entire cross-section at midheight. Nevertheless, they did not trigger failure. Fig. 23 is an internal view of the cylinder after failure. The first buckle that formed is at the far end of the panel most centrally placed in the photograph; the stiffeners adjacent to it do not show any evidence of tripping.

Comparison with design predictions

The experimental results were compared with the failure loads predicted by various design methods. The results are summarized in Table 3, in which all stresses have been non-dimensionalized by the yield stress. The values given in brackets are for cases for which the methods do not strictly apply.

The first three columns give predictions using the DnV rules [3]. The first value is the buckling stress for local buckling and the other two are the stresses for plate induced and stiffener induced failure, both calculated neglecting the effect of curvature. The next column was obtained using a more accurate method developed by Valsgard and Steen [6] for narrow panelled cylinders to account for the effect of curvature. Finally, a draft proposal by the ECCS [7] and a paper by Samuelson and Vestergren [8] were used in combination to obtain the last two columns, the local buckling stress for a curved unstiffened panel and the stress for overall buckling of an orthotropic shell. In both cases the additional partial safety factor, $\gamma = 4/3$ was omitted.

The flat plate approximation of the DnV rules underestimated the failure load considerably, especially for broad panelled or long cylinders. The refinements introduced by Valsgard and Steen gave the best predictions of all methods for the cylinders to which the method is applicable. However, the assumed failure mode in all cases was overall buckling, which did not generally occur. The ECCS predictions, on the other hand, gave more conservative results, but always distinguished correctly between local and overall buckling.

Both Valsgard and Steen's method, and Samuelson's method predicted the correct mode shape for cylinder C, i.e. eight waves around the circumference.

Table 3. Strength Predictions

CYLINDER NO.	DnV [3]			Valsgard & Steen [6]	ECCS [7]			ρ_{DnV}	ρ_{VAL}	ρ_{ECCS}
	LOCAL BUCKLING	PLATE INDUCED	STIFFENER INDUCED	CURVED PANEL	LOCAL BUCKLING	OVERALL BUCKLING	TEST			
1	.9215	.905	.800	.893	.819	.954	.96	1.20	1.08	1.17
2	.6256	.577	.532	[.6157]	.723	[.916]	.96	1.80	[1.56]	1.33
3	.9215	.885	.791	.872	.819	[.917]	.95	1.20	1.09	1.16
4	.9215	.905	.800	.893	.819	.954	1.10	1.38	1.23	1.34
5	.6256	.577	.532	[.6157]	.723	[.916]	1.00	1.88	[1.62]	1.38
6	.9215	.885	.791	.872	.819	[.917]	1.09	1.38	1.25	1.33
A	.9215	.905	.800	.893	.820	.956	.93	1.16	1.04	1.13
B	.9215	.778	.567	.879	.820	.833	.99	1.76	1.13	1.21
C	.9215	.441	.354	.877	.820	.772	1.00	2.82	1.14	1.30
IC4	.7785	.658	.643	[.755]	.804	[.887]	.87	1.35	[1.15]	1.08

All strength values are non-dimensionalized by the yield stress
Underlined values are for critical modes
Bracketed values are for situations to which the method does not strictly apply.

$\rho = \sigma test/\sigma prediction$

Conclusions

A series of tests on large and small scale stringer stiffened cylinders has been described in this paper. The models were carefully manufactured and attention was paid to the detailed measurement of initial imperfections and residual stresses. This allowed correlations to be made between the model imperfections and the collapse modes. The effect of boundary restraint on the collapse was also investigated.

The two series of small models had very similar failure loads, which were all close to squash. This suggests that the out-of-roundness imperfections, which were significantly larger for the first series, do not have an important effect on collapse. Analytical information suggests that the out-of-straightness, which was very small in all the models, would have had a more significant influence on failure.

The radial deflection readings which were taken during the tests allowed the determination of overall buckling mode shapes and the points of initiation of failure.

The design rule comparisons showed the DnV and ECCS rules to be conservative. The former predicts an incorrect mode of failure in cases. Use of the latter was complicated, especially in the case of orthotropic shell buckling.

Acknowledgements

The authors wish to thank the U.K. Department of Energy, the Science and Engineering Research Council and Det norske Veritas for funding the work described.

References

1. Dowling, P.J., Harding, J.E., Fahy, W and Agelidis, N. "Report on the testing of large and small scale stiffened shells under axial compression" Dept. of Energy Report OT-R-8107, August, 1981

2. Agelidis, N., Harding, J.E. and Dowling, P.J. "Final report on buckling tests on stringer stiffened cylinder models subject to load combinations" Vol. 1. LCMT Report, Jan. 1982

3. Det norske Veritas "Rules for the design, construction and inspection of offshore structures" 1977

4. Agelidis, N., Harding, J.E. and Dowling, P.J. "Axial load tests of three small scale stringer stiffened cylinder models" Imperial College Report, to be published.

5. Denston, R.J. and White, J.D. "An electrical demountable extensometer" Cambridge Uni., Dept. of Eng. Report CUED/C-Struct/TR61 1977

6. Valsgard, S. and Steen, E "Simplified strength analysis of narrow panelled stringer stiffened cylinders under axial compressive and lateral load" Progress Report No. 4, DnV Rep. 80-0590, August 1980

7. European Convention for Constructional Steelwork "Draft edition of shell rules" 1982

8. Samuelson, A. and Vestergren, P. "Analysis of axially loaded stringer stiffened circular cylinders" Aeronautical Research Institute of Sweden, Tech. Note. FFA-2134, Dec. 1979

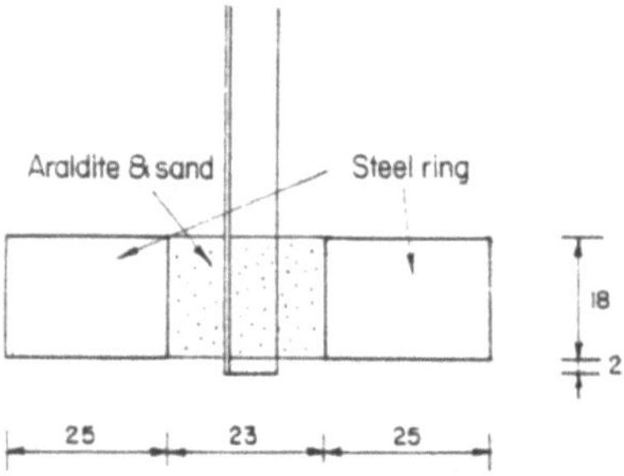

Fig. 1 End ring detail
- Series 1

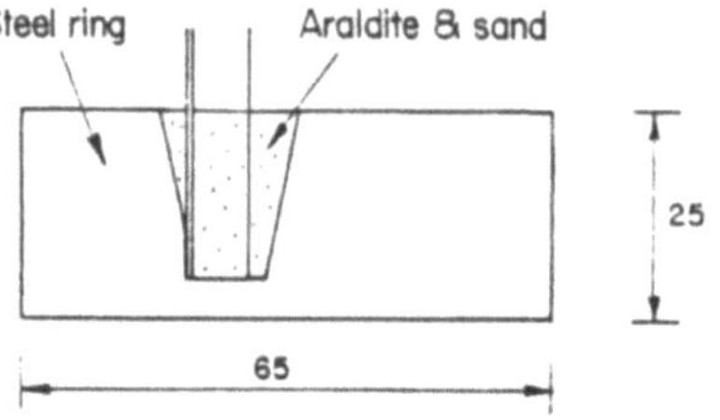

Fig. 2 End ring detail
- Series 2

Fig. 3 Imperfection measuring frame

Fig. 4 Stiffener imperfection measuring equipment

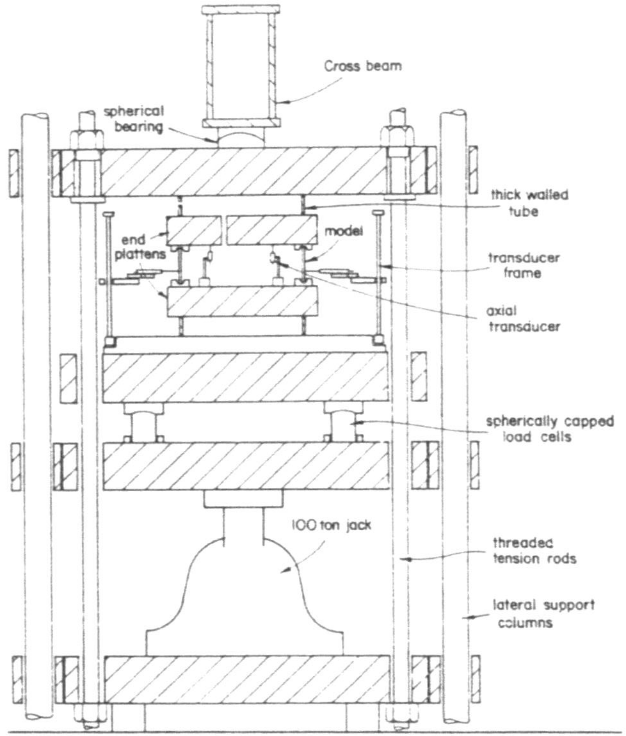

Fig. 5 Test rig

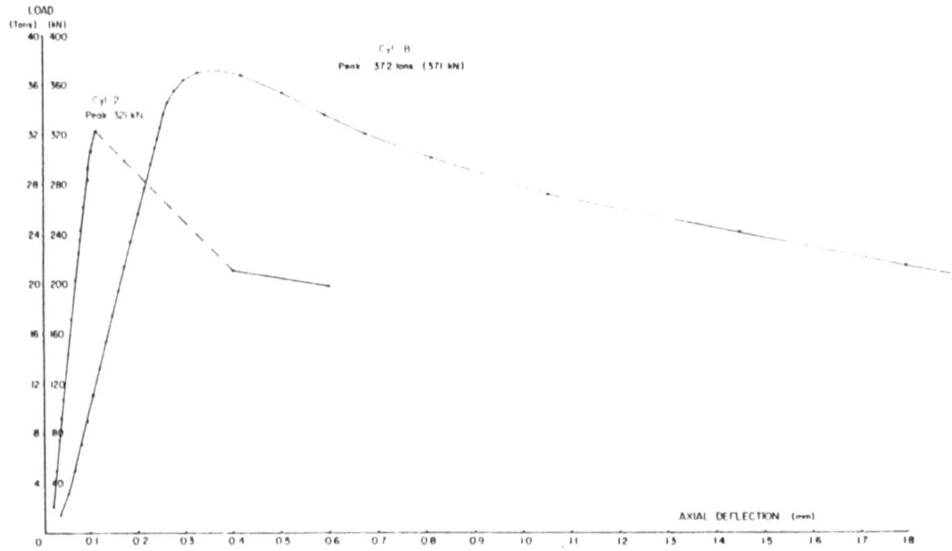

Fig. 6 Load-deflection plots of cylinders 2 and B

Fig. 7 Cylinder C after failure

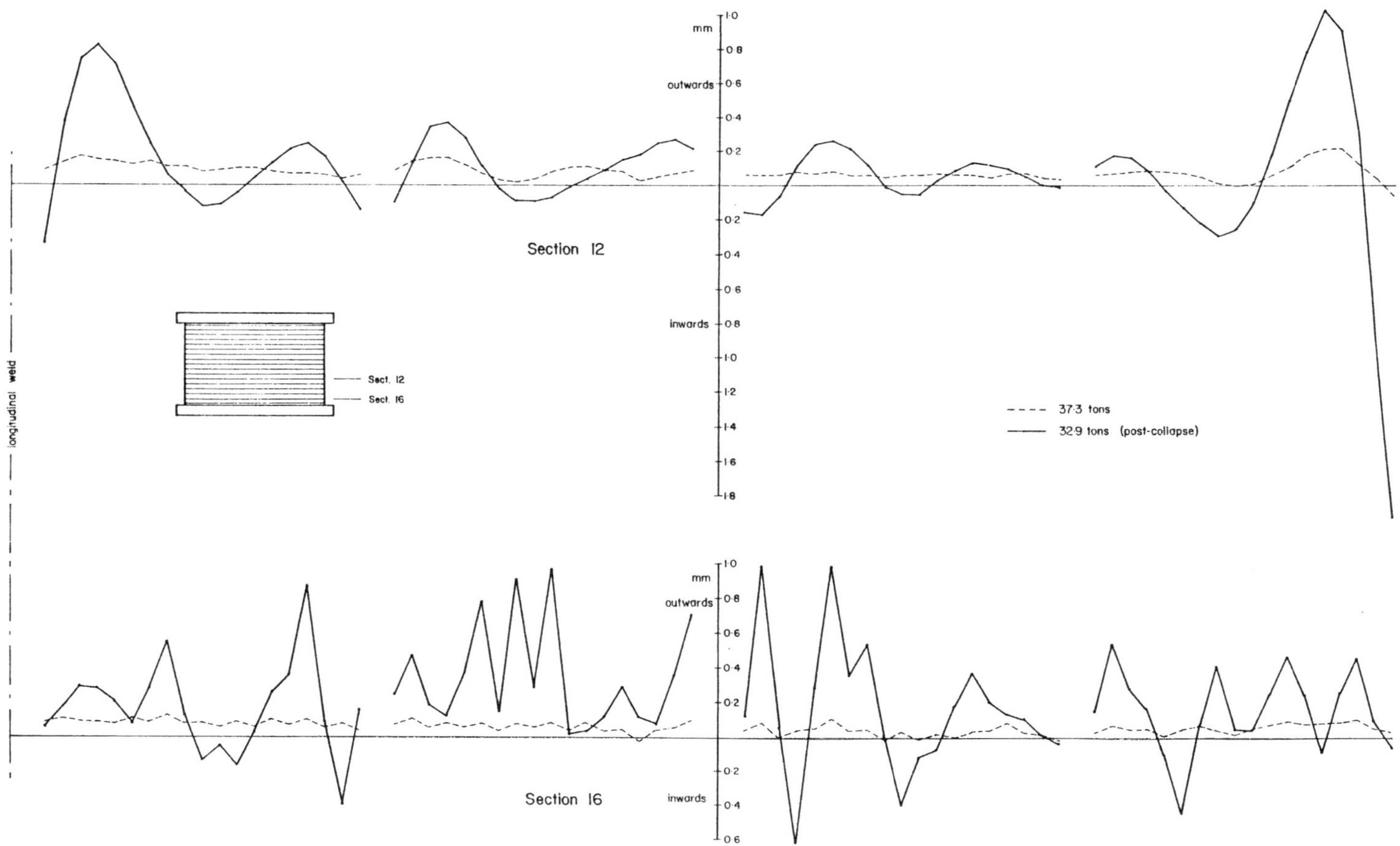

Fig. 8 Radial deflection data - cylinder C

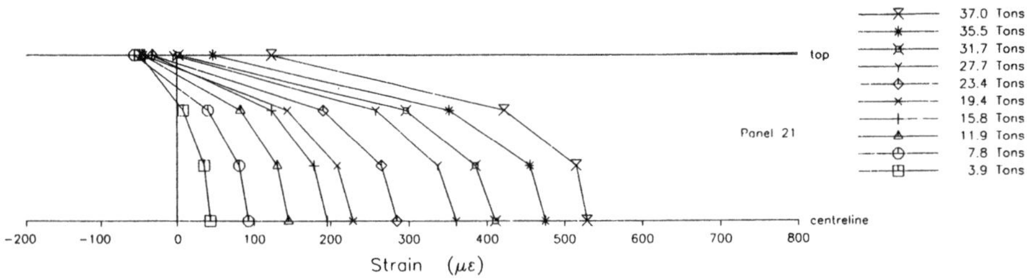

Fig. 9 Hoop strain - cylinder B

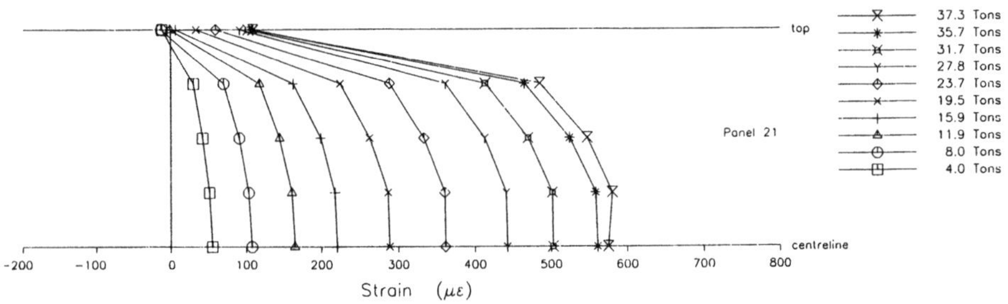

Fig. 10 Hoop strain - cylinder C

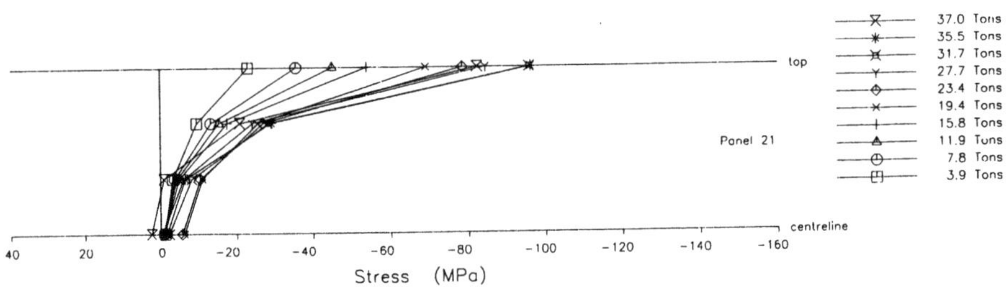

Fig. 11 Hoop stress - cylinder B

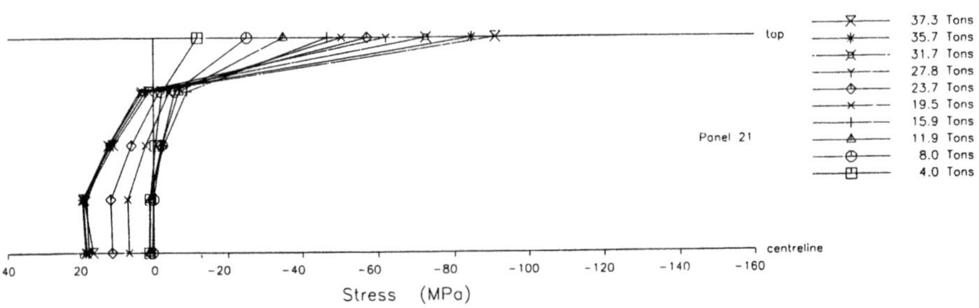

Fig. 12 Hoop stress - cylinder C

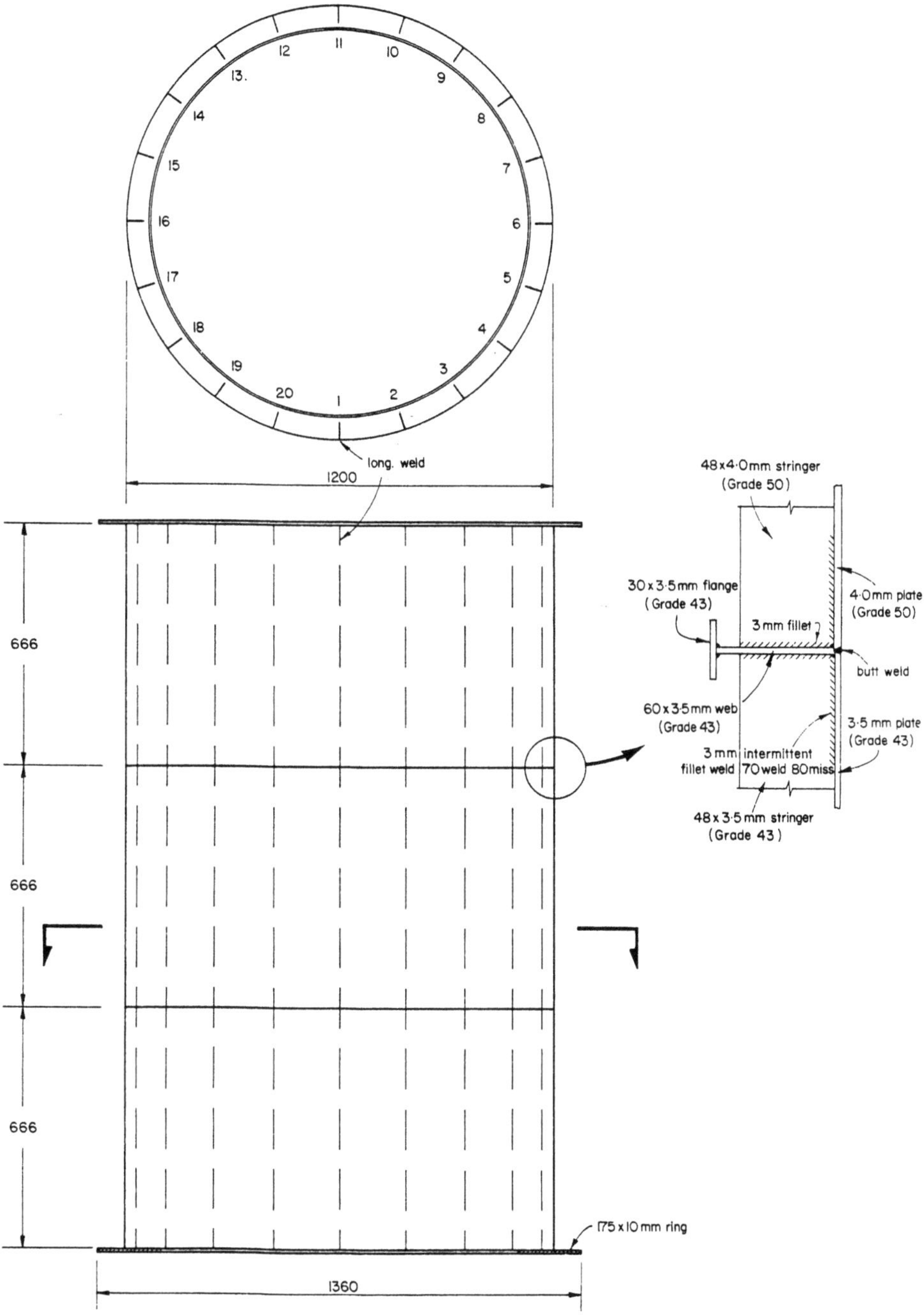

Fig. 13 Fabrication details - cylinder IC4

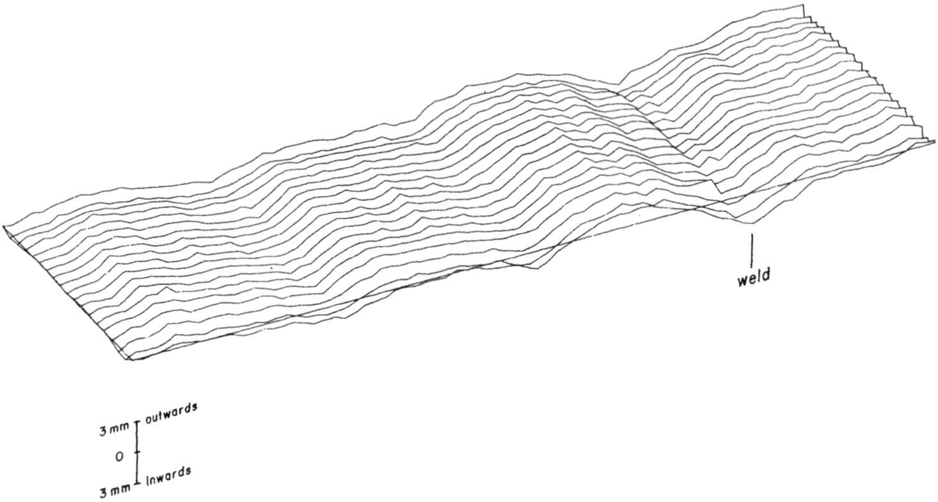

Fig. 14 Imperfection surface - cylinder IC4

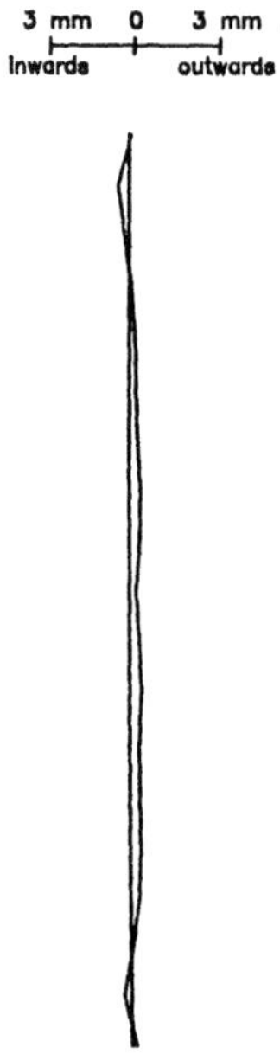

Fig. 15 Imperfections along typical vertical section

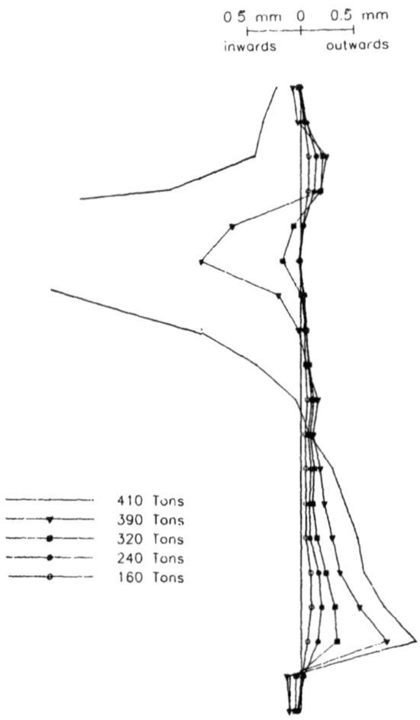

Fig. 16 Deflections along typical vertical section

Fig. 17 Test rig

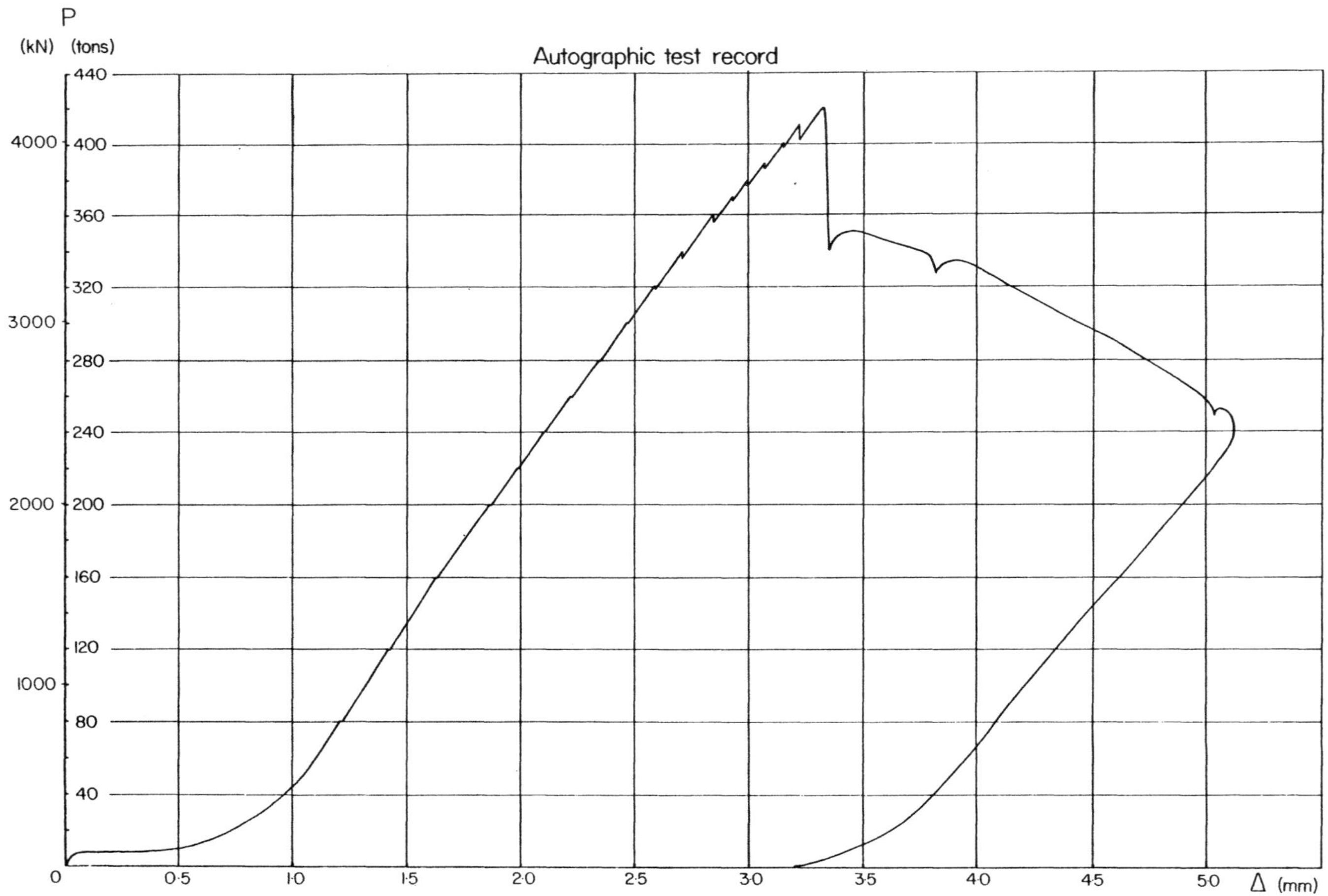

Fig. 18 Load-deflection plot - cylinder IC4

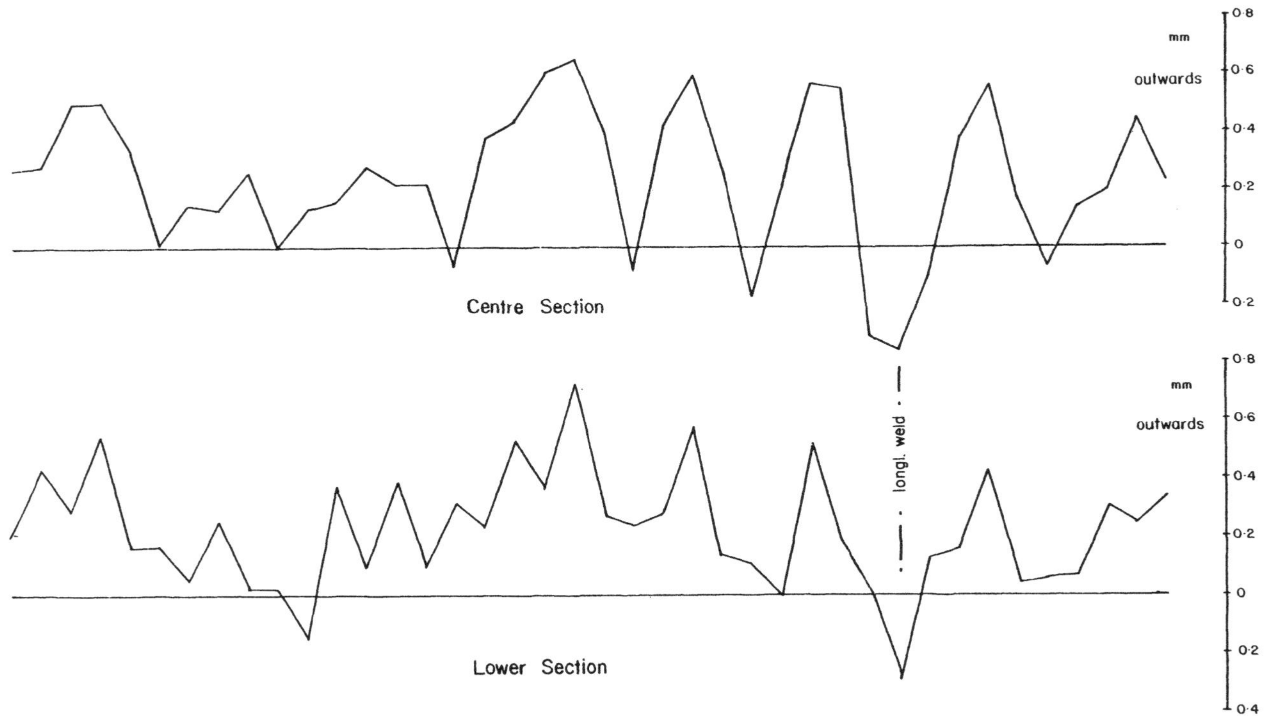

Fig.19 Radial deflection data - cylinder IC4

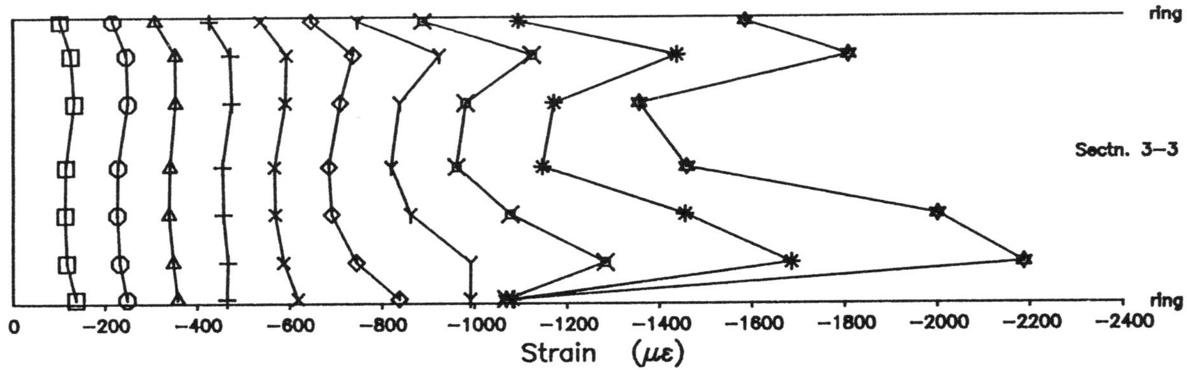

Fig. 20 Longitudinal strain along length - cylinder IC4

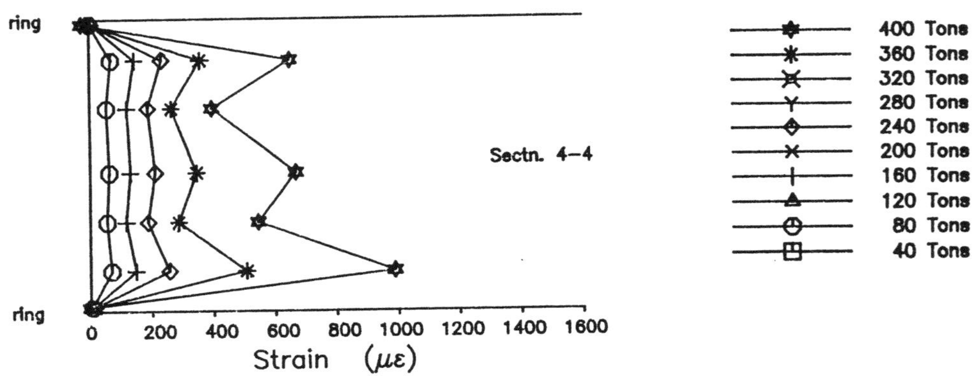

Fig. 21 Hoop strain along length - cylinder IC4

Fig. 22 Interior buckled shape - cylinder IC4

THE BUCKLING OF THIN-WALLED RING-STIFFENED STEEL SHELLS

A C Walker; Y Segal; S McCall

Department of Mechanical Engineering

University of Surrey

GUILDFORD

United Kingdom

SUMMARY

The buckling of thin-walled cylindrical shells subject to various types of combined loading is of current interest to engineers engaged in offshore engineering practice. This paper considers the inter-ring buckling of cylinders when subject simultaneously to pressure and axial loading. Some example results from an experimental research programme on steel models are presented and compared to the interaction diagrams proposed by other researchers.

1. INTRODUCTION

The buckling behaviour of circular cylindrical shells has intrigued engineers for the past half century, both from the intellectual requirement of the theory and the demand for necessary rigour in the experimental treatment. So many aspects have to be considered; including, types of load, shell geometry including initial imperfections, material characteristics boundary conditions, linear and non-linear mathematical formulation etc. that it is not surprising a comprehensive and complete treatment is as yet unavailable and that gaps in our knowledge and understanding of shell buckling still exist. Occasionally this lack of information can become evident when industrial use of thin walled shells requires design data which is not available. Such a situation is developing in offshore engineering where very thin walled shell components are being designed to form parts of floating tethered platforms. These components are subject simultaneously to axial and pressure loading. Although some information exists to guide designers it would seem that it is based on little experimental data relevant to offshore construction materials and practice and on insufficient theoretical background.

In this paper we consider the behaviour of ring stiffened cylinders

Buckling of Shells, Ed.: E. Ramm
Proceedings of a State-of-the-Art Colloqium

subject to axial compression and pressure loading. Particularly, it is assumed that the rings will remain unbuckled and that the deformations will be confined to the portions of shell between the rings. The paper reviews briefly the previous work in this field and presents some preliminary results from a new programme of small scale model testing.

2. Review of Inter-ring buckling

2.1 Theoretical Background

Classical approaches with either Donnell or Flugge approximations in the mathematical formulation have been used to study the buckling of shells subject to load combinations. Timoshenko (1) suggested an interaction equation (Appendix A) based on the general equilibrium equations for axial load together with lateral external pressure. These equations were solved by assuming simple-support boundary conditions and had been previously used by Flugge (2) as the linear part of a 3rd degree polynomial solution. Sharman (3) used this interaction equation (restricted to the linear part) to suggest a lower and upper bound approach (Appendix A) that was originally developed by Terebushko (4) for other load combinations.

Brush and Almroth (5) used the Donnell eighth order differential equation to determine a linear interaction equation (Appendix A) for various buckling modes and referred to interaction curved contained in the handbook by Gerard & Becker (6). Similarly employing the Donnell equation, Lakshmikantham (7) reported the effects of various compressive loading combinations on the stability of stiffened circular cylinders in conjunction with the Batdorf parameter ($Z = (L^2/Rt)(1-\nu^2)^{1/2}$). These equations were used by Krishnamoorthy (8) to derive a solution of the critical stresses relevant to the clamped boundary condition. The effects of including the characteristics of composite materials were studied by a number of researchers, including Baker (9), Levi and Cheng (10) and Hess (11). These studies show similar trends to the isotropic case in the elastic region for the same loading combination.

The above theoretical analyses do not take into account the influence of initial imperfections, boundary conditions and elasto-plastic material behaviour which are particularly important in offshore structures. The effect of initial imperfections have been discussed by Hutchinson (12) in a preliminary manner and refer only to axisymmetric imperfections.

There is interest in this subject in Russia and Eastern Europe, notable among the available publications is that by Zyczkowski (13) who applied a generalised power series method to analyse the stability of circular cylinders under combined axial and external pressure loading. Other reports include Mushtari and Satchenkow (14), Immerman (15), Kurshin and Pustovoi (16) and Kabanov (17).

Recent developments have made great use of computers and there are available a number of general and special purpose programmes. Among the former are STAGS (18) and BOSOR (19) but these tend to be expensive for design analyses. A programme which incorporates better estimates of elasto-plastic behaviour is being developed in the UK (20). This makes use of Sander's non-linear shell equation (21) solved by the Dynamic Relaxation method. The programme includes the effect of asymmetric initial imperfections and has been applied to biaxial loading systems (22). A more special purpose programme restricted to axisymmetric modes of deformation, relevant to very short cylinders, has been developed (25, 26) but like the others although it is valuable for particular geometries it has not been used to generate a wider understanding of the phenomena and interaction effects across the necessary range of geometric parameters.

Abramovich and Singer (23), Midgley and Johnson (34), have studied the effect of combined loading on the buckling of integrally stiffened aluminium cylinders and Croll (24) has developed a lower bound approach for this general problem.

2.2 EXPERIMENTAL BACKGROUND

The comparison of theoretical linear and non-linear analyses to experimental results on "as manufactured" models is very limited. The root to the problem is the small amount of experimental data which could enable such a comparison to be made. Interaction equations based on linear theory were used to verify experimental results on "perfect" elastic models. Donnell equations were employed by Tennyson (27), and together with Booton (28) he described the correlation between experiments carried out on epoxy models subjected to various loading combinations and a proposed interaction equation (Appendix A). Weingertan and Seide (29) and later Foster (30) carried out similar tests on Myler models and suggested a linear interaction equation as a conservative bound. Foster (30) introduced an empirical equation (Appendix A) as a result of tests on fifteen elastic models for axial compression combined with external pressure or torsion. The application of these tests using low modulus materials for practical design is poor and can only give a basic guide for direction in further investigations.

Testing models made from high modulus materials even on a small scale (about 1/20) is a complex effort. Miller (31) pointed out only a few of those problems. His summary of American research includes an interaction equation (Appendix A) which is valid only for certain cases. Among the twenty two models reported just six were tested under axial compression combined with lateral external pressure while the rest were subjected to axial tension and pressure.

Another experimental scheme devoted to this purpose is reported by Abramovich and Singer (23) using aluminium models in their tests while the recent study by Walker and McCall (32) on welded stiffened stringer steel cylinders was reported to DnV but not yet released for publication.

Adding external pressure loads to axial compression acting on cylinders requires some form of hyperbaric type testing rig. The design of such a pressure facility may exhibit increasing complexity as the scale of the models increases or as the level of pressure load increase.

One can compare two testing systems designed for different loading levels and similar model geometry. The first described in Ref.23 and the other in this paper. It is easy to notice the increased problems that the higher loading system provokes. This recent development of the hyperbaric experimental rig described below gives the opportunity for a wider investigation of steel models and some preliminary results are described in the present study. However, this will not be sufficient in itself to provide designers with all the necessary information.

A move to bigger scale modelling (about 1/5) is essential to verify using a small number of tests, the results obtained from a more comprehensive study on the smaller scale models to provide a reliable design code.

3. EXPERIMENTAL INVESTIGATION

As part of a continuing experimental programme at the University of Surrey, eight models were tested to investigate the inter-ring buckling phenomenon. The development of the test procedure and apparatus included two preliminary models; however, because of the uncertain accuracy of the measurements during these tests the results are not included in this paper. The geometries of the remaining six models, S3 - S8 are shown in Table 1, the scale can be regarded as about 1/20 of that contemplated for the outer skin of the legs of practical tethered leg platforms. Since we are considering here the buckling between ring stiffeners, this stage of the research programme was restricted to the consideration of a single tier of a ring stiffened cylinder. Thus we have in effect a plain cylinder between end plattens, as shown in Fig 1. The boundary conditions are assumed to be effectively clamped, and certainly in these, and other tests on stringer stiffened shells, the failed shells indicated that the end supports had maintained the initial form even in the presence of localised plasticity.

3.1 MANUFACTURE OF MODELS

The models were fabricated from specially prepared cold reduced steel strip 0.84mm thick. The manufacturing procedure consisted of rolling the strip round a mandrel of the appropriate diameter and then welding the longitudinal seam using the TIG process. To remove all residual stresses and to obtain a stress-strain diagram representative of the full-scale counterpart, the shell was clamped between the mandrel and an external sleeve and the assembly heat treated in a moist atmosphere oven. The stress relieved shell then had its ends lapped parallel and perpendicular to its length. The end location rings were fitted and the shell cast into place using a sand-Araldite mixture, following which the end parallelity and shell out-of roundness were measured.

TABLE NO 1 SURREY UNIVERSITY MODELS/GEOMETRICAL AND MATERIAL PROPERTIES

Model	Description	Length (mm) L	Maximum Initial Imp. wo(max)/t
S3	Long	115	0.3
S4	"	130	0.3
S5	"	130	0.18
S6	Short	50	0.08
S7	"	50	0.06
S8	"	50	0.14

$r = 160mm$ $t = 0.84mm$ $\sigma_o(av) = 313\ N/mm^2$

$E(av) = 203\ KN/mm^2$

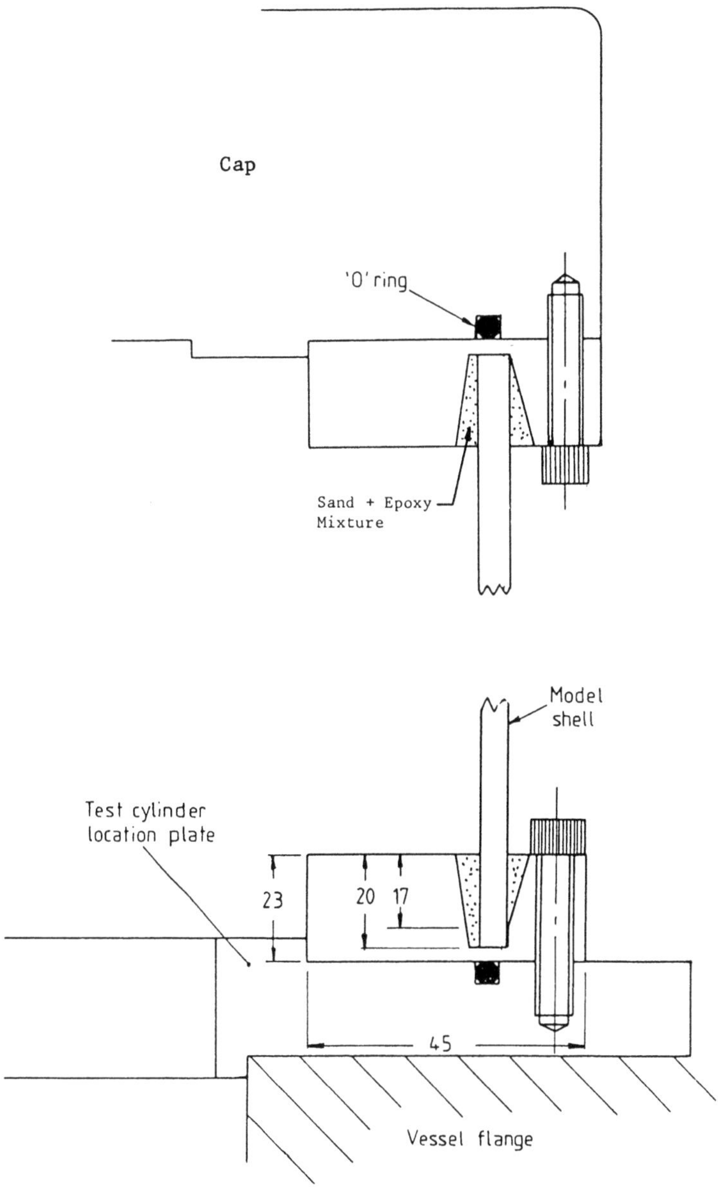

Fig.1. Shell boundary conditions and end-plate connection

Coupon plates were cut from the steel sheet used to manufacture the model and were heat treated simultaneously with the model. Tensile test specimens were prepared and it was found that the average material properties were E = 203 KN/mm^2; $\nu = 0.3$; $\sigma_o = 313\ N/mm^2$. The transition from elastic to plastic behaviour was quite distinct and a yield plateau extending to several yield strains was noted in the heat treated sheet.

3.2 MEASUREMENT OF INITIAL IMPERFECTIONS

The initial imperfections were measured using a turntable rotating relatively to a fixed L.V.D.T. connected electrically to a Solatron 'Orion' data logger (see Fig 2). The resulting best fit cylinder, and deviations from that shape were determined using a statistically based analysis (33). Readings of the L.V.D.T. were taken at 256 mesh points around the shell and at 0.8cm intervals along the generator. A typical output from the computer programme (33) is shown in Fig 3 and the corresponding coefficients from a Fourier analysis of the initial imperfections are summarised in Table 2, where

$$wo = \sum_{i=1,2,} \sum_{j=0,1,} (Aij \cos\theta j + Bij \sin\theta j) \sin \frac{i\pi x}{L}$$

3.3 HYPERBARIC FACILITY

The test facility is shown diagrammatically in Fig 4. It combines a self reacting frame with a very heavy flange to which a hyperbaric chamber is fitted. Axial load is applied by an Instron 500 kN servo-hydraulic actuator on which the ram displacement is controlled using a signal from a 1mm L.V.D.T. mounted on the frame of the rig. The sensitivity of the L.V.D.T. is better than 0.12% hence the end displacement can be controlled to 10^{-3} mm which for the 50mm long model represents 1% yield strain.

The model was located on a machined mounting ring fitted to the heavy flange. The axial load was transmitted to the model from the actuator by means of a spherical seating which could accommodate any small inevitable lack of parallelity of the end plates. The spherical seating can be inverted to facilitate either axial tension or compression modes of loading.

Fig.2. Initial imperfections measurements set up

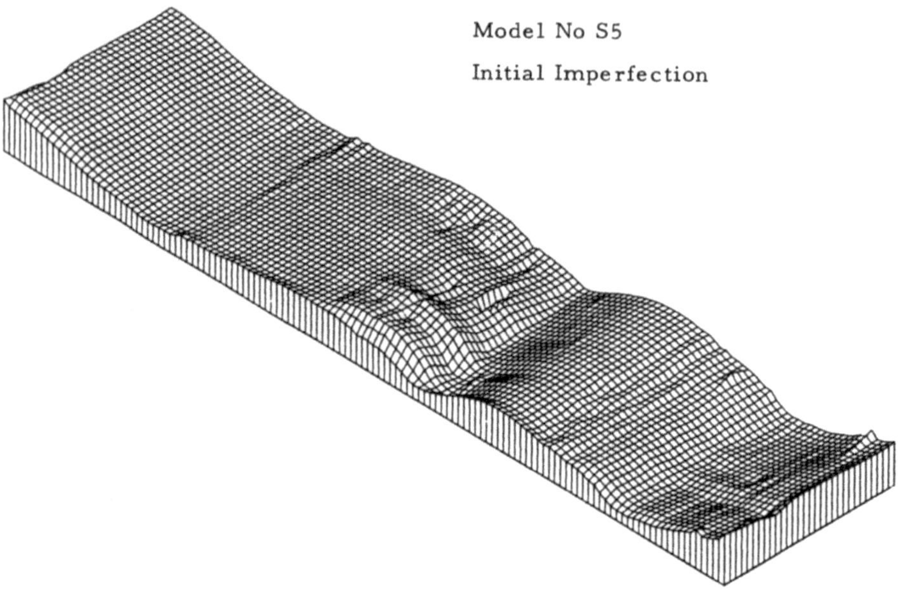

Fig.3. Model S5 - Initial imperfections plotting

TABLE NO 2 MODEL S5 FOURIER COEFFICIENTS

FOURIER COEFFICIENTS

A(M,N.

M =	1	2	3	4	5
N = 0	-0.0007	0.0045	-0.0012	0.0027	-0.0008
N = 1	-0.0001	-0.0025	-0.0011	0.0007	0.0031
N = 2	0.0286	0.0142	0.0073	0.0077	0.0024
N = 3	0.0385	-0.0132	0.0114	-0.0049	0.0086
N = 4	0.0329	0.0066	0.0109	0.0027	0.0045
N = 5	-0.0078	-0.0072	-0.0014	-0.0029	0.0002
N = 6	0.0141	0.0035	0.0047	0.0012	0.0023
N = 7	-0.0044	-0.0019	-0.0020	-0.0001	-0.0011
N = 8	-0.0025	-0.0013	-0.0004	-0.0003	0.0001
N = 9	0.0028	0.0012	0.0003	0.0008	0.0003
N = 10	-0.0003	-0.0006	-0.0000	-0.0003	0.0000
N = 11	0.0051	-0.0006	0.0015	-0.0004	0.0011
N = 12	-0.0022	0.0002	-0.0005	0.0003	-0.0007
N = 13	0.0013	0.0001	0.0005	-0.0006	0.0007
N = 14	-0.0013	0.0004	-0.0012	0.0009	-0.0009
N = 15	0.0010	-0.0008	0.0005	-0.0006	0.0010
N = 16	0.0010	0.0002	-0.0001	0.0009	-0.0002
N = 17	0.0007	0.0008	0.0004	0.0001	0.0005
N = 18	0.0014	0.0003	0.0003	0.0006	-0.0001
N = 19	-0.0004	0.0000	-0.0000	-0.0002	0.0001
N = 20	0.0004	-0.0002	0.0005	-0.0001	0.0004
N = 21	0.0007	0.0007	-0.0004	0.0005	0.0002
N = 22	-0.0002	0.0001	0.0005	-0.0001	0.0001
N = 23	0.0008	-0.0006	-0.0006	0.0003	-0.0002
N = 24	-0.0010	-0.0000	0.0006	-0.0007	0.0001
N = 25	0.0011	0.0000	-0.0004	0.0005	0.0002
N = 26	-0.0003	0.0006	0.0008	-0.0003	0.0001
N = 27	0.0011	-0.0006	-0.0002	0.0004	-0.0001
N = 28	-0.0002	0.0003	0.0005	-0.0005	-0.0000
N = 29	0.0004	-0.0000	-0.0002	0.0002	0.0002
N = 30	0.0005	0.0001	0.0001	-0.0002	0.0003
N = 31	0.0004	-0.0004	0.0002	-0.0003	0.0004
N = 32	0.0006	-0.0004	-0.0002	0.0002	-0.0002
N = 33	-0.0006	0.0005	0.0001	-0.0002	0.0001
N = 34	0.0002	-0.0003	0.0000	0.0002	0.0000
N = 35	-0.0003	0.0000	0.0003	-0.0004	0.0002
N = 36	0.0006	-0.0006	0.0000	0.0005	-0.0002
N = 37	-0.0005	0.0007	0.0001	-0.0001	-0.0001
N = 38	0.0002	-0.0004	-0.0001	0.0003	0.0000
N = 39	0.0002	0.0001	0.0005	-0.0007	0.0004
N = 40	0.0003	-0.0004	-0.0001	0.0002	-0.0000

1

B(M,N)

M =	1	2	3	4	5
N = 0	0.0000	0.0000	0.0000	0.0000	0.0000
N = 1	-0.0007	0.0003	0.0014	-0.0020	0.0013
N = 2	-0.0082	0.0007	-0.0040	0.0019	-0.0014
N = 3	-0.0600	0.0015	-0.0186	-0.0002	-0.0095
N = 4	0.0067	-0.0064	0.0004	-0.0003	0.0020
N = 5	-0.0041	0.0010	-0.0012	-0.0005	-0.0011
N = 6	-0.0019	-0.0044	-0.0000	-0.0010	0.0013
N = 7	0.0085	0.0023	0.0026	0.0002	0.0009
N = 8	-0.0098	-0.0037	-0.0030	-0.0007	-0.0014
N = 9	-0.0009	-0.0002	-0.0008	0.0004	-0.0002
N = 10	-0.0018	-0.0006	-0.0007	0.0002	-0.0002
N = 11	-0.0001	-0.0011	-0.0001	-0.0003	0.0002
N = 12	0.0011	-0.0004	0.0002	-0.0002	-0.0000
N = 13	-0.0027	-0.0008	-0.0005	-0.0006	-0.0002
N = 14	0.0010	0.0007	0.0001	0.0001	-0.0000
N = 15	-0.0030	-0.0011	-0.0011	-0.0001	-0.0007
N = 16	-0.0010	-0.0005	-0.0003	-0.0004	-0.0000
N = 17	-0.0010	-0.0006	-0.0008	0.0004	-0.0001
N = 18	-0.0004	-0.0002	0.0002	-0.0002	0.0004
N = 19	0.0001	-0.0003	-0.0007	0.0008	-0.0005
N = 20	-0.0021	-0.0002	-0.0000	-0.0006	-0.0005
N = 21	0.0006	-0.0002	-0.0002	0.0002	0.0008
N = 22	-0.0008	0.0001	0.0001	-0.0003	0.0001
N = 23	-0.0000	-0.0002	-0.0006	0.0006	0.0002
N = 24	-0.0011	0.0000	-0.0005	-0.0000	-0.0007
N = 25	-0.0002	-0.0005	-0.0005	-0.0001	0.0004
N = 26	0.0003	-0.0001	0.0003	0.0001	0.0000
N = 27	-0.0009	-0.0001	-0.0002	-0.0000	0.0001
N = 28	0.0001	-0.0002	-0.0001	0.0002	-0.0005
N = 29	-0.0007	0.0003	-0.0003	-0.0002	0.0000
N = 30	0.0004	-0.0004	0.0001	0.0002	0.0001
N = 31	-0.0006	0.0002	-0.0001	-0.0004	0.0001
N = 32	0.0005	-0.0005	0.0001	0.0003	-0.0000
N = 33	-0.0000	-0.0000	-0.0004	-0.0003	0.0001
N = 34	-0.0002	-0.0001	-0.0003	0.0004	-0.0002
N = 35	-0.0004	0.0001	-0.0000	-0.0004	0.0002
N = 36	0.0001	-0.0002	0.0000	0.0001	-0.0001
N = 37	0.0001	-0.0001	-0.0000	-0.0001	0.0003
N = 38	-0.0004	0.0000	-0.0002	0.0002	-0.0004
N = 39	-0.0003	0.0000	0.0001	-0.0000	0.0002
N = 40	-0.0003	-0.0001	0.0002	-0.0004	-0.0001

N = circumferential direction
M = longitudinal direction

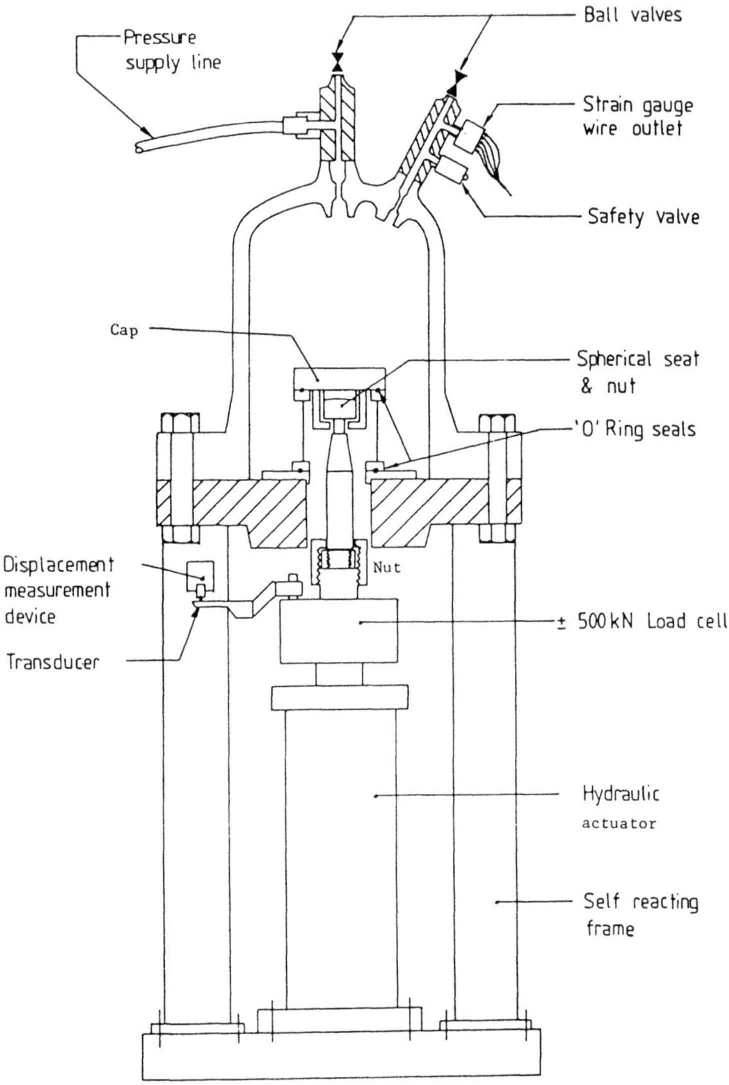

Fig.4. General view of hyperbaric chamber

The hyperbaric chamber is serviced by its own variable speed positive injection pump and pressuring monitoring equipment, with digital read-out. The chamber also incorporates a particularly conventional facility to take strain gauge lead wires from the high pressure interior to the external data logging system.

3.4 TEST PROCEDURE

The objective of each test required that the combined effect of axial load and external pressure should be such that the ratio of $\eta = \sigma_x/\sigma_\theta$ should be constant at a predetermined value throughout the test. Unfortunately, it was not possible to control the application of the pressure and axial load as a simultaneous continuous function so, instead, it was decided to apply small step increments of pressure and then appropriately adjust the level of axial load.

3.5 EXPERIMENTAL RESULTS

The models were tested with various load combinations; models S4 and S8 were fitted with waterproofed strain gauge prior to mounting in the test rig (see Fig 5). The buckling load was defined to be the value just existing just before a drop in the indicated pressure. This reduction occurred because the walls of the chamber were massive, relative to the shell thickness, and since water at the test pressures is effective incompressible, any change in the shell volume due to onset of the buckling mode would result in a corresponding pressure change. In all the tests, with the exception of S8, the change in indicated pressure was abrupt and therefore the buckling load was clearly discernable. In the case of S8 the collapse load was determined from the measured strain versus load plot, Fig 8, as the point of sudden increase in bending strain.

Table 3 summarises the test results; Fig 6 shows a photograph of a typical failed mode and a typical distribution of buckles is shown in Fig 7. The longer models (L/R = 0.813) were tested with a load combination of $\eta = 1.0$. The models S3 and S4 have slightly different geometries and the slightly shorter of the two (S3) buckled at a higher load as expected, compared to S4. Both had very similar levels of imperfection $w_o(max)/t \approx 0.3$. Model S5 had a lower level of initial

TABLE NO 3 SURREY UNIVERSITY MODELS/EXPERIMENTAL RESULTS

Model	$\eta = \frac{\sigma x}{\sigma \theta}$	Applied Pressure p N/mm^2	Applied Load P KN	N*
S3	1.0	0.483	43	10
S4	1.0	0.476	39.5	10
S5	2.0	0.494	119.5	12
S6	0.5	1.282	0.0	16
S7	2.0	0.772	186.5	15
S8	0.0	1.482	^{T}119.0	20

T - In tension to resist the pressure

* - Estimated number of waves from the observed region of buckles

Fig.5. Model S8 - Ready for testing

Fig.6. Model S7 - Failure mode

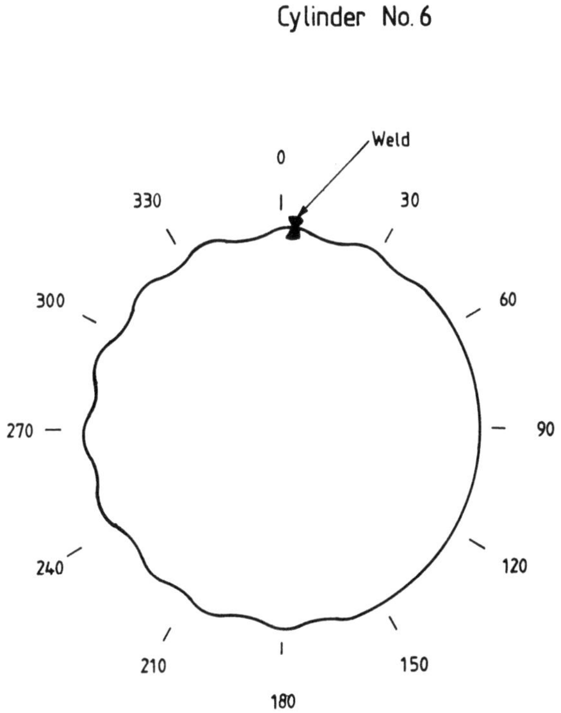

Fig.7. Model S6 - Distribution of buckles around circumference

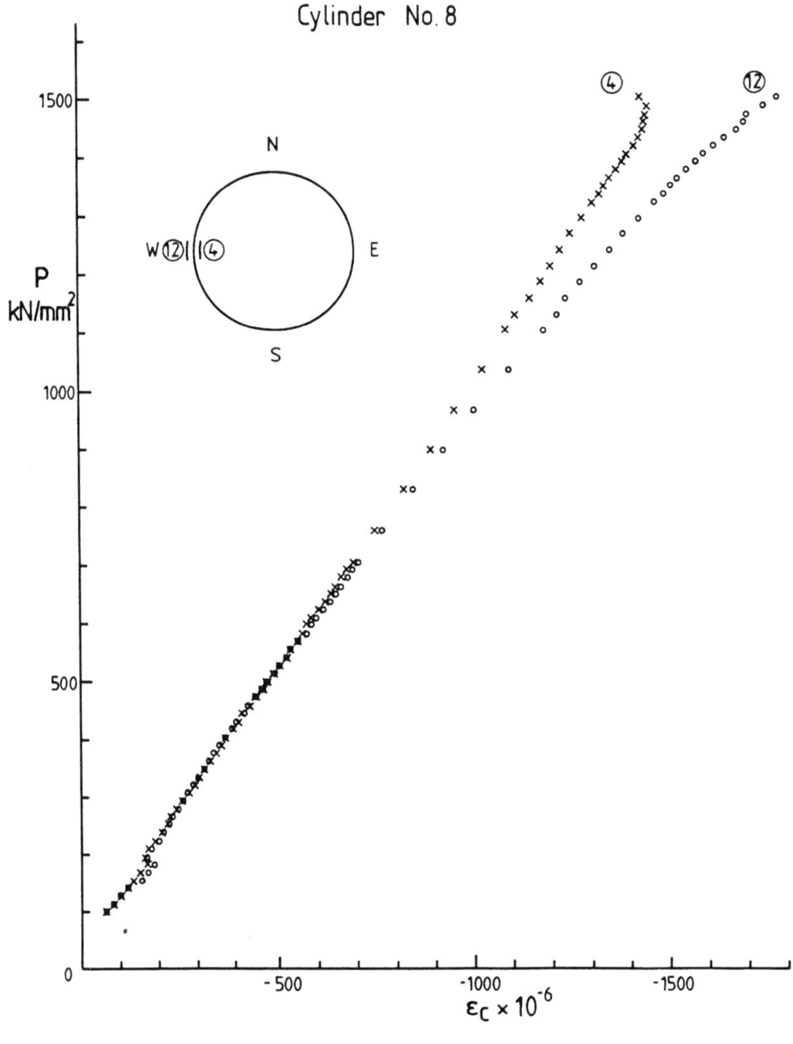

Fig.8. Model S8 - Load vs Strain

imperfection (wo(max)/t = 0.18, see Fig 3) and was subject to a stress ratio of $\eta = 2.0$. The apparently large magnitude of the applied stress at failure seems to reflect the degree of the perfection of the shell.

The shorter models (S6,S7 and S8) were nominally identical (L/r = 0.3125) and the tests were designed to generate information with which to prove a complete interaction graph; S6 was subjected to hydrostatic loading ($\eta = 0.5$), S7 was subjected to greater levels of axial loading $\eta = 2.0$ and S8 had external pressure only $\eta = 0.0$ Models S6 and S7 exhibited sudden failure in contrast to S8 which, as mentioned above, failed "gently". In fact by increasing the rate of pumping the applied pressure could be increased to 1.55 N/mm^2, whereas failure is considered to be at 1.48 N/mm^2, according to Fig 8. It is noticeable that, as may be expected, the greatest imperfections are localised to the weld zone. This factor is reflected in the buckling mode which showed that deformations predominated in that zone. This feature occurred irrespective of the location of the weld relative to the mounting ring in the test equipment.

4. DISCUSSION

It is evident from the diversity of the interaction curves shown in Figs 9 and 10 that the analysis of thin-walled cylindrical shells subject to interactive loading is by no means thoroughly understood. Indeed the end points of the lines have themselves been the subject of considerable activity throughout this century and even now great reliance has to be placed on experimental results. It is widely accepted that in general the buckling pressure of shells subjected to pressure alone has a low sensitivity to initial imperfections. This is in contrast to the situation for axial compressive loading acting alone and when the two types of loading act in conjunction there is little information on the imperfection sensitivity or of the effect of change in mode shape with the relative magnitude of the loading types. Therefore in the derivation of the interaction curves it has been difficult to define a buckling criterion. In the present study it was decided to follow the linear assumption in the various equations used by researchers to define the critical load, as described in Appendix A as equations (1) and

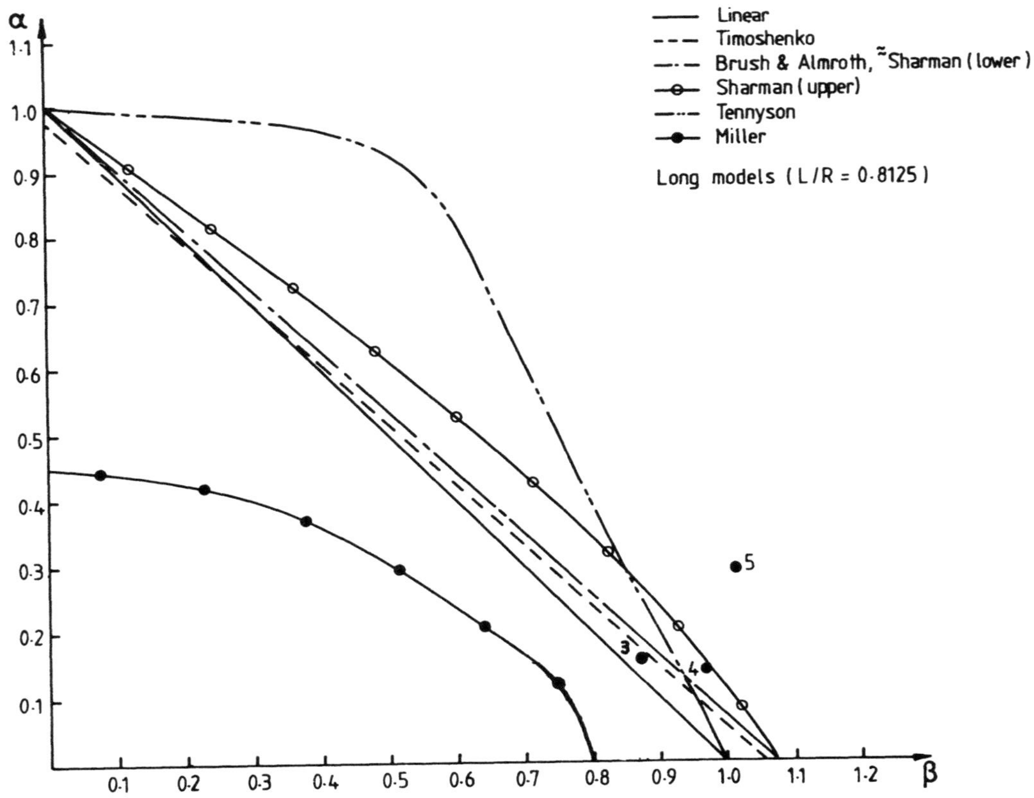

Fig. 9. Interaction curves for long models

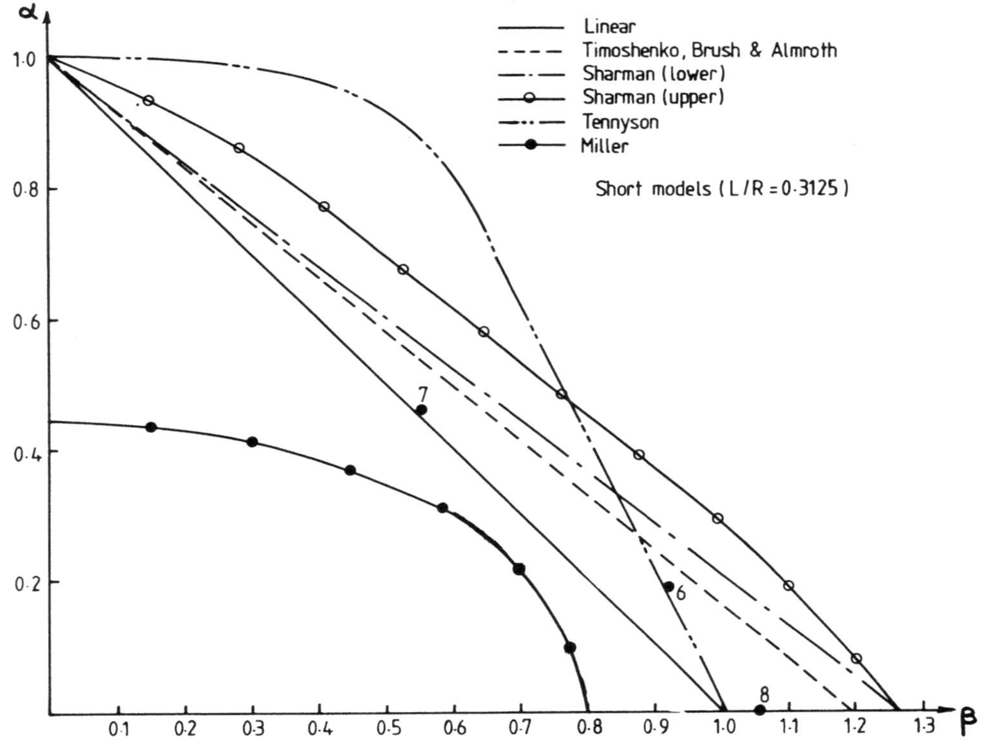

Fig. 10. Interaction curves for short models

(2). The results quoted in the Appendix have been basically determined either from a linearised buckling approach (1,3,5), or on an empirical basis based results from elastic models (27,30). An exceptional case is the work produced by Miller (31) which includes the effects of geometric variation. The buckling criteria in Appendix A have been plotted in Figs 9 and 10 to correspond to the model test results, discussed above, the authors could find no other references to coherent test programmes attempting to prove the full range of the interactive curve for elastic-plastic engineering materials, although of course a considerable body of information exists for shells subject to hydrostatic loading. It is interesting to see in Figs 9 and 10, the differences between the various interaction curves especially in the purely external pressure domain. Contrarily, in the purely axial compression domain most of the curves converge to $\alpha = 1$, because they do not take into consideration the effect of initial imperfections, which is most unfortunate. The exception to this remark is the curve proposed by Miller. An obvious approach, but one which lacks a theoretical basis is to take a linear interaction between the empirically derived failure mode, including plasticity, for axial compression alone and the corresponding pressure loading, also acting alone. This is under consideration as a design recommendation in Europe.

The experimental results shown in the Figures appear not to agree well with any of the proposed curves, although the shape seems to coincide more nearly with Millers' proposals. Hence, if these are the only relevant test data it is difficult to arrive at any useful practical recommendations for design analyses. This comment is reinforced by Fig 11, which would show that there is little correlation between the test results and the corresponding theoretical predictions.

The major conclusion which may be drawn from the present study is that much more experimental work has to be undertaken if a rational set of interaction curves are to be obtained. This work must be designed to cover the possible range of practical geometries and to examine the effect of boundary conditions and of the possibility of various buckling modes at different relative magnitude of axial and pressure loading.

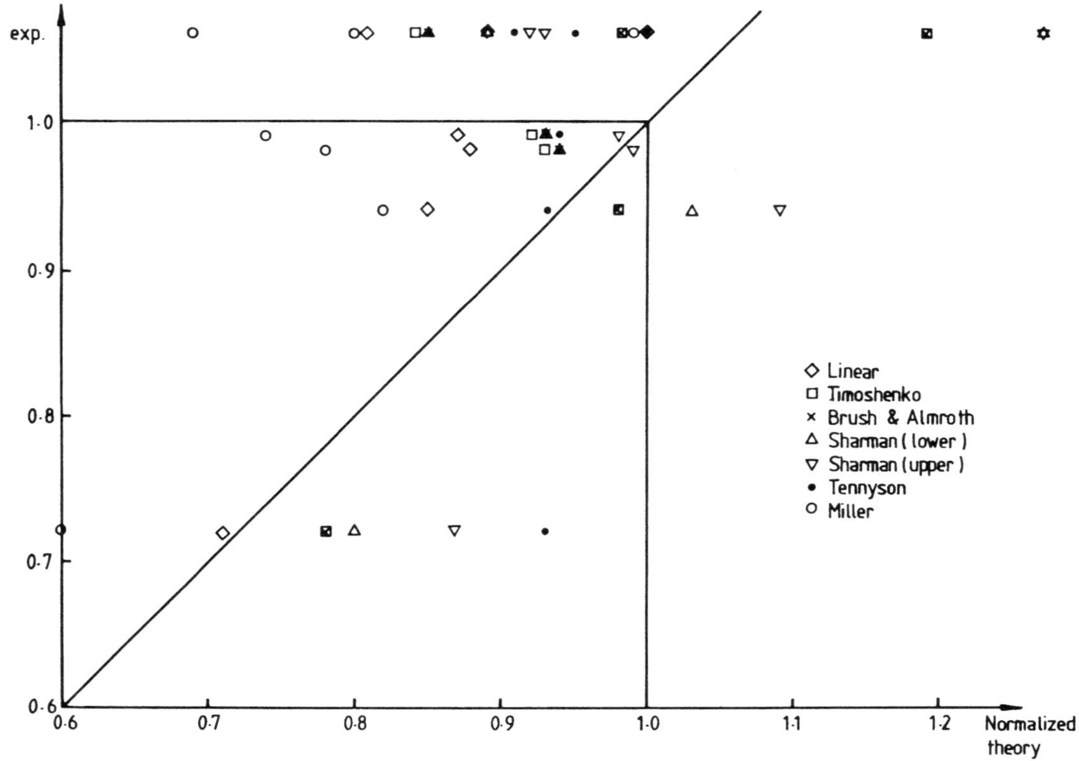

Fig.11 Correlation between experimental results and theoretical predictions

Appendix A

1 $$\sigma_{x_{cr}} = 0{\cdot}606\, E \frac{t}{R}$$

2 $$\sigma_{\theta_{cr}} = \frac{\pi^2}{12(1-\nu^2)} k_{min}\, E\left(\frac{t}{L}\right)^2$$

$$k_{min} = \frac{1}{0{\cdot}5 + q}\left[(1+q)^2 + \frac{12Z^2}{\pi^4(1+q)^2}\right]$$

$$q = \left(\frac{nL}{\pi R}\right)^2 \; ; \quad Z = \frac{L^2}{Rt}(1-\nu^2)^{1/2}$$

3 $$\alpha = \frac{\sigma_x}{\sigma_{x_{cr}}} \quad ; \quad \beta = \frac{\sigma_\theta}{\sigma_{\theta_{cr}}}$$

a. Timoshenko (Ref 1)

4 $$T1\cdot\alpha + T2\beta = 1{\cdot}0$$

$$T1 = \frac{\sigma_{x_{cr}}}{C_I} \qquad T2 = \frac{\sigma_{\theta_{cr}}}{C_{II}}$$

$$C_I = \frac{C_1 + C_2\left(\frac{t}{12R}\right)^2}{C_4(1-\nu^2)} E \qquad C_{II} = \frac{C_1 + C_2\left(\frac{t}{12R}\right)^2}{C_3(1-\nu^2)} E$$

$$C_1 = (1-\nu^2)\, q^4$$

$$C_2 = (q^2+n^2)^4 - 2\left[\nu q^6 + 3q^4n^2 + (4-\nu)q^2n^2 + n^6\right]$$

$$C_3 = n^2(q^2+n^2)^2 - (3q^2n^2 + n^4)$$

$$C_4 = q^2(q^2+n^2)^2 + n^2q^2$$

$$q = \frac{\pi R}{L}$$

b. Sharman (Ref 3)

5. $$\alpha + S\beta = 1{\cdot}0 \quad \text{Lower}$$

$$\alpha^{5/4} + S\beta = 1{\cdot}0 \quad \text{Upper}$$

$$S = \frac{\sigma_{\theta cr}}{C_I} \qquad C_I = \frac{0{\cdot}92\,E\,\frac{t}{R}}{L/R\,\sqrt{R/t} - 1{\cdot}4}$$

c. Brush & Almroth (Ref 5)

6. $$B1\,\alpha + B2\,\beta = 1{\cdot}0$$

$$B1 = \frac{\sigma_{x cr}}{C_{I\,min}}\,q^2 \qquad B2 = \frac{\sigma_{\theta cr}\,n^2}{C_{I\,min}}$$

$$C_{I\,min} = \frac{E}{12R^2(1-\nu^2)(q^2+n^2)^2}\left[(q^2+n^2)^4 t^2 + q^4\,12(1-\nu^2)R^2\right]$$

$$q = \frac{\pi R}{L}$$

d. Tennyson (Ref 27)

7. $$\alpha(1{\cdot}5 - 0{\cdot}5\alpha) + \beta(1-\alpha) = 1{\cdot}0$$

No information about the critical loads used to establish this formula, hence, equations 1 and 2 were used in the present case.

e. Miller (Ref 31)

8. $$M1\cdot\alpha + M2\cdot\beta = 1{\cdot}0 \quad \text{for } \sigma_x \leq 0{\cdot}5\,\sigma_\theta$$

$$M1 = \frac{\sigma_{x cr}(\sigma_{re} - \sigma_{he})}{0{\cdot}5\,\sigma_{re}\,\sigma_{he}} \qquad M2 = \frac{\sigma_{\theta cr}}{\sigma_{he}}$$

9. $$M1 \cdot \alpha + M2 \cdot \beta^2 = 1{\cdot}0 \quad \text{for } \sigma_x > 0{\cdot}5\,\sigma_\theta$$

$$M1 = \frac{\sigma_{x\,cr}}{\sigma_{xe}} \qquad M2 = \sigma_{\theta cr}^{\;2}\left[\frac{1}{\sigma_{ne}^{2}} - \frac{0{\cdot}5}{\sigma_{ne}\,\sigma_{xe}}\right]$$

$$\sigma_{xe} = CE\,\frac{t}{R}$$

C is the greater of C_1 and C_2

$C_1 = 0{\cdot}125 \quad$ for $M > 10$

$C_1 = 0{\cdot}5\,M^{-0{\cdot}6} \leqslant 0{\cdot}392 \quad$ for $M \leqslant 10$

$C_2 = 0{\cdot}919 - 0{\cdot}286 \log R/t \quad$ for $R/t > 146$

$C_2 = 0{\cdot}3 \quad$ for $R/t \leqslant 146$

$$\sigma_{he} = CE\,\frac{t}{D}$$

$C = 0{\cdot}88\,\frac{t}{D} \quad$ for $M \geqslant 0{\cdot}75\ D/t$

$C = \dfrac{1{\cdot}472}{M - 0{\cdot}636} \quad$ for $3{\cdot}5 \leqslant M < 0{\cdot}75\ D/t$

$C = \dfrac{1{\cdot}51}{M - 0{\cdot}559} \quad$ for $1{\cdot}5 \leqslant M < 3{\cdot}5$

$$\sigma_{re} = CE\,\frac{t}{D}$$

$C = 0{\cdot}88\,\frac{t}{D} \quad$ for $M \geqslant 0{\cdot}75\ D/t$

$C = \dfrac{1{\cdot}472}{M - 1{\cdot}17} \quad$ for $3{\cdot}5 \leqslant M < 0{\cdot}75\ D/t$

$C = \dfrac{4{\cdot}88}{M^{1{\cdot}64}} \quad$ for $1{\cdot}5 \leqslant M < 3{\cdot}5$

$$M = \frac{L}{\sqrt{Rt}}$$

f. Foster (Ref 30)

$$\left\{\left(\frac{\sigma_x}{\sigma_{x_c}}\right)^{1.5} + \left(\frac{\sigma_\theta}{\sigma_{\theta_c}}\right)^{1.5}\right\}^{2/3} = 1.0$$

σ_{x_c} Axial critical stress

σ_{θ_c} External pressure critical stress

List of Symbols

Aij,Bij	-	Fourier coefficient for initial imperfection
E	-	Young's modulus
L	-	Length of the shell
N	-	Number of circumferential buckle halfwaves
p	-	Applied pressure
P	-	Applied axial load
R	-	Radius of shell (ave)
t	-	Thickness of the shell (ave)
wo	-	Amplitude of initial imperfection
Z	-	Batdorf parameter
ν	-	Poisson's ratio
σ_x	-	Axial stress
σ_θ	-	Hoop stress
η	-	The ratio between axial stress and hoop stress

References

1. Timoshenko, S.P.; Gere, J.M. Theory of Elastic Stability. McGraw Hill, 495-499 1961.

2. Flugge, W. Sresses in Shells. Springer Verlag, Berlin 1960.

3. Sharman, P.W.: A theoretical interaction equation for the buckling of circular shells under axial compression and external pressure. J. of the Aeronaut Sci. 29, 7, 878 July 1962.

4. Terebushko, O.I. Stability of a cylindrical shell under twisting, external pressure and compression. ARS J. 31, 3, 378-386 (Russian Suppl.) March 1961.

5. Brush, D.O.; Almroth, B.O. Buckling of Bar, Plates and Shells. McGraw Hill, 172-174 1975.

6. Gerard, G.; Becker, H. Handbook of Structure Stability Part III, Buckling of Curved Plates and Shells. NACA TN 3783 1957.

7. Lackshmikantham, C.; Gerard, G. Elastic stability of cylindrical shells under axial and lateral loads. J. of the Royal Aeronaut Soc. 68, 773-775 1964.

8. Kirshnamoorthy, G. Buckling of thin cylinders under combined external pressure and axial compression. J. of Aircraft 11, 2, 65-68 Feb. 1974.

9. Baker, E.H. Stability of circumferential corrugated sandwich cylinders under combined loads. AIAA 2142-2149 Dec. 1964.

10. Levi, M.M.; Cheng,S. Buckling of composite and homogeneous isotropic cylindrical shells under biaxial compression. AIAA 2259-2263 Nov 1971.

11. Hess, T.E. Stability of orthotropic cylindrical shells under combined loading. ARS J. 31, 237-246 1961.

12. Hutchinson, J. Buckling of imperfect cylindrical shells under axial compression and external pressure. AIAA 1968-1970 Oct. 1965.

13. Zyczkowski, M.; Buck, S. A method of stability analysis of cylindrical shells under biaxial compression. AIAA 2259-2263 Nov. 1971.

14. Mushtari, K.M.; Satchenkow, A.V. Ob Vstoytchivosti cilindritcheskikh i konitcheskikh obolotchek kruglogonego normalnogo davlenya. Prikladnaya Matematika i Mechanika 18, 6, 667-674 1954.

15. Immerman, A.G. Proverkie ustoytchivosti krugovykh obolotchek pri sovmestnom deystvu prodolnoy poperetchnoy nagruski. Rastchot prostranstrennykh konstrukeiy W, Gosstroyizdat, Moscow 525-530 1958.

16. Kurshin, L.M.; Pustovoi, N.V. Stability of a long cylindrical shell subjected to axial compression and annular load. Soviet Aeromautics 17, 1, 50-53 1974.

17. Kabanow, V.V. Stability of inhomogeneous cylindrical shells under external pressure and axial stress. Mech. of Solids 10, 2, 132-137 1975.

18. Almroth, B.D.; Brogan, F.A.; Marlowe, M.B. Collapse analysis for shell of general shape, Vol.I, Analysis. AFFDL-TR-71-78 Aug. 1972 Air Force Flight Dynamics Lab. Wright-Pallerson Air Force Base, Ohio (STAGS - A computer program).

19. Bushnell, D. Stress, stability and vibration of complex shells of revolution. Analysis and User Manual for BOSOR 5.

20. Harding, J.E. The elasto-plastic analysis of imperfect cylinders. Proc. Instn. Civil Eng. 65, 2, 875-892 1978.

21. Sanders, J.L. Nonlinear theory for shells. Q. Appl. Math. 21, 21-36 1963.

22. Harding, J.E. Ring-stiffened cylinders under axial and external pressure loading. Proc. Inst. Civil Eng. 2, 863-878 Sept. 1963.

23. Abramovich, H.; Singer, J. Correlation between vibration and buckling of stiffened cylindrical shells under external pressure and combined loading. Israel J. of Tech. 16, 1/2, 34-44 Dec. 1978.

24. Croll, J.G. Elastic-plastic buckling of pressure and axial loaded cylinder. Report of UCL.

25. Walker, A.C.; Andronicou, A.; Sridharan, S. Local plastic collapse of ring stiffened cylinders. Proc. Inst. Civil Eng. Vol 71, 2, 341-367 June 1981.

26. Walker, A.C.; Andronicou, A. Buckling of ring stiffened shells subject to pressure and axial loading. To be published.

27. Tennyson, R.C. Interaction of cylindrical shell buckling experiments and theory. Theory of Shells, Proc. of the 3rd IUTAM, Ch. 1 1980.

28. Tennyson, R.C.; Booton, M. Buckling of short cylinders under combined loading. Trans. ASME (E) 45, 574-578 Sept. 1978.

29. Weingertan, V.I.; Seide, P. Elastic stability of thin walled cylindrical and conical shells under combined external pressure and axial compression. AIAA 913-920 May 1965.

30. Foster, C.G. Interaction of buckling modes in thin walled cylinders. Exp. Mech. 124-129 March 1981.

31. Miller, C.D. Summary of buckling tests on fabricated steel cylindrical shells in U.S.A. Imperial College Conference on offshore structure April 1981.

32. Walker, A.C.; McCall, S. Buckling test on stringer stiffened cylinder models subjected to load combination. Report for DnV, Surrey University Dec. 1981.

33. Nelson, H.M.; Green, D.R.; Phillips, D.V. Buckling studies of large diameter stiffened tubes. Pilot test, University of Glasgow, Civil Eng. Dept. June 1978.

34. Midgley, W.R.; Johnson, A.E.Jr. Experimental buckling of internal integral ring-stiffened cylinders. Exp. Mech. 145-154 1967.

BUCKLING OF WIND LOADED CYLINDRICAL SHELLS -APPLICATION TO UNSTIFFENED AND RING-STIFFENED TANKS

F.RESINGER and R.GREINER

Institut für Stahlbau, Holzbau und Flächentragwerke
Technische Universität Graz, Austria

1. Introduction

Shells of cylindrical storage tanks may buckle due to wind loading in the empty or partially filled state (Fig.1) -and in particular when they are subjected additionally to an internal operating vacuum or to large snow loads on the roof. Since on one hand the size of these tanks has been increased up to diameters of about 110 m and on the other hand the plate thickness of

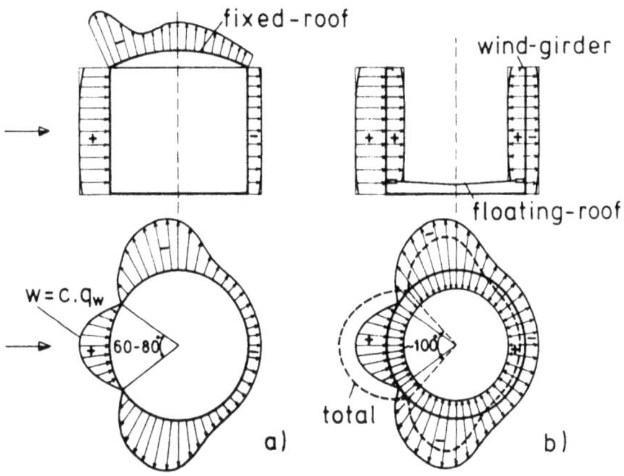

Fig.1. Wind pressure on closed-top tank (a) and open-top tank (b)

the tank-shells has been decreased gradually through use of high tensile steels, the stability problem of tank-shells became a problem of major importance in the recent years. Tank-shells damaged considerably by wind pressure are recorded in [1], [2].
It is a wide spread opinion that wind buckling has not to be considered as a problem of safety, because local dents in the shell as a result of wind pressure will not cause collapse of the tank. Although this will prove true

Buckling of Shells, Ed.: E. Ramm
Proceedings of a State-of-the-Art Colloquium

for wind loads alone, wind loads in combination with internal vacuum or axial loads may induce by far larger damages. On the other hand repairs of large storage tanks will be very expensive since the tanks will be out of operation for a long time and might have to be made gasfree and to be cleaned. Therefore from economic considerations reliable design rules will be of particular importance in practice.

Relevant design Codes, as there are the British Standard BS 2654 and the German Standard DIN 4119, lead to quite different results for buckling due to wind pressure.

The aim of the following investigation is to obtain a more exact solution for this problem as well as to present an easy to handle design concept. Easy to handle for that reason, because the stability calculation has to be considered as an additional check of a tank-shell only, whose wallthickness was already designed due to internal hydrostatic pressure of the liquid. The investigation comprises cylindrical shells subjected to wind loads, as well as its combination with internal vacuum, internal pressure and additional axial forces, the latter simulating snow loading on the roof a.o. It was performed by means of wind tunnel tests as well as analytical investigations. For application in practice the design concept was extended to tank-shells of varying wallthickness, i.e.stepped cylinders. Some information is given on tanks stiffened by secondary rings and also on the sizing of the ring stiffeners.

2. The Problem

The typical pressure distribution of vertical cylindrical shells is shown in Fig.1 for a fixed-roof tank as well as for a floating-roof tank. A tank with closed top is subjected to wind pressure on the windward side over an arc of $6o^{0}$ to $8o^{0}$ and over the rest of the shell to a suction load. In front of the shell the pressure coefficient is $C = 1{,}o$; the pressure distribution is approximately assumed as constant over the height of the tank. In the case of an empty floating-roof tank the wind -in addition to the above mentioned pressure from outside- causes an uniform suction load inside the shell. In good accordance to wind tunnel tests [2] this suction will be covered by $C = o{,}6$, which can be considered as a conservative value. As a result of this assumptions the arc of the pressure zone on the windward side extends up to $1oo^{0}$.

In the pressure zone of a wind loaded tank the shell is subject to buckling, while in the suction zone there is no risk of instability. Since a theoretical stability analysis for nonuniform external pressure is rather involved and too time-consuming for an additional check, the problem will be approached by more practice-orientated considerations [1], [8], which allow a simplified solution.

Provided that the buckling pattern due to uniform external pressure will form a great number of circumferential waves, so that the wind pressure acts over a number of several waves (Fig.2a), it is evident that the shell in this area

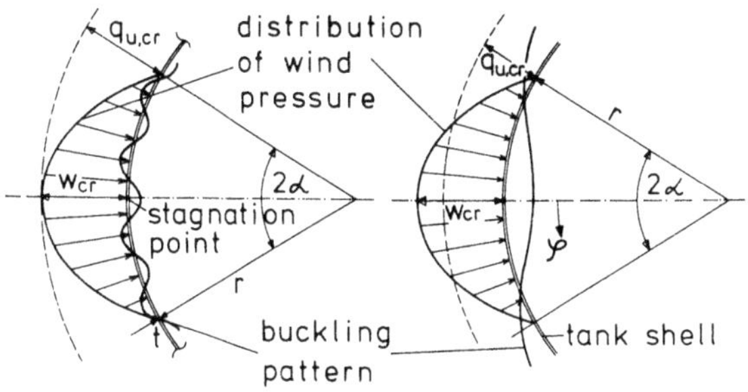

Fig.2. Uniform buckling pressure and windbuckling pressure of shells with different wave-patterns

will respond very similar, in effect, as to an uniform pressure equal the stagnation pressure, since the wind pressure upon the waves at the center of this area is nearly constant.

In the case of a pattern with a small number of waves (Fig.2b) one might argument that the average wind pressure over one buckle will be smaller than the maximum of the stagnation pressure and that the adjacent suction zone in addition will contribute a stabilizing affect too. An uniform pressure that is equivalent in effect, therefore will obviously be of a lower value compared to the stagnation pressure. This leads to the conclusion that the nonuniform wind pressure could be substituted by an "equivalent uniform external pressure $q_{u\,e}$" of certain amount for the purpose of determining buckling in a localized area.

Following this concept the pressure $q_{u\,e}$ is defined in the BS 2654 with a conservative value equal the stagnation pressure q_w and in the DIN 4119 with half of this value only, which might be too small for special cases(Fig.3).

Theoretical investigations of this subject [3],[4]are of little help in solving the problem, since the derived results lead to values $q_{u\,e}$ of 32% to 62% of the stagnation pressure for closed-top tanks, but with an inconsistent relation to the buckling pattern. For open-top tanks these values increase

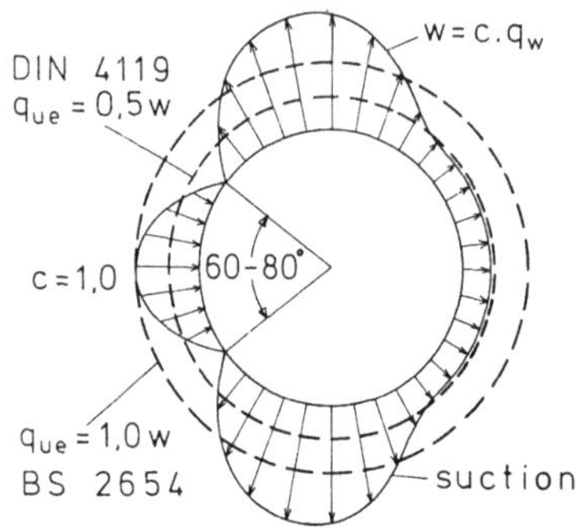

Fig.3. Design concepts for windbuckling in relevant Codes

even to 1,8 of the total pressure in the stagnation point, which is completely incompatible with practical experience.

In order to obtain a reliable solution for this problem wind tunnel tests seemed to be the most appropriate approach. By varying the dimensions of the models and the loading conditions a test series representative for tanks used in practice could be investigated.

3. Wind tunnel tests

The tests were performed in the wind tunnel of the Institut für Strömungslehre und Gasdynamik of the TU Graz (Fig.4). The dimensions of the Aluminium-models had to be designed in close accordance to the characteristics of the tunnel and are shown in Fig.5. In order to obtain clear boundary conditions a fixed connection by glueing was provided between the sheet of the cylinder and the very stiff end-plates. The series of tests under the above mentioned

Fig.4. Wind tunnel with test-cylinder

d (mm)	l (mm)	t (mm)	Model Nr.
440	440	0,105	1
	320	0,105	2
	180	0,102	3(I)
	180	0,100	3(II)
620	320	0,105	4(I)
	320	0,102	4(II)
	250	0,105	5
	180	0,105	6

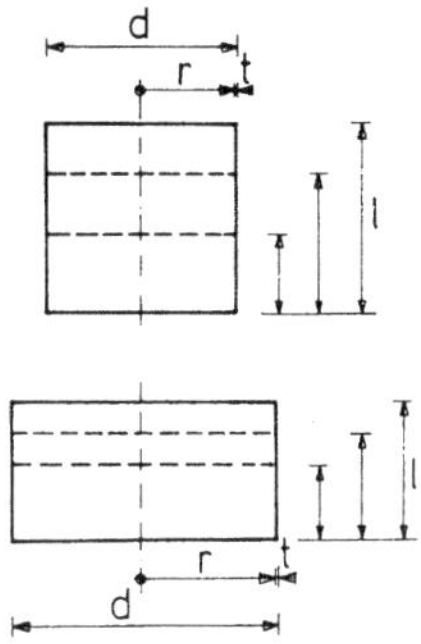

Fig.5. Dimensions of the models

load-combinations could be performed on the same model because of its fully elastic behaviour during buckling. A more detailed report is given in [5], [7].

Dynamic effects, that means the interaction of vibration and buckling of the wind loaded shell, turned out to be of negligible influence since a stationary buckling pattern appeared.

4. Results of the wind tunnel tests

4.1 Distribution of wind pressure

In order to establish a relation to the buckling tests and for comparison with test data already published, the wind pressure distribution was measured for all models. The pressure coefficients taken from [5] are compared with results of [9],[1o],[11] in Fig.6. It turned out that the maximum wind pressure at the center of the windward side didn't reach the full stagnation pressure, which is due to the special velocity profile of the wind tunnel. This has to be considered in comparing the buckling test data, but not for tanks in reality, where pressures with C = 1,o act in the stagnation area.

4.2 Results of buckling tests

The test results, taken from [5], are shown in Fig.7. The critical stagnation pressure $q_{w,cr}$, which caused buckling, is presented in relation to the simultaneously acting internal vacuum q_u and internal pressure q_i in form of interaction-curves for certain values of the axial force P. The critical wind pressure w_{cr} actually acting on the model has to be calculated from $q_{w,cr}$ by multiplying with the coefficient C. The figures obviously show -what already has been expected- that internal vacuum and axial pressure forces reduce the

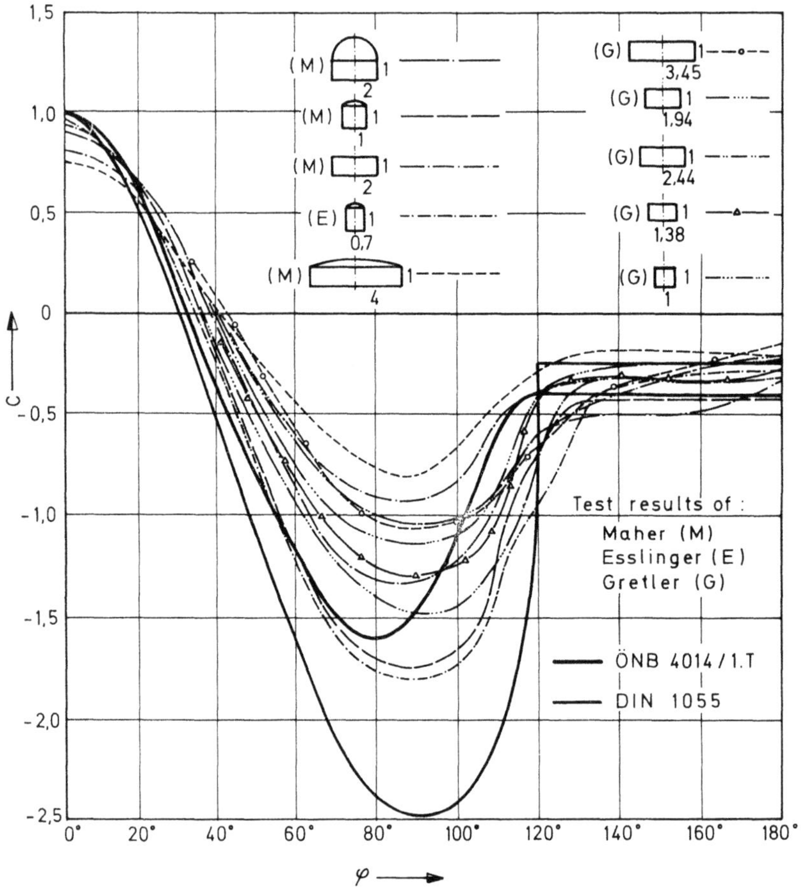

Fig.6. Distributions of wind pressure according to test results

critical wind loads, while internal pressure and axial tension will increase them.

During the tests in particular the buckling pattern was observed (Fig.8). In the case without any wind load, i.e. pure internal vacuum, the buckles are distributed evenly along the circumference. For combined loading of wind and internal vacuum buckling appears on parts of the shell only, namely on the front and back half in case of low wind pressures and on the front side only in case of high wind pressures. Here the observed wave-length due to wind pressure alone was approximately equal to that for internal vacuum, i.e. uniform external pressure. Fig.9 shows as example two buckling-patterns due to wind load and vacuum.

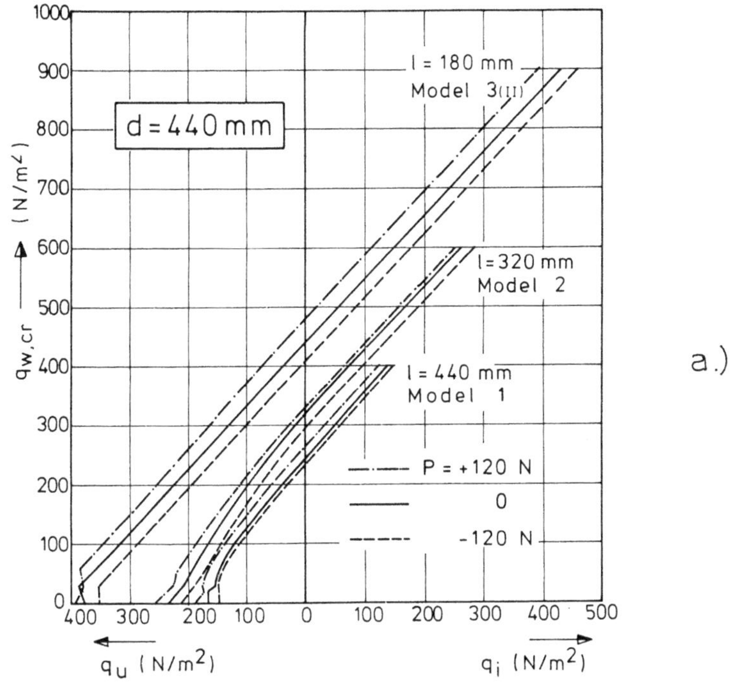

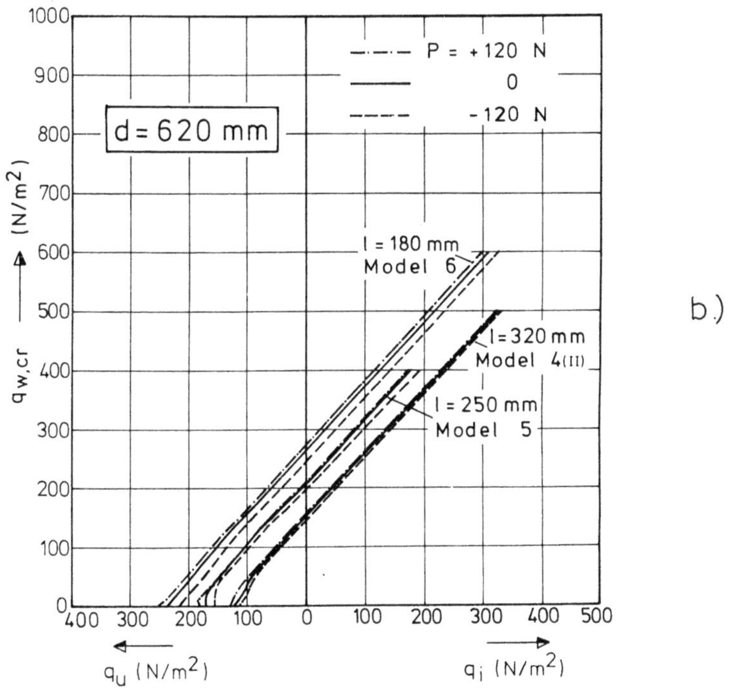

Fig.7. Results of the buckling tests

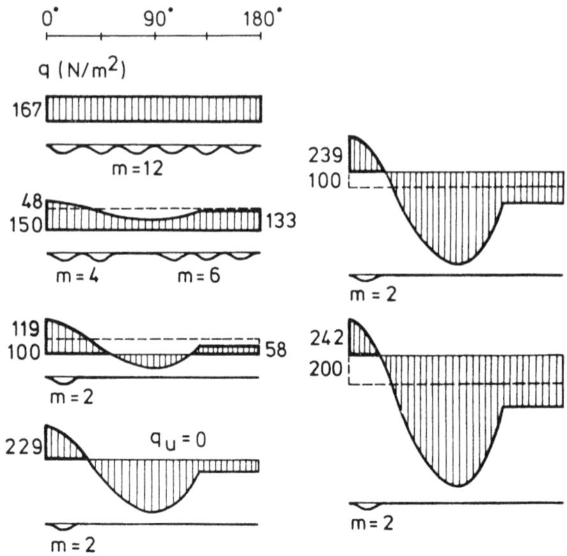

Fig.8. Pressure distribution and buckling pattern of model

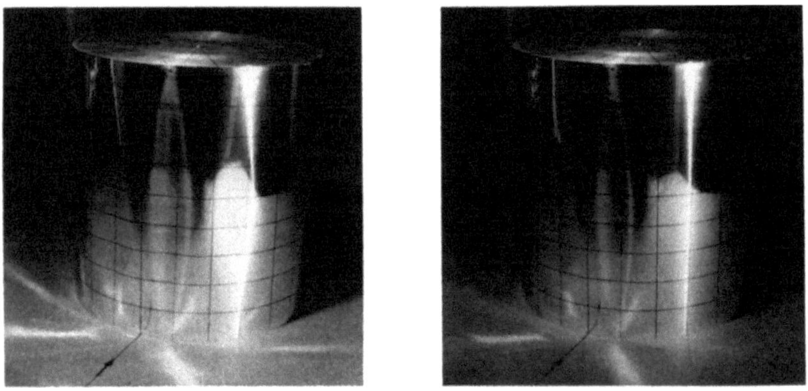

Fig.9. Buckling pattern of model 1 due to uniform external pressure(a) and wind pressure(b)

5. Buckling due to uniform external pressure

The theoretical buckling pressure $q_{u,th}$ and the wave-number $m_{u,th}$ of the test-cylinders may be determined analytically on the basis of the classical stability theory by using the well-known formulas

$$q_{u,th} = 1,5 \cdot 0,92\ E\ \frac{r}{l}\ \left(\frac{t}{r}\right)^{2,5}$$

$$m_{u,th} = \sqrt{1{,}5 \cdot 7{,}52 \frac{r}{l} \sqrt{\frac{r}{t}}}$$

The factor 1,5 in these formulas takes into account the effect of the axial constraint on the boundary conditions, which assume the shell to be completely fixed at the rigid end-plates. In [13] some influences on the theoretical buckling pressure were investigated, as there are the flexibility of the end plates in vertical direction and the weight of the upper plate. Both effects turned out to be negligible. The elasticity modulus E of the extremely thin Aluminium-sheets used for the cylinders was assumed with 7o.ooo N/mm^2. This is of influence on the relation between theoretical and critical buckling pressures only.

The critical buckling pressure $q_{u,cr}$ of the real, imperfect shell is derived from the theoretical pressure by multiplying with a reduction factor. This factor represents mainly the imperfection sensitivity of the shell and its

Table 1

Model				internal vacuum						wind load			
Nr	d	l	t	$q_{u,cr}$	$q_{u,th}$	$\frac{q_{u,cr}}{q_{u,th}}$	m_{cr}	m_{th}	$\frac{m_{cr}}{m_{th}}$	$q_{w,cr}$	C	w_{cr}	$\varkappa = \frac{q_{u,cr}}{w_{cr}}$
1	440	440	0,105	167	240	0,695	12	16,1	0,75	241	0,95	229	0,73
2	440	320	0,105	235	331	0,711	14	18,9	0,74	320	0,93	298	0,79
3(I)	440	180	0,102	398	546	0,728	20	25,3	0,80	486	0,92	447	0,89
3(II)	440	180	0,100	377	520	0,725	20	25,5	0,78	436	0,92	401	0,94
4(I)	620	320	0,105	131	198	0,693	20	24,4	0,82	158	0,94	149	0,88
4(II)	620	320	0,102	124	184	0,675	20	24,6	0,82	154	0,94	145	0,86
5	620	250	0,105	168	253	0,664	21	27,6	0,76	206	0,93	192	0,88
6	620	180	0,105	238	351	0,677	27	32,5	0,83	261	0,93	243	0,98

d, l, t (mm), q (N/m^2)

value is given in the literature with o,7 based on a great number of buckling tests. The reduction factors of the 8 test cylinders-calculated from the ratio $q_{u,cr}/q_{u,th}$ in Table 1- have values of o,664 up to o,728 and confirm the above mentioned value o,7. The ratio of the wave numbers m_{cr}/m_{th} is in good accordance with the value o,8 also, which is recorded in the literature (see Table 1).

6. Evaluation of the test results

6.1. Wind buckling pressure

The considerations made above suggest that the wind buckling pressure w_{cr} may be related to the buckling pressure $q_{u,cr}$ due to uniform external pres-

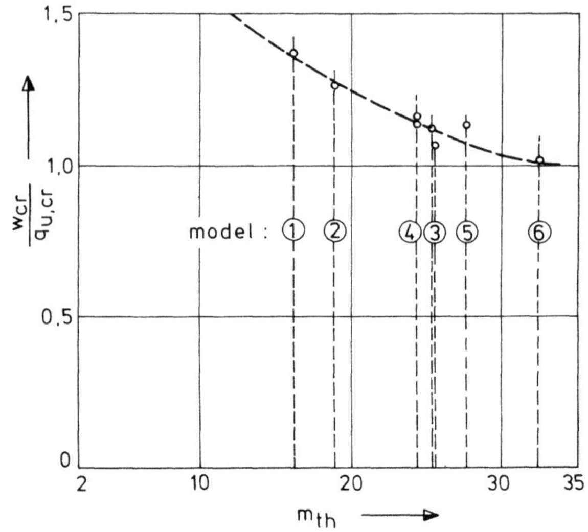

Fig.1o. Wind buckling pressures of models

sure. For this reason the maximum wind pressure w_{cr}, effectively acting on the wall of the model, has to be derived from the stagnation pressure $q_{w,cr}$ by $w_{cr} = C_o \cdot q_{w,cr}$. In this formula C_o means the pressure coefficient of the models at the stagnation point, which are somewhat smaller than 1,o. These reduced wind buckling pressures are plotted in Fig.1o versus the wave numbers m_{th} of the models due to uniform pressure. The wave numbers contain the radius-height-thickness relationships of the shells as well as the effect of the boundary conditions and can therefore be used to transfer the test results to shells of different geometry or different supports.As could be expected w_{cr} in any case is higher than $q_{u,cr}$ and may be expressed by

$$w_{cr} = \frac{q_{u,cr}}{\varkappa} ,$$

$\varkappa$ being tabulated in Table 1. It is obvious, that with increasing number of buckles acted upon by the wind pressure, the stabilizing effect of the parabolic pressure distribution as well as of the adjacent suction forces will be reduced and the wind buckling pressure decreases to the value $q_{u,cr}$.

6.2 Interaction of windload with internal vacuum or internal pressure

Fig.11 shows the test data, approximated by curves. $q_{cr(\varphi=o)}$ means the total critical pressure in the stagnation point consisting of wind pressure

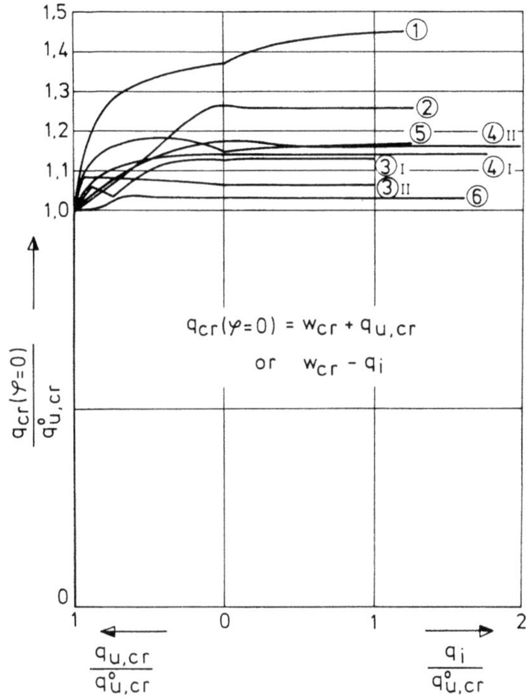

Fig.11. Interaction of wind load with internal vacuum or internal pressure

w_{cr} plus internal vacuum $q_{u,cr}$ or minus internal pressure $q_i \cdot q^o_{u,cr}$ and w^o_{cr} denote the buckling pressures due to isolated acting vacuum or wind loads. It is obvious that the various curves represent a quite different behaviour and that in some cases the increase of the total pressure is quite large, when the vacuum decreases for a very small amount only. This can be explained by the fact, that for internal vacuum buckling starts at the worst dent (pre-buckle) somewhere on the surface of the shell, while windbuckling depends on the imperfections in the windward area only. The form of the curves therefore is probably arbitrary and the interaction curve for wind and vacuum was generally chosen as a straight line. For a combination of wind load and internal pressure the total pressure $q_{cr(\varphi=o)}$ remains nearly constant, with exception of model 1, where a small increase occurs.

Therefore the entire interaction diagram may be simplified for practical application as shown in Fig.12. Since at real tanks $C_o = 1$, the value w_{cr} was substituted by $q_{w,cr}$. From this diagram follow the formulas for combination of

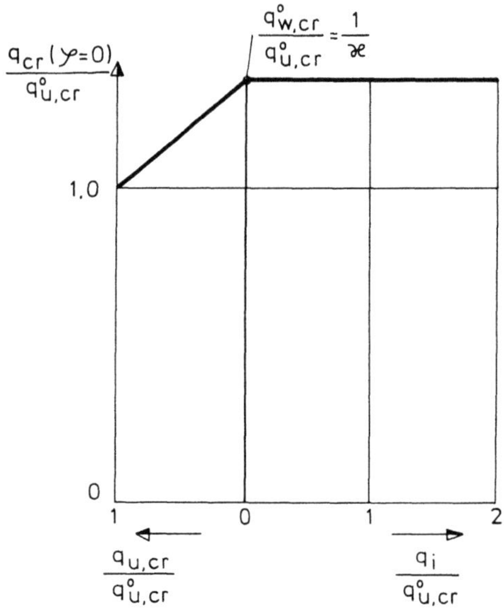

Fig.12. Simplified interaction-diagram

wind and internal vacuum $\frac{q_{cr(\varphi=0)}}{q^o_{u,cr}} = \frac{1 + q_u/q_w}{\varkappa + q_u/q_w}$

or wind and internal pressure $\frac{q_{cr(\varphi=0)}}{q^o_{u,cr}} = \frac{1}{\varkappa}$

6.3 Effect of additional axial forces

The influence of additional axial forces is shown in Fig.13 by means of the test data in the usual related presentation. In the absence of own test data the value P^o_{cr} was calculated according to the relevant rules [12].It is obvious that even for wind pressure the results agree well with the interaction lines proposed in design codes.

6.4 Comparison with analytical investigations

Comparative investigations for one of the recorded test models subjected to wind loads, which were based on the linear and nonlinear stability analysis, show good accordance with the test results and are reported in detail in [6], [15].

The analytical results in [3],[4] -although based on practical wind pressure

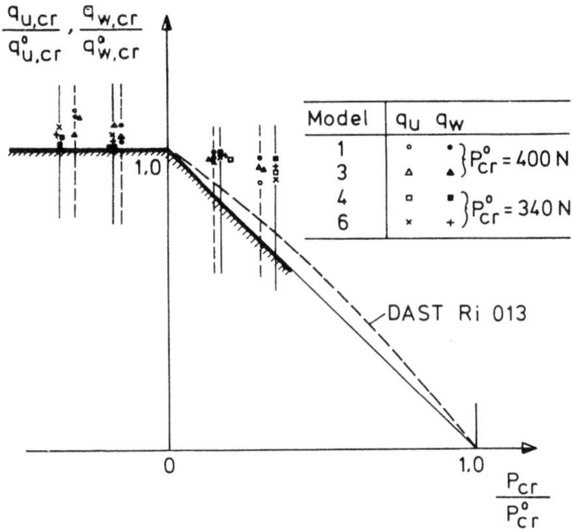

Fig.13. Effect of additional axial forces

distributions -do not agree with the wind tunnel tests as already indicated before.

In Fig.14 the results of a basic study of Almroth of the buckling-problem of cylindrical shells subjected to nonuniform external pressure [14] are set in relation to the results of the wind tunnel tests. Although the pressure dis-

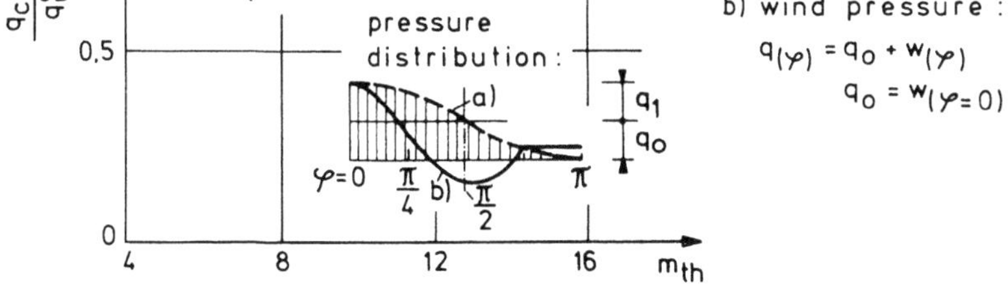

Fig.14. Comparison with results of Almroth [14]

tribution considered by Almroth is quite different in detail from that of the wind pressure, especially at the sides of the cylinder, the results show the same tendency: decreasing buckling pressure for increasing wave-number and smaller buckling pressure due to the more severe loading.

7. Design concept based on an equivalent uniform pressure

The concept of substituting the nonuniform wind pressure by an "equivalent uniform external pressure q_{ue}" can now be established on the basis of the results of the wind tunnel tests.

For wind loads alone the value q_{ue} may be expressed -as shown above- according to the wind buckling pressures by the formula

$$q_{ue} = \varkappa \cdot q_w \quad ,$$

assuming the shape factor $C_0 = 1$, valid for applications to tanks used in practice.

The factor $\varkappa$ is derived from Fig.15 in relation to the wave number m_{th} and may be given in a simplified form for practical application:

$$\varkappa = 0{,}46 + 0{,}017\ m_{th} \leq 1{,}0$$

This formula is confirmed by the tests for wave-numbers $m_{th} \geqq 16$, but may be extrapolated down to $m_{th} = 10$ without any risks. Tests which will give information about wind buckling of shells with lower wave-number are difficult to perform, since the models would become quite long for this purpose. But the range covered by the tests seems to be wide enough in order to include tank-sizes usual in practice.

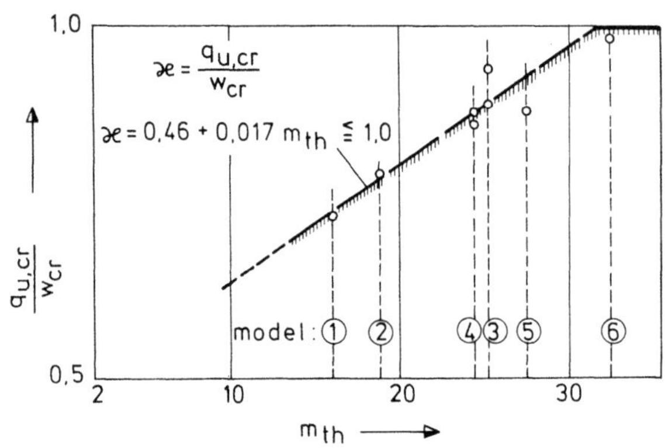

Fig.15. Factor $\varkappa$ as a function of the wave-number

The equivalent pressures q_{ue} reach values of o,7 to 1,o of the stagnation pressure, which indicates that the value $q_{ue} = 0{,}5\ q_w$ of the DIN 4119 is on the unsafe side.

By means of the equivalent pressure q_{ue} the buckling calculation of a wind loaded tank may be reduced to a buckling calculation due to uniform external pressure. In order to determine the factor $\varkappa$,the wave-number has to be calculated. This can be done by the formula:

$$m_{th} = 2{,}74\sqrt{C_R \frac{r}{l}\sqrt{\frac{r}{t}}}\ ,$$

which takes into account the boundary conditions by the factor C_R. Here $C_R = 1{,}0$ belongs to tanks without axial constraint at the ends and rises to 1,5 for the case of full axial constraint.

For combined loadings the equivalent pressure may be determined as follows:

for wind and internal vacuum: $q_{ue} = q_u + \varkappa \cdot q_w$

and for wind and internal pressure: $q_{ue} = \varkappa (q_w - q_i)$.

Such load combinations occur frequently in the case of fixed-roof tanks, where vacuum valves create internal vacuum, or in the case of floating-roof tanks, where under windloads alone a suction load of the value $q_u \cong 0{,}6 \cdot q_w$ is caused inside the empty tank. Since the top edge of a floating-roof tank is stiffened radially by a windgirder, the buckling analysis of such a tank may be performed similar to a closed-top tank using the equivalent pressure $q_{ue} = q_w\ (\varkappa + 0{,}6)$.

So far the design concept is applicable to cylindrical tanks of variable geometry provided that the shell is of constant wallthickness. Since tank-shells in general are designed due to the hydrostatic pressure of the liquid their

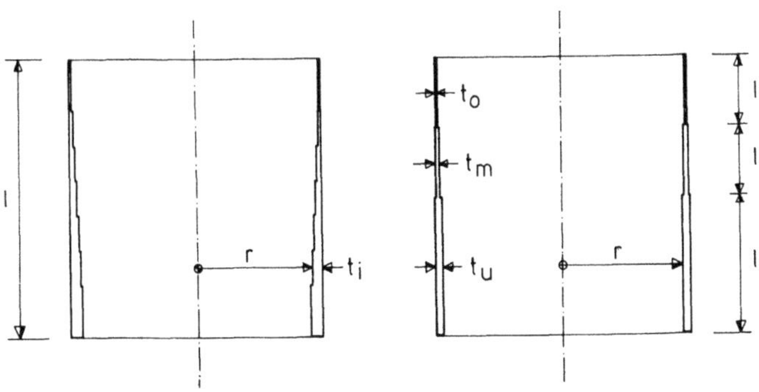

Fig.16. Tank with varying wallthickness and "substitute cylinder"

wallthickness gradually increases downwards from course to course, which leads to a stepped shape of the wall.

8. Design Concept for tanks with varying wallthickness

Based on the understanding of the behaviour of shell buckling due to wind loads the design concept established above can be extended to tanks with varying wallthickness too. In order to obtain the equivalent pressure q_{ue} appropriate for such tanks, the theoretical wave-number m_{th} has to be determined for a stepped wallthickness. The buckling analysis then reduces to that of a stepped cylinder under the uniform external pressure q_{ue}. This may be performed according to a simplified method -given in detail in [16], [12]- introducing a "substitute cylinder"(Fig.16). This "substitute cylinder" represents a hypothetical shell with diameter and height equal to those of the real tank, but built up of three courses only, each of them having a constant wall-thickness equal to the average plate thickness of the tank in the respective zone.

8.1 Buckling behaviour of cylindrical shells with varying thickness

In principle the general buckling behaviour of a stepped cylinder may be derived from the example of a shell composed of two courses. In Fig.17 and 18 the theoretical buckling pressures and wave-numbers are plotted as a function of the ratio of the wallthicknesses and of the heights of the courses. The buckling pressures q_{th} and wave-numbers m_{th} are related to the buckling pressures q_o and wave-numbers m_o of a cylinder, which consists of the top course only [17]:

$$q_o = 0{,}92\, E\, \frac{r}{l_o} \left(\frac{t_o}{r}\right)^{2,5} \quad \text{and} \quad m_o = 2{,}74 \sqrt{\frac{r}{l_o} \sqrt{\frac{r}{t_o}}} \;.$$

Depending on the distribution of the wallthicknesses the shell can buckle in the thinwalled upper part only or it might buckle over the whole length of the cylinder. The first buckling mode will be called partial buckling, the latter one total buckling. In general total buckling takes place in the case of a thin-walled bottom course, which buckles due to its thin wall together with the upper course. Partial buckling is caused by a thick bottom course, which will not take part in overall buckling but on the contrary has a stabilizing effect. While the change from one mode to the other results in a gradual transition of the buckling pressures, the wave-numbers show a strongly marked jump at the transition point. This typical behaviour of the wave-numbers affords to determine the wave-numbers seperately with regard to

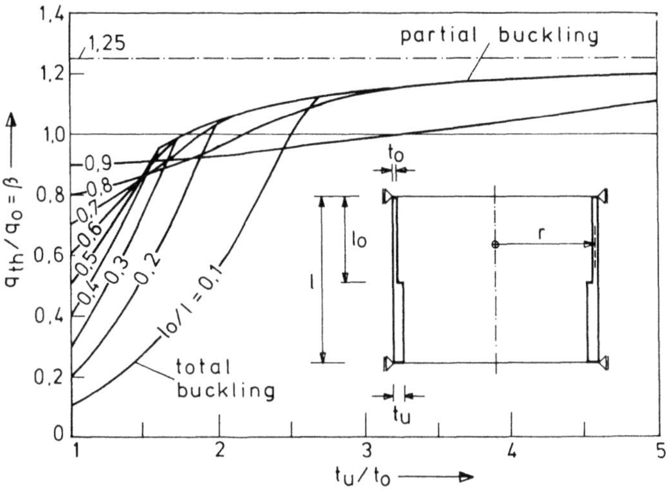

Fig.17. Buckling pressure of cylinders with 2 courses

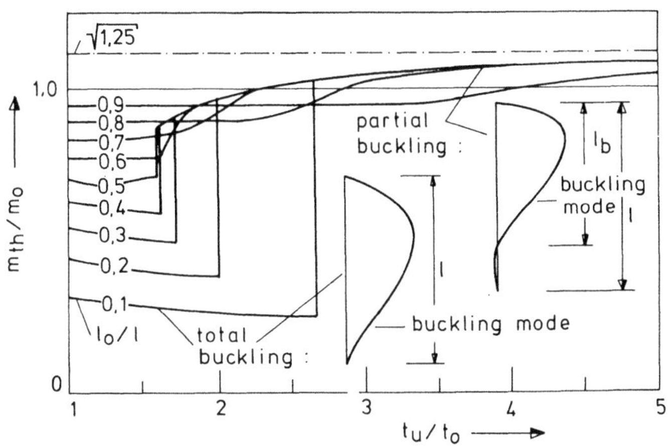

Fig.18. Wave-numbers and buckling modes
of cylinder with 2 courses

the different buckling modes.

8.2 Simplified method based on the "substitute cylinder"

The approximate method in [16] reduces the tank with its actual plate thicknesses to a substitute cylinder composed of three courses of the thicknesses t_o, t_m and t_u(Fig.16).

In the case of total buckling the wave-number may be calculated in good approximation for a shell with an average thickness:

$$t_m^* = (t_o \cdot l_o + t_m \cdot l_m + t_u \cdot l_u)/l$$

$$m_{tot} = 2{,}74 \sqrt{\frac{r}{l}\sqrt{\frac{r}{t_m^*}}} \qquad (1)$$

In the case of partial buckling the wave-number increases considerably, because due to the restraining effect of the lower courses the effective buckling length is shortened from l to l_B. The value $l_B = \frac{l_o}{\beta}$ can be derived from the buckling pressure of the substitute cylinder:

$$q_{th} = q_o \cdot \beta = 0{,}92E\,\frac{r}{l_o}\,\left(\frac{t_o}{r}\right)^{2,5} \cdot \beta$$

β includes the restraint of the lower courses and is tabulated in [16],[12]. It was shown by comparative calculations [18], that the introduction of the value l_B alone doesn't give satisfactory results, but that good accordance was obtained by use of an additional factor δ.

$$m_{part} = m_o \cdot \beta \cdot \delta = 2{,}74 \sqrt{\frac{r}{l_o}\sqrt{\frac{r}{t_o}}} \cdot \beta \cdot \delta \qquad (2)$$

with m_{part} being the wave-number for partial buckling. The factor δ is given in Fig.19.

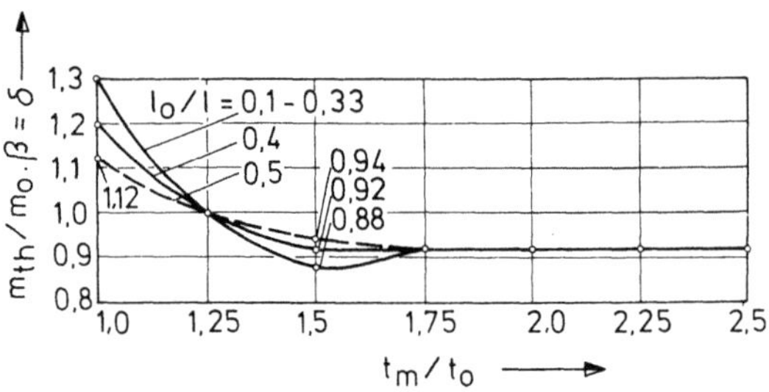

Fig.19. Wave-number for partial buckling: factor δ

In order to decide wether total buckling or partial buckling should be assumed the transition points between the two buckling modes are plotted in Fig.2o. By consideration of the geometry of the substitute cylinder it can be decided, whether the tank in question will buckle on the whole or only partially.

By means of this simplified method the additional amount of calculations for the practical design calculation of a stepped tank subjected to wind loads can be such reduced to an extremely small amount.

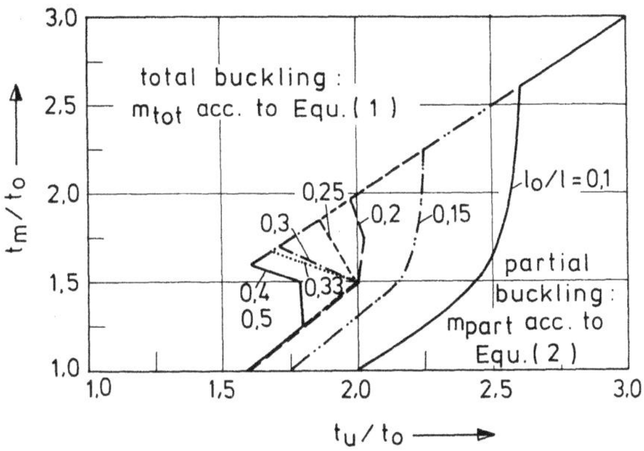

Fig.2o. Definition of the transition between total and partial buckling

9. Ring stiffened shells

Since the plate thickness of tank-shells in general is designed due to internal pressure the most effective measure to prevent shell-buckling because of external pressure is through fixing by secondary stiffening rings. According to the concept of the "equivalent pressure q_{ue}" described above, wind loads may be considered as uniform external pressure also. For the sizing of such rings only few rules are given in design Codes. While the BS 2654 has established a list of sizes for ring-stiffeners, depending on the tank diameter, the DIN 4119 doesn't give any rule. Although a theoretical stability analysis of a ring-stiffened shell due to uniform external pressure based on a linear shell theory is no problem in principle, since relevant Computer-programs are available,[19] e.g., this approach is too time-consuming for design practice.

A simplified approach to this problem will be given in the following as well as some information about "limiting values" of the stiffness of secondary rings. This investigation is based on a simplified shell theory -the so called Semi-membrane-theory (SMT). Since the main simplification of this method is to neglect the axial bending stiffness of the wall, the results are limited to thin-walled shells. In addition the results were compared to those

of an exact theory [19].
The example considered in detail is a cylindrical shell subjected to uniform external pressure, stiffened by a ring at the half of its height(Fig.21). The cross-sectional properties of this ring are its flexural rigidity J_R and its area A_R. Depending on the rigidity of the ring the well-known response of this shell may be a symmetric buckling mode or an antisymmetric mode with respect to the axial direction. At the value J_R^* both configurations are equally possible. The investigations therefore may be limited to the study of the symmetric buckling and to the determination of the limiting value J_R^*, which corresponds to the minimum value of J_R that just guarantees antisymmetric buckling of the shell. Since it turned out that the value J_R^* practically will lead to very small ring-sizes, there is not much sense in the designing smaller stiffening rings even in case of small external loads, which would not require the full buckling strength of the antisymmetric mode.

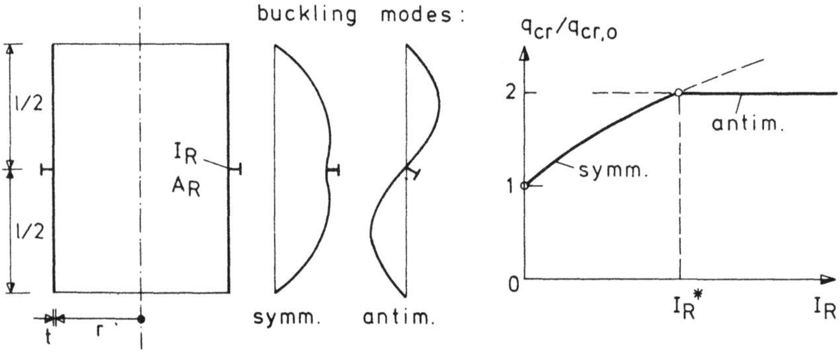

Fig.21. Ring stiffened cylindrical shell: buckling modes and "limiting value J_R^* "

This is due to the fact, that the wave-number corresponding to the symmetric buckling pattern is nearly as high as that of the unstiffened shell and that therefore the span from node to node becomes small. By use of the buckling formulas described above the buckling pressure for the antisymmetric mode reaches double the value of the buckling pressure $q_{cr,o}$ of the unstiffened Cylinder. According to the SMT the stability equation of the symmetric buckling mode can be expressed in the following form [2o] :

$$\frac{16.E.F(\lambda)}{m^4(m^2-1)} \cdot \left(\frac{t}{r}\right)\left(\frac{r}{l}\right)^3 + \frac{E\,J_R}{r^4}\,(m^2-1) - \frac{A_R}{r^2}\cdot\sigma_R = 0 \qquad (3)$$

$$\text{with } F(\lambda) = \lambda^3 \frac{2\cosh\lambda \cdot \cos\lambda}{\cosh\lambda\cdot\sin\lambda - \sinh\lambda\cdot\cos\lambda} ,$$

$$\lambda = \frac{l}{2r} \sqrt[4]{m^4(m^2-1)\left(\frac{\sigma}{E} - \left(\frac{t}{r}\right)^2 \cdot \frac{m^2-1}{12\,(1-\mu^2)}\right)} ,$$

$$\text{and } \sigma = q\,\frac{r}{t} .$$

In the pre-buckling state the stress of the ring-stiffener σ_R was assumed equal to the stress σ in the wall, the latter being supposed constant over the height of the shell.

Fig.22 shows the effect of the rigidity of the stiffening rings on the buckling pressure considering symmetric buckling modes only. The curve for J_R^* touches the line for $q_{cr} = 2.q_{cr,o}$; the curve for $J_R = o$ shows its minimum slightly below $q_{cr,o}$, which is due to the assumption of an area A_R only, neglecting the rigidity J_R.

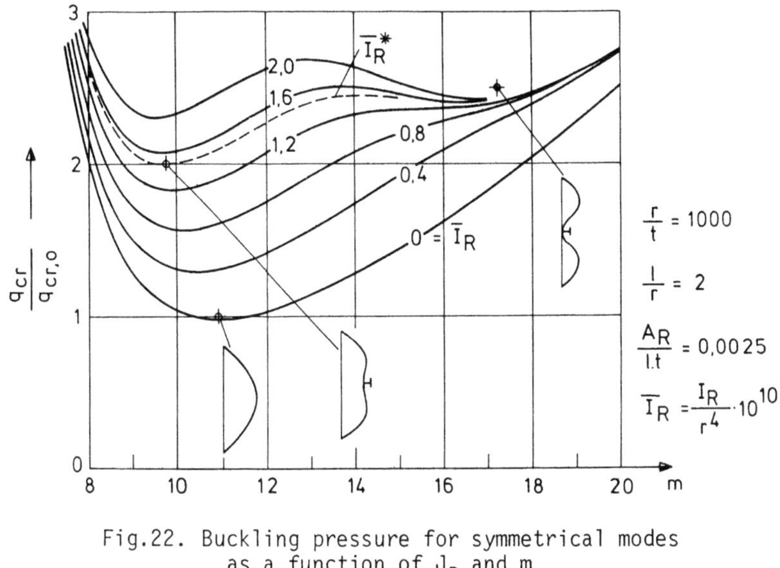

Fig.22. Buckling pressure for symmetrical modes as a function of J_R and m

Using the SMT it is not necessary to perform these calculations of the buckling pressure as function of the value J_R in order to obtain the limiting value J_R^*, since the stability equation allows to express J_R directly. Fig.23 shows the results of the calculation of the reduced values $\bar{J}_R$, which lead to an increase of the buckling pressure up to $2.q_{cr,o}$. The maxima of the curves

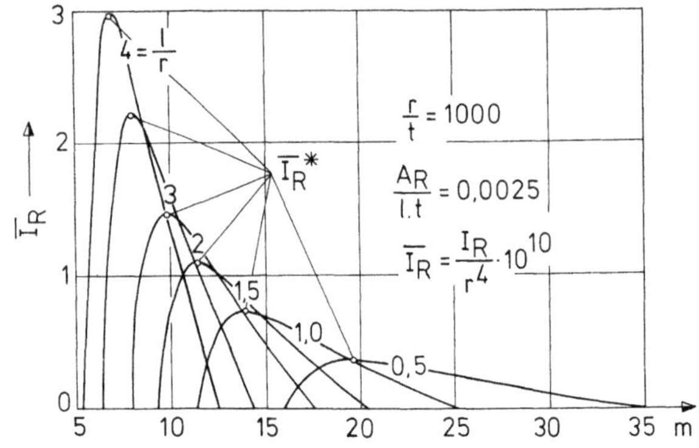

Fig.23. Reduced rigidities $\bar{J}_R$, which afford $q_{cr} = 2.q_{cr,o}$ for various lengths, as a function of m

for $\bar{J}_R$ as a function of the wave-number m give as a result the limiting value $\bar{J}_R^*$. Since it is obvious from this figure, that the values J_R^* depend significantly on the length of the cylinder a more suitable parameter was chosen to relate the flexural rigidity J_R of the rings. The value γ_R represents the ratio of the rigidity of the ring to that of the wall over the whole cylinder-length, the value δ_R the analogous ratio of the sectional areas.
Fig.24 gives the results of this related presentation, which shows that the limiting value γ_R^* is nearly independent of the ratios r/t and l/r. Consequently the rigidity J_R^* turns out to be practically independent of the radius of the shell -a result that doesn't verify the dependence of the tank-diameter, as stated in the BS 2654. On the other hand the sizes of the stiffe-

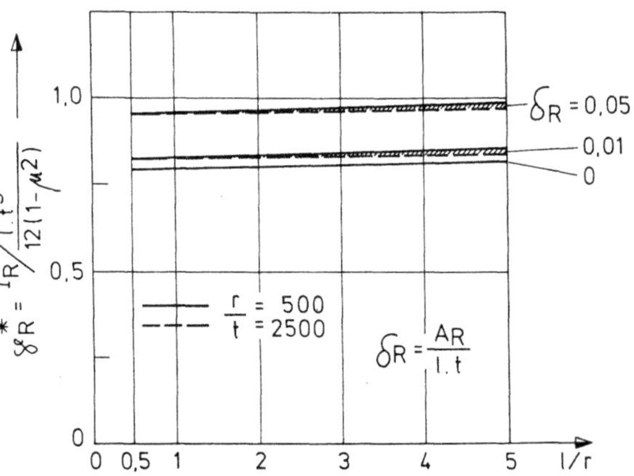

Fig.24. "Limiting stiffness γ_R^*", which affords $q_{cr} = 2.q_{cr,o}$

ning rings as required by BS 2654 -if applied to the dimensions of practical tanks- turn out to be very large, with values of about 1o to 2o times the values given in Fig.24.

Since the determination of these sizes in the BS 2654 doesn't depend on the length of the shell or on the wall-thickness a complete comparison cannot be made.

Relatively good accordance of the results derived above was found with a design proposal published in [8] .There it is proposed, that the stiffener alone should be adequate to carry the entire buckling pressure from half the span on each side of the stiffener, assuming the wave-number equal to that of the unstiffened shell. This leads to a constant value $\delta_R^* = 1{,}33$, which is about 5o% higher than the values obtained from the SMT.

In order to confirm the validity of the results of the SMT comparative calculations were performed using the CAD-Computerprogram [19]. Fig.25 shows as example the comparison for two cylindrical shells. Considering the simplifications of the SMT the results agree quite well with those of the exact theory except the fact, that the buckling pressure q_{cr} reaches values higher than $2 \cdot q_{cr,o}$. But this could have been expected, since it is well-known, that -compared to the results given by the simplified buckling formula used above- the buckling pressure increases for decreasing length or increasing wallthickness. This increase is due to the bending stiffness of the wall in axial direction, which was neglected in the SMT, and is worth to be considered

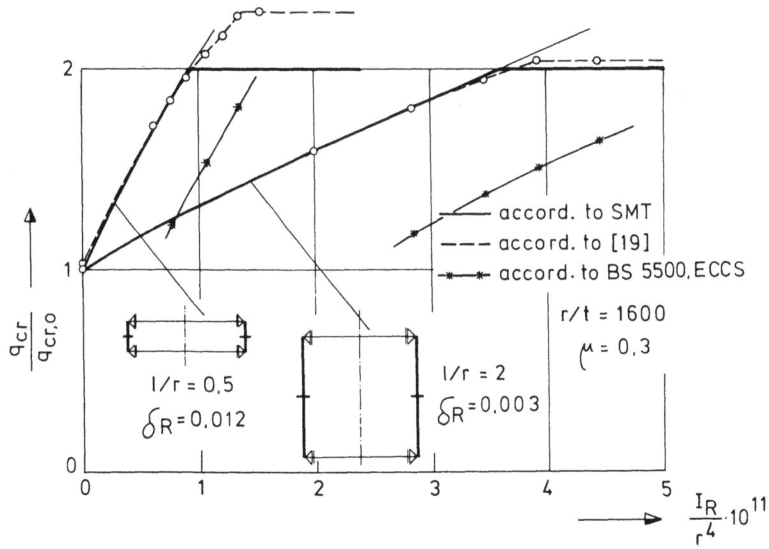

Fig.25. Comparison with other results

only for short cylinders with ratios $\bar{l} = \frac{l}{r}\sqrt{\frac{r}{t}} \leq 2o$, where l is meaning the wave-length in axial direction. Since in the antisymmetric buckling mode l is to replace by l/2 according to Fig.21, the value $\bar{l}$ amounts to 1o for the short cylinder of Fig.25 and to 4o for the longer one. Accordingly the discrepancy shown in Fig.25 is of practical interest only for the short shell. But the amount of this discrepancy would be of the half value only when taking into account also the axial component of an external pressure loading, which is assumed to act all around the tank, while in the comparison of Fig.25 the external pressure on the wall alone was considered. More rigorous investigations [2o] show, that even in the case of short shells there will be achieved close accordance with exact results when introducing the extensional stiffness of the ring-girder in the second term of the stability equation (3). By this the stiffness of the ring decreases significantly for high wave-numbers as well as for small values δ_R compared to the inextensional stiffness-value.

Furthermore there was tried a comparison of the results of the SMT and of the Computer-calculation [19] with the rules for the stability of ringstiffened shells in the British Standard BS 55oo, which was also included in the ECCS-Recommandations, chapter 1o. But it turned out that these rules don't allow an adequate buckling analysis in the case of such thin-walled structures with wave-numbers of about 1o to 2o, since the code limits the wave-number by 6. Fixing the wave-number with 6 in the analysis naturally would bring about by far too high buckling pressures of the ring-stiffened shell or on the other hand too small ring-sizes. But it should be noted that the Standard certainly isn't meant to be applied in this way to thin-walled tank-structures. Obviously it was established to be applied for heavily loaded shells, closely stiffened by several rings, as used in the chemical industry or for underwater structures. But this should be emphasized strongly in the Standard. In Fig.25 therefore only the theoretical buckling pressures,upon which the Standard is based, were plotted for comparison. The difference to the other results is rather large. This is due to the fact, that the effect of the circumferential bending-stiffness of the wall was neglected in the buckling formula of the Code.

As a consequence of the higher buckling pressure of short shells in the antisymmetric mode the limiting value J_R^* increases when calculated according to the exact theory, as shown in Fig.25. The discrepancy of the results partially can be eliminated by calculating the values J_R^* on the basis of the buckling pressures according to the well-known Windenburg-Trilling-formula instead of using, as before, the value $2.q_{cr,o}$. The results of this analysis are presented in Fig.26. As could be expected due to the limitation $\frac{l}{r}\sqrt{\frac{r}{t}} \leq 2o$, the values increase when the shells become thicker and shorter.

In summary these results give some information about the sizing of stiffening rings and point out in particular the dependence on the geometry of the tanks. Accordingly it may be stated that the limiting stiffness of secondary rings of thin-walled tanks is nearly independent of the tank-diameter and further that the required sizes of such stiffening rings turn out to be quite small for practical tank dimensions. This is due to the fact, that the wave-number of the stiffened shell is nearly as high as that of the unstiffened shell and that therefore the span of the wall from node-line to node-line becomes rather small.

Since the limiting values γ_R^* are based on the assumption of a perfect shell, buckling tests would be of special interest in order to find out an appropriate reduction factor, which will take into account the imperfection sensitivity of the stiffened shell.

By means of the "limiting stiffness" γ_R^* of secondary rings the stability analysis of tanks is reduced to a simple check, easy to handle in design practice. Further information on cylindrical shells stiffened by more than one ring or by rings of varying spacing will be given in [2o].

Although these results were derived for uniform external pressure they are applicable to windloads also, when the calculation of q_{ue} is based on the wave-number of the antisymmetric buckling mode.

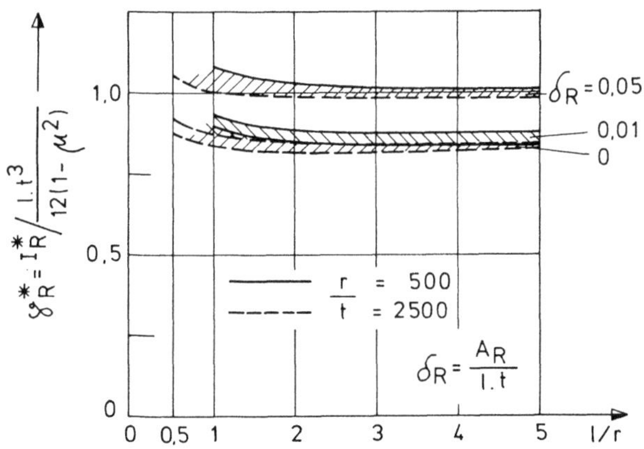

Fig.26. "Limiting stiffness γ_R^* "

1o. Summary

Cylindrical shells subject to buckling due to wind loads in engineering practice generally are thin walled storage tanks, either with closed top or with a floating roof. In the design procedure of such tanks the stability analysis is an additional check only. For that purpose the buckling analysis should be reduced to a calculation not too theoretical or time-consuming, since in addition stabilizing measures by means of stiffening rings will be rather inexpensive.

This paper presents a design concept for the buckling analysis of wind-loaded tanks, based on results of wind tunnel-tests. Included in this concept is further on the interaction of wind loads with internal vacuum or internal pressure as well as additional axial forces, since such loading conditions may occur frequently in practice. Although the test data refer to models of constant wall-thickness the design concept could be extended to tanks with varying wallthickness.

More over some information is given about the required stiffness of secondary rings based on theoretical investigations. These result in quite small sizes of stiffening rings compared to those proposed in some design Codes.

References

1. De Wit, J.: How to calculate the Stability of Empty Storage Tanks. Oil and Gas International, Vol.11,1971,H.8,S.74-8o.
2. Ziólko, J.: Modelluntersuchungen der Windeinwirkung auf Stahlbehälter mit Schwimmdach. Der Stahlbau 47(1978),H.11,S.321-329.
3. Maderspach, V.; Gaunt, J.T., and Sword, J.H.: Buckling of Cylindrical Shells due to Wind Loading. Der Stahlbau 42(1973,H.9, S.269-277.
4. Maderspach, V.; Kamat, M.: Buckling of open cylindrical tanks due to wind-loading. Der Stahlbau 48(1979),H.2,S.53-56.
5. Gretler, W.: Beulversuche an zylindrischen Modellen im Windkanal. Institut für Strömungslehre und Gasdynamik, TU Graz (1978).
6. Fischer, D.; Rammerstorfer, F.G.; Fastner, R.E.: Vergleichsberechnungen. EDV-Dokumentation VOEST-ALPINE (1978).
7. Resinger,F.; Greiner,R.: Auswertung der Windkanalversuche. Forschungsbericht des Forschungsförderungs-Fonds der Gewerblichen Wirtschaft Zl.1/367--I/P
8. Mc Grath, R.V.: Stability of API Standard 65o Tank Shells. Session on Oil Storage Tanks of the API, Philadelphia, Pa., May 16, 1963.
9. Maher, F.J.: Wind Loads on Dome-Cylinder and Dome-Cone Shapes. Proc.of the ASCE, Journ.of Struct.Div. 92(1966),No.St.5(Oct),S.79-96.
1o. Purdy, D.M.; Maher, F.J. and Frederick, D.: Model Studies of Wind Loads on Flat-Top Cylinders. Proc. of the ASCE, Journ.of Struct.Div. 93(1967), No.St2(April), S.379-395.

11. Esslinger, M.; Ahmed, S.R., and Schroeder, H.H.: Stationäre Windbelastung offener und geschlossener kreiszylindrischer Silos. Der Stahlbau 4o(1971) H.12,S.361-368.

12. DAST-Richtlinie o13, Beulsicherheitsnachweise für Schalen (Juli 198o)

13. Resinger,F., und Greiner,R.: Kreiszylinderschalen unter Winddruck. Anwendung auf die Beulberechnung oberirdischer Tankbauwerke. Der Stahlbau 5o(1981), H.3, S.65-72 .

14. Almroth, B.O.: Buckling of a Cylindrical Shell Subjected to Nonuniform External Pressure. Journal of Appl.Mech. 29(1962),No.4 (Dec.),S.675-682.

15. Ramm,E.; Rammerstofer, F.G.; Brendel, B.; Fastner, R.E.: Stabilität windbelasteter dünner Kreiszylinderschalen. Schalenbeultagung in Darmstadt am 6./7.Juni 1979. Sonderheft der DFVLR Braunschweig.

16. Resinger,F. und Greiner, R.: Praktische Beulberechnung oberirdischer zylindrischer Tankbauwerke für Unterdruck. Der Stahlbau 45(1976), H.1, S.1o-15 .

17. Resinger, F. und Greiner, R.: Zum Beulverhalten von Kreiszylinderschalen mit abgestufter Wanddicke unter Manteldruck. Der Stahlbau 43(1974), H.6, S.182-187.

18. Greiner, R.: Zum Beulnachweis von Zylinderschalen unter Winddruck bei abgestuftem Wanddickenverlauf. Der Stahlbau 5o(1981), H.6, S.176-179.

19. Esslinger, M. und Kerkhoff, H.: CAD-Berichte. Berechnung der Beullasten von ringversteiften Rotationsschalen unter axialsymmetrischer Belastung im elastischen Bereich. KFK-CAD 137, (Mai 1979) .

2o. Greiner, R.: Ringversteifte Zylinder unter Außendruck, to be published.

Buckling of Long Liquid-Filled Cylindrical Shells

H. Saal

Lengfeld i. Odw., Germany

1. Introduction

Pipelines are widely used for the transport of material. While pressure is the dominant sustained design load with gaseous materials, the weight of the contents becomes important with liquid transport and hydraulic solids transport.

In the case of the customary saddle support the support reactions cause high stresses in addition to the stresses from the tube acting as a beam across the supports /1/. These additional stresses reduce the load carrying capacity of the favourable circular cross section. To eliminate these stresses ring girders, first introduced in Sweden 70 years ago /2/, are incorporated with the support. In the case of horizontal vessels the same effect may also be attained by moving the saddle support to the stiffened ends of the vessel. Because of the advantages of this type of support this investigation is restricted to cylinders with the corresponding boundary conditions.

The stresses due to dead load of the shell and weight of the fluid when the cylinder is completely filled agree with beam theory and will not be discussed here. However, if the shell is only partially filled, the discontinuous variation of the normal pressure along the circumference causes circumferential bending stresses and deviations of the axial membrane stresses from those according to beam theory. It has been stated many years ago that these stresses for the partially filled pipe (Fig. 1) may exceed those for the full pipe /2/, /3/.

The plot of the minimum axial membrane stresses due to a linear analysis versus the load (Fig. 2) demonstrates the nonlinearity of the stress - load-relation increasing with decreasing slenderness of the pipe, which is expressed by the parameter β . The nonlinearity is even more pronounced with the equivalent stress at midspan according to Tresca (Fig. 3). This is due to the circumferential bending stresses of the partially filled pipe.

Buckling of Shells, Ed.: E. Ramm
Proceedings of a State-of-the-Art Colloqium

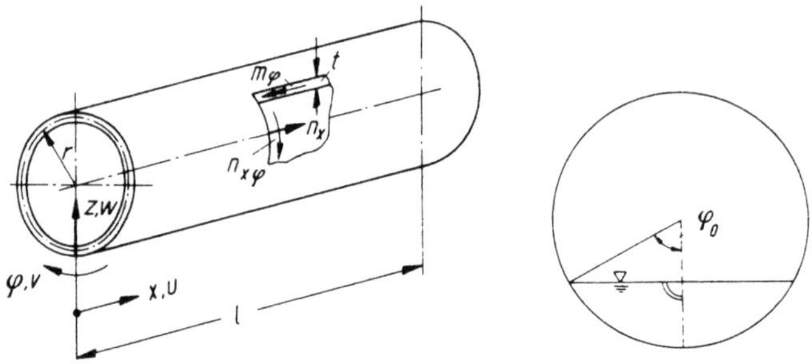

Fig. 1. Definition of dimensions, coordinates, stress resultants, displacements and the degree of fill φ_0

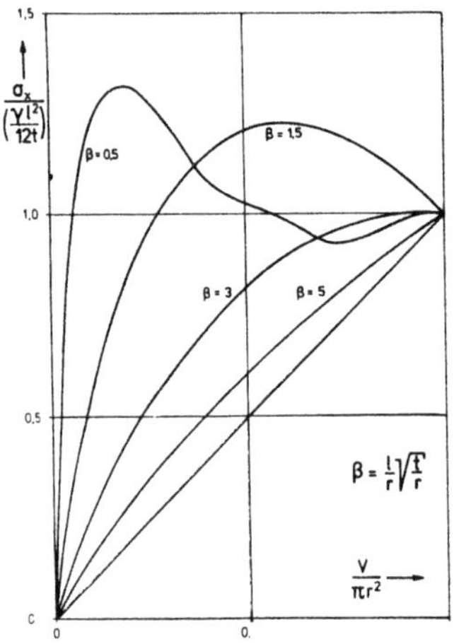

Fig.2. Minimum axial membrane stresses σ_x at the clamped edge versus the volume of water contents

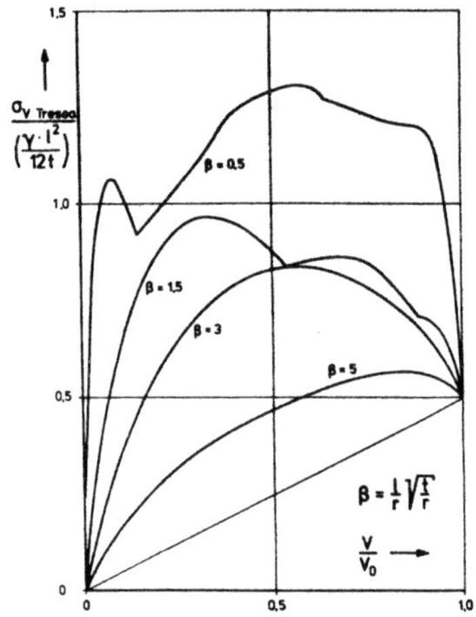

Fig.3. Equivalent stresses (Tresca) σ_v at midspan versus the volume of water contents

In the theoretical investigation of the buckling load the deformations of the cross section resulting from these bending stresses have to be considered as well as the increased axial membrane stresses. Both phenomena which disappear with the full pipe may reduce the buckling strength. If at all the liquid filled pipe will therefore buckle before it is full. This has been demonstrated by the experiments of Stern /4/ and Tooth and Fernandez /5/. In their tests which were compared with their theoretical results for the full cylinder they applied a slight internal pressure to avoid buckling of the partially filled pipe.

In practice the pipe with liquid or solid contents will be partially filled during startup and shutdown in any case and sometimes even under normal operating conditions. Therefore it was desired to provide stress concentration factors which give the actual maximum stresses of the pipe

when they are applied to the stresses according to beam theory. Also an appropriate tool for the proof of buckling safety was required which should account for various boundary conditions and any degree of fill.
The lack of experimental results for the partially filled pipe - neither strain measurements (except from /6/) nor buckling loads available - demanded for an experimental verification of theoretical strains and buckling loads. Because of its destructive effect on the buckling load a slight vacuum in the pipe should also be investigated in the experiments.

2. Theoretical Investigations

It is obvious that because of the deformations of the cross section of the partially filled pipe a nonlinear analysis of the shell is necessary. However, on the other hand the great number of calculations, which had to be performed for the parametric investigation demanded for an utmost reduction of the numerical work. This reduction was based on the experience that

a) circumferential extension and shear strain of the middle surface of the shell are very small compared to unity.
b) buckling is essentially due to the axial membrane stresses if there is no vacuum.

2.1 Nonlinear Stress Analysis

The nonlinear equations of equilibrium of the cylindrical shell are derived from the principal of virtual displacements

$$\begin{aligned}-\delta\Pi = \oint\Big\{\int_0^{\ell}\Big[\, & n_x\,\delta\varepsilon_x + n_\varphi\,\delta\varepsilon_\varphi + n_{x\varphi}\,\delta\gamma_{x\varphi} \\ & - m_x\,\delta\varkappa_x + m_\varphi\,\delta\varkappa_\varphi + m_{x\varphi}\,\delta\varkappa_{x\varphi} \\ & - p_z\Big(w'\,\delta u + \frac{\dot{w}-v}{r}\,\delta v - (1+u')\,\delta w\Big)\Big]\,dx \\ & + \Big[n_x\,\delta u + n_{x\varphi}\,\delta v\Big]_0^{\ell}\Big\}\, r\,d\varphi\end{aligned} \tag{1}$$

with the nonlinear strain-displacement-relations

$$\begin{aligned}\varepsilon_x &= u' + (v'^2 + w'^2)/2 \\ \varepsilon_\varphi &= (\dot{v} + w)/r + (\dot{u}^2 + \dot{w}^2 + v^2 - 2v\dot{w})/(2r^2) \\ \gamma_{x\varphi} &= (\dot{u} + rv')/r + (\dot{u}u' + \dot{v}v' + \dot{w}w' - vw' + wv')/r \\ \varkappa_x &= w'' \\ \varkappa_\varphi &= (\ddot{w} + w)/r^2 \\ \varkappa_{x\varphi} &= (2r\dot{w}' - rv' + \dot{u})/r\end{aligned} \tag{2}$$

With the assumptions that the linear component of

1. the circumferential strain of the middle surface of the shell is zero

$$\varepsilon_\varphi = \frac{\partial v}{r\,\partial\varphi} + \frac{w}{r} \equiv 0 \qquad (3a)$$

2. the shear strain of the middle surface of the shell is zero

$$\gamma_{x\varphi} = \frac{\partial u}{r\,\partial\varphi} + \frac{\partial v}{\partial x} \equiv 0 \qquad (3b)$$

the three displacement functions u, v, w are no longer independent. If we choose a Fourier-representation for the longitudinal displacement

$$u = \sum_{m=-\infty}^{+\infty} {}^{m}V'(x) \cdot {}^{m}\Phi(\varphi)$$

we get from (3a) and (3b) (4)

$$v = -\frac{1}{r} \sum_m {}^{m}V(x) \cdot {}^{m}\dot{\Phi}(\varphi)$$

$$w = \frac{1}{r} \sum_m {}^{m}V(x) \cdot {}^{m}\ddot{\Phi}(\varphi)$$

with the trigonometric functions

$${}^{m}\Phi(\varphi) = \begin{cases} r\cdot\cos(m\varphi) & m = 0, -1, -2, \dots \\ r\cdot\sin(m\varphi) & m = 1, 2, 3, \dots \end{cases} \qquad (5)$$

Because of the assumptions (3a) and (3b) only the stress resultants

$$n_x = D\,(\varepsilon_x + \mu\,\varepsilon_\varphi)$$

$$m_x = K\,(\varkappa_x + \mu\,\varkappa_\varphi) \qquad m_\varphi = K\,(\varkappa_\varphi + \mu\,\varkappa_x) \qquad (6)$$

$$m_{x\varphi} = K\,(1-\mu)\,\varkappa_{x\varphi}/2 = m_{\varphi x}$$

can be expressed by the elastic law and the stress resultants $n_{x\varphi}$ and n_φ have to be determined from the equations of equilibrium. For this purpose we apply the equations of the linear shell theory

$$r\,n_x' + \dot{n}_{x\varphi} = 0$$

$$r\,n_{x\varphi}' + \dot{n}_\varphi - \dot{m}_\varphi/r + m_{x\varphi}' = 0 \qquad (7)$$

$$n_\varphi + r\,m_x'' + 2\,\dot{m}_{x\varphi}' + \ddot{m}_\varphi/r = p_z \cdot r$$

Thus by the introduction of (2), (4), (5), (6) and (7) into (1) we get a set of differential equations of 4th order for the unknown functions ${}^{m}V(x)$. With this reduction compared to the 8th order of the exact differential equation for the cylindrical shell we loose the rapidly decaying solutions which describe the transverse edge load of the shell. Nevertheless there is a good agreement of the results from this analysis with experimental strain measurements /6/, /7/, /8/ as long as the length of the pipe is not less than its diameter. Thus the computation saving assumtions (3a) and (3b) are justified.

The solution of equation (1) is by iteration of the linearized equations where the derivatives are expressed in terms of finite difference approximations. The iteration starts from the linear solution (First Order Theory). For the linear problem we get a set of independent, linear differential equations of 4th order, which correspond to those for the beam on elastic foundation.

2.2 Boundary Conditions

At the boundaries x = const. according to equation (4) v is proportional to w and u is proportional to $\partial w/\partial x$. This is due to the loss of the rapidly decaying solutions of the shell equation. Therefore at the boundaries x = const. only the axial displacement u or the axial stress resultant n_x and the circumferential displacement v or the shear stress resultant $n_{x\varphi}$ may be described. - If the boundary conditions for w and/ or $\partial w/\partial x$ are not yet satisfied with the aforementioned boundary conditions this may be achieved by the superposition of the decaying solution /9/. However, in this investigation as in many other cases this additional computation effo rt was not necessary.

The analysis was performed for a single span pipe with equal boundary conditions at both supports (Fig. 4).

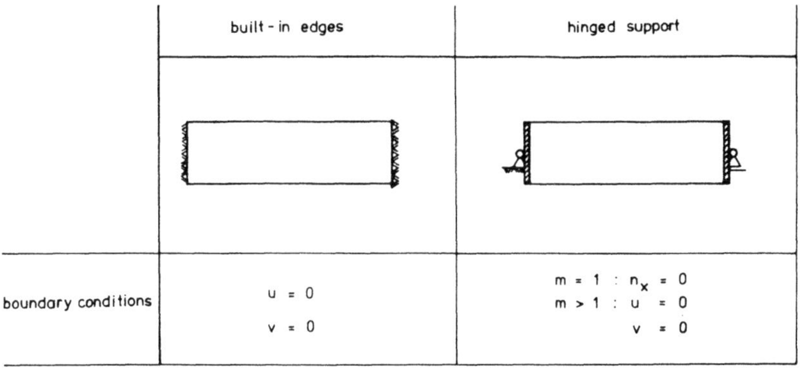

Fig. 4. Support and boundary conditions of the cylinders

The single span pipe with built-in edges (u=0 ; v=0) is representative of a continuous pipe with equally spaced supports.

The hinged support is characterized by undeformable plane cross sections which only can rotate about the horizontal axis through its centroid. This type of support may be assumed at pipe ends with heavy flanges or at vessel heads. Its mathematical verification is by the boundary condition $n_x = 0$ for the first harmonic and $u = 0$ for the higher harmonics while $v = 0$ for all terms of the Fourier-representation.

2.3 Stability Criteria

From the theoretical investigations and experiments /4/, /5/ it is known that the buckles of the full pipe (Fig. 5) are very similar in shape and size to those of the cylinder under axial compression /10/ and/or bending /7/. This buckling wave length is very short compared to the characteristic length of the prebuckling state of the cylinder: the prebuckling state of stress and deformations changes over the whole cylinder length.

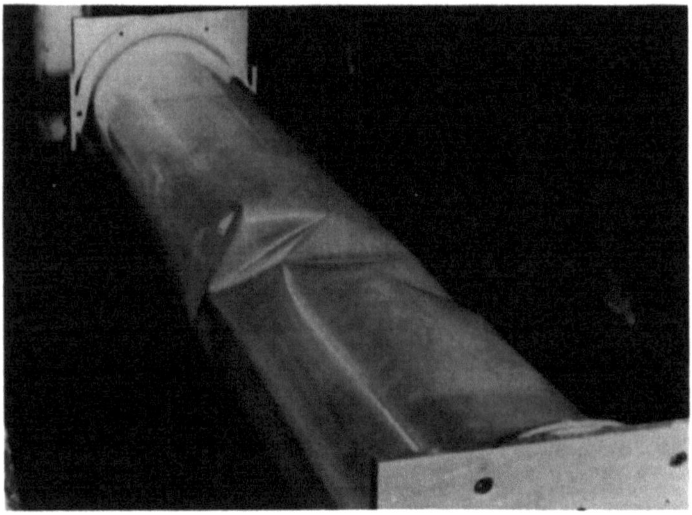

Fig. 5. Postbuckling pattern of a liquid filled cylinder

Because of the short buckling wave length the assumtions (3) are not justified for the buckling analysis. Furthermore this short wave length would require a much closer distance of the mesh points for the finite difference solution and more terms of the Fourier representation in the circumferential direction. All this would considerably increase the computational effort of the buckling analysis compared to the stress analysis. On the other hand as a consequence of these differences between the prebuckling state and the buckling state the prebuckling stresses and deformations are nearly constant over the area of a buckle.

Therefore it is assumed that the cylinder buckles when

$$\alpha = \frac{|\sigma_{x,c}(x,\varphi)| \cdot r_\varphi(x,\varphi)}{E \cdot t} = 0{,}605 \qquad (8)$$

where $\sigma_{x,c}$ is the axial compressive membrane stress, r_φ is the radius of the deformed cross section and E is Young's modulus. This assumption which has already been used for cylindrical shells under axial compression and/or bending /11/, /12/, /13/ corresponds to the representation of the zone which is prone to buckle by an equivalent cylinder under axial compression. This model also has been verified for the liquid filled cylinder by an exact stability analysis as will be shown later on.

The criterion (8) only gives realistic buckling loads as long as buckling is mainly due to the axial membrane stresses. In the case of a pure vacuum loading, where it obviously fails, the buckling loads from the simplified analysis according to section 2.1 are very close to the exact result despite the assumptions (3) because of the long buckling wave length in both directions /14/. However, for the sake of simplicity we will confine our investigation of the partially filled pipe with vacuum pressure to plotting the experimental results in an interaction diagram. This requires the theoretical buckling pressure of the cylinder with built-in edges which according to /15/ is given by

$$q_{ki} = 1.50 \cdot 0.92 \cdot \quad (r/l) \cdot (t/r)^{2.5} \qquad (9)$$

3. Experimental Investigations

The cylinders were manufactured from "Hostaphan Typ RN", a polyester film which exhibits linear behaviour beyond the range of stress at which buckling was likely to occur. The thickness of the film used in the tests was 0.250mm (RN250) and 0.350mm (RN350) respectively. Four ranges of r/t ratios were examined, namely 286, 357, 400 and 500. The slenderness parameter was in the range $0.59 \leq \beta = (l/r) \cdot \sqrt{t/r} \leq 1.91$.

The cylinders were made by lapjointing along the longitudinal seam and attaching aluminum cover plates along the ends. This procedure as well as the determination of Young's modulus E, Poisson's ratio μ and the shell thicknes t is described in detail in /16/ and will not be reported here.

Two different test set ups were used. Fig. 6 shows the partially filled pipe with built-in (clamped) ends which simultaneously is subjected to a slight vacuum pressure. The hinged support of the pipe according to Fig. 4 is to be seen Fig. 7. The details of the support are described in /8/.

Fig. 6. Test set up for a cylinder with clamped (built-in) ends with manometer for vacuum pressure

Fig. 7. Test set up for a cylinder with hinged ends with equipment for strain measurements

Fig. 7 also shows the test set up for the strain measurements. Strain gauges in the longitudinal and circumferential direction were applied to both surfaces of the cylinder at eight different positions of the cross section at midspan. The strain gauges on the inside surface were protected from the water by covering putty. Due to the low thermal conductivity of the polyester film and to the effect of the covering putty and the water the influence of the temperature on the strain measurements had to be eliminated. This was achieved with the application of the dummy gauges on the surface of a vertical cylinder (Fig. 7) of the same material and thickness as the test cylinder. The water level in this cylinder was always as high above the bottom as in the test cylinder.

By turning the test cylinder about its axis the positions for the strain measurements were increased and the influence of the longitudinal seam was investigated.

A special device has been developped for measuring the radial displacements of the shell surface (cross section deformation) /7/. This was necessary because on the one hand the small bending stiffness of the pipe wall did not allow any contact between the gauge and the pipe wall and on the other hand rather large deformations had to be measured. The change of the horizontal diameter was considered to be representative of the deformation of the cross section.

4. Experimental Results and Theoretical Comparisons

The experimental results of 37 buckling tests, deformation measurements and strain measurements with 3 cylinders are described in detail in /8/, /16/ and will only be summarized in the following.

4.1 Stresses and Strains

The axial membrane stresses at midspan of cylinder no. 31 /16/ are shown in Fig 8 to 10. The large differences between the experimental results and the stresses from the beam formula are obvious. The experimental stress values are in good agreement with the analytical results according to the nonlinear analysis. However, there are only small differences between these values and those according to the linear analysis. The consideration of the deformations in the equilibrium conditions generally results in a reduction of the axial membrane stresses. The maximum reduction is about the height of the water level.

The circumferential bending stresses at the inside surface of cylinder no. 31 are shown in Fig. 11 to 13 for the cross section at midspan. As in the case of the axial membrane stresses the linear analysis overrates the stresses. However, this difference is more pronounced so that the results from the nonlinear analysis are clearly confirmed by the experimental

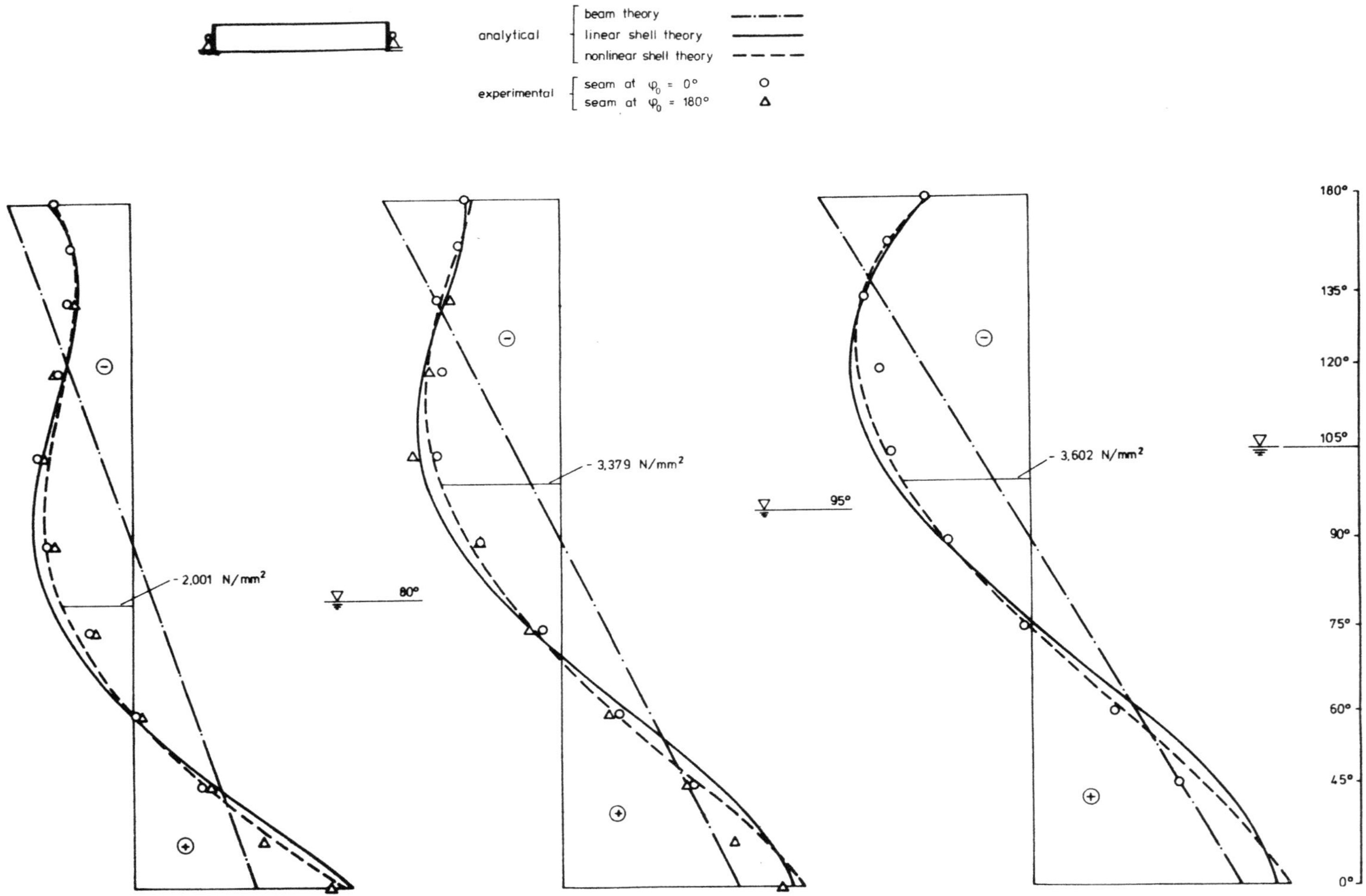

Fig. 8 to 10. Longitudinal membrane stresses σ_x of cylinder no. 31 (β = 0.65) for various degrees of fill φ_0

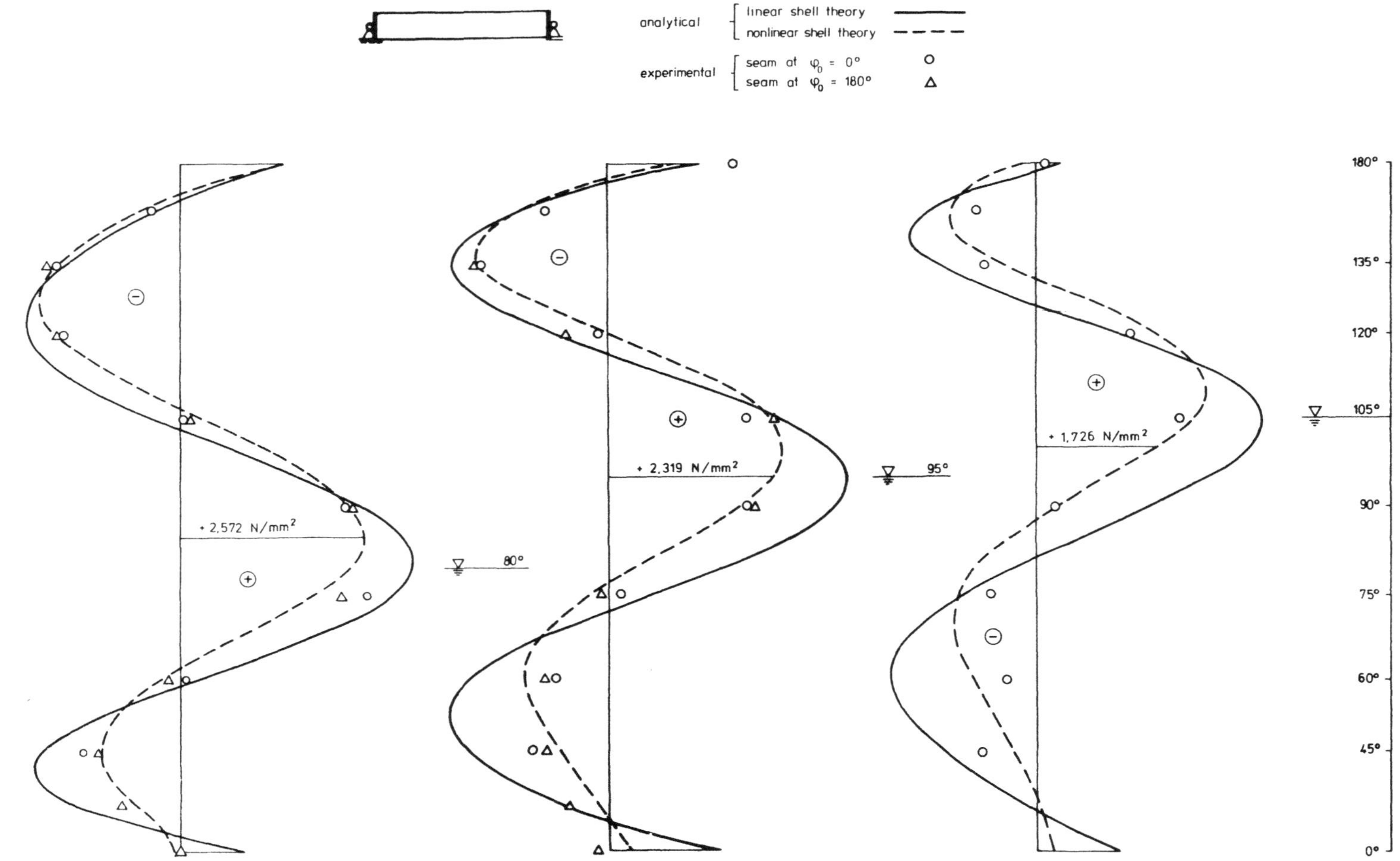

Fig. 11 to 13. Circumferential bending stresses $\sigma_{\varphi i}$ at inside surface of cylinder no. 31 for various degrees of fill

stress values.
The strain measurements at the highest load ($\varphi_0 = 105^\circ$) are very close to the ultimate load since buckling of this cylinder occured at $\varphi_0 = 113^\circ$.
The comparison of experimental and analytical stresses for two more cylinders ($\beta = 0.60$, $\beta = 1.25$) in /8/ gives similar results.
Thus the assumptions (3a) and (3b) of the analysis are justified by the experiments.

4.2 Displacements

The change of the horizontal diameter of the cylinders was chosen to characterize the deformations of the cross section (Fig. 14 to 16). In the initial stage of loading this change may be positive or negative. However, with increasing load the horizontal diameter decreases. It has a minimum value when the water level is at its height and from then on increases. With the decrease of the diameter the maximum radius r_φ of the cross section increases.
The displacement measurements from 20 cylinders confirmed the nonlinear analysis - the results are improved with respect to the linear analysis - /8/ as may also be seen from the examples given in Fig. 14 to 16. The linear analysis overestimates the flattening of the cross section. The differences between the results of the linear and the nonlinear analysis are of the same order of magnitude as with the circumferential bending stresses. The displacement measurements of some cylinders were taken at load steps which were very close to the buckling load.

4.3 Buckling Loads

Fig. 5 shows the typical postbuckling pattern of a liquid filled cylinder. This diamond shaped postbuckling pattern is well known from the cylinder under axial compression.
The degree of fill φ_{0k} at which according to equation (8) buckling was expected is plotted in Fig. 17 to 21 versus the shell parameter β . The buckling loads which result from the introduction of the stresses and deformations from the nonlinear analysis into equation (8) are higher than those which are obtained in the same procedure with the stresses and deformations from the linear analysis if φ_{0k} is less than 120° . This is due to the fact that the linear analysis overestimates the flattening of the cross section.

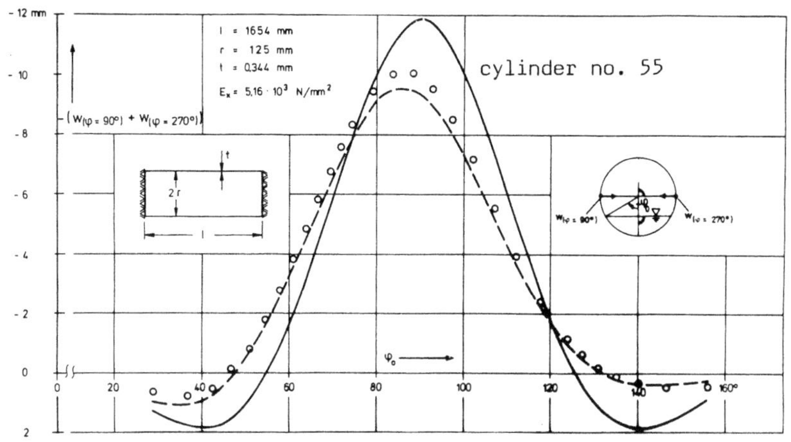

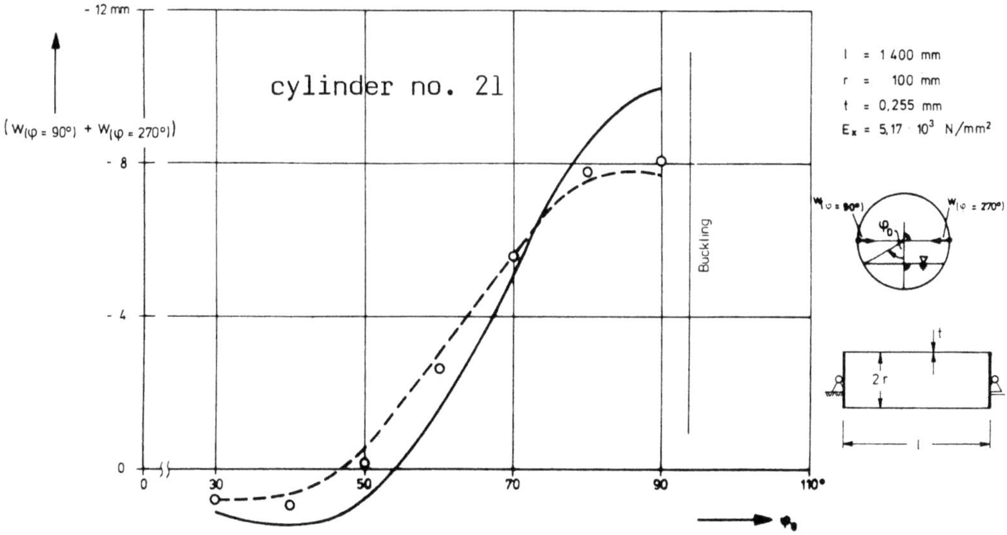

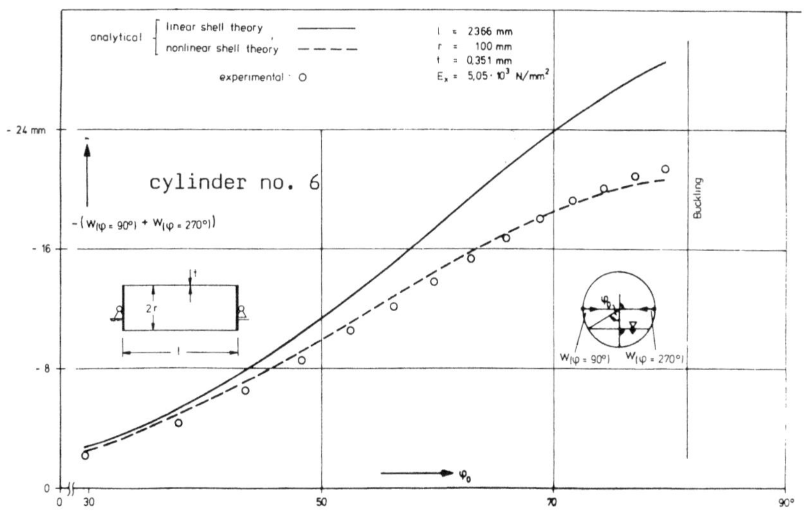

Fig. 14 to 16. Change of the horizontal diameter at midspan with φ_0

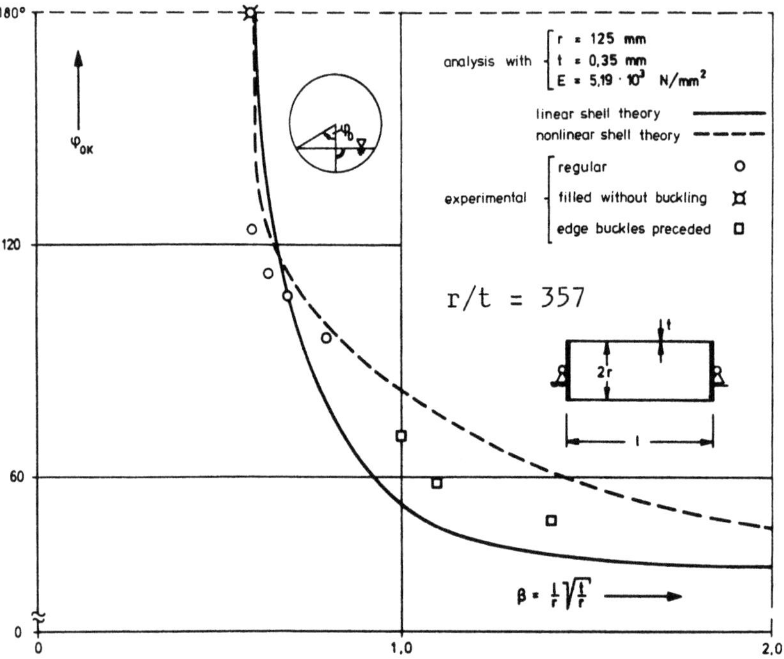

Fig. 17. Buckling at midspan of cylinders with hinged support and r/t=286

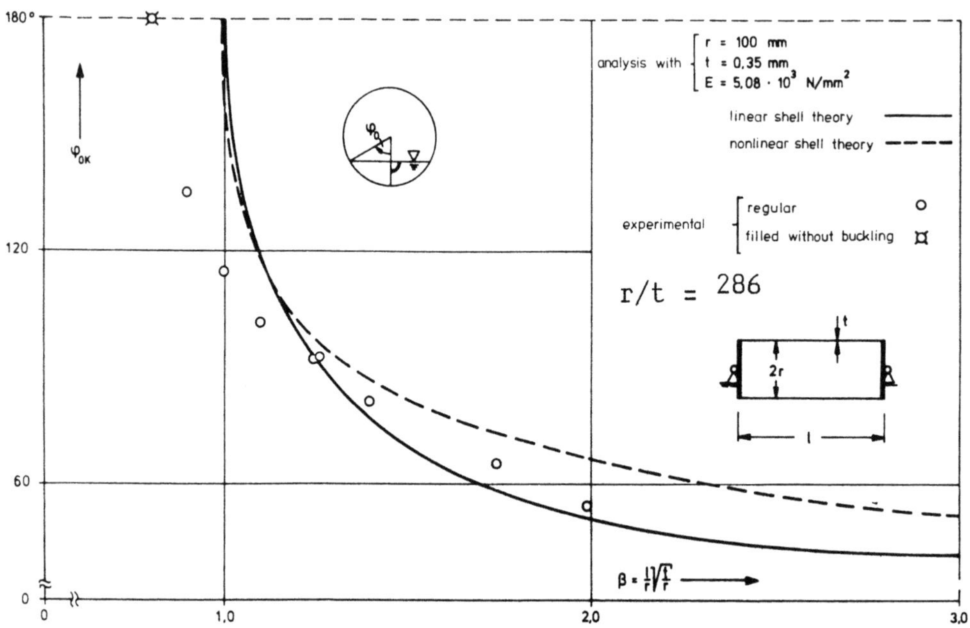

Fig. 18. Buckling at midspan of cylinders with hinged support and r/t=357

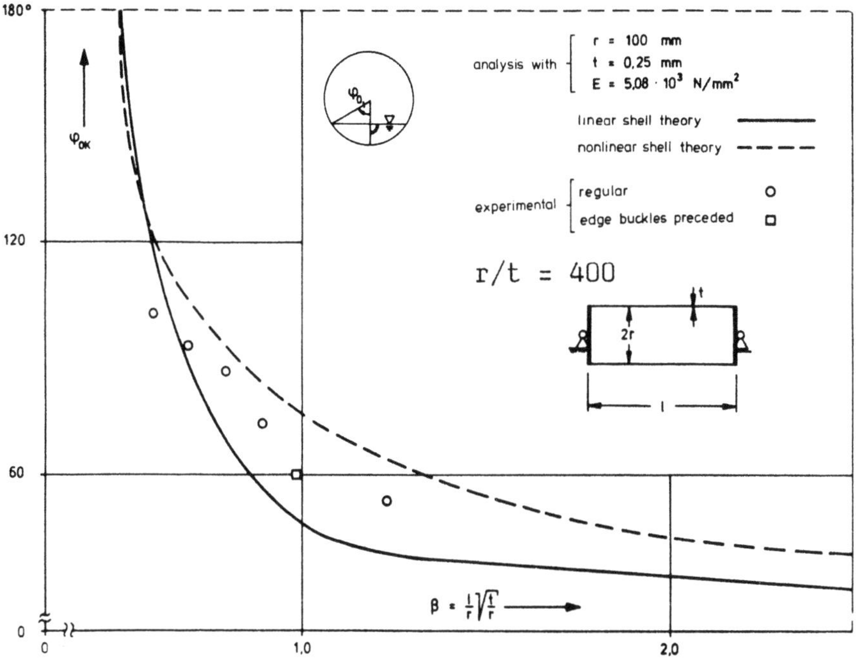

Fig. 19. Buckling at midspan of cylinders with hinged support and r/t=400

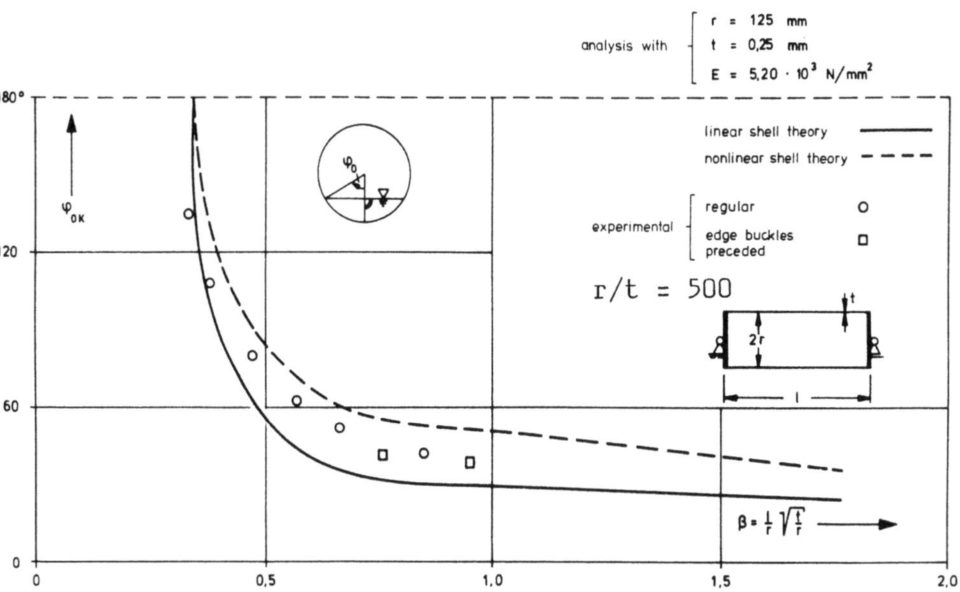

Fig. 20. Buckling at midspan of cylinders with hinged support and r/t=500

The small scatter of the experimental buckling loads (φ_{0k}) in Fig. 17 to 21 indicates that with this type of loading the cylindrical shell is not very sensitive to imperfections. However, one must have in mind that the buckling parameter α according to equation (8) is a nonlinear function of the load parameter φ_0. Thus the experimental result $\varphi_{0k} = 115^{\circ}$ which is far below the theoretical value $\varphi_{0k} \approx 180^{\circ}$ for $\beta = 1.0$ in Fig. 17 has a buckling parameter $\alpha = 0.502$, i.e. 83% of the theoretical buckling load. Apart from very few exceptions the buckling parameter α for the experimental buckling loads was greater than 0.43, i.e. the effect of the imperfections on the buckling load was less than 30%.
With increasing r/t and β the cylindrical shell is prone to buckle at the bottom near the cover plates. These "edge buckles" are due to compressive membrane stresses which are caused by the restraint of the warping deformations by the cover plates. They are also predicted by equation (8) as has been demonstrated in /16/: The buckling parameter α at midspan is less than that at the edges and that at the edges is of the same order of magnitude as that at midspan in the experiments where buckling occured at midspan ($\alpha = 0.43$ to 0.60). The prediction of the edge buckles on the basis of the results of the linear analysis is on the safe side.
However, the edge buckles do not cause failure of the cylinder. The water contents at failure /16/ demonstrate that there is still a considerable load carrying capacity after the appearance of the edge buckles. From Fig. 18 to 21 it seems that the failure loads of cylinders with edge buckles are within the scatter band of those without edge buckles. The analytical results in these figures do not account for edge buckling because this would be rather difficult.
It must be emphasized that edge buckling occurs not only with the clamped support (built-in edges) but also with the hinged support.

Combined liquid loading and vacuum pressure was applied to cylinders with clamped support only. Fig. 22 shows the postbuckling pattern of cylinder no. 40 ($\varphi_0 = 110^{\circ}$, $p = 2.2 \cdot 10^{-3}$ bar). This buckling pattern was observed with all other cylinders of this series except of cylinders no. 42, 43 and 46, where the postbuckling pattern according to Fig. 5 appeared. Fig. 23 shows that a linear interaction on the basis of stresses, deformations and buckling pressures from the linear analysis gives reasonable approximations of the buckling under combined loading. However, it must be admitted that this combined loading needs further investigation.

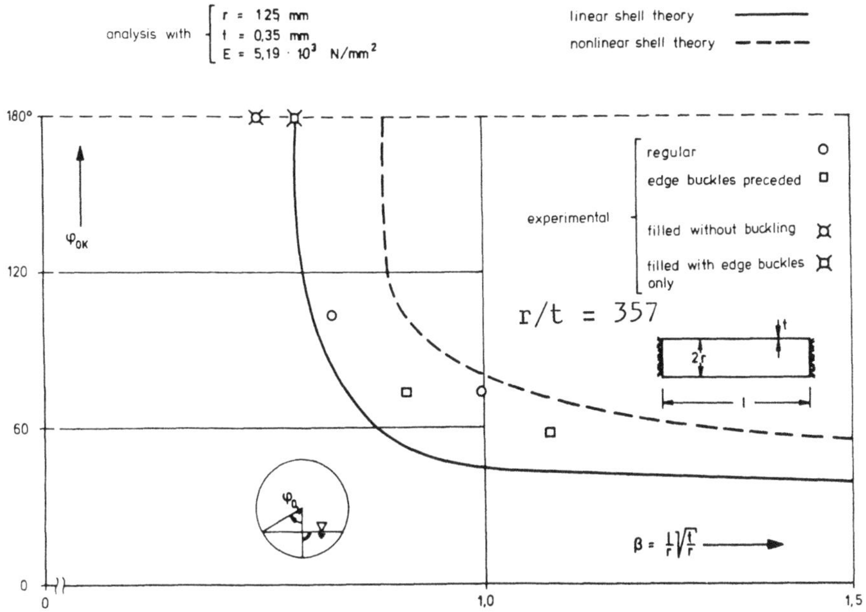

Fig. 21. Buckling at midspan of cylinders with clamped ends and r/t=357

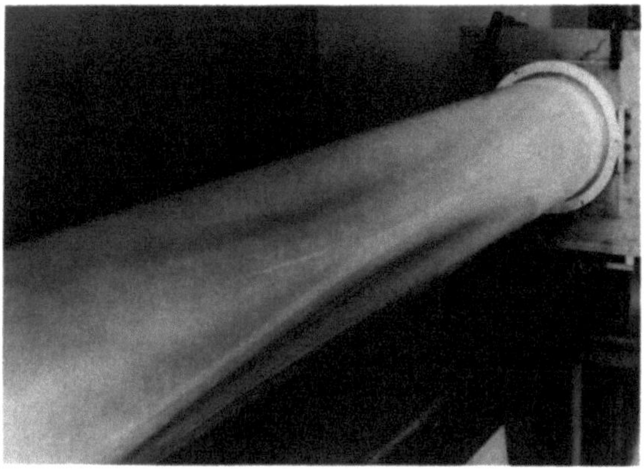

Fig. 22. Postbuckling pattern of a liquid filled cylinder (partially) with vacuum pressure (cylinder no. 40)

Finally we come to the justification of equation (8) by the results of an exact stability analysis. The buckling load of a partially liquid-filled cylinder was determined from the stability equations of shell theory in /17/ among other cases. For the numerical values of the physical properties (length 2366mm, radius 100mm, wall thickness 0.35mm, Younǵs modulus $5050 N/mm^2$, Poissoń ratio 0.30 and specific weight of the liquid $10^{-5} N/mm^3$)

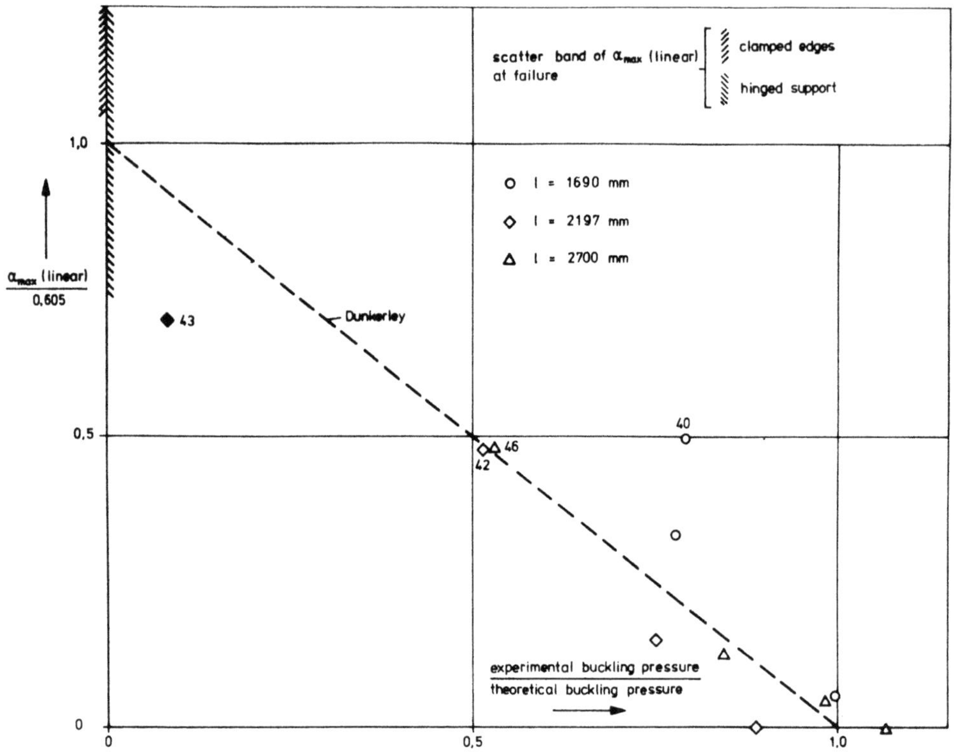

Fig. 23. Interaction diagram for the buckling of partially liquid-filled cylinders with vacuum pressure

the degree of fill at which buckling had to be expected according to the shell stability equations was $\varphi_{0k} = 80^o$. The corresponding value which results from the application of equation (8) is $\varphi_{0k} = 86^o$ which is rather close to the exact result. The experimental buckling load of this cylinder was at $\varphi_{0k} = 81.5^o$.

5. Conclusion

The experimental investigation confirms the stresses and displacements which result from a nonlinear analysis of the shell. The linear analysis gives a good approximation for the stresses and deformations which significantly deviate from those according to beam theory. This approximation is to the safe side - (remarkably only for the displacements and circumferential stresses).
The application of an equivalent cylinder model to the determination of the buckling load gives rather good agreement with the experimental results. There is only little imperfection sensitivity in this load case as the ex-

periments show. Again the theoretical buckling load which is based on the stresses and displacements from linear shell theory is on the safe side. This encourages a design procedure (stresses and buckling loads) which is based on the results from linear shell theory and is outlined in detail in /16/.

References

1. Del Gaizo,R.I.: Liegende zylindrische Behälter und Rohre auf Sattellagern endlicher Breite. Dissertation. Universität Stuttgart. 1980.

2. Karlson,K.I.: Über Schwerkraftspannungen in Rohrleitungen von großen Durchmessern und deren rationelle Konstruktion. Schweizerische Bauzeitung 80 (1922). S. 105 - 109.

3. Schorer,H.: Design of Large Pipe Lines. Transactions of the ASCE, Vol. 98 (1933). pp. 101 - 191.

4. Stern,J.: Zur Stabilität zylindrischer Lagertanks. Dissertation. Technische Universität Hannover. 1969.

5. Tooth,A.S., a. Fernandez,J.A.: A Study of the Buckling Behaviour of Horizontally Supported Thin-walled Cylindrical Storage Vessels Which Contain Fluid. In: Stability Problems in Engineering Structures and Components. Richards,T.H., a. Stanley,P. (eds.). pp. 315 - 340. Essex. Applied Science Publishers.

6. Schardt,R., u. Steingaß,J.: Eine Erweiterung der Technischen Biegelehre für die Berechnung dünnwandiger, geschlossener Kreiszylinderschalen. Der Stahlbau 39 (1970). S. 65 - 70 u. S. 146 - 150.

7. Saal,H., u. Kahmer,H.: Experimentelle Bestimmung der Beullasten axial zentrisch oder exzentrisch gedrückter, langer Kreisrohre. Zwischenbericht zum DFG-Forschungsvorhaben Sa241/2. Darmstadt. 1978.

8. Saal,H., u. Reif,A.: Experimentelle und theoretische Bestimmung der Beullasten von kreiszylindrischen, liegenden Rohren mit Flüssigkeitsteilfüllung. Abschlußbericht zum DFG-Forschungsvorhaben Sa241/3. Darmstadt. 1980.

9. Duddeck,H., u. Niemann,H.: Kreiszylindrische Behälter. Tabellen und Rechenprogramme für allgemeine Lastfälle. Wilhelm Ernst & Sohn. Berlin. 1976.

10. Eßlinger,M., u. Geier,B.: Postbuckling Behavior of Structures. Springer Verlag. Wien - New York. 1975.

11. Axelrad,E.L.: Refinement of the Upper Critical Loading of Pipe Bending Taking Account of the Geometrical Nonlinearity (In Russian). Izvestiia, Akademiia Nauk, SSSR, Otdelenie Teknicheskikh Nauk, Mekhanika, No. 4, 1965. pp. 123 -139.

12. Saal,H., Kahmer,H., u. Hein, J.-C.: Experimentelle und theoretische Untersuchung an beulgefährdeten, langen Kreisrohren. Der Stahlbau 48 (1979). S. 353 - 359.

13. Eßlinger,M., u. Geier,B.: Berechnung der Spannungen und der endlich großen Deformationen von Rotationsschalen unter großflächiger, axialsymmetrischer und nicht-axialsymmetrischer Belastung. Der Stahlbau 50 (1981). S. 263 - 270.

14. Steingaß,J.: Ein Beitrag zur Klärung des Tragverhaltens von geschlossenen, isotropen Kreiszylinderschalen. Dissertation D 17. Darmstadt. 1972.

15. DASt-Richtlinie 013. Beulsicherheitsnachweise für Schalen. Deutscher Ausschuß für Stahlbau. Juli 1980.

16. Saal,H., u. Reif,A.: Der Spannungs- und Stabilitätsnachweis für horizontale Zylinder mit beliebiger Flüssigkeitsteilfüllung. Der Stahlbau 50 (1981), S. 1 - 13.

17. Kahmer,H.: Zum Tragverhalten der Kreiszylinderschale mit endlichen, elastischen Formänderungen. Dissertation D 17. Darmstadt. 1981.

Part III :

Conical and Spherical Shells

Buckling of Thin Conical Shells under Axial Loads with and without Internal Pressure

M. ESSLINGER, J. CIPRIAN

Deutsche Forschungs- und Versuchsanstalt für Luft- und Raumfahrt, Braunschweig/Hoechst Aktiengesellschaft, Frankfurt am Main

Summary

Our lecture on conical shells subject to axial load consists of two parts:

The first author gives an account of the basic research work done at DFVLR.

The second author reports on the applied research work performed at Hoechst AG.

Part I

We consider the cone as a degenerated cylinder. Fig. 1 shows two isotropic cylinders of medium length, with equal wall thickness and different radii. Their theoretical buckling stresses are different. But the theoretical buckling load F is independent of the radius, and hence the two cylinders buckle under the same axial load.

Fig. 2 presents an isotropic cone of medium length under axial load. Its buckling stress depends on the wall thickness, the radius and the inclination angle. The buckling load F is independent of the radius and hence it is the same for all cross sections.

The cylinders buckle with a chess board pattern, theoretically extending over the entire surface. The circumferential wave number of that pattern is smaller for the cylinder with the smaller radius, Fig. 3.

The cone cannot adapt the circumferential wave number to the varying radius. It has to decide for one wave number and it

Buckling of Shells, Ed.: E. Ramm
Proceedings of a State-of-the-Art Colloqium

chooses the number corresponding to the smallest radius. As a consequence the buckling mode is most pronounced near the upper edge and decays towards the region of the larger radii. Fig. 4 shows the variation of the amplitude along the meridian for a theoretical buckling pattern.

The cone's choice of the smallest circumferential wave number can again be understood from the buckling behavior of cylinders. For a buckled cylinder having one or two half waves in longitudinal direction, in experiment and theory, numerous circumferential wave numbers can be obtained which are smaller than that of the theoretical buckling pattern /1/,/2/. But it is impossible to obtain, in theory or experiment, greater circumferential wave numbers.

High speed motion pictures exhibit that carefully manufactured test cones buckle according to their theoretical buckling mode. Only in the exceptional case of substantial initial imperfections in the vicinity of the largest radius the cone buckles there. This can be seen in a film /3/.

But before considering cones subject to simultaneous axial compression and internal pressure it is useful to demonstrate the membrane forces due to pure internal pressure. As it can be seen in Fig. 6 the meridional as well as the circumferential tensile force grow with increasing radius.

Due to the augmentation of the circumferential tension towards the largest radius the stiffening effect of the internal pressure is greatest in the lower part of the cone. From this it follows that with combined axial compression and internal pressure, the buckling pattern will be more concentrated to the upper edge, than with pure axial compression.

Fig. 6 shows the superposition of the meridional membrane forces due to axial load and internal pressure. It can be seen that in the cone (in contrast to the cylinder) the internal pressure influences, not only the magnitude of the meridional membrane force, but also its distribution. In the presence of internal pressure the compressive meridional stress is more concentrated towards the edge with the smaller radius and this has an additional effect on concentrating the buckling pattern.

The buckling behavior of the cone under simultaneous axial compression and internal pressure is also shown in the film /3/.

Part II

Our chemists and chemical engineers, who have developed a new process for chemical waste water treatment, have set us design engineers the task of designing the shell illustrated in Fig.7 for a water filling with a height of 20 metres. The outer shell consists of a cone and a toric section. The diameter at the top edge is 45 meters and at the bottom edge it is 23 metres. The height is 21.3 meters.

For technical reasons of manufacture, we have replaced the toric section by five short conical courses and have fixed the wall thickness as 30 mm in the toric area, then gradually diminishing towards the top to a thickness of 10 mm at the top rim (Fig. 8).

Fig. 9 shows the distribution of the membrane forces due to the hydrostatic load. It can be seen that the shell is subjected to meridional pressure because it has to transmit the effective weight of water in the area between the greatest (45 m) and the smallest diameter (23 m) into the foundation. In the torus, the meridional compressive forces produce inward-acting radial components at the mitre joints between the conical courses; these are counteracted by the horizontal components of the internal pressure, so that there are only low circumferential stresses. The design is a problem of stability. It is the toric section that is most endangered by buckling as a result of the meridional compressive forces. Since this is a novel type of structure for which there exist no specified standard design rules, in order to analyse and prove the stability it was necessary to carry out model tests with the goal of determining the reduction factor α , which is the relationship between the experimental and the theoretical buckling loads.

In the test models (Fig. 10), the bottom, buckling-prone section of the shell was produced true-to-scale by turning from a thick PVC plate. The top section was a smooth cone made of ®Hostaphan (polyester sheet). The longitudinal seam of this cone and the joint between the cone and the torus were glued together with an overlap. The bottom edge of the shell was placed in the groove of a rigid base plate and cast firmly in position with epoxy resin. The top edge was affixed to a narrow PVC ring using the same glue.

Fig. 11 shows a part of the PVC torus after the test, where permanent deformations of the postbuckling pattern can be clearly realized.

The numerical calculation took into account the individual conical courses with the mitre joints and the overlaped circumferential seam. The top edge was assumed to be free of forces; the boundary conditions at the bottom edge were taken as

$$u = v = w = M = 0.$$

Neglecting the lowest as well as the highest value the evaluation of the first six tests gave as a result α values of 1.

In order to explain these unusually high α values, the following possible causes were considered:

- First, the discontinually curved meridian of the torus represents effective prebuckles that are exactly taken into account in the non-linear prebuckling calculation.
- Secondly, the stabilising effect of the internal pressure.

In order to prove the correctness of theses assumptions, further investigations were performed.

1. Thus two buckling tests were carried out with toriconical models with the same dimensions, whereby the torus was not

approximated by short conical courses but the meridian was a continuous curve. The result: the theoretical buckling loads were the same and the experimental buckling loads also lay within the scatter range of the values obtained for models with a discontinuous meridian in the torus area. This result was surprising because in contrast to this, according to a report by Ebel /4/, in the case of spherical shells subjected to external pressure the discontinuous meridian has a considerable influence on the breakdown load. The different prebuckling sensitivity of the two shells is probably not due to the different loads - axial force in one case and external pressure in the other - but to the fact that in the case of an inward-convex torus the stiffening effect deriving from membrane forces is considerably smaller than in the case of the outward-convex spherical shell.

2. The stabilising effect of the internal pressure in the torus area was seriously questioned, also among experts, because in this case the internal pressure does not produce any circumferential tensile forces. For this reason, we felt ourselves called upon to produce evidence that the internal pressure always has a stiffening effect quite independently of whether it produces circumferential tensile forces or counteracts circumferential compressive forces. Fig. 12 shows a seamless toriconical steel model under axial load, on the left without internal pressure and on the right with internal pressure. The boundary conditions and the connections of the shell with the top and bottom base plates were the same as for the PVC models: the shell edges rested in narrow grooves that were cast with epoxy resin. Using these models, Prof. Schulz has carried out buckling tests at the Research Centre for Steel, Wood and Stone of the University of Karlsruhe. He has kindly placed at our disposal the dimensions of the models and the experimental buckling loads obtained by two tests, each with and without internal pressure. A comparison of the experimentally determined buckling loads with the theoretical buckling loads we had calculated with a non-linear prebuckling state /5/,/6/

produced α values of 0.636 for the two tests without internal pressure and α values of 0.98 for the two tests performed with different internal pressures of 0.0675 N/mm² and 0.173 N/mm².

Fig. 13 concerns the model with an internal pressure of 0.0675 N/mm² and shows the stress distribution along the meridian in a state of equilibrium immediately before buckling; the discontinuous torus area is easily recognised by the high stress peaks of the bending components. It is obvious that the circumferential as well as the meridional membrane stresses are compressive stresses of the same order of magnitude.

The α values attained in these tests (Fig. 12) clearly prove that there is a stiffening effect of the internal pressure in this approximated torus built up of conical courses, although there are membrane compressive stresses acting in the circumferential direction.

Over the last few years, there have been many discussions on the subject of the stiffening effect of internal pressure in water-filled cones. Low buckling loads known from the relevant literature were the reason why, according to the DASt Guideline 013 /7/, the possibility of taking into account the stabilising effect of the internal pressure for cones is not admissible. This induced us to clarifay this question by a series of tests with cones made of ®Hostaphan, PVC and steel.

In our opinion the disputed test results could in particular be traced back to an incorrect assessment of the boundary conditions. The following is a report on the results of six buckling tests with water-filled cones of PVC with different types of connection to the base plate.

Fig. 14 shows the PVC model and the distribution of the membrane forces along the meridian for the version with rigid edge clamping and a state of equilibrium immediately before buckling. A comparison with the membrane forces in the torus (Fig. 9) shows that the membrane forces in the cone rise more sharply towards the bottom edge.

The different types of connection with the base plate and the evaluated test results are shown in Fig. 15. In Versions a and b, the cone rests in a groove; in the case of Version c, it stands on a smooth plate. On Version a, the edge of the cone is firmly cast in the groove with hard epoxy resin whereas on Versions b and c it is only sealed with a soft glue. On comparing the buckling stresses, it can be seen what a decisive influence the boundary conditions have on the buckling load. Whereas with the bottom edge fitted in a groove, the use of a soft glue (Version b) instead of a hard resin (Version a) already results in a clear reduction of the buckling load, when the bottom edge rests on a smooth plate and is provided with a seal of soft glue (Version c), the buckling load drops to 30 % of the value obtained with Version a. The soft glue clearly permits displacements so that on Versions b and c the cone can buckle directly on the edge of the shell, i.e. at the point subject to the highest load.

Despite the big differences in the buckling loads, the differences in the α values are small. This depends on the choice of the corresponding boundary conditions when calculating the theoretical buckling load. The calculation was based on the following boundary conditions:

-Version a) $u = v = w = \beta = 0$
-Version b) $u = v = w = M = 0$
-Version c) $u = T = R_x = M = 0$

Our experience with buckling tests on conical shells made of ®Hostaphan (polyester sheet) - which will not be covered in more detail in this report - has produced less satisfactory results, because these sheets are irregular and orthotropic. The measured values of the moduli of elasticity fluctuate between 4000 and 6000 N/mm². In the case of the cone, this orthotropic behaviour disturbs the symmetry of rotation.

Finally, we also investigated a further variant version of the load applied. The model shown in Fig. 16 was subjected to a constant water pressure and an increasing axial force. The evaluation of two tests on models with rigid edge clamping - on the left in the diagram - produced α values of 1. Two further tests with slightly altered boundary conditions, i.e. using a soft silicon glue instead of epoxy resin - on the right in the diagram - produced a 24 % lower buckling load. These results confirm the stiffening effect of the internal pressure as well as the sensitivity of the buckling behaviour to small displacements of the edge of the shell.

Consideration of the test results presented here show how important it is to observe the following basic principles: Data on the reduction factor can only then be admitting but one interpretation if all test parameters and the procedure used for calculating the theoretical buckling load are known. It is only permissible to transfer the reduction factors determined in tests to the large-scale structure when the test parameters are in agreement with the actual conditions existing in the real object and the theoretical buckling load was in both cases calculated by the same method.

We have strictly adhered to these basic principles when calculating the dimensions and carrying out the construction of the Bioreactor mentioned at the beginning. It goes without saying that during the construction work we have also taken care to guarantee flawless manufacture and to make sure in particular that both, global and local initial imperfections could be kept to a minimum. Thus for example we have kept well within the tolerances laid down in the DASt Guideline 013 for local imperfections. Fig. 17 shows the reactor under construction and Fig. 18 shows two completed reactors during the test filling. During these, the water level was set at a height at which the membrane stresses were 33 % higher than those under operating conditions. The calculated safety factor against buckling under operating conditions is S = 4.48. It is as high as this because no test results were available when the reactor was designed.

In the meantime, three further reactors of this type and size have been completed and two more are currently under construction.

Aknowledgment

It is gratefully acknowledged that the film was sponsored by the Deutsche Forschungsgemeinschaft.

The second autor wants to thank Dipl.-Ing. W. Grein, who guided the performance of all test series, as well as Dr.-Ing. H. Hey, who carried out the numerical computations, for their kind support.

References

1. Esslinger, M.: Beulen und Nachbeulen exzentrisch versteifter dünnwandiger Kreiszylinder unter axialsymmetrischer Belastung. DLR-FB 70-48 (1970) 40 S., 9 Bild., 11 Lit.

2. Esslinger, M.; Geier, B.: Calculated Postbuckling Loads as Lower Limit for the Buckling loads of Thinwalled Circular Cylinders. Ed. B. Budiansky, Springer Verlag Berlin, Heidelberg, New York (1976) S. 274-290, 17 Bild., 5 Lit.

3. Esslinger, M.; Klein, H.: Beulen und Nachbeulen dünnwandiger Schalen, isotroper Kegelstumpf unter Axiallast ohne und mit Innendruck, halber Öffnungswinkel des Kegels 40 °. (Buckling and Postbuckling of Thinwalled Shells, Isotropic Circular Truncated Cone subject to Axial Load combined without and with Internal Pressure, Semiapertural Angle of the Cone 40 °). Inst. Wiss. Film, Göttingen, Tonkommentar deutsch und englisch, schwarzweiß, Vorführdauer 11.40 Min.

4. Ebel, H.: Bericht über Durchschlagversuche an unversteiften Kugelschalenkalotten aus verschiedenen Werkstoffen. Technische Hochschule Darmstadt, Institut für Statik und Stahlbau (1979) 10 S., 4 Bild., 4 Tab., 6 Lit.

5. Esslinger, M.; Kerkhoff, H.: Berechnung der Beullasten von ringversteiften Rotationsschalen unter axialsymmetrischer Belastung im elastischen Bereich. KFK-CAD 176 (1980), 84 S., 10 Bild, 24 Tab., 10 Lit.

6. Esslinger, M.; Kerkhoff, H.; Melzer, H.-W.; Taelmann,E.-W.: Theoretische Grundlagen des Programms FO4BO7 "Berechnung der Beullasten von dünnwandigen Rotationsschalen unter axialsymmetrischer Belastung im elastischen Bereich". KFK-CAD 176 (1980),84 S.,10 Bild., 24 Tab., 10 Lit.

7. Beulsicherheitsnachweise für Schalen, Deutscher Ausschuß für Stahlbau, DASt Richtlinie 013, Juli 1980, 16 S.

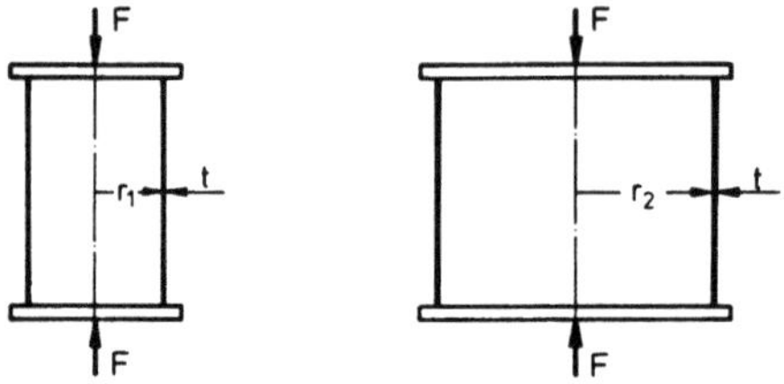

buckling stress $\sigma = 0{,}6E\,\frac{t}{r_1}$ $\qquad \sigma = 0{,}6E\,\frac{t}{r_2}$

buckling load $F = \sigma \cdot A = (0{,}6E\,\frac{t}{r})(2\pi r t) = (0{,}6 \cdot 2\pi)\,E t^2$

Fig. 1. Two cylinders with equal wall thickness under axial load

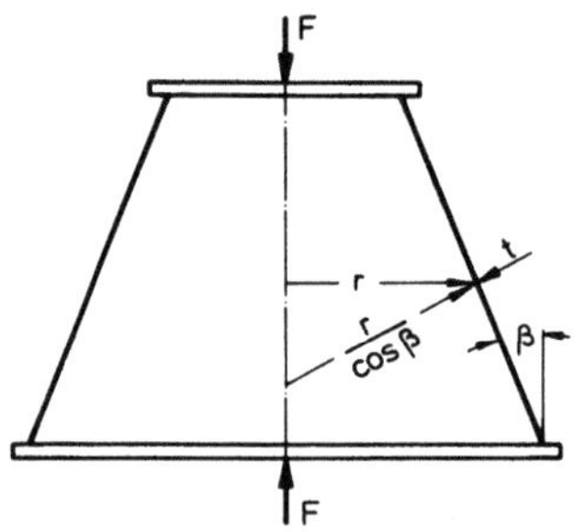

buckling stress $\sigma = 0{,}6E\,\frac{t \cdot \cos\beta}{r}$

buckling load $F = \sigma \cdot A \cdot \cos\beta = (0{,}6E\,\frac{t \cdot \cos\beta}{r})(2\pi r t)\cos\beta = (0{,}6 \cdot 2\pi) E t^2 \cdot \cos^2\beta$

Fig. 2. Cone under axial load

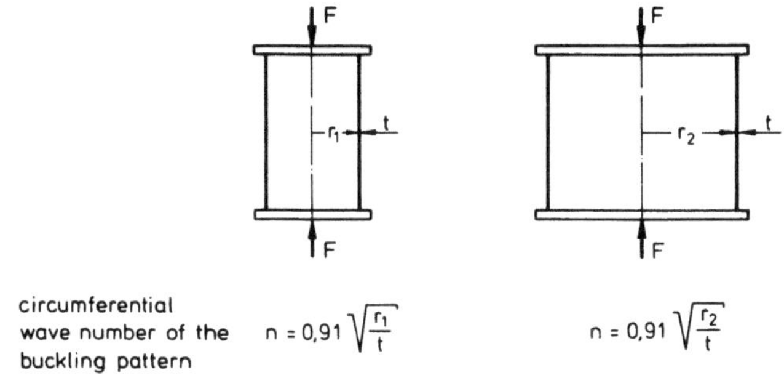

Fig. 3. Two cylinders with equal wall thickness under axial load

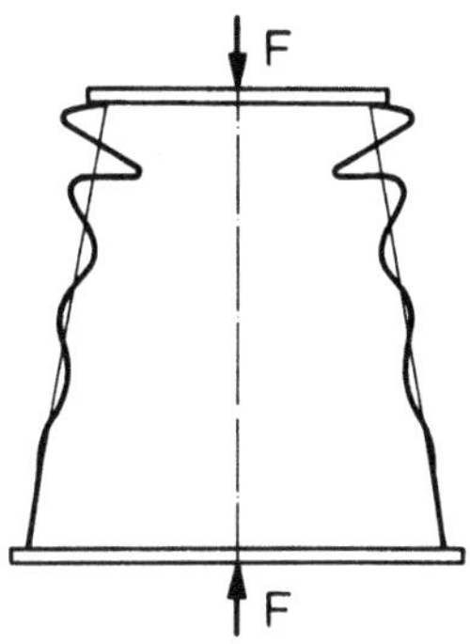

Fig. 4. Variation of the buckling pattern along the meridian

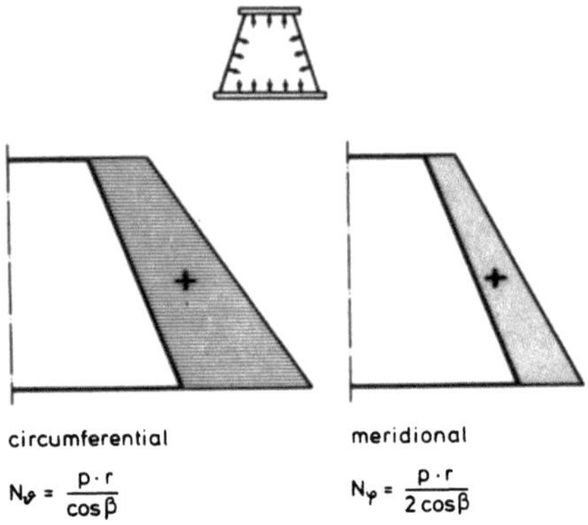

Fig. 5. Membrane forces of a cone under internal pressure

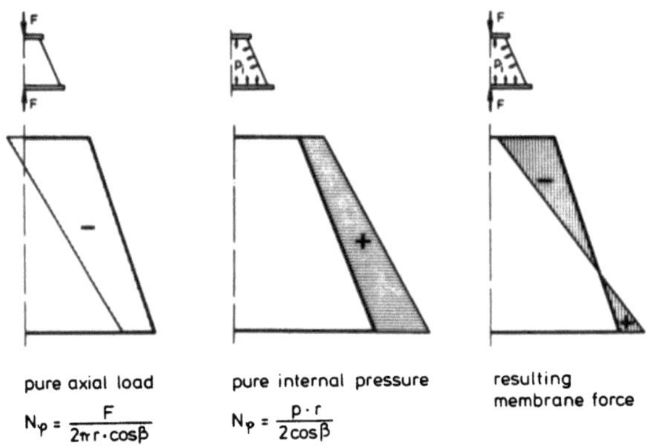

Fig. 6. Resulting meridional membrane forces in a cone under axial load and internal pressure

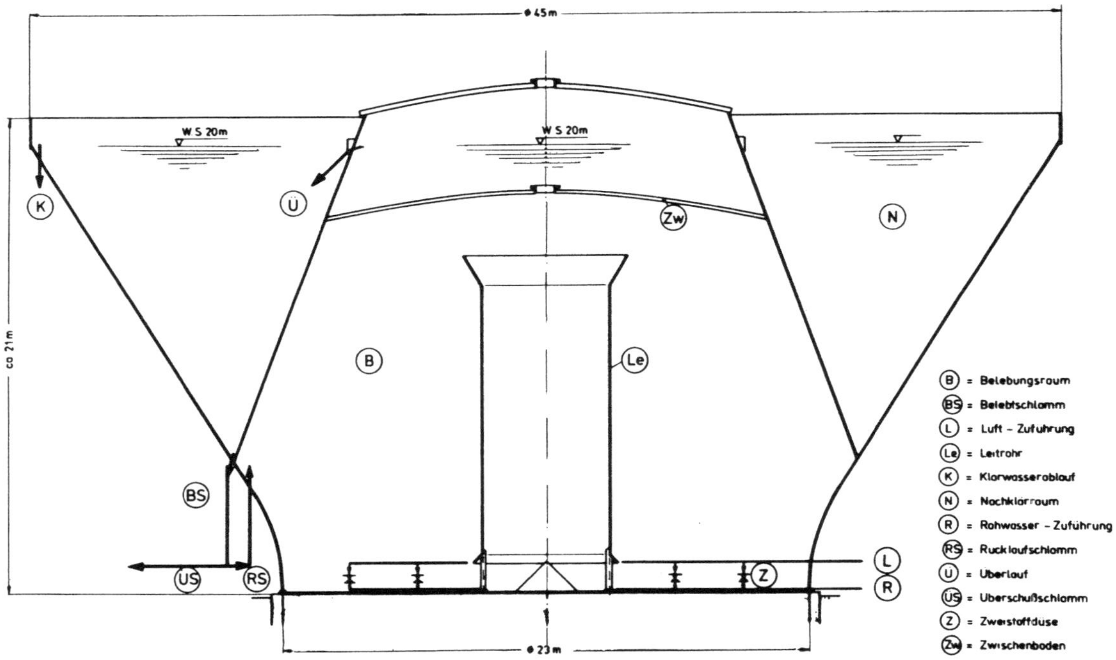

Fig. 7. Vertical-type biological reactor

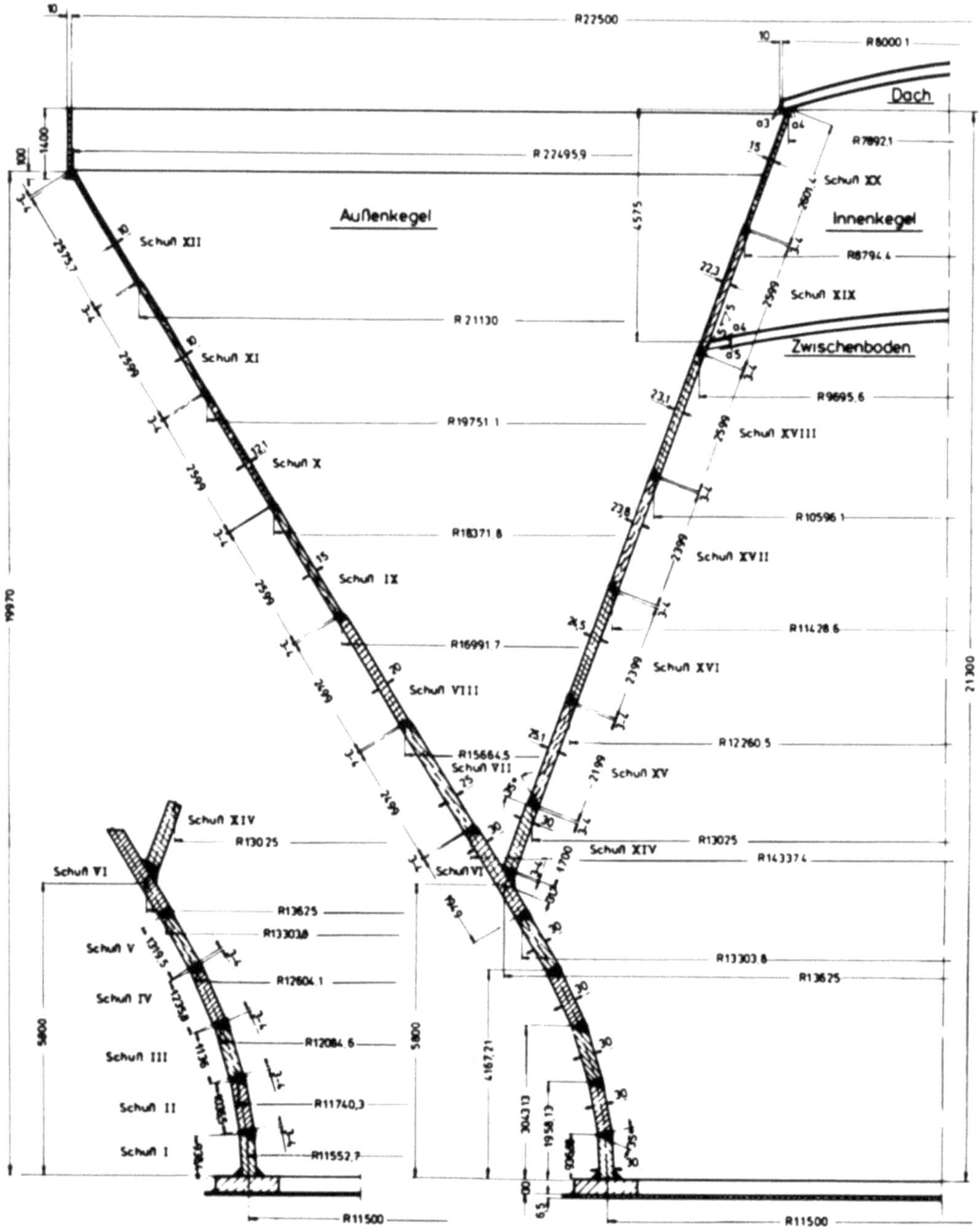

Fig. 8. The structure of the reactor

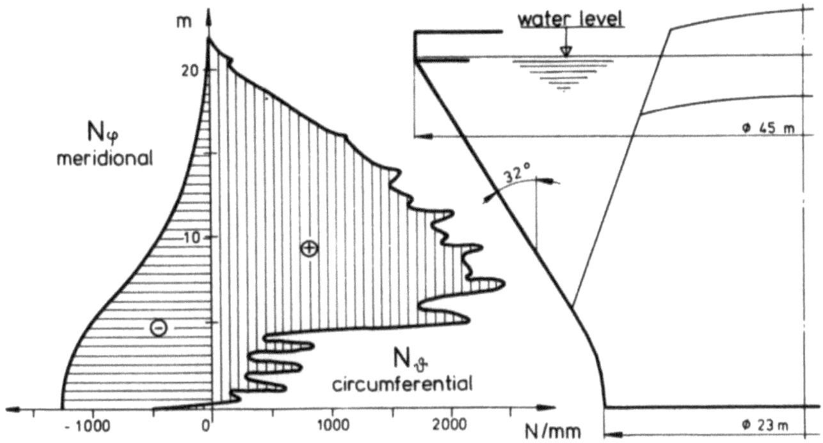

Fig. 9. Membrane forces in the hydrostatically loaded outer shell of the reactor

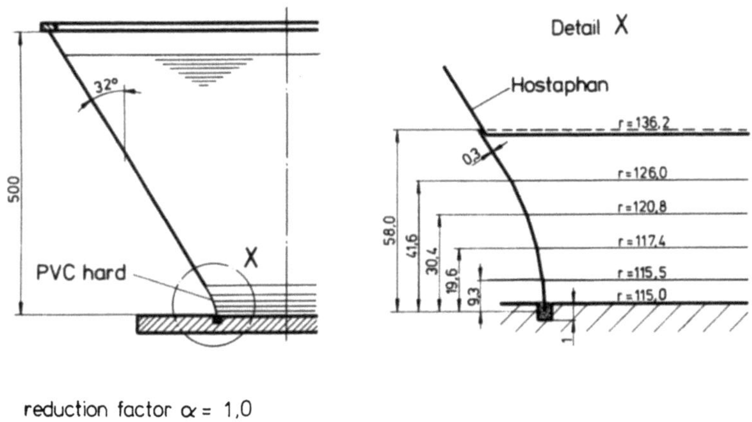

Fig. 10. Hydrostatically loaded toriconical model

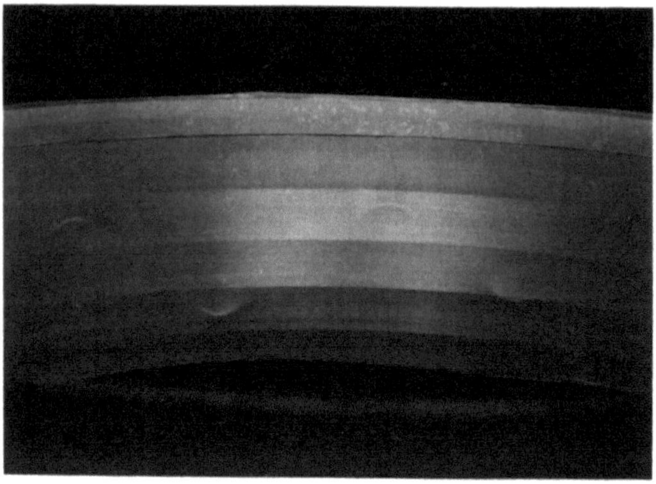

Fig. 11. Section of the PVC torus after the test

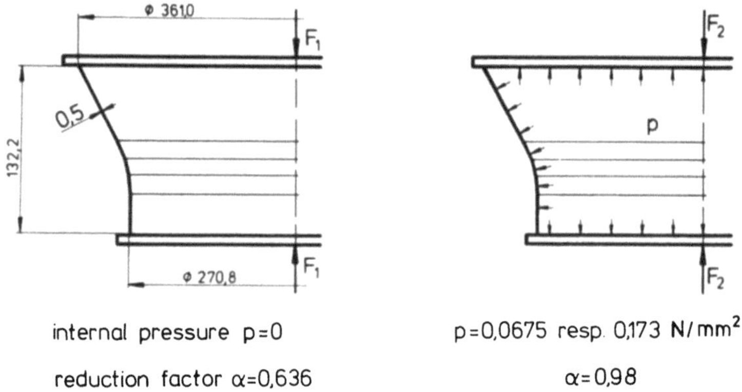

Fig. 12. Toriconical steel model under axial load without and with internal pressure

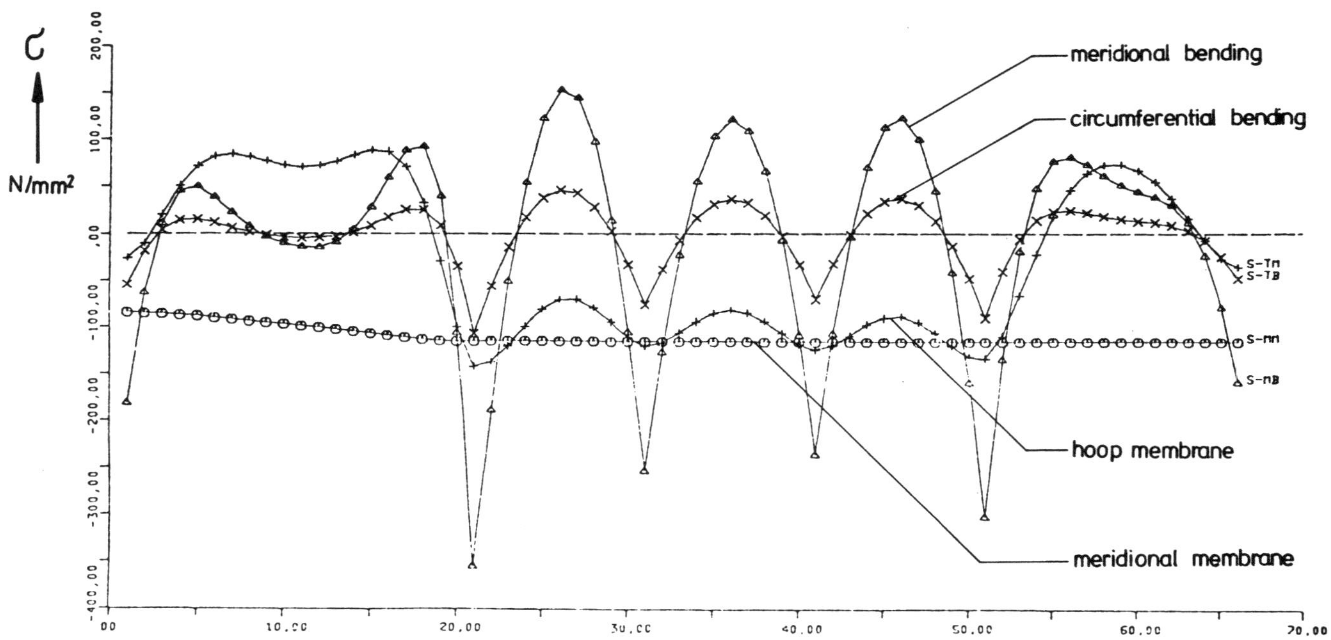

Fig. 13. Stress distribution along the meridian of the toriconical steel model under simultaneous axial load and internal pressure; the figures at the abscissa represent the node numbers between the upper edge (left) and the lower edge (right)

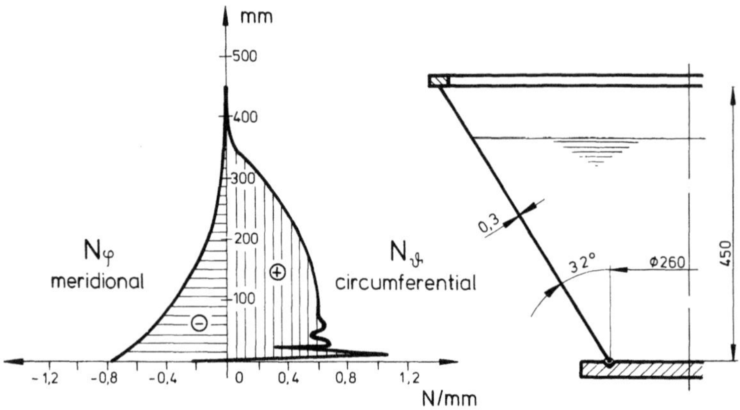

Fig. 14. Membrane forces in the hydrostatically loaded conical PVC model

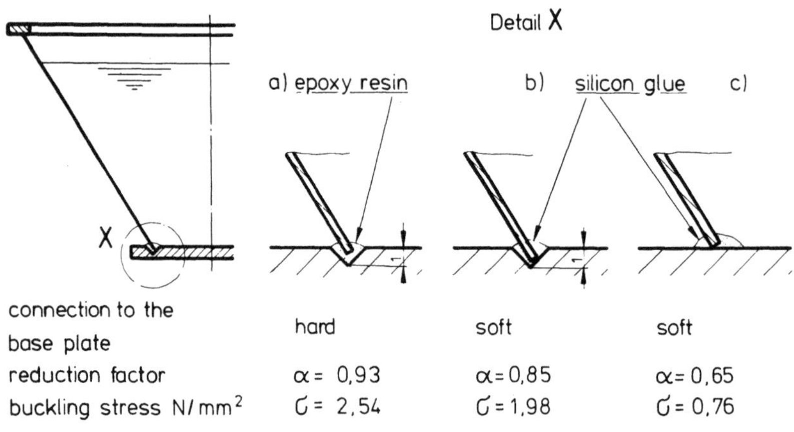

Fig. 15. Hydrostatically loaded conical model - effect of different boundary conditions

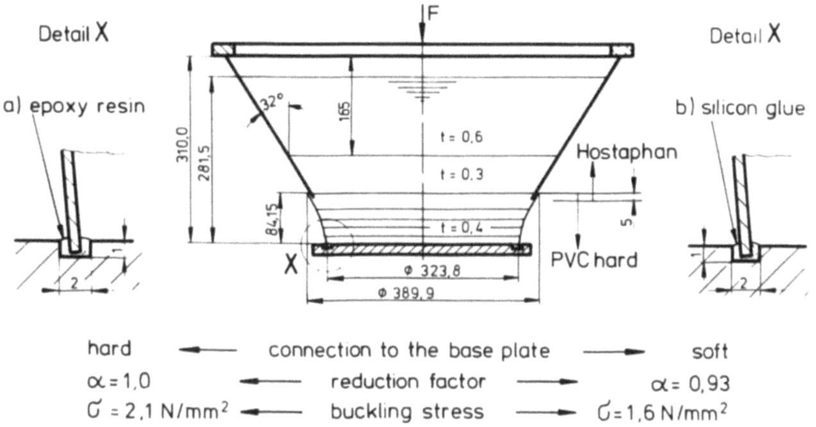

Fig. 16. Toriconical model under axial load and hydrostatic pressure - effect of different boundary conditions

Fig. 17. The reactor under construction

Fig. 18. Two reactors during the test filling

EXPERIMENTAL INVESTIGATION OF BUCKLING OF HYDROSTATICALLY LOADED, CONICAL SHELLS AND PRACTICAL EVALUATION OF THE BUCKLING LOAD

D. VANDEPITTE, J. RATHE, B. VERHEGGHE, R. PARIDAENS and C. VERSCHAEVE

State University at Gent, Belgium

Synopsis

An extensive experimental investigation of buckling of conical shells that are supported only along their lower edge and that are loaded by the weight of a liquid leads to diagrams and formulas which can be used directly in the design of such shells. The margin of safety to be applied, the influence of welding stresses and the limitations on the validity of the formulas for structures with different boundary conditions are discussed.

1. Introduction

Consider a conical shell with a vertical axis. Its lower edge has the smaller radius, r, and is supported. Its upper rim is free (fig. 1). When the

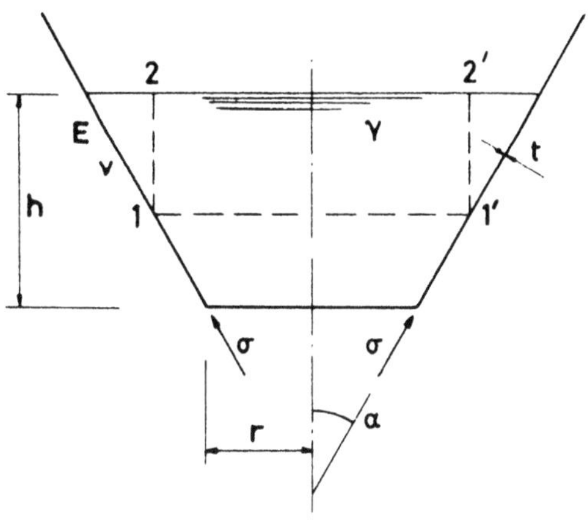

Fig. 1

cone is partly filled with a liquid of specific weight γ, compressive stresses originate in the shell in the meridional direction and tensile stresses develop in the circumferential direction. The compressive meridional stres-

Buckling of Shells, Ed.: E. Ramm
Proceedings of a State-of-the-Art Colloqium

ses acting along the parallel circle *1-1'* upon the part of the shell located above *1-1'* equilibrate the weight of that part and the weight of the liquid contained in the toroidal space of triangular cross-section bounded by the wall of the shell and by the cylindrical surface defined by the vertical generatrices *1-2* and *1'-2'*. The weight of the toroidal mass of liquid increases more rapidly than according to a linear law when we lower the horizontal cross-section *1-1'* considered. Since, moreover, the perimeter of the parallel circle then decreases, the compressive membrane stress increases very rapidly along the generatrices of the cone between the surface of the liquid and the base of the shell. The meridional membrane stress along the bottom edge of the shell is given by

$$\sigma = \frac{\gamma h^2 (r + \frac{h}{3}\, tg\alpha)\, tg\alpha}{2rt \cos\alpha} \qquad (1)$$

when the weight of the shell itself is assumed to be negligible as compared with that of the liquid. t is the thickness of the wall and α is the angle between its generatrices and the vertical. *(1)* does not hold when $\alpha \to 90^o$ because plate action then supersedes membrane action.

When the level of the liquid in the conical vessel rises gradually, there comes a moment when the compressive stresses cause the bottom part of the shell to buckle, in spite of the stabilizing effect of the circumferential tensile stresses , and since the load is a gravity load , the shell fails suddenly and catastrophically.

Certain water towers and bioreactors [1] may have a conical shape and may be endangered by the type of instability described above. A few years ago, a steel water tower with a capacity of more than 1000 tons collapsed while the tank was being filled for the first time [2 – 3] and before the water had reached overflow level.

The stability of conical shells under axial load and under certain other loading conditions has been studied theoretically by SEIDE [4] and by others, and also experimentally [5] . Partly on account of the failure mentioned above we have undertaken to investigate systematically the stability of hydrostatically loaded, conical shells. To the best of our knowledge, this had not been done before. On previous occasions we have reported tentatively and succinctly on the results obtained [6 – 7 – 8], but our investigation was still in full swing and is continuing even now. The present paper is more comprehensive than [7] and, especially, than [6], is based on more complete results and also contains a better interpretation and presentation of the results, enabling a designer to use them in an easier and more direct manner.

2. Model materials and model geometry

We have tested hundreds of model cones made of mylar, brass, aluminium or steel and covering a wide range of the parameters α and $r/(t\cos\alpha)$ (table 1).

Table 1 - "Simply supported" cones

Model material	Mylar	Brass	Aluminium	Steel	
Number of tests	530	33	15	30	2
t (mm)	0,125-0,350	0,20	0,20	0,31	1,50
α	10° 20° 30° 40° 50° 60° 75°	20° 30° 40° 50° 60°	20° 30° 40° 50° 60°	20° 30° 40° 50° 60°	40°
$r/(t\cos\alpha)$	100 - 3600	200 - 1500	200 - 1500	200 - 1500	520

Angles α of 75° and even of 60° occur probably rarely, if ever, in actual structures. All models were filled with water until they collapsed. The thickness of the wall of each specimen was constant and its upper rim was free and unstiffened. Some models, mostly mylar models, consisted of a cylinder superimposed on a cone (fig. 2), but this arrangement had no noticeable effect on the critical value of the meridional membrane stress in the supported edge ; this is understandable : the stiffening effect of the transition zone between cone and cylinder did not extend as far as the critical bottom region of the cone.

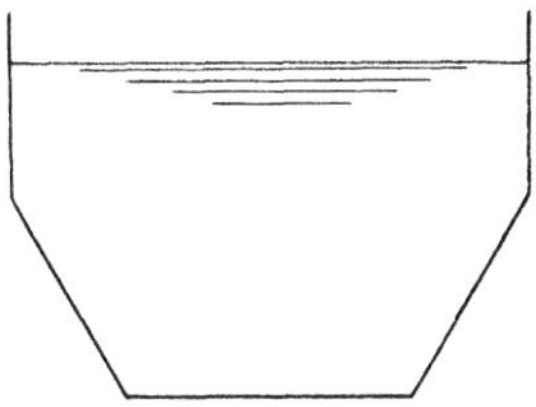

Fig. 2

Each mylar model was cut from a single sheet of the plastic material and had one meridional overlap joint. Except for the two steel models with wall thickness $t = 1,5\ mm$ mentioned in table 1, all the joints between sheets of brass, aluminium or steel were soldered.

3. Boundary conditions

3.1. In a small number of tests, the lower edge of the models was clamped. The edge of the mylar cones was cast into resin and that of the metal cones was clamped between an outer female and an inner male steel cone having the same angle α as the test specimen. The boundary condition under consideration may lead to an overestimation of the buckling strength of real conical shell structures, whose base is usually less constrained rotationally.

3.2. We therefore attempted to simulate real conditions more truly and carried out a number of tests on mylar specimens supported by the steel device shown in figure 3. In a circular groove B machined into the bottom of the bearing, resin was cast around the lower edge of the model cone. Besides, the wall of the cone was supported laterally at point A by the rim of the device and pressed against that rim by a rubber ring surrounding the bottom plate on the model reservoir. Thus the lower part of the hydrostatically loaded wall of the vessel was elastically restrained against rotation by the dry part AB of the shell. We experimentally determined the length of AB in

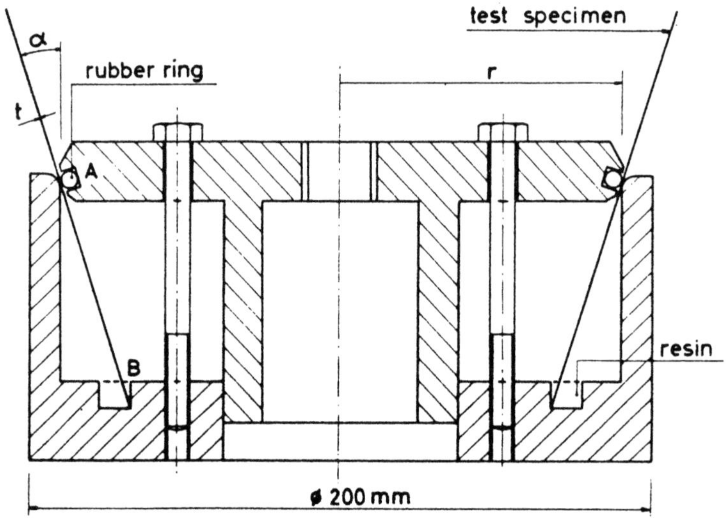

Fig. 3

such a manner that increasing it did not further diminish the buckling load of the cone. We feel that by exceeding this limit length in a test arrangement we achieved a boundary condition which was tantamount to a simple support.

3.3. A similar device was employed to support a few aluminium and brass models, but for steel specimens *AB* became so long that buckling occurred between *A* and *B* and that the tests did not produce usable results. We therefore supported the steel cones and most of the aluminium and brass cones in a different way. They were placed on a thick conical steel bearing plate whose generatrices were perpendicular to those of the test specimen and a soft resin was poured into the groove between the two conical surfaces in order to prevent leakage (fig. 4). This type of support hardly impedes rotation of the lower edge of the model, but also barely prevents it from sliding in the direction of the generatrices of the support cone, thus allowing it to assume a sinusoidal shape along the circumference. This condition, which is more unfavourable and results in a lower buckling load than the boundary conditions normally obtaining in full scale structures, existed in some tests which are discussed in section 12 of this paper.

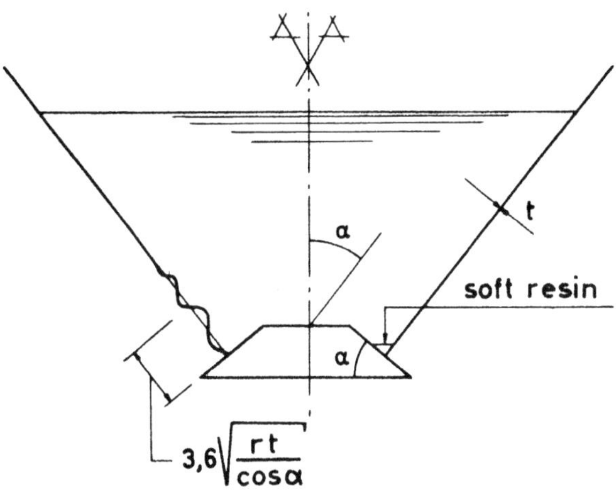

Fig. 4

3.4. In our other tests on metal models we prevented sliding of the bottom edge of the test cone by placing a large number of studs around and against its perimeter and by fastening them to the bearing plate.

This arrangement comes as close to a simple support as we think feasible and is possibly the best simulation of the connection between most full scale, hydrostatically loaded, conical shells and their supporting structures. The latter may not always preserve strictly the circularity of the lower edge of the conical tank, but on the other hand they usually provide some rotational restraint.

3.5. The numbers of tests mentioned in table 1 pertain exclusively to cones having either the simple support described in 3.4 or the weak elastic rotational restraint described in 3.2 and almost equivalent to simple support. The models with a clamped (3.1) or sliding (3.3) lower edge are not included in table 1 and will be discussed in sections 13 and 12, respectively.

When a conical bearing plate was used it was not easy to achieve good contact between it and the test cone all along their circumference. In fact, in a few tests there was over a portion of the base a slit, whose maximum width was as high as *0,35 mm* in one case, between the test specimen and its support, and when nothing was done about it, the model collapsed prematurely and unsymmetrically, more or less falling to one side while the water in the vessel fluctuated to and fro. We disregarded the results of those tests. When calibrated shims, not just one shim, but a series of shims of different thicknesses, were inserted into and over the length of any aperture wider than *0,05 mm* , the model behaved normally at failure.

4. Registration of shape imperfections

Since shape imperfections may affect the carrying capacity of a shell considerably we have measured them before each test. Mapping of the shape imperfections was achieved by moving a needle along a number (usually equal to 16) of generatrices of the cone and registering the displacement of the tip of the needle in the direction perpendicular to the generatrix. The force exerted by the pick-up on the very flexible wall of the mylar models had to be very slight, of the order of magnitude of *0,01 N* , lest the needle itself depress the wall. The critical buckling region is the lower part of the cone, and we therefore limited shape measurements to that part. Figure 5 shows a set of registered shape curves, the ordinates representing the magnified deviations from the theoretical straight generatrices. The kinks in the curves are likely to coincide with joints between the sheets of metal used to fabricate the model. For further details about the registration of shape imperfections, see [6].

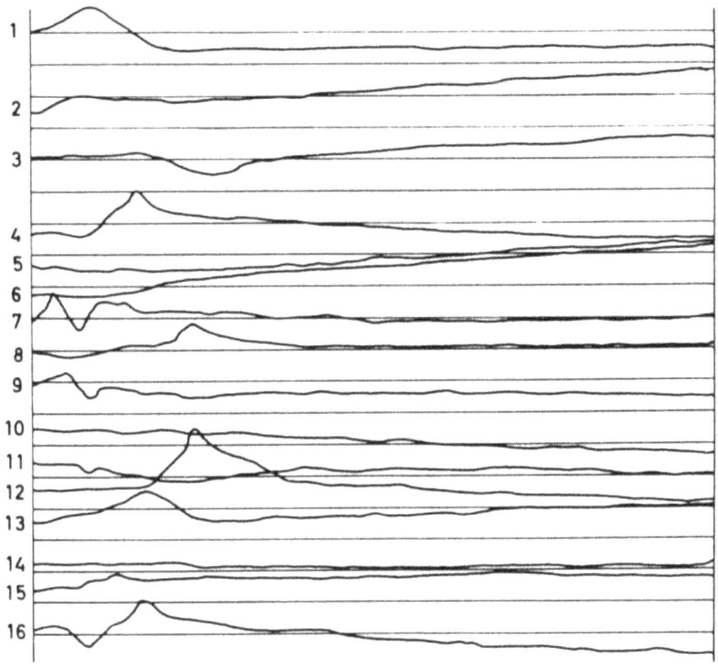

Fig. 5

FIG. 6

Figure 6 is a photograph of a collapsing mylar model. [7] contains a photograph of a collapsing brass model.

5. Calculation of the theoretical buckling load - Definition of "good" cones and "poor" cones

For each test material and for many model geometries a numerical calculation of the critical level of the liquid was made by means of the BOSOR 4 computer program [9]. This program is based on minimizing the total potential energy of the system and on replacing differentials by difference expressions. It assumes linear elastic behaviour of the material and is applicable only to shells of revolution. In the prebuckling stage it takes account of non-linear relationships between deformations and displacements. It allows consideration of buckling displacements that are either axially symmetrical or not.

The mean calculated total meridional wave length of the buckles adjoining the support of the cone (fig. 4) is $\ell = 3{,}6\sqrt{rt/\cos\alpha}$. We relate the measured shape imperfections to this theoretical buckling wave length. From the real shape of the test cone generatrices registered before loading we deduced the largest inward amplitude, $\bar{w}$, of any initial dent or depression, taken from a straight rod of length $\ell = 3{,}6\sqrt{rt/\cos\alpha}$ placed anywhere against the generatrices in the lower part of the cone (fig. 7). When

$$\bar{w} \leq 0{,}008\ell = 0{,}008 \times 3{,}6\sqrt{rt/\cos\alpha} \qquad (2)$$

we call the cone a "good" one. When $\bar{w} > 0{,}008\ell$, we classify the test cone as a "poor" one.

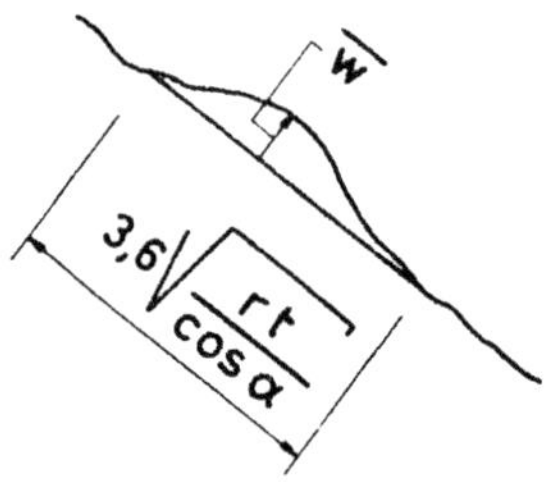

Fig. 7

6. Identification of the main system parameter

The many hundreds of numerical and experimental results constituted a rather bewildering array of data. The following pragmatic reasoning enabled us to bring some order into the chaos.

The parameters to be considered (fig. 1) are : r, t, α, γ, h, σ, all defined previously, E and ν, the modulus of elasticity and Poisson's coefficient of the test cone material. h and σ are connected by (1) and represent more specifically the height of the liquid content and the meridional membrane stress in the lower edge of the cone at the moment of failure by buckling under the hydrostatic load. Dimensional analysis provides us with the following complete set of independent non-dimensional parameters which must normally appear in relations that should be valid for full scale structures as well as for models : $\frac{\sigma}{E}$, $\frac{E}{\gamma r}$, $\frac{r}{t}$, α and ν. Any combination of these 5 parameters is also allowed. In principle the critical value of σ/E, as well as that of h/r, is determined by the values of the other 4 parameters. We know that E and ν appear in the combination $\frac{E}{\sqrt{1-\nu^2}}$ for shell structures. Hence a relation of the kind

$$\frac{\sigma\sqrt{1-\nu^2}}{E} = f\left(\frac{E}{\gamma r\sqrt{1-\nu^2}}\, , \frac{r}{t}\, , \alpha\right) \tag{3}$$

between 4 dimensionless parameters presumably exists for perfect elastic cones. We now rearrange equation (1) :

$$\frac{\sigma\sqrt{1-\nu^2}}{E} = \frac{\gamma r\sqrt{1-\nu^2}}{2E\, tg\alpha} \cdot \frac{r}{t\cos\alpha}\left(1 + \frac{h\, tg\alpha}{3r}\right)\cdot\left(\frac{h\, tg\alpha}{r}\right)^2 \tag{4}$$

and we combine (3) and (4) :

$$f\left(\frac{E}{\gamma r\sqrt{1-\nu^2}}\, , \frac{r}{t}\, , \alpha\right) = \frac{\gamma r\sqrt{1-\nu^2}}{2E\, tg\alpha} \cdot \frac{r}{t\cos\alpha}\left(1 + \frac{h\, tg\alpha}{3r}\right)\cdot\left(\frac{h\, tg\alpha}{r}\right)^2 \tag{5}$$

The 3 variables determining the numerical value of the function f appear only in the combinations $\frac{E\, tg\alpha}{\gamma r\sqrt{1-\nu^2}}$ and $\frac{r}{t\cos\alpha}$ in the second member of (5), and h appears only in the combination $\frac{h\, tg\alpha}{r}$. We conclude that for perfect cones some relationship

$$\zeta = f_1(\xi,\eta) \tag{6}$$

between the 3 dimensionless parameters

$$\xi = \frac{r}{t\cos\alpha}\quad , \quad \eta = 10^{-6}\,\frac{E\, tg\alpha}{\gamma r\sqrt{1-\nu^2}}\quad , \quad \zeta = \frac{h\, tg\alpha}{r} \tag{7}$$

may possibly exist.

We rewrite (1) in yet another form :

$$\psi = \frac{\sigma t\sin\alpha}{\gamma r^2} = \frac{1}{2}\left(1 + \frac{h\, tg\alpha}{3r}\right)\cdot\left(\frac{h\, tg\alpha}{r}\right)^2 = \frac{\zeta^2}{2}\left(1 + \frac{\zeta}{3}\right) \tag{8}$$

If (6) is true, then the non-dimensional quantity ψ should also be a function of ξ and η.

We calculated ξ, η and ψ, inserting the experimentally obtained buckling values of σ into $\psi = \frac{\sigma t \sin\alpha}{\gamma r^2}$, for the 610 tests referred to in table 1 and plotted the logarithms of these 3 numbers in a three-dimensional cartesian coordinate system. We noticed that all the representative points lay fairly close to the plane defined by the equation

$$\log\psi = 5{,}036 - 1{,}777\log\xi + 0{,}8525\log\eta \qquad (9)$$

Hence, equation $\psi = 108\,600\,\xi^{-1{,}777}\eta^{0{,}8525} = 108\,600(\xi\eta^{-0{,}48})^{-1{,}777}$ reflects roughly the test results on simply supported models (a similar argument presented in [7] bears upon a larger set of results, partly obtained with other boundary conditions). It seemed natural to assume that the exponent $-0{,}48$ should in fact be $-0{,}5$ and hence to conclude that

$$\omega = \frac{\xi}{\eta^{0{,}5}} = 1\,000\left(\frac{2\gamma r^3\sqrt{1-\nu^2}}{Et^2\sin 2\alpha}\right)^{\frac{1}{2}} \qquad (10)$$

is probably a decisive parameter.

7. Diagram and formula summarizing the calculated values of the buckling strength

We plotted the ψ-values, which result when the critical stresses σ obtained by means of the BOSOR 4 computer program are inserted into the second member of (8), as a function of ω in an ω-ψ diagram and found that most of them lie very nearly on a single curve. This finding confirms a posteriori the significance of the dimensionless parameter ω.

The curvature of the theoretical ω-ψ curve in a doubly logarithmic diagram is extremely slight and the curve is represented with a good approximation by the upper dashed straight line (short dashes) in figure 8 and by the equation $\log\psi = 5{,}638 - 1{,}946\log\omega$, obtained by a least sum of squares method, of that line. Hence the relationship

$$\psi = 434\,000\,\omega^{-1{,}946}, \quad \text{with} \quad \psi = \frac{\sigma t \sin\alpha}{\gamma r^2} \qquad (11)$$

and with ω given by (10), summarizes reasonably well the results of our numerical calculations. (11) gives the theoretical critical value of σ for perfect, perfectly elastic cones.

Because the shape of a vessel containing a liquid may be conical over only part of its height and be different elsewhere, and also because the relative influence of the dead load of a full scale structure on σ may be greater than in the case of our test specimens, where it was insignificant, it is more practically useful to express the buckling load in terms of the critical stress σ or of the number ψ than in terms of the critical height h or of the number ζ.

8. Lower bounds of the experimental results

The results of the 610 tests to which table 1 pertains are also plotted in figure 8. As might be expected the experimental dots lie below the theoretical, upper dashed line. A dozen test points are, however, located very slightly above the curve ; they all represent mylar cones ; the shape of these specimens was quite good and their buckling strength may have benefited after all by their being continuous with the part *AB* inside the bearing device shown in figure 3.

The relative narrowness of the scatter band also vindicates the use of the parameters ω and ψ. Although, owing to the logarithmic scale of the ordinates, the scatter is wider than one might infer from a quick glance at diagram 8, the band is in fact quite narrow, as buckling tests on various types of shell go.

ω did not exceed *400* for any of the test specimens made of brass, aluminium or steel. For the mylar models ω went up to *5300*.

The coordinates of the small circle marked *S* in figure 8 characterize the collapse of the water tower discussed in [2]. The circle represents a full scale buckling test (not meant to be that kind of test !) and is seen to fall well inside the cloud of experimental points.

The lower bound of the test points is well-nigh straight. If we a priori assume a straight line and give it the slope which is the most plausible from the look of the lower bound, and when a statistician's method is then used to so determine the position of the straight line that its ordinates ψ have a *99* % probability of being surpassed by those of the experimental points for either the poor or the good cones, the lower dashed line (long dashes) in figure 8 is obtained. Its equation is

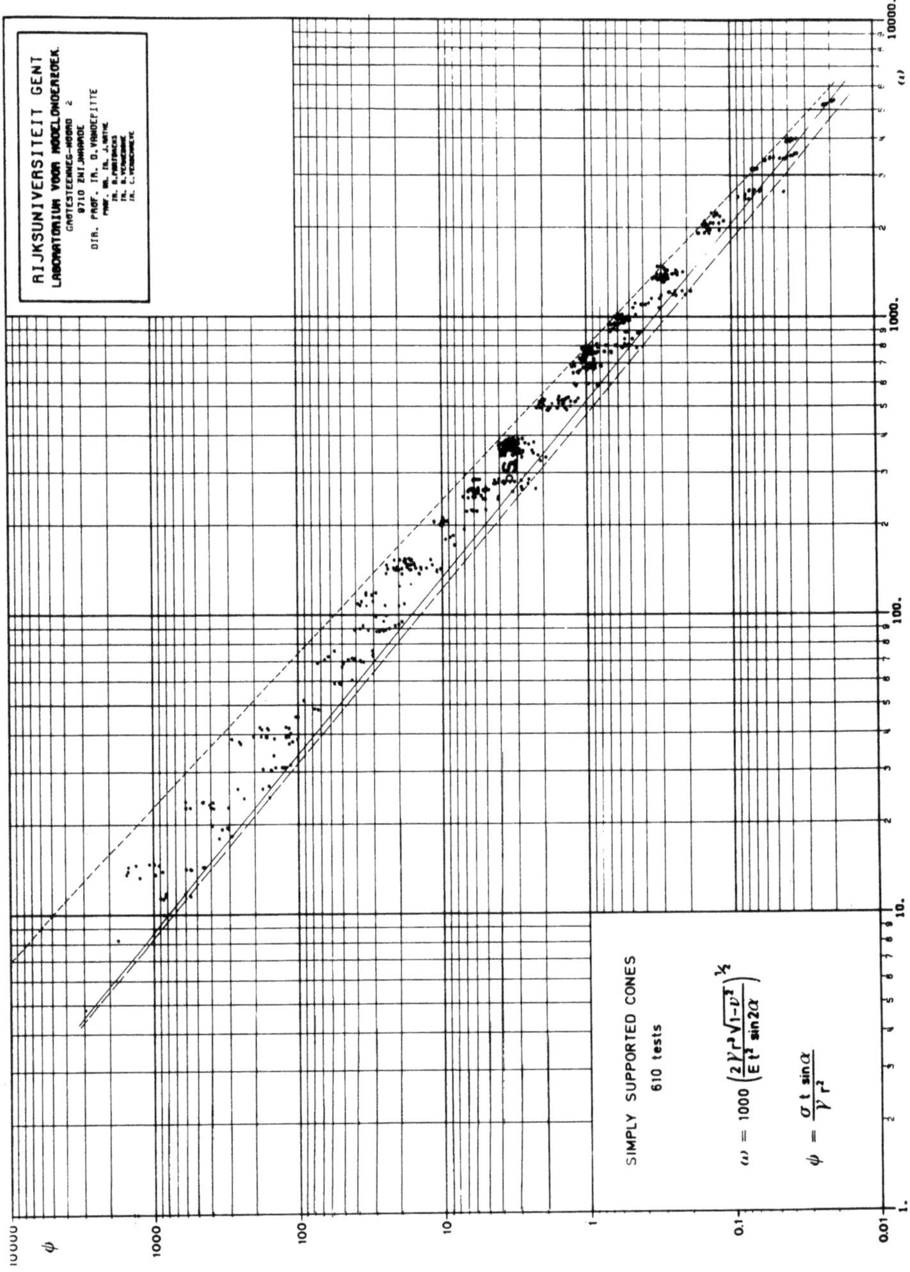

Fig. 8 - Cones with simply supported lower edge

$$\psi = 37500\omega^{-1,710} \qquad (12)$$

Six test points are located somewhat below the dashed straight line.

The equation of the straight line answering to the same description as the lower dashed line, except for the replacement of the 99 % by the 99,9 % probability, is

$$\psi = 30600\omega^{-1,710} \qquad (13)$$

Not a single test point falls below the latter straight line, which is not shown in figure 8.

When we eliminate the poor models and consider only the tests on the good models satisfying condition (2) - there were 373 such tests - and proceed exactly as in the preceding two paragraphs, we find for the 99 % line, which is the fully drawn line in figure 8, the equation

$$\psi = 37900\omega^{-1,686} \qquad (14)$$

and for the 99,9 % line, which is omitted from figure 8, the equation

$$\psi = 31500\omega^{-1,686} \qquad (15)$$

Only one good cone provided a test point actually located below the 99 % boundary line.

Many poor models are represented by points above the boundary line for the good models. This may be due partly to the fact that their largest imperfection did not happen to be located in the immediate vicinity of their supported edge. It may also indicate that we should have confined the registration of imperfections to a narrower annular region, perhaps to a $15\sqrt{rt/\cos\alpha}$ wide region, measured along the generatrices. This is not to say that the imperfections elsewhere are immaterial, but only that they are not predominant.

Since the next test which one carries out may by chance yield a lower result than all previous ones made under the same circumstances, no definitive lower bound or 100 % curve exists. Therefore, although the 99 % line is almost, but not truly a lower bound, we believe it is justified to use that line (for the good models) and equation (14) as a basis for design, as if it were a lower bound, provided, of course, that the imperfections of the actual structure are within the limitation defined by (2).

If $\bar{w}$ exceeds the limit (2), equation (12) should be used as a basis for design. The largest measured imperfection $\bar{w}$ of some of our test models considerably surpassed $0{,}008\ell$ and in one exceptional case even reached $0{,}10\ell$. Imperfections $\bar{w}$ exceeding $0{,}05\ell$ occurred exclusively in mylar cones which had been tested to failure previously and which we had subsequently straightened out with the purpose of subjecting them to a second test ; of course, the first buckling collapse had constituted a very rough treatment and had produced considerable imperfections. Metal models were not tested twice and only 12 of them were characterized by a value of $\bar{w}$ within the interval $0{,}02\ell - 0{,}05\ell$. Since their largest imperfection $\bar{w}$ may have been located at some distance from their lower edge, while the imperfections which mainly influenced their buckling behaviour may have been smaller, we tentatively suggest, pending a more detailed review of the location of the deepest dents, that (12) be used only when $0{,}008\ell < \bar{w} \leq 0{,}02\ell$.

Generally speaking, test points corresponding with small values of α tend to segregate in the lower strip of the experimental scatter band. If we retained only angles α greater than, say, $15°$ or $25°$, the 99 % boundary would be situated somewhat above the lines defined by (12) and (14), at any rate in the range of high values of ω.

9. Partial safety factor accounting for imperfection sensitivity Design formula

Writers of codes and designers not guided or bound by codes must decide the essential practical question which margin they should keep between the working load and the lower bound of the experimental buckling loads or what we have taken to be the lower bound. The margin should depend on the imperfection sensitivity of the kind of shell considered under the type of load considered. In order to answer this question as rationally as possible we have determined the two curves which in a diagram with linear (not logarithmic) ordinates ψ present a probability of respectively 50 % and 90 % of being surpassed by the test points. The ratio μ of the ordinate of the 50 % curve to the ordinate of the 90 % curve is 1,40 if no distinction is made between good and poor cones, and it amounts to 1,36 when only the good cones are retained. The number μ is a criterion for the degree of scatter of the test points, which itself results mainly from the imperfection sensitivity.

For cylindrical shells subjected to uniform external radial pressure μ has been found to be 1,27, and for axially loaded, cylindrical shells, which are more imperfection sensitive, μ varies between 1,65 and 1,8 [10, p. 287]. Hence the scatter is from 1,65/1,27 = 1,30 to 1,8/1,27 = 1,42 times wider for the latter loading case than for the former. The European Convention for Constructional Steelwork (ECCS) adopts the same over-all factor of safety for cylinders under external pressure as for most structural steel components, such as columns, beams and frames. The ECCS deemed reasonable to assume that, the wider the scatter, the lower the buckling strength of an actual shell might fall below the experimental lower bound curve used as a basis for its design and the farther the representative acting load should be kept removed from that curve. The ECCS therefore divides the ordinates of the experimental lower bound for axially loaded cylinders buckling in the elastic range by $\gamma' = 1,333$ in order to obtain their "design" buckling strength [11, p. 8]. $\gamma' = 1,333$ lies within the interval 1,30 - 1,42 of ratios of μ mentioned above.

Using the same argument for hydrostatically loaded cones, we propose that the buckling stress σ_u evaluated for good cones by means of equation (14) be divided by approximately 1,40/1,27 , say by $\gamma' = 1,1$, and that the design buckling stress σ_{ud} thus obtained not be permitted to be exceeded by the "design" acting stress. The design acting stress is the membrane stress produced by the factored working loads which result from the multiplication of the working loads by the usual load factors specified generally (in the relevant norm or code) for structures that are not particularly imperfection sensitive. Hence, in the final analysis, σ_{ud} should be calculated for good cones by means of the equations

$$\sigma_{ud} = \frac{\sigma_u}{\gamma'} \tag{16}$$

and

$$\frac{\sigma_u t \sin\alpha}{\gamma r^2} = 37\,900\,\omega^{-1,686} \tag{17}$$

with

$$\gamma' = 1,1\ , \quad \omega = 1\,000\left(\frac{2\gamma r^3\sqrt{1-\nu^2}}{Et^2 \sin 2\alpha}\right)^{1/2} \quad \text{and} \quad 15^\circ < \alpha < 65^\circ \tag{18}$$

Because none of the models with $\alpha < 15^\circ$ or $\alpha > 65^\circ$ was made of metal and, furthermore, minding the last sentence of the first paragraph of section 1 and the last paragraph of section 8, we feel it is judicious to restrict α to the interval $15^\circ < \alpha < 65^\circ$ when applying (17).

10. Influence of welding stresses

We mentioned before that all our brass, aluminium and steel test cones, except two steel cones, had soldered joints. Apparently, soldering causes lesser residual stresses than welding. Since welding stresses are known to affect the buckling load of steel columns quite considerably, the question arises whether our specimens may not be unrepresentative of actual steel shells, which are welded, and whether our test results may consequently not be too optimistic. Our steel specimens were made of sheets which were too thin (t = 0,31 mm) for welding. Therefore we fabricated two steel models whose wall was t = 1,5 mm thick and could be welded. Their other dimensions were also greater, in more or less the same proportion, than those of the cones with t = 0,31 mm. The diameter $2r$ of their base was 1,20 m, that of their upper rim was 7,50 m and their height, measured vertically, was 3,75 m. We designed them intentionally with a welded circumferential seam located quite close to their supported edge, more specifically : 9,25 cm from the edge. Welding produced noticeable shape imperfections : for one cone $\overline{w}$ was $0{,}0178\ell$ and for the other $\overline{w} = 0{,}0130\ell$. Nevertheless, the two corresponding points are well inside the scatter band of all the test points and are even slightly above the 99 % curve for good cones. Although only two welded, simply supported cones were tested - fabricating and testing a great many shells of comparable size is not within our financial means - we think that welding of the joints does not significantly affect the ultimate load of hydrostatically loaded, conical shells.

11. Elastic-plastic buckling

At the moment of failure by buckling, the meridional membrane stress σ at the base of our test models amounted to half or less than half of the yield stress f_y of the model material. Hence formula (17) for σ_u applies only when $\sigma_u \leqq f_y/2$.

When elastic-plastic buckling comes into play, which we assume to occur whenever the value of σ_u resulting from (17) exceeds $f_y/2$, σ_{ud} may possibly be given by the equation

$$\frac{\sigma_{ud}}{f_y} = 1 - 0{,}3061\left(\frac{f_y}{\sigma_u}\right)^{0{,}8333} \qquad (\text{when} \quad \sigma_u \geqq \frac{f_y}{2}) \qquad (19)$$

Equation (19) is so devised that the curve representing σ_{ud}/f_y as a function of f_y/σ_u has the ordinate 1 and a horizontal tangent for $f_y/\sigma_u = 0$, that it yields the same value $\sigma_{ud} = 0{,}5f_y/1{,}1$ as *(16)* for $\sigma_u = 0{,}5f_y$, and that it has the same tangent as the curve $\sigma_{ud}/f_y = \sigma_u/(1{,}1f_y)$ for $\sigma_u = 0{,}5f_y$. The same line of reasoning leads to a similar equation in [11, p. 8]. When the ratio σ_u/f_y increases indefinitely, *(19)* leads to $\sigma_{ud} \to f_y$. Hence, when σ_u keeps increasing, equation *(19)* gradually eliminates the additional partial safety factor $\gamma' = 1{,}1$, which accounts for imperfection sensitivity. This is logical, for a very high value of σ_u is characteristic for a cone whose thickness is great, given its other dimensions, and such a cone would not fail by buckling, but by yielding of the material. We must emphasize, however, that *(19)* represents at present only a guess. We intend to perform a few tests on rather thick cones in order to either substantiate or invalidate equation *(19)*.

12. Cones with sliding lower edge

Figure 9 is a plot, again with logarithmic abscissas ω and logarithmic ordinates ψ, of the results of 122 tests on cones whose lower edge could rotate freely, but could also slide more or less freely along the generatrices of the conical base plate. Friction and the resin used to prevent leakage did, however, impede the latter movement to some extent. 33 of the models involved were made of brass, 22 were made of aluminium, 66 were made of *0,31 mm* thick steel and one had a *1,5 mm* thick steel wall, was welded and was as large as the two specimens referred to in section 10. In figure 9 (and in figure 10) no distinction is made between good and poor cones.

It stands to reason that allowing the lower edge of the cone to move almost freely in the direction perpendicular to its surface lowers the critical level of the loading liquid. This is confirmed by the considerably lower general elevation of the cloud of test points in figure 9 with respect to the three straight lines, which are exactly the same as in figure 8. After buckling, the lower edge of at least one model was visibly deformed more or less sinusoidally along the perimeter of the bearing plate. The single large welded test specimen had no circular weld near its lower edge ; its representative point in figure 9 is well inside the scatter band. Many test points are markedly below the lower bound for good or poor, simply supported cones, and a somewhat larger number are above it. For the test specimens represented by the latter points, the detrimental effect of the sliding support was evidently eclipsed by the normal scatter.

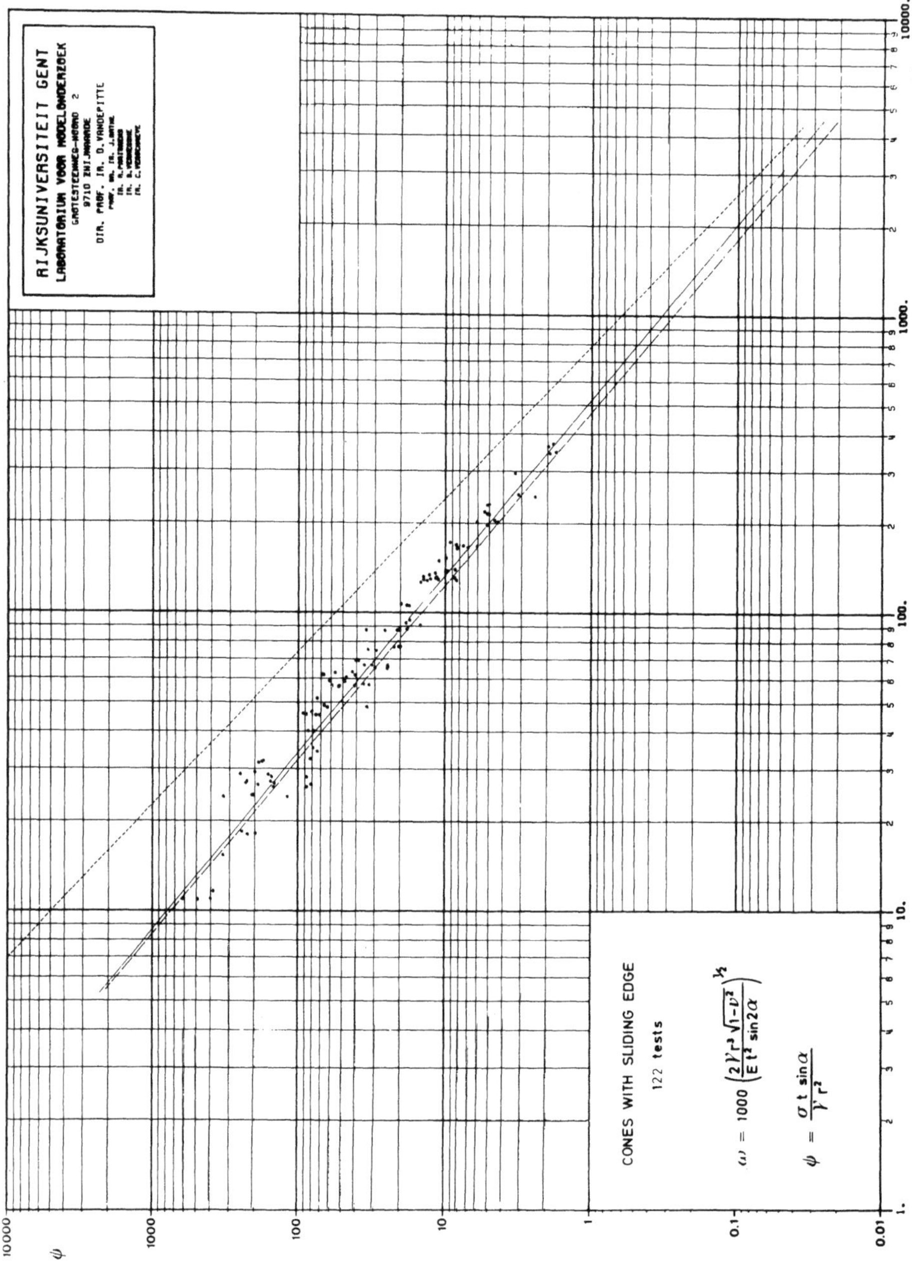

Fig. 9 - Cones with sliding lower edge

13. Cones with clamped lower edge

Figure 10 contains the same three straight lines as the figures 8 and 9, as well as the points representing 36 tests on cones with a clamped lower edge. 25 of these models were made of mylar, 4 of brass, 3 of aluminium and 4 of steel. Taken as a whole, the rather small set of experimental points for clamped cones rises above the set for simply supported cones. The difference in mean elevation is particularly conspicuous, again bearing in mind the logarithmic scale of the ordinates ψ, for larger values of ω.

14. Conical vessels of varying thickness

Since the meridional compressive stress in a hydrostatically loaded, conical shell decreases rapidly with increasing distance from the supported edge, it is natural to design the shell with two or more different thicknesses. The conical bottom of the water tower discussed in [2] was *15 mm* thick in a narrow annular region around its supporting structure and was *8 mm* thick elsewhere. Its failure was actually initiated by buckling of the lowest portion of its thinnest part.

When one wants to check the stability of such a thinner annular part, one is tempted to apply equation *(17)* to that part. But this equation does not automatically hold in the present situation, since the thicker part of the shell wall does not guarantee the circularity of the lower edge of the thinner part. In order to find out whether equation *(17)* might nevertheless be valid, we plan to test several models made of sheet metal of two different thicknesses.

15. Influence of the supporting structure

Another problem designers may be faced with and which is related to the one broached in section 14 arises from the fact that a conical reservoir does not always rest on a well-nigh rigid support. The tank of a steel water tower usually sits on top of a cylindrical shaft, and between the shaft and the bottom of the tank proper there may be a transition consisting of one or more conical shells of increasing conicity, which are not in contact with water. The question is whether the cylinder or the intermediate conical shell is stiff enough to maintain the circularity of the supported edge of the hydrostatically loaded shell sufficiently and, if not, how much additional stiffening of the horizontal junction between the latter shell and

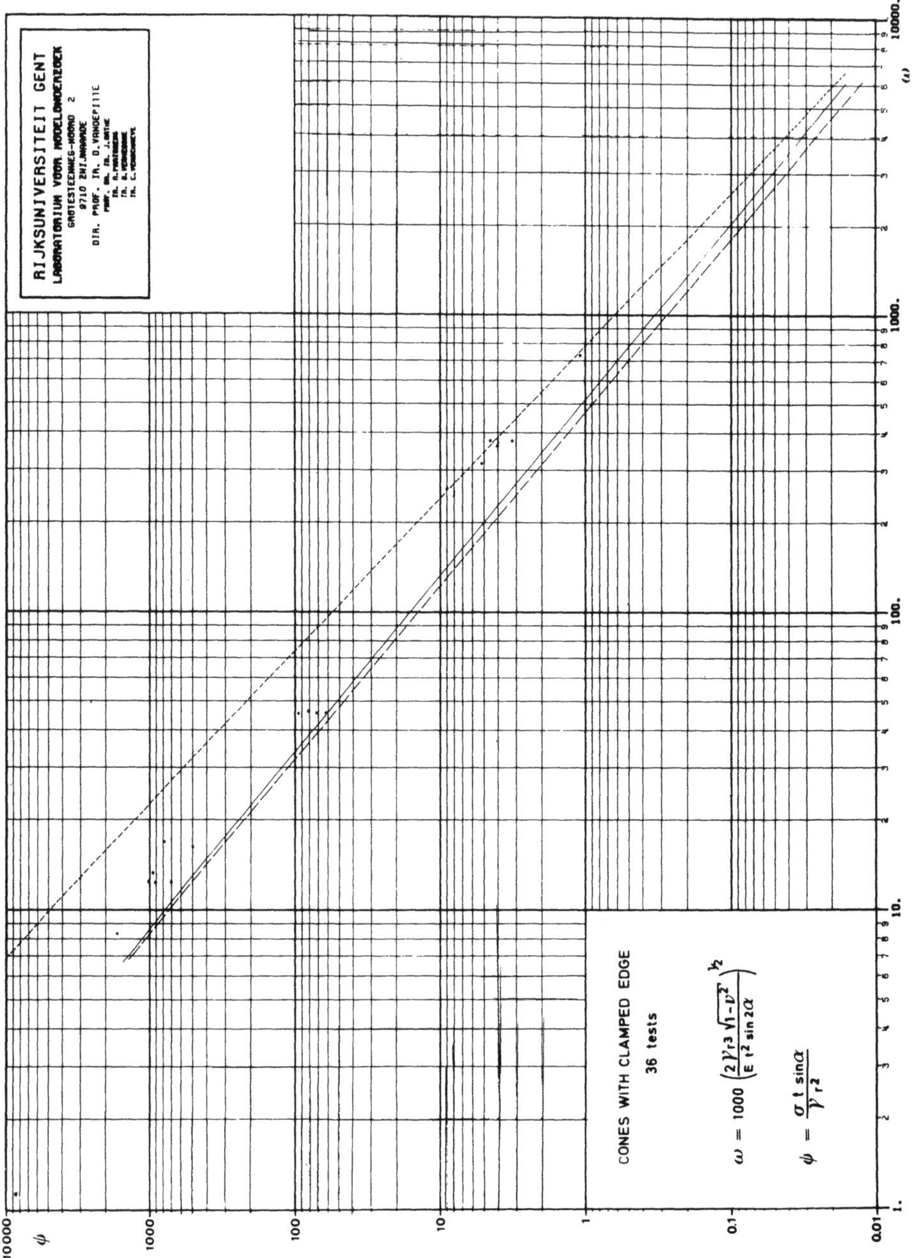

Fig. 10 - Cones with clamped lower edge

its supporting structure is required. We shall try to find at least a partial answer to this question by carrying out some tests on models similar to the one outlined in figure 11.

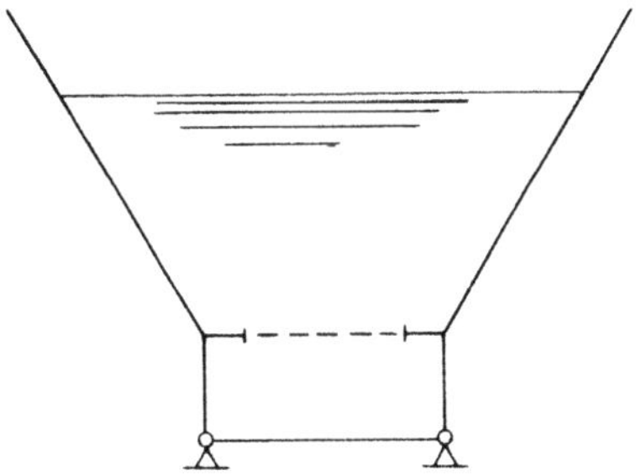

Fig. 11

16. Conclusions

1) The buckling load of a hydrostatically loaded, conical shell depends on the system parameter ω defined by equation *(10)*.

2) The design value of the meridional membrane stress acting along the lower edge of a simply supported shell should not exceed the design value σ_{ud} of the buckling stress. We evaluate σ_{ud} prudently by means of the equations *(16)*, *(17)* and *(18)*, provided that $\sigma_u \leq f_y/2$ and that the largest shape imperfection $\bar{w}$ of the cone in the vicinity of the support fulfil condition *(2)*.

3) The imperfection sensitivity of the shell is moderate and the partial safety factor $\gamma' = 1{,}1$ in *(16)* accounts for it.

4) If $0{,}008 \times 3{,}6\sqrt{rt/\cos\alpha} < \bar{w} < 0{,}02 \times 3{,}6\sqrt{rt/\cos\alpha}$, σ_u should be calculated by means of equation *(12)* instead of *(17)*.

5) Equation *(19)* may allow σ_{ud} to be evaluated when $\sigma_u > f_y/2$, but it is not yet substantiated experimentally.

6) Welding stresses do not seem to affect the buckling strength of the shell significantly and we believe that their effect is covered by the design procedure outlined above. The experimental evidence supporting this opinion is admittedly not overwhelming.

7) Allowing the supported edge of the cone to move in the direction perpendicular to the wall decreases the buckling strength of the shell. Clamping of the supported edge increases the buckling strength somewhat.

8) Equations (*11*) and (*12*) hold when the supported edge of the cone is so restrained that it must remain circular. Further tests are needed to find out whether they are valid also when circularity is not strictly preserved, either because the cone is supported by an annular portion of a thicker cone, which is also loaded hydrostatically, or because the cone is supported by a laterally unloaded cylinder or cone of different conicity.

17. Acknowledgement

The investigation described in this report was supported financially by the State University at Gent, by the Instituut tot Aanmoediging van het Wetenschappelijk Onderzoek in Nijverheid en Landbouw (IWONL), by the Centrum voor Wetenschappelijk en Technisch Onderzoek der Metaalverwerkende Nijverheid (CRIF), by the Ateliers de Construction de Jambes-Namur, by Nobels-Peelman N.V. at Sint-Niklaas, and by the bureau SECO for technical checking of building and civil engineering structures in Brussels. Sidmar N.V. has donated the sheet steel used to fabricate the steel test models. Professor BORNSCHEUER made pertinent comments about our test program.

18. Addendum

Just prior to forwarding the present contribution to the organizers of the Stuttgart colloquium we became aware of a paper [12] in which ESSLINGER and GEIER comment on the results summarized in [6]. A comparison with values given by a formula for axially loaded and internally pressurized cylinders [11, p. 13] elicits the remark that the lower bound curves in [6] are quite low, which is true. Trying to explain this, as well as the wide scatter of the experimental results shown in [6], ESSLINGER and GEIER list several hypotheses. We believe that two of those are not operative and for reasons that will emerge below we deem a third one groundless, to wit : that our method of supporting the test cones was perhaps "fehlerhaft". For one thing, as we have detailed in section 3, we supported them in four different ways, two of which we consider to be practically equivalent. The real explanation is simple : as is quite clear from the account given in our tentative report [6], all the test results were plotted together in a single diagram, irrespective of the various boundary conditions employed, which ranged from sliding support to full clamping. Consequently the scatter in that diagram was wide indeed, and the lower bound was naturally determined by the most disadvantageous of the boundary conditions used and conservative for the other ones. The effect of the bourdary condition at the supported edge is more perceptible for a hydrostatically loaded cone than for an axially loaded cylinder, because it influences directly the adjacent buckling region of the cone.

In the present paper the experimental results, classified in three groups, dependent upon the boundary condition, are plotted separately in the figures 8, 9 and 10. The two lower bounds in figure 8, corresponding with the boundary condition that is probably most often representative of the situation of full scale structures, give notably higher values of σ_u than the lower bounds in [6]. Moreover, the scatter band in figure 8 is not wide at all : on the contrary, μ is only *1,36* or *1,40*, while μ varies between *1,65* and *1,8* for axially loaded cylinders. That μ is so low must be attributed to the fact that the liquid pressure perpendicular to the wall of the shell alleviates the detrimental effect of its imperfections.

In order to establish the fully drawn lower bound in figure 8 we made a distinction between cones with $\bar{w} \leqq 0,008\ell$ and cones with $\bar{w} > 0,008\ell$. It does not seem to be too difficult for fabricators and erectors to achieve $\bar{w} \leqq 0,008\ell$ and even to achieve a higher degree of perfection. We could easily determine the lower bound of the test points representing cones with, for instance, $\bar{w} \leqq 0,005\ell$. It should be higher than the fully drawn line in figure 8 and builders guaranteeing $\bar{w} \leqq 0,005\ell$ would be justified in using it as a basis for design. As indicated at the end of section 8, part of the lower bound may climb even higher if we exclude small angles α.

We are aware of the results of some model tests carried out by Hoechst A.G. in Frankfurt and related to a specific bioreactor design [13]. The model material used by Hoechst was not isotropic (the moduli of elasticity E in the longitudinal and in the transverse direction of the sheets were quite different) and the thickness t varied somewhat. When we use the mean of the extreme values given in [13] for E and for t, we find that all the experimentally obtained ψ values are definitely below the theoretical upper bound in figure 8 and that all the test points, except one, are located in the lower third of the band between the lower bound for good cones and the upper bound. The one exception pertains to a test in which the "Beulform war durch leichtes diagonales Vorbeulen per Hand initiiert" ; the corresponding test point lies precisely on the lower bound for either poor or good cones. As shown in figure 8 some of our own test results were considerably higher and even reached the level of the approximate theoretical upper dashed line calculated for perfect shells.

19. Second addendum

In the present paper we have determined experimental lower bounds without using knock down or reduction factors. The theoretical buckling stress for an axially loaded, perfect elastic cone of medium height and not subjected to internal pressure is $\sigma_{cr} = \frac{E}{\sqrt{3(1-\nu^2)}} \cdot \frac{t\cos\alpha}{r}$.

We write this formula in nondimensional form :

$$\psi_{cr} = \frac{\sigma_{cr} t \sin\alpha}{\gamma r^2} = \frac{1\,000\,000}{\sqrt{3}} \cdot \frac{Et^2 \sin 2\alpha}{1\,000\,000 \times 2\gamma r^3 \sqrt{1-\nu^2}}$$

or $\psi_{cr} = 577\,350\,\omega^{-2}$ (20)

The ratio of the ψ_B value, resulting from the BOSOR 4 computer program for a hydrostatically loaded, simply supported, perfect elastic cone and given approximately by expression (11), to ψ_{cr} is

$$\frac{\psi_B}{\psi_{cr}} = 0{,}7517\,\omega^{0,054} \qquad (21)$$

This ratio increases from $0{,}884$ for $\omega = 20$ to $1{,}133$ for $\omega = 2000$. ψ_B exceeds ψ_{cr} when ω is higher than 197. Given the geometry, i.e. given r/t and α, a high ω represents a high ratio $\gamma r/E$ and hence corresponds with stabilizing liquid pressures that are high with respect to the stiffness of the cone material.

If we express the buckling pressure σ_u given by (17) for imperfect, but good cones or by (12) for either poor or good cones in terms of σ_{cr} for an axially loaded, perfect cone , we obtain the following expressions for the knock down factor $\alpha' = \sigma_u/\sigma_{cr} = \psi/\psi_{cr}$:

$$\alpha' = 0{,}066\,\omega^{0,314} \quad \text{when} \quad \overline{w} \leq 0{,}008 \times 3{,}6\sqrt{rt/\cos\alpha} \qquad (22)$$

$$\alpha' = 0{,}065\,\omega^{0,290} \quad \text{when} \quad \overline{w} < 0{,}02 \times 3{,}6\sqrt{rt/\cos\alpha} \qquad (23)$$

Table 2 contains a few numerical values resulting from the equations (22) and (23). The beneficial effect of high liquid pressures against the bottom part of an imperfect conical shell wall, mentioned in section 18, manifests itself in values of the knock down factor α' that increase together with ω.

Table 2 - Knock down factors α'

ω	20	100	400	2000
Formula (22)	0,169	0,280	0,433	0,718
Formula (23)	0,155	0,247	0,369	0,589

20. References

1. W. THIESSEN und A. GREIN - Bau von Bio-Hochreaktoren aus Stahl zur Abwasserreinigung - Schweißen und Schneiden 33, 1981, Heft 9, S. 465-469.

2. D. VANDEPITTE - Model investigation of the collapse of a steel water tower - Preliminary Report of the Second International Colloquium on Stability of Steel Structures, Liège, April 1977, p. 599-607.

3. R. BALTUS and Ch. MASSONNET - Use of computer programs BOSOR 4 and 5 in the stability analysis of two civil engineering steel shell structures - Preliminary Report of the Second International Colloquium on Stability of Steel Structures, Liège, April 1977, p. 609-618.

4. P. SEIDE - Axisymmetrical buckling of circular cones under axial compression - Journal of Applied Mechanics, December 1956, p. 625-628.

5. Column Research Committee, Japan - Handbook of structural stability, 1971, p. 4-235-259 - Corona Publishing Company, Tokyo.

6. D. VANDEPITTE, J. RATHE und G. WEYMEIS - Experimentelle Beullasten von Kegelschalen unter hydrostatischer Belastung - Schalenbeultagung Darmstadt, Juni 1979 - Vorträge und Diskussionsbeiträge, herausgegeben von M. ESSLINGER, S. 151-161.

7. D. VANDEPITTE, J. RATHE, G. WEYMEIS, B. VERHEGGHE and E. TAELMAN - Buckling of hydrostatically loaded conical shells - Final Report of the 11th Congress, Vienna, 1980, p. 885-890 - International Association for Bridge and Structural Engineering.

8. D. VANDEPITTE - Instabiliteit van betonnen bolkappen en van stalen kegelschalen - Symposion Schaalconstructies en Ruimtestructuren, Leuven, 1981, p. 1-21.

9. D. BUSHNELL - Stress, stability and vibration of complex, branched shells of revolution - Computers & Structures, Vol. 4, 1974, p. 399-435.

10. Shells - Chapter 10 of Manual on the Stability of Steel Structures - Introductory Report, Second International Colloquium on Stability - Tokyo, 1976 ; Liège, 1977 ; Washington, 1977.

11. European Recommendations for Steel Construction ; Section 4.6 ; Buckling of Shells, 1981 - European Convention for Constructional Steelwork - The Construction Press, London and New York.

12. M. ESSLINGER und B. GEIER - Bemerkungen zur DASt-Richtlinie 013 - Der Stahlbau, 2/1982, S. 37-40

13. HOECHST A.G. ; Abteilung Eigenüberwachung - Aktennotiz : Beulversuche an wasserbelasteten Kegelstumpfschalen - 13 November 1980 (private communication).

BUCKLING OF COMPLETE SPHERICAL SHELLS AND SPHERICAL CAPS

L. KOLLÁR

H-1122 Budapest
Karap utca 9
Hungary

Summary

The buckling behaviour of complete spherical shells and spherical caps is explained and test results are extensively reviewed. The development of the theory up to the point where theoretical and experimental results agree is presented. The treatment is confined to completely elastic material.

Introduction

Among the doubly curved surfaces with positive Gaussian curvature it is the sphere which is simplest to treat mathematically, due to its constant curvature. In fact, its buckling problems have been extensively clarified in the literature.

We shall present the results of elastic stability analysis for complete spherical shells and for spherical caps. In our treatment we shall survey the field of the linear buckling theory, the results of the model tests and, finally, the development and achievements of the nonlinear theory /post-buckling analysis/.

Buckling of complete spherical shells and spherical caps subjected to uniform overall radial pressure

It was Zoelly who first established the critical load of the spherical shell, assuming an axisymmetric buckling pattern and using the linear theory [Zoelly 1915]:

$$p_{cr}^{lin} = \frac{2E}{\sqrt{3(1-\nu^2)}} \frac{t^2}{R^2} . \qquad /1/$$

Later van der Neut showed [van der Neut 1932] that assuming an

Buckling of Shells, Ed.: E. Ramm
Proceedings of a State-of-the-Art Colloqium

asymetric buckling pattern yields the same result.

This "classical" buckling theory shows some peculiar features of the buckling phenomenon. First, the shell develops small, local buckles, so that the theory of shallow shalls can also be used for deriving the critical load, since, inside of a buckle, the shell can be considered shallow indeed.

The second remarkable characteristic of the phenomenon can be most easily shown by using the shallow shell theory. Assuming a reticulated buckling pattern we find [Wolmir 1962], [Bürgermeister et al. 1963], [Dulácska 1965] that the side lengths of the buckle remain undetermined, only a relation between them is given. This circumstance means that several buckling modes belong to the same critical load, i.e. we have to deal with a "compound buckling" [Thompson and Hunt 1973]. Consequently, a sharp drop in the post-buckling load-bearing capacity, i.e. a high imperfection sensitivity is to be expected.

Indeed, experiments gave substantially lower values than the linear critical load /1/. In Figs. 1 and 2 we compiled the results of some test series plotted against the ratio of radius to wall thickness R/t and the geometric parameter λ respectively.

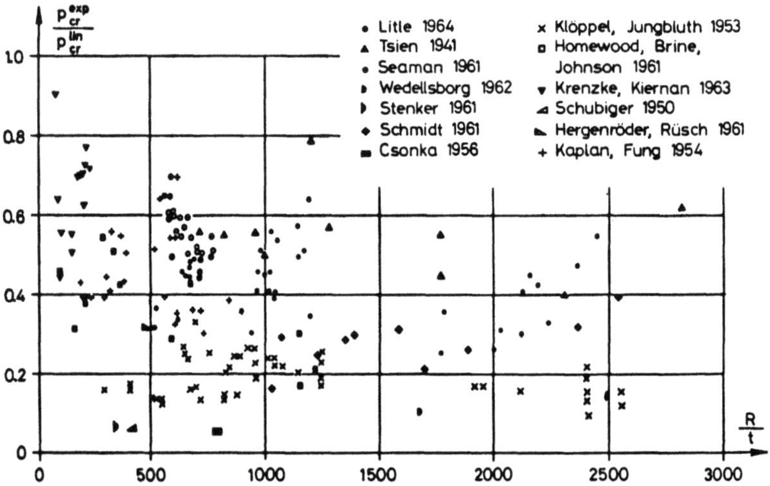

Fig.1. Experimental results on spherical shells versus R/t

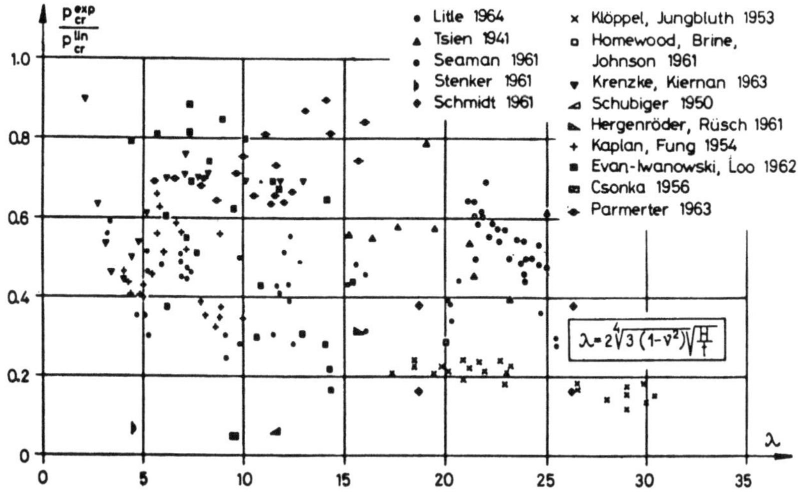

Fig.2. Experimental results on spherical shells versus the geometric parameter λ

The parameter λ is a characteristic of the spherical cap:

$$\lambda = 2\sqrt[4]{3\left(1 - \nu^{2}\right)}\sqrt{\frac{H}{t}}\ , \qquad /2/$$

with H being the rise of the cap, see Fig.7. /The radius of curvature R does not appear in λ./

From the results of [Schmidt 1961] we plotted in Fig.2 only those representing the four extreme cases; the others lie in between. Some results for non-spherical shells were also plotted in these figures /see item d./ below/.

To these seemingly completely irregular results some sense can be given by the following viewpoints:

a./ For some experiments "n e a r - p e r f e c t" s h e l l s with practically no initial imperfections were manufactured by special methods. These were the specimens of [Parmerter 1963], [Krenzke and Kiernan 1963], [Evan-Iwanowski and Loo 1962]. Their results are, in fact, higher than those for the other, less perfect, shells. The models of [Adam and King 1965] and of [Carlson, Sendelbeck and Hoff 1967], not shown in the figures, also belong to this group; they reached 80% or sometimes even 86% of the p_{cr}^{lin}.

b./ In some experiments the supports of the shell's edges were not stiff enough: either the base ring was too weak or - in the case of polygonal shells - the edges were supported by vertical diaphragms having no rigidity perpendicular to their planes. To this group belong the shells of [Csonka 1956], [Schubiger 1950], [Schmidt 1961], [Hergenröder and Rüsch 1961], [Stenker 1961], as well as some of the experiments of [Klöppel and Jungbluth 1953].

The supports of the models of [Kaplan and Fung 1954], [Homewood, Brine and Johnson 1961], [Seaman 1961] and of [Litle 1964] were also weaker than a rigid clamping, see the evaluation in [Wang 1966]. These buckled at a lower critical load than the others.

c./ In the following cases the experimental shells had non-circular ground plans. This caused deviations from the axisymmetric distribution of the internal forces resulting in not exactly definable local disturbances, in addition to the effect mentioned under b./. Here belong the shells of Csonka, Schubiger, Schmidt, Hergenröder and Rüsch.

d./ In the following researchers' experiments the shell surfaces were not spherical: Csonka, Schubiger, Schmidt.

e./ The lowest test results were yielded by reinforced concrete shells /Csonka, Schubiger, Stenker/. One can account for these by recognizing that concrete has cracks, it creeps and probably has greater initial imperfections than metallic shells.

f./ Finally, it should be noted that every experimental shell was shallow except those of [Tsien 1941] and [Litle 1964], as well as some of the shells of [Klöppel and Jungbluth 1953], [Homewood, Brine and Johnson 1961], [Seaman 1961] and [Krenzke and Kiernan 1963]. The study of [Wang 1966], which evaluates a great number of experiments, points to the fact that only the critical loads of the shallow shells depend on the parameter λ /2/, while those of deep ones do not.

The kind of loading /whether it maintains its direction like gravity loading, or whether it remains perpendicular to the buckled surface like fluid pressure/ is, according to [Simitses and Cole 1968], practically irrelevant for the magnitude of the critical stress of the complete sphere, as contrasted with the case of the cylinder under lateral pressure. This is most probably due to the fact that spherical shells always buckle in small, local, shallow waves. Therefore, we will not deal with the kind of loading here any more.

The discrepancy between the linear critical load and the test results necessitated the development of a buckling theory for spherical shells which takes l a r g e d e f o r m a t i o n s into account /"nonlinear" theory/. Its essential feature is that - maintaining the assumption concerning the shallowness of the shell surface - in the expressions for the deformations the second powers of the first derivatives of the displacement w, perpendicular to the shell surface, are also taken into account. This makes possible the description of the buckling process beyond the equilibrium position close to the undeformed state, up to displacements several times the wall thickness. For geometrically p e r f e c t , c o m p l e t e s p h e r i c a l s h e l l s this calculation was first made by [Kármán and Tsien 1939]. They assumed the shape of the buckling pattern, characterizing its extension by a free parameter, with respect to which they minimized the load at every loading step. Thus, they determined the lowest point of the post-critical load--displacement curve, obtaining /for $\nu = 0$/ the result

$$p_{cr}^{lower} = 0.365 \frac{E\, t^2}{R^2} , \qquad /3/$$

see Fig. 3. They used as the abscissa the ratio of the average displacement f of the load to the displacement f_{cr}^{lin} pertaining to the linear critical load. As ordinates we plotted the ratio p/p_{cr}^{lin}, which, for Eq./3/, assumes the value 0.315.

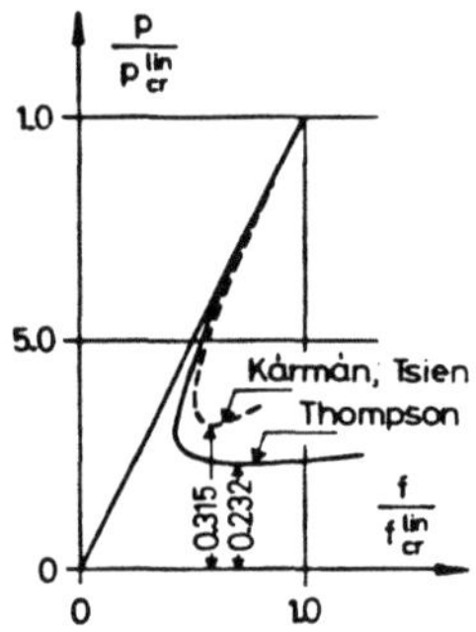

Fig.3. Post-critical behaviour of the perfect spherical shell

Later, improving the accuracy of the calculation, [Tsien 1941] arrived at the value of

$$p_{cr}^{lower} = 0.312 \frac{Et^2}{R^2} , \qquad /4/$$

i.e. 0.27 times the linear critical load.

Thompson assumed four free parameters in the expression for the buckling pattern [Thompson 1962], Thus, he was able to show that both the shape and the extension of the buckling pattern greatly vary during snapping, that is, the buckling half-wavelength increases with the decreasing load. He assumed the deformation to be axisymmetric during the entire buckling process. Thus, he obtained, with $\nu = 1/3$, the second curve of Fig.3. Its lowest point is

$$p_{cr}^{lower} = 0.283 \frac{Et^2}{R^2} ,$$

i.e. 0.232 times the linear critical load.

Other scientists performed further calculations on a similar basis in order to find the "true" value of the numerical factor appearing in the expression for p_{cr}^{lower}. So [Pozo Frutos and Pozo Vindel 1979] got /with $\nu = 0$/ 0.23, [Mushtari 1955] 0.22, [Gabril'i͡ant͡s and Feodos'ev 1961] 0.13, and [Dostanova and Raĭzer 1973] obtained /with $\nu = 0$/ the value

$$p_{cr}^{lower} = 0.126 \frac{Et^2}{R^2} , \qquad /5/$$

i.e. 0.108 times the linear critical load.

Rauch, Jacobs and Marz found that, using Reissner's equations of small finite deflections, p_{cr}^{lower} depends to some extent on the R/t ratio as well. They got 0.07 and 0.05 times the linear /classical/ critical pressure for R/t = 100 and 200 respectively [Rauch, Jacobs and Marz 1978].

All these results show that, although a perfect sphere theoretically buckles at the linear critical load, the shell is indeed very sensitive to initial imperfections, as is indicated by the shape of the curve, and as was to be expected because of the "compound buckling" mentioned earlier.

Some researchers state that the lower critical load p_{cr}^{lower} /Fig. 3/ can practically be considered as a lower bound for the value of p_{cr}^{upper}, which causes snapping of an imperfect spherical shell /Fig.4a/. The reason for this statement is that the dashed lines, that describe the behaviour of imperfect shells, osculate the full line of the perfect shell /Fig.4a/. Consequently, if their peak would lie lower than p_{cr}^{lower}, then, in fact, they have no peak, they rather have a steadily ascending character.

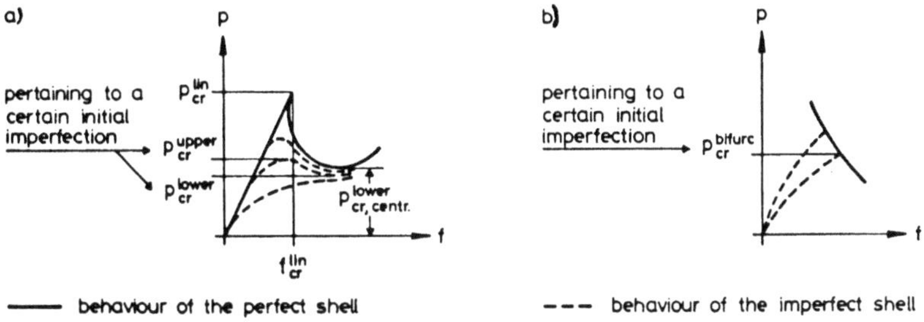

Fig.4. Two characteristic cases of post-buckling load-bearing behaviour plotted against the displacement in the direction of the load

The trouble with this reasoning is, however, that we do not exactly know the magnitude of p_{cr}^{lower}. The aforementioned results of several authors and the character of the p_{cr}^{upper} - curves /see Figs.5 and 6/ makes it likely that the "true" value of p_{cr}^{lower} is only slightly higher than /or even equal to/ zero. Consequently, in the following we have to directly deal with the determination of the snapping load of imperfect shells.

The investigations of i m p e r f e c t c o m p l e t e s h e l l s are mostly based on Koiter's method [Koiter 1945]. Hutchinson, starting from the shallow-shell equations, and also taking asymmetric buckling modes into account [Hutchinson 1967] obtained snapping loads p_{cr}^{upper} as given by the two upper curves in Fig. 5, plotted against the ratio of the initial imperfection amplitude w_0 to the wall thickness t /calculated for $\nu = 1/3$/. The two curves differ according to the shape of the assumed initial imperfection.

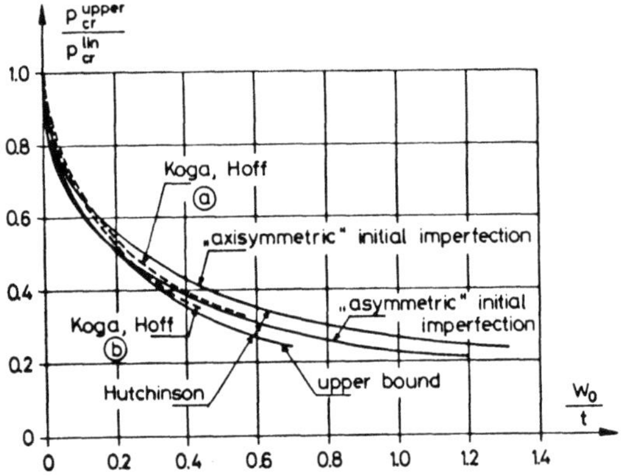

Fig.5. Influence of the initial imperfection amplitude on the snapping load of the spherical shell

The upper curve is for a "symmetrical" initial imperfection, while the initial imperfection assumed for the lower curve has an "asymmetric" shape. The investigation of Hutchinson can be considered as being completely general, since it is not confined to axisymmetric imperfections. It becomes inaccurate only for large imperfection amplitudes due to the approximations inherent to Koiter's theory. This is illustrated by the lowest curve of Fig.5, which represents - according to the special theory of Koiter - an exact solution /i.e. also correct for larger imperfection amplitudes/ taking a chosen imperfection shape as a basis. Since this does not represent the most onerous case, the truly exact curve has to lie somewhat lower.

Koiter did not make use of the approximations of the shallow-shell theory, but he confined his investigation to axisymmetric imperfections and buckling shapes [Koiter 1969]. His results are very close to those of Hutchinson.

Since the accuracy of Koiter's method - at least in its original form - becomes less and less with increasing imperfection amplitudes, due to omission of the higher terms, other investigators chose a different approach: they confined themselves to axisymmetric imperfections and buckling shapes but tried to solve this problem exactly. This was also motivated by the experimental observation /see e.g. in [Koga and Hoff 1969] / that spherical shells always deform - at least at the onset of the buckling process - axisymmetrically.

On the basis of these principles, Bushnell computed the load intensity p_{cr}^{upper} causing symmetric snapping [Bushnell 1967a]. He assumed axisymmetric initial imperfections the diameter of which were characterized by the geometric parameter λ, Eq. /2/, in such a way that H denotes the height of the original /perfect/ sphere over the imperfection circumference. When plotting these p_{cr}^{upper}-values against the ratio of the imperfection amplitude to the wall thickness /w_o/t/, he obtained different curves for the different values of λ. These curves can be combined to give a "festoon curve" /Fig. 6/.

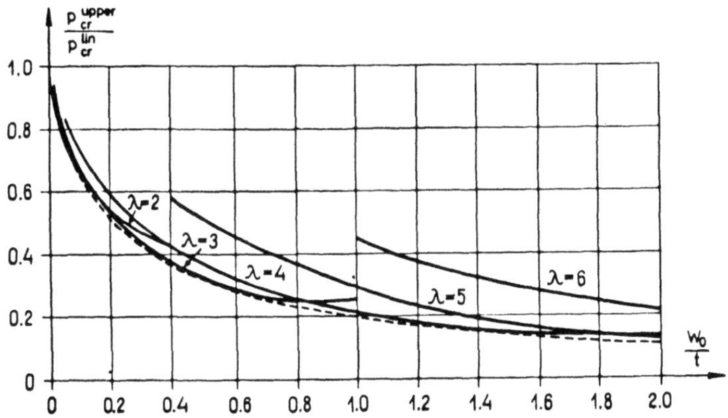

Fig.6. Influence of the initial imperfection according to Bushnell

The envelope of this festoon curve, indicated by the dashed line, can be used for design.

The development of the festoon curve can be explained by two factors. On the one hand, spherical shells with initial imperfections of different λ-s may behave differently because of the nonlinear character of the whole phenomenon, which means that, depending on the extent to which the diameter of the initial imperfection differs from the "natural" half buckling wave length /given by the linear theory/, the shell exhibite different behaviour patterns during buckling. On the other hand, that part of the spherical shell which is inside the initial imperfection behaves in a similar way to a spherical cap clamped along a circle with the diameter of the initial imperfection, and for that reason it is quite natural to obtain a festoon curve /see below/.

Fig. 6 shows that for a given ratio of w_o/t the minimum value of p_{cr}^{upper} is given by the curves with λ between 2 and 4. This approximately corresponds to $\lambda_{lin} = 2.35$ [Hoff 1969], which characterizes the half buckling wevalength of the axisymmetric mode given by the linear theory.

Koga and Hoff solved the axisymmetric buckling problem of the complete sphere [Koga and Hoff 1969] by a numerical method different from Bushnell's. They considered two somewhat different imperfection shapes a and b, and found that the minimum values of p_{cr}^{upper} were obtained for $\lambda \approx 4$ and $\lambda \approx 3$ respectively. This latter result approximately corresponds to that of Bushnell /see Fig.6/.

These minimum values of p_{cr}^{upper} of Koga and Hoff are plotted by dashed lines, marked with a and b, versus w_o/t in Fig.5. The curve marked with b practically coincides with the envelope of Bushnell in Fig.6, providing a good check.

It is worth noting that the two curves of Koga and Hoff hardly deviate from each other. Consequently, for practical purposes, it is sufficient to characterize the initial imperfection by its amplitude w_o only, independent of its shape.

It seems that the close agreement of these curves with those of

Hutchinson, who also took asymmetric imperfections and buckling shapes into account, is due to two factors compensating for each other. On the one hand, Koga and Hoff - as contrasted to Hutchinson - did not make use of the approximations of Koiter's theory, so that they obtained somewhat lower curves. On the other hand, as mentioned before, the buckling shape is, in fact, - at least in the initial stage of the buckling process - axisymmetric, so this assumption might increase the calculated p_{cr}^{upper} only to a small degree. In summary, we may conclude that for the critical load of a spherical shell practically the same result is obtained whether only symmetric imperfection and buckling deformation, or also asymmetric imperfection and deformation, are considered; see also in [Koiter 1969].

Another series of theoretical investigations refers to the spherical cap with a clamped edge. The notations to be used are indicated in Fig. 7. For the understanding of the behaviour of the clamped spherical cap, we first consider a cap whose contour points may roll freely on a conical surface having the vertex angle φ_0, but cannot rotate /see Fig. 8b/. Hence, it is in a situtation similar to a partial sphere: when subjected

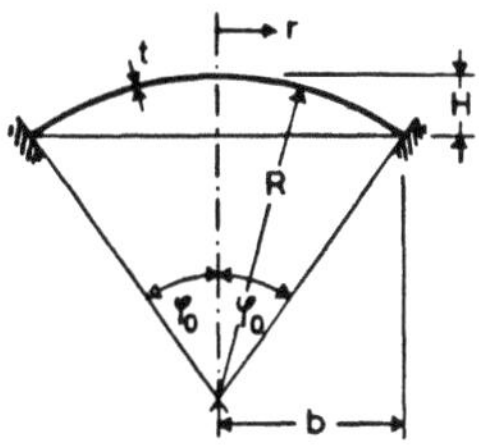

Fig.7. Clamped spherical cap

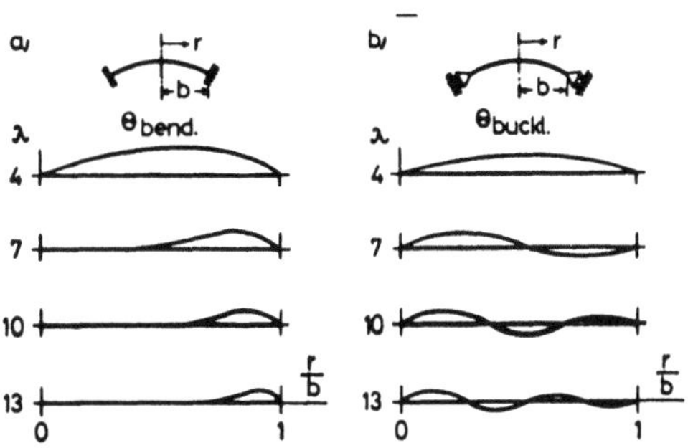

Fig.8. Change in slope of the bending and buckling deformations

to external pressure it can undergo compression without bending. The linear theory furnished for the critical load of this cap a festoon curve with a horizontal lower tangent, which coincides with the linear critical load /1/ of the perfect complete sphere [Budiansky 1959], [Huang 1964]. Hence, the spherical cap supported in this way tends to buckle almost exactly like the complete shpere.

However, if the edge of the spherical cap is clamped, the shell - even if it was geometrically perfect - undergoes bending under the load before buckling, since the spherical cap gets compressed, while the support, being considered as infinitely rigid, does not. In certain cases this bending deformation has a shape of the same character as the buckling deformation, while in other cases it has the opposite character. Computations show that this behaviour depends on the geometric parameter λ /2/.

The diagrams of Fig. 8a show, on the basis of [Budiansky 1959], the change in the slope of the c l a m p e d cap /Fig.7/ due to bending for some values of λ. On the other hand, Fig.8b represents the diagrams for the change in the slope of the cap, supported by rollers, due to axisymmetric buckling deformation. /The cap is "most willing" to buckle in this latter way./ It can be seen that the two diagrams have similar forms for $\lambda = 4$ and 10, while for $\lambda = 7$ and 13 they have opposite characters.

Due to these bending deformations even the geometrically perfect spherical cap behaves as if it had an initial imperfection. Thus, its deformation corresponds to the "eccentric" curves of Fig.4a. Using the results of [Budiansky 1959], [Weinitschke 1965] and [Thurston 1961], in Fig. 9 we plotted with full lines the p_{cr}^{upper} and p_{cr}^{lower} values of these curves against λ. The curve p_{cr}^{upper} has the peculiarity of oscillating about the line $p_{cr}/p_{cr}^{lin} = 1$. The comparison with Fig.8 shows that at λ-values, where bending deformations due to the load are similar to the buckling deformation, the shell buckles at a load lower than the linear critical value, while where they are of opposite character, the buckling load is greater than the linear one [Budiansky 1959].

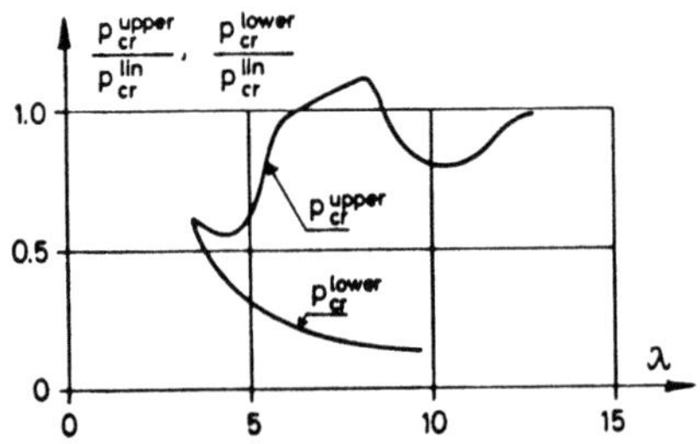

Fig.9. Snapping load of the clamped perfect spherical cap /assuming axisymmetric buckling shapes/

However, the p_{cr}^{upper}-curve of the clamped spherical cap according to Fig.9 gives considerably higher values for the critical load than the experimental ones /Fig.3/. Consequently, in order to explain this discrepancy, the initial imperfections of the cap have also to be taken into account. First [Budiansky 1959], then, refining the calculation, [Thurston and Penning 1966] developed a method for the investigation of the stability of clamped spherical caps with an axisymmetric initial imperfection. They described the deformation of the shell by the nonlinear theory and determined the p_{cr}^{upper} according to Fig.4a, assuming an axisymmetric buckling mode. The last-mentioned authors investigated experimentally and theoretically forty shell models which were manufactured with great care in order to produce predetermined initial imperfections. Theory and experiments corresponded fairly well except for the very thin and the "near-perfect" shells /about half of the models/. Later Thurston, further improving the calculation, obtained somewhat lower p_{cr}^{upper}-values [Thurston 1971].

However, all these investigations had the shortcoming of disregarding the possibility of asymmetric buckling modes, although high-speed motion-picture recording of the buckling process of spherical caps /see e.g. [Klöppel and Jungbluth 1953]/ showed that they often buckle asymmetrically, at least in one phase of the process. Thus, the phenomenon should correspond to Fig.4b. Consequently, further theoretical investigations aimed at finding out the load intensity at which the axisymmetric deformation of the perfect spherical cap, described by the large-deflection /nonlinear/ theory, bifurcates into an asymmetric mode. This bifurcation itself can be treated by the linear theory, so that a linear eigenvalue problem has to be combined with the nonlinear axisymmetric deformation. This problem was solved independently by [Huang 1964] and [Weinitschke 1965], both arriving at the same result: p_{cr}^{bifurc} ist represented by a festoon curve when plotted against λ. Fig. 10 shows the curve for the clamped spherical cap /n denotes the circumferential full-wave number of the asymmetric mode/. The characterictic feature of this curve as contrasted to that describing the axisymmetric buckling of perfect spherical caps Fig.9/ is that it does not oscillate, being rather constant and practically independent of λ.

Archer and Famili obtained the same curve when investigating the asymmetric vibrations of axisymmetrically deformed, clamped, spherical caps. The critical load intensity was reached when the eigenfrequency became zero [Archer and Famili 1965].

Fig. 10 shows that the clamped spherical cap is prone to bifurcation into asymmetric buckling for $\lambda > 5.5$ only. At smaller values of λ it buckles axisymmetrically.

This theory, considering also asymmetric buckling, agrees fairly well with the experiments made on "nearly-perfect" shells /the results of Krenzke and Kiernan, of Evan-Iwanowsky and Loo, and of Parmerter in Fig.2/. Still, there remained a considerable discrepancy with respect to the experiments on "less perfect" shells, suggesting that asymmetric initial imperfections should also be taken into account. These calculations, involving great mathematical difficulties, were first performed by [Kao and Perrone 1971]. Later Kao improved on the results using a more exact method [Kao 1972]. They assumed an initial imperfection

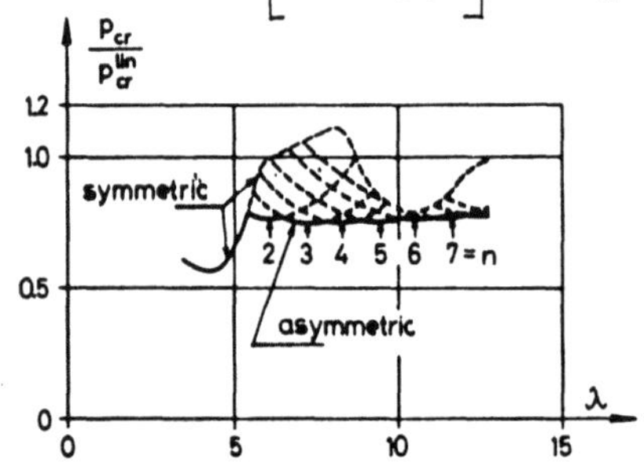

Fig.10. Critical load pertaining to the asymmetric buckling which bifurcates from the axisymmetric deformation of the clamped perfect spherical cap

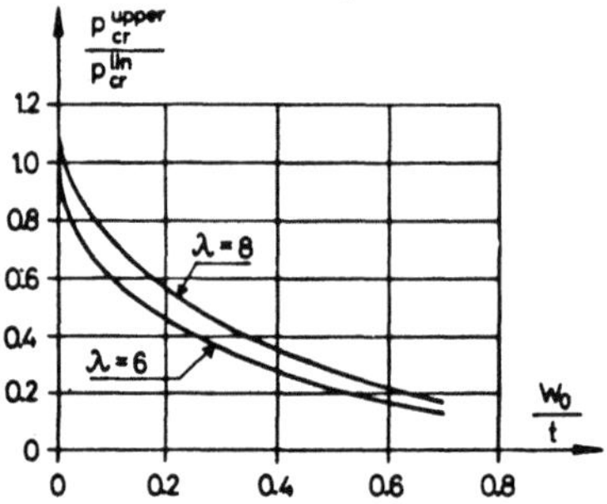

Fig.11. Influence of the asymmetric imperfection on the snapping load of the clamped spherical cap

extending in ground plan to one quarter of the clamped cap, and they described the deformation by the nonlinear theory of shallow shells. The maximum value of the load gave p_{cr}^{upper}. The ratio $p_{cr}^{upper}/p_{cr}^{lin}$ as a function of w_o/t was calculated for two values of λ /Fig.11/. These curves apparently yield the critical loads corresponding to the test results, so that they can be regarded as the final solution of the problem. /The curve pertaining to $\lambda = 8$ starts at $w_o = 0$ from a value of $p_{cr}^{upper}/p_{cr}^{lin}$ greater than 1, corresponding to Fig.9./

If we compare the theoretical results on the complete sphere and on the clamped spherical caps we arrive at the following conclusions:

When investigating the buckling of a complete sphere, the shell may be divided into two parts [Koga and Hoff 1969]: a shallow cap that buckles and the remaining part that undergoes only slight deformation due to the joining conditions. Hence, the clamped cap differs from the complete sphere only in being supported against displacement and rotation along the edge somewhat more rigidly, which results in a critical load greater than /or at least equal to/ that of the complete sphere. This explains why the curves for the snapping loads of the complete sphere and of the spherical cap are nearly identical. A detectable discrepancy is to be expected only if the extension of the cap is less than the buckle which develops on the complete sphere, i.e. for small λ-values.

The similarity between the behaviour of the complete sphere and that of the cap also explains why complete spherical shells buckle axisymmetrically, and why the results of the axisymmetric buckling investigations on complete shells can be used for design /Fig.5/. It was shown that the complete sphere is most sensitive to imperfections with λ-values around 3. These spherical shells behave similarly to caps with approximately the same λ-s. As Fig.10 shows, caps are not prone to buckle asymmetrically if $\lambda < 5.5$. Bushnell found, in fact, that complete spherical shells are inclined to buckle asymmetrically if λ is greater than about 6 [Bushnell 1967c], [Hoff 1969].

It is also obvious from the foregoing that if the support of the spherical cap is weaker than that provided by the adjacent part of the shell /when imagining that the cap is part of a complete sphere/, then the critical load of the cap is less than that of the complete sphere. Shells with a polygonal ground plan and supported by diaphragms with no bending rigidity belong to this group.

Figs. 5 and 11 enable us to design spherical shells against buckling. We only have to assume in our computation the imperfection amplitudes to be expected. However, these diagrams do not contain the circumstances mentioned under items b, c, d, and e in the description of the model tests /Figs. 1 and 2/, i.e. the weakness of the supports, the non-circular ground plan, the non-spherical shape, and the material properties of the reinforced concrete. Nevertheless, the test results of Figs. 1 and 2 could sufficiently be explained only by taking all these circumstances into account. These effects cannot be analysed exactly yet, so we try to make allowance for them by half-empirical methods based on model tests and approximate calculations, to be surveyed in what follows.

Of the approximate methods we first present the h a l f - e m pir i c a l f o r m u l a established on the basis of the already mentioned test series by [Klöppel and Jungbluth 1953]. For the critical uniform radial load of the spherical cap they write /see Fig.7/:

$$p_{cr} = k\left(\varphi_0\right) \times k\left(\frac{R}{t}\right) \times 0{,}3\,E\,\frac{t^2}{R^2} \qquad /6/$$

where

$$k\left(\varphi_0\right) = 1 - 0{,}175\,\frac{\varphi_0 - 20^\circ}{20^\circ} \qquad /7a/$$

and

$$k\left(\frac{R}{t}\right) = 1 - 0{,}07\,\frac{R}{400t}\,. \qquad /7b/$$

This formula is valid between the limits

$$20^\circ \leqq \varphi_0 \leqq 60^\circ$$

$$400 \leqq R/t \leqq 2000.$$

Formula /6/ actually completes that of Kármán and Tsien /3/, or that improved by Tsien /4/, by the factors $k(\varphi_0)$ and $k(R/t)$.

Eq. /6/ theoretically gives the $p^{lower}_{cr,centr}$ of the perfect spherical cap /Figs. 3, 4a/. In fact, we may regard it as the p^{upper}_{cr} of the imperfect shell, since this latter value was measured in the experiments. The imperfection amplitude corresponding to the factor 0.3 is, according to Fig.11, $w_0/t \approx 0.4$.

The scatter of the test results around the value of Eq./6/ was at maximum ± 20%. The metal shells were supported by squeezing their plane edges between two rings to be regarded as infinitely rigid.

In order to assess the influence of the support stiffness on the critical load, [Klöppel and Roos 1956] performed a series of experiments in which the spherical caps - like the reservoirs - are connected to cylinders welded at the joints without edge rings /Fig.12/. The pressure acted only on the spherical cap.

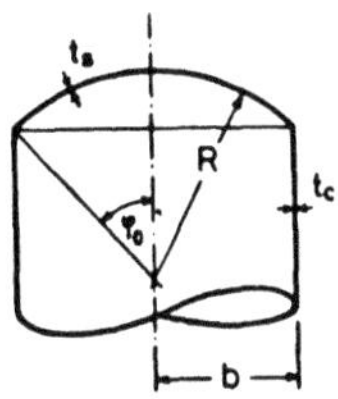

Fig.12. Spherical cap joining a cylindrical shell

Varying the ratio of the wall thicknesses t_s of the sphere and t_c of the cylinder, they found that the critical load /6/ still has to be multiplied by the factor

$$\lambda_1 = 0.65\left(1 + \frac{R/t_s}{400 + R/t_s}\right)\left(\Phi_M + 0.5\right)\cos\varphi_0, \qquad /8/$$

where

$$\Phi_M = \frac{\varkappa}{1 + \varkappa}, \qquad /9a/$$

$$\varkappa = \sqrt{\frac{b}{R}}\sqrt{\left(\frac{t_c}{t_s}\right)^3}, \qquad /9b/$$

and the restriction

$$t_c/t_s \leqq 2.5$$

has to be valid.

The deviation of the experimental results from Eq. /6/ multiplied by /8/ was at maximum + 23% ~ -11%.
These data give valuable information on the influence of the supporting edge: if the edge has no rigidity against rotation at all $\left(t_c/t_s = 0\right)$, then this reduces the critical load of the cap supported as in the experiments of Klöppel-Jungbluth to about half its value. The clamping effect of the cylindrical wall reaches that of the support used by Klöppel and Jungbluth at about $t_c/t_s = 1$, while thicker cylinder walls obviously further increase p_{cr}. This effect is, in fact, obvious when we consider that shell models in most cases started to buckle asymmetrically near the edges.

The influence of support stiffness on the critical load of spherical caps was also theoretically investigated by several researchers. A simple, approximate method was developed by [van Koten and Haas 1970-71], which yielded a result for the test shell of [Schubiger 1950] very close to that obtained in the experiment, although this latter was exceptionally low /see Figs. 1 and 2/.

A more exact analysis was performed by [Bushnell 1967b] who took into account the extensional rigidity of the ring as well. He followed the /initially/ axisymmetric deformation of the spherical cap by the nonlinear theory and determined the load intensity at which asymmetric bifurcation occurs. For the cases of built-in, hinged and roller-supported edges /i.e. zero displacement and rotation, zero displacement and unprevented rotation, and unprevented displacement and rotation of the edge/ he obtained upper critical load intensities of 0.790 p_{cr}^{lin} , 0.690 p_{cr}^{lin} and 0.160 p_{cr}^{lin} respectively. Here p_{cr}^{lin} again means the value defined by Eq. /1/.

In their study [Wang, Rodriquez-Agrait and Litle 1966] investigated experimentally the influence of the stiffnesses of these three kinds of support on the critical load on fifty-two spherical PVC caps. They obtained 0.635 p_{cr}^{lin}, 0.530 p_{cr}^{lin} and 0.190 p_{cr}^{lin} as average values for the clamped, hinged and roller-supported edges respectively, these being in acceptable agreement with the results of Bushnell. The discrepancy between the experimental and theoretical values might be accounted for by the fact that Bushnell did not take initial imperfections into account. The imperfections affected the experimental critical loads considerably in the first two cases but much less in the third case because of the much greater bending deformation.

The above results show that the edge rigidity against rotation has far less influence on the critical load then its extensional rigidity, effective against displacement.

Spherical caps under one-sided load

The buckling of spherical caps under one-sided uniform load was investigated experimentally by [Klöppel and Roos 1956]. We shall not be dealing here with this problem in detail, but only state that the intensity of the one-sided load causing buckling is close to that of the uniform load. A similar conclusion can be drawn from the calculation of [Famili and Archer 1965].

However, we may remark that, according to [Dickie and Broughton 1971], the load intensity causing symmetric snapping of a flat arch becomes lower if it is arranged partially instead of over the whole arch; i.e. if we leave unloaded the two extreme section of the arch that displace upwards. This reduction of the critical load intensity of the arch, if it is loaded by a single concentrated load applied at the centre, can even be 30%./In this case the concentrated load distributed over the central portion of the arch, displacing downwards during buckling, is to be regarded as load intensity./ Since we shall not deal with the most onerous arrangement of the partial load acting on a spherical cap in detail, the above information may be used for assessing this effect.

Buckling of spherical caps under a centrally applied concentrated load

The buckling problem of a clamped spherical cap under a centrally applied c o n c e n t r a t e d l o a d was solved by [Fitch 1968]. The critical load p_{cr} was made dimensionless as follows /see Fig. 13/:

$$Q_{cr} = \frac{P_{cr} \ R}{2 \pi B} , \qquad /10/$$

R again being the radius of the sphere and $B = Et^3/12(1-\nu^2)$ the bending rigidity of the shell.

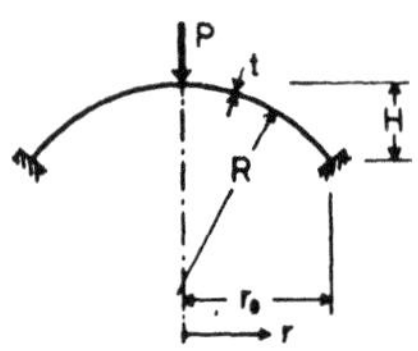

Fig.13. Spherical cap under a concentrated load

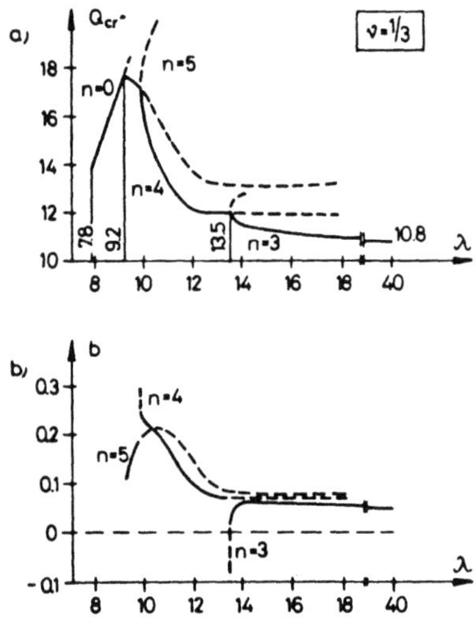

Fig.14. Critical concentrated load and initial post-buckling behaviour of the spherical cap

The value of Q_{cr} as plotted against the geometric parameter of the shell /2/ is given for $\nu = 1/3$ by the diagram of Fig. 14a, where n denotes the wave number of the buckling mode in the circumferential direction. /The dashed lines are theoretical values, the full lines are relevant./ Accordingly, n = 0 denotes axisymmetric, and n = 1 asymmetric, buckling modes. In the case of $\lambda < 7.8$ neither snapping nor bifurcation develop: the load--deflection curve has a steadily increasing character. In the range $7.8 \leqq \lambda \leqq 9.2$ the curve has a maximum, i.e. snapping occurs, but these maxima are quite gentle and approach a point of inflexion as λ tends to 7.8 /Fig.15a/. For $\lambda > 9.2$ asymmetric bifurcation accours before the snapping load intensity is reached /Fig.15b/. These bifurcation points are indicated as critical loads at $\lambda > 9.2$ in Fig.14a. Finally, if $\lambda \rightarrow \infty$, than $Q_{cr} \rightarrow 10.8$. For such large λ-s the buckling deformation is confined to the environment of the apex, so that this asymptotic value can be taken as the critical load of a complete spherical shell under two diametrically opposite concentrated loads.

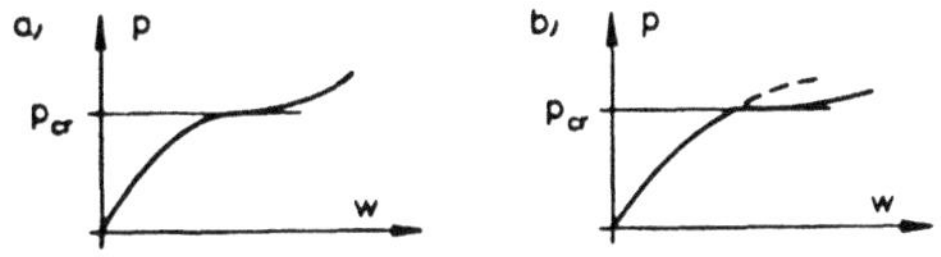

Fig.15. Two possible load-deflection paths of the spherical cap

The quantity b, which is characteristic of the initial tangent to the post-buckling load-deflection curve, is plotted in Fig. 14b against λ. It can be seen that at $\lambda > 9.2$ b is always positive.

Fitch found that his results qualitatively agree with those of the model tests reported on in the literature.

References

1. Adam, H.P.; King, P.A. /1965/: Experimental Investigation of the Stability of Monocoque Domes Subjected to External Pressure. Experimental Mechanics 5, 313-320.

2. Archer, R.R.; Famili, J. /1965/: On the Vibration and Stability of Finitely Deformed Shallow Spherical Shells. Journ. Appl. Mech. 32, 116-120.

3. Budiansky, B. /1959/: Buckling of Clamped Shallow Spherical Shells. Proc. Symp. Theory of Thin Elastic Shells /Delft/. North-Holland Publ., Amsterdam, 1960.

4. Bushnell, D. /1967a/: Nonlinear Axisymmetric Behavior of Shells of Revolution. AIAA Journ. 5, 432-439.

5. Bushnell, D. /1967b/: Buckling of Spherical Shells Ring-Supported at the Edges. AIAA Journ. 5, 2041-46.

6. Bushnell, D. /1967c/: Symmetric and Nonsymmetric Buckling of Finitely Deformed Eccentrically Stiffened Shells of Revolu-

tion. AIAA Journ. 5, 1455-62.

7. Bürgermeister, G.; Steup, H.; Kretschmar, H. /1963/: Stabilitätstheorie, Teil II. Akademie-Verlag, Berlin.

8. Carlson, R.L.; Sendelbeck, R.L.; Hoff, N.J. /1967/: Experimental Studies of the Buckling of Complete Spherical Shells. Experimental Mechanics 7, 281-288.

9. Csonka, P. /1956/: The Buckling of a Spheroidal Shell Curved in Two Directions. Acta Techn. Acad. Sci. Hung. 14, 425-437.

10. Dickie, J.F.; Broughton, P. /1971/: Stability Criteria for Shallow Arches. Journ. Eng. Mech. Divis. /Proc. ASCE/ 97 EM3, 951-965.

11. Dostanova, S.Kh.; Raĭzer, V.D. /1973/: Issledovanie ustoĭchivosti pologih obolochek polozhitel'noĭ krivizny metodom lokal'nykh variat͡siĭ. Stroit. Mekh. i. Rasch. Sooruzh. 34-39.

12. Dulácska, E. /1965/: Experimental Determination of the Critical Load of Shell Structures. Proc. Symposium IASS /Budapest, 1965/. Épitéstudományi Intézet, Budapest.

13. Evan-Iwanowski, R.M.; Loo, T.C. /1962/: Deformations and Stability of Spherical Shells under Action of Concentrated Loads and Uniform Pressure. Techn. Report 834 /11/, No. 4. Syracuse University Research Institute.

14. Famili, J.; Archer, R.R. /1965/: Finite Asymmetric Deformation of Shallow Spherical Shells.AIAA Journ. 3, 506-510.

15. Fitch, J.R. /1968/: The Buckling and Postbuckling Behavior of Spherical Caps under Concentrated Loads. Int. Journ. Solids Struct. 4, 421-446.

16. Gabril'i͡ant͡s, A.G.; Feodos'ev, F.I. /1961/: Ob osesimmetrichnykh formakh ravnovesii͡a uprugoĭ sfericheskoĭ obolochki, nakhodi͡ashcheisi͡a pod deĭstviem ravnomerno raszpredelennogo davlenii͡a. Prikl. Mat. i Mekh. 25, No. 6.

17. Hergenröder, A.; Rüsch, H. /1961/: Recent Findings in the Testing of Models. Proc. Symp. Shell Research, /Delft 1961/, North-Holland Publ., Amsterdam, 1961

18. Hoff, N.J. /1969/: Some Recent Studies of the Buckling of Thin Shells. Aeronaut, Journ. /London/ 73, 1057-70.

19. Homewood, R.H.; Brine, A.C.; Johnson, A.E.Jr. /1961/: Experimental Investigations for the Buckling Instability of Monocoque Shells. Proc. Soc. Experimental Stress Analysis 18, 88-96.

20. Huang, N.C. /1964/: Unsymmetrical Buckling of Thin Shallow Spherical Shells, Journ. Appl. Mech. 31, 447-457.

21. Hutchinson, J.W. /1967/: Imperfection Sensitivity of Externally Pressurized Spherical Shells. Journ. Appl. Mech. 34, 49-55.

22. Kao, R.; Perrone, N. /1971/: Asymmetric Buckling of Spherical Caps with Asymmetrical Imperfections. Journ. Appl. Mech. 38, 172-178.

23. Kao, R. /1972/: A Note on Buckling of Spherical Caps with Initial Asymmetric Imperfections. Journ. Appl. Mech. 39, 842-844.

24. Kaplan, A.; Fung, Y.C. /1954/: A Nonlinear Theory of the Bending and Buckling of Thin Elastic Shallow Shells. NACA TN 3212.

25. Kármán, Th. von; Tsien, H.S. /1939/: The Buckling of Spherical Shells by External Pressure. Journ. Aeronaut. Sci. 7, p. 43.

26. Klöppel, K.; Jungbluth, O. /1953/: Beitrag zum Durchschlagproblem dünnwandiger Kugelschalen. /Versuche und Bemessungsformeln./ Stahlbau 22, 121-130.

27. Klöppel, K.; Roos, E. /1956/: Beitrag zum Durchschlagproblem dünnwandiger, versteifter und unversteifter Kugelschalen für voll- und halbseitige Belastung. Stahlbau 25, 49-60.

28. Koga, T.; Hoff, N.J. /1969/: The Axisymmetric Buckling of Initially Imperfect Complete Spherical Shells. Int. Journ. Solids Struct. 5, 679-697.

29. Koiter, W.T. /1945/: Over de stabiliteit van het elastisch evenwicht. /On the stability of the elastic equilibrium./ Thesis, Delft. H.J. Paris, Amsterdam. English translation: NASA, Techn. Transl. F 10, 833, 1967.

30. Koiter, W.T. /1969/: The nonlinear Buckling Problem of a Complete Spherical Shell under Uniform External Pressure. I, II, III and IV. Proc. Kon. Nederl. Akad. Wet. Series B. 72, 40-123.

31. Koten, H. van; Haas, A.M. /1970-71/: The Stability of Doubly Curved Surfaces Having a Positive Gaussian Curvature Index. Heron, 17, No. 4.

32. Krenzke, M.A.; Kiernan, T.J. /1963/: Elastic Stability of Near-Perfect Shallow Spherical Shells. AIAA Jorun. 1, 2855-57.

33. Litle, W.A. /1964/: Reliability of Shell Buckling Predictions. Research Monograph No. 25. Cambridge, Mass., M.I.T. Press

34. Mushtari, Kh.M. /1955/: K teorii ustoǐchivosti sfericheskoǐ obolochki pod deǐstviem vneshnego davleniia͡. Prikl. Mat. i

Mekh. 19, No. 2.

35. Neut, A. van der /1932/: De elastische stabiliteit van de dunwandige bol. /The elastic stability of the thin-walled sphere./ Thesis. Delft.

36. Parmerter, R.R. /1963/: The Buckling of Clamped Shallow Spherical Shells under Uniform Pressure. Thesis. California Institute of Technology.

37. Pozo Frutos, F. del; Pozo Vindel, F. del /1979/: Elastic Buckling of Non-perfect Spherical Shells of Constant Thickness. Proc. IASS World Congress, Madrid, 1, 1.173-1.179.

38. Rauch, H.E.; Jacobs, N.H; Marz, J.L. /1978/: Buckling of a Complete Spherical Shell Under Uniform External Pressure. Studies in Appl. Math. 58, 141-158.

39. Schmidt, H. /1961/: Ergebnisse von Beulversuchen mit doppelt gekrümmten Schalenmodellen aus Aluminium. Proc. Symp. Shell Research /Delft, 1961/, North-Holland Publ. Amsterdam.

40. Schubiger, E. /1950/: Die Schalenkuppel im vorgespannten Beton der Kirche Felix und Regula in Zürich. Schweizerische Bauzeitung, 68, 223-228.

41. Seaman, L. /1961/: The Nature of Buckling in Thin Spherical Shells. Thesis. Cambridge, Mass. Department of Civil Engineering, M.I.T.

42. Simitses, G.J.; Cole, R.T. /1968/: Effect of Load Behavior on the Buckling of Thin Spherical Shells under Pressure. Journ. Appl. Mech. 35, 420-422.

43. Stenker, H. /1961/: Gesamtbericht von Grossversuchen an doppelt gekrümmten Montageschalen für raumabschliessende Geschossdecken. Wissensch. Zeitschr. Hochschule für Architektur und Bauwesen Weimar, 8, 181-202.

44. Thompson, J.M.T. /1962/: The Elastic Instability of a Complete Spherical Shell. Aeronaut. Quart. 13, 189-201.

45. Thompson, J.M.T.; Hunt, G.W. /1973/: A General Theory of Elastic Stability. John Wiley & Sons, London - New York - Sydney - Toronto.

46. Thurston, G.A. /1961/: A Numerical Solution of the Nonlinear Equations for Axisymmetric Bending of Shallow Spherical Shells. Journ. Appl. Mech. 28, 557-568.

47. Thurston, G.A. /1971/: A New Method for Computing Axisymmetric Buckling of Spherical Caps. Journ. Appl. Mech. 38, 179-184.

48. Thurston, G.A.; Penning, F.A. /1966/: Effect of Axisymmetric Imperfections on the Buckling of Spherical Caps under Uniform Pressure. AIAA Journ. 4, 319-327.

49. Tsien, H.S. /1941/: A Theory for the Buckling of Thin Shells. Journ. Aeronaut. Sci. 9, 373-384.

50. Wang, L.R.L. /1967/: Discrepancy of Experimental Buckling Pressures of Spherical Shells. AIAA Journ. 5, 357-359.

51. Wang, L.R.L.; Rodriquez-Agrait, L. /1966/; Litle, W.A.: Effect of Boundary Conditions on Shell Buckling. Journ. Eng. Mech. Divis. /Proc. ASCE/ 92 EM6, 101-116.

52. Wedellsborg, B.W. /1962/: Critical Buckling Load on Large Spherical Shells. Journ. Struct. Divis. /Proc. ASCE/ 88 ST1, 111-121.

53. Weinitschke, H.J. /1965/: On Asymmetric Buckling of Shallow Spherical Shells. Journ. Math. Phys. 44, 141-163.

54. Wolmir, A.S. /1956/: Biegsame Platten und Schalen. Verlag für Bauwesen, Berlin, 1962.
Vol'mir, A.S.: Gibkie plastinki i obolochki. Gostekhizdat, Moskva.

55. Zoelly, R. /1915/: Über ein Knickproblem an der Kugelschale. Thesis. Zürich.

Part IV :

Torispherical and Hyperboloidal Shells

THE BUCKLING OF FABRICATED TORISPHERICAL SHELLS UNDER INTERNAL PRESSURE

G. D. GALLETLY

Department of Mechanical Engineering,
University of Liverpool, Liverpool, U.K.

SUMMARY
Some simple approximate equations for predicting the buckling pressures of internally-pressurised perfect torispheres are given and compared with experimental results obtained on (i) small machined aluminium alloy models and (ii) large 'as-manufactured' stainless steel heads. The agreement between theory and test was reasonably satisfactory. Based on these results, a possible procedure for the design of these 'as-fabricated' heads is outlined and evaluated. The axisymmetric plastic collapse mode in these heads is also discussed briefly.

INTRODUCTION

End closures on internally-pressurised cylindrical containers are often dished for economic or functional reasons and the shapes normally employed are either ellipsoidal or torispherical. Other shapes have been suggested, e.g. the Biezeno head of constant strength [1], but they do not seem to have found general application.

Dished ends are employed in various applications, e.g. the fermentation tanks in breweries, oil or gas storage tanks, water towers, fluid cokers in the chemical industry, internal bulkheads in some missiles, the primary tanks of LMFBR's, the containment structures of PWR's, the food processing industry, etc.

As is well-known, many dished ends develop large compressive hoop stresses in their knuckle regions when they are subjected to internal pressure. One consequence of this is that the possibility of buckling in the circumferential direction must be considered in the design of the head. Either elastic or plastic buckling may occur, depending on the geometric ratios of the shell and the material from which it is constructed. For steel

Buckling of Shells, Ed.: E. Ramm
Proceedings of a State-of-the-Art Colloqium

dished ends of the usual geometric proportions, plastic buckling is more likely to occur than elastic buckling.

In addition to avoiding circumferential buckling in the design of these end closures (in which waves or wrinkles form around the circumference - see Fig. 1) it is necessary to ensure that large axisymmetric deformations are not produced in service. This is accomplished by checking that there is an adequate margin of safety between the plastic collapse (or limit) pressure and the operating pressure [2], [3].

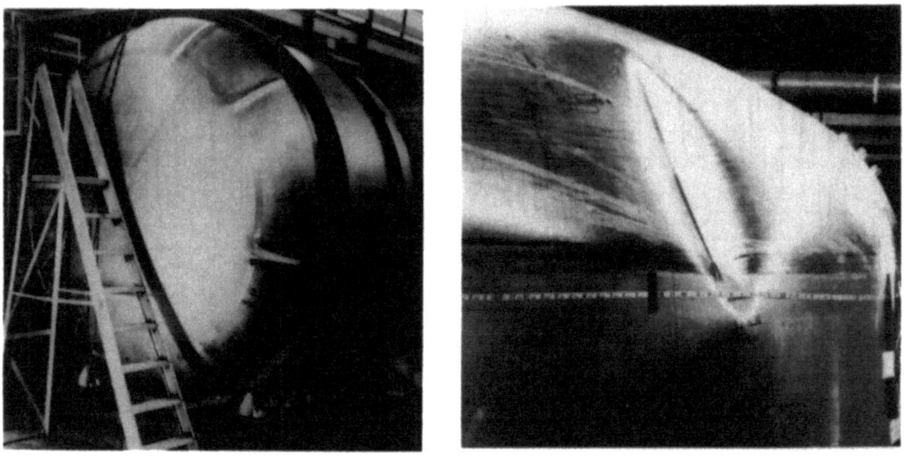

Fig. 1. Buckling due to internal pressure in a 3 m dia. stainless steel torispherical shell
a. Outward buckles b. Inward buckle

In practice, the dished ends are usually manufactured in one of two ways, i.e. either they are pressed and spun or they are made by welding a spherical cap onto a toroidal portion made from several segments. This latter construction is known as the 'crown and segment' method and it involves considerable amounts of hammering and welding. The spun ends undergo significant plastic distortion and the final thickness in the knuckle (or toroidal) portion of these ends may sometimes be only about 70% of the starting plate thickness. Spun ends may also have meridional welds in them, depending on the diameter of the vessel. More details on the two methods of manufacture may be found in [4].

Dished ends fabricated as described above will contain residual stresses due to the forming and welding operations and the magnitude and distribution of these residual stresses will not be known in detail. For spun ends made from strain-hardening materials (e.g. austenitic stainless steel) another factor which has to be considered is the change in the mechanical properties of the material (e.g. the 0.2% proof stress) brought about by the thinning of the shell during manufacture. For example, 15% thinning can change the magnitude of the 0.2% proof stress by a factor of 2.0 (cf. Fig. 10 in [5]).

Insofar as Codes for the prevention of plastic buckling in fabricated dished ends are concerned, there do not appear to be any available at the moment although groups in several countries are working on the problem. The situation is such that tentative design recommendations for this buckling problem should be forthcoming in the near future.

In subsequent sections of the paper, the main type of dished end to be discussed will be the torispherical one (the Korbbogen and Klöpper heads used in Germany belong to this class). However, reasonably up-to-date results on ellipsoidal dished ends may be found in a recent publication of the author [5].

The principal objective of the present work is the presentation and assessment (via test results) of some simple equations for predicting the occurrence of asymmetric plastic buckling in fabricated torispherical shells (made from either strain-hardening or perfectly plastic materials). Before considering these simple equations, a brief review of the problem is given and this is followed by a discussion of some characteristics of internal pressure buckling in torispheres. The experimental/theoretical correlation obtained for some machined models is reviewed and then the buckling results for 17 large fabricated stainless steel models are considered. Several failures of dished ends in service are also taken into account. A simple design procedure to prevent circumferential buckling in torispheres is then proposed and its predictions are compared with the test results on fabricated torispherical heads. The axisymmetric plastic collapse mode for these internally-pressurised heads (a mode similar to the formation of yield circles in a limit analysis) is also reviewed briefly.

BRIEF HISTORICAL REVIEW

In February 1956, a 45 ft. diameter x 78 ft. high fluid coker ($^D/t$ = 430) failed in a brittle manner while undergoing its hydrostatic proof test in Avon, California [6,7]. The bottom head of this vessel was a torispherical one and it was supported by a cylindrical skirt. A crack propagated right around the circumference of the torispherical head (just below its attachment with the skirt) causing the head to separate from the cylindrical body. The resulting outflow of the large volume of water in the vessel put a partial vacuum on the cylindrical body of the vessel and caused it to collapse - see Fig. 2.

Fig. 2. Buckling of the 15 m cylindrical body of the Avon fluid coker after brittle failure of the lower torispherical head

As this vessel had been designed in accordance with the then current edition of the ASME VIII Unfired Pressure Vessel Code, some concern was naturally felt over the failure. The author was working for Shell Development Corporation, Emeryville, California at that time and he was asked to investigate the problem. He carried out a rigorous small-deflection elastic stress analysis of the vessel and found that the circumferential stresses were compressive and very high, exceeding the yield point of the material over a considerable portion of the head [6]. It was also clear that circumferential buckling of torispheres was a possible failure mode, although there did not seem to be any Code provisions which catered for its occurrence. Another interesting point about this work is that it was one of the very first analyses of axisymmetric shells (toroidal and spherical)to employ the Runge-Kutta numerical integration technique on the digital computer. Prior to this, one usually had to use solutions involving Bessel, Kelvin, Legendre, etc., functions in order to make accurate analyses.

The elastic analysis in [6] was confirmed, in its essentials, by W. Flügge of Stanford University. Due to the high levels of the elastic stresses in parts of the toroidal portion of the bottom head, it was apparent that an elastic-plastic stress analysis was required in order to be able to assess the strength of the head properly. Such an analysis would have been difficult to perform at that time but Drucker and Shield, who were consulted about the problem, managed to carry out a limit analysis of the head. They confirmed that the margin of safety was small.

Ref. [6] was not published in the open literature but the main conclusions of the analysis were given in [8]. In Ref. [8] a 'hypothetical' vessel was referred to, but it was actually a half-scale version of the one analysed in [6].

One result of the foregoing investigation into the failed Avon fluid coker was that it prompted Drucker and Shield into carrying out limit analyses on a range of torispheres [2],[9]. Their work on this problem has since been incorporated into several Codes. Another effect was that it caused Watertown Arsenal (U.S. Army) to carry out some experiments on model torispheres. Their interest in the problem was the prediction of possible elastic buckling mentioned in [8] - the U.S. Army was, at that time, using torispherical shells as internal bulkheads in one of its missiles. Their experiments confirmed that buckling did indeed occur and one of their researchers, Mescall, carried out a small-deflection elastic buckling analysis of torispheres using the Rayleigh-Ritz technique [10]. Further experiments were carried out by Adachi and Benicek [11] - mainly on PVC models (which usually buckle elastically) but they also tested some metallic ones (aluminium and titanium) which usually buckle plastically.

The next buckling analysis to appear was by Thurston and Holston in 1966 [12] and this utilised Reissner's finite-deflection shell equations. They analysed both ellipsoidal and torispherical elastic end closures and they produced some charts for p_{cr} - although the ranges of parameters they covered were not wide.

In 1975, three experimental investigations on internally-pressurised torispheres were reported. One [13] was on large fabricated stainless steel models (1.37 m dia. to 2.74 m dia.) and the other two [14],[15] were on small (0.137 m dia.) machined heads made from aluminium alloys. At about the same time, the BOSOR shell buckling programs written by Bushnell[16], [17] were starting to be used for the analysis of these shells. Some elastic buckling results appeared in [18] while one or two plastic buckling results were given in [15]. Additional experimental results on the small machined torispherical shells were given in [19] and [20] and BOSOR predictions of p_{cr} for them appeared in [21] and [22].

The BOSOR programs use a finite-difference energy formulation of the problem and BOSOR 5 incorporates large deflections and plastic effects. There are also several finite element programs which will analyse these heads. Among these are FESIA, developed by the Cornell University group [23], CEASEMT which comes from the nuclear centre at Saclay [24] and ROT B from Bochum[25].

The Liverpool University group has published several papers in the last 3-4 years, both on the analysis of torispheres and on the correlation of tests with theory [3,5,26 to 29]. The most comprehensive of these papers is probably Ref. [5], which also includes an assessment of the experimental/theoretical correlation for ellipsoidal heads.

SOME TORISPHERICAL SHELLS WHICH BUCKLED IN SERVICE

It is, of course, when investigating vessels which have failed in service (or during the hydrostatic proof test) that one sometimes comes across inconsistencies in design rules. The fluid coker at Avon [6,7] is a well-known example of this. While that failure was ascribed to a weld defect, and the fact that the proof test was carried out below the transition temperature of the material, much useful information on the buckling resistance and the limit pressures of torispheres has been gained in subsequent years.

With regard to vessels which actually buckled in service, there is information available on only a few of these. Buckles appeared along meridional welds in the knuckle region during the proof test of a large (18.3 m diameter) carbon steel head [30]. Stennett [31] reported some buckles which appeared in the torispherical ends of two 2 m diameter stainless steel storage tanks;these failures occurred due to a safety valve in an air line becoming blocked. Similar failures in 3 m diameter fermentation tanks in breweries, due to the mal-operation of valves, have also been reported.

Kemper, in 1972, gave details of a number of stainless steel heads [4] (whose diameters varied from about 2.5 m to 4 m). Some of these heads buckled in service whereas a number of similar ones did not. Ref.[13] also mentions an early failure of a torispherical head in Holland.

A number of the above service failures will be taken into account later in the paper when a simple design method is being assessed.

GEOMETRIC CHARACTERISTICS OF TORISPHERICAL SHELLS

As may be seen from Fig. 3, a torisphere consists of a spherical cap joined to a toroidal part (frequently called 'the knuckle'). At the junction J there is no discontinuity in slope but there is one in the meridional curvature. For a given $^{D}/H$-ratio, there are many torispheres which will satisfy the slope requirement, as may be seen from the following.

From Fig. 3, it may be seen that:

$$H = R_s - (R_s - r)\cos\alpha \qquad (1)$$

$$^{D}/2 = r + (R_s - r)\sin\alpha \qquad (2)$$

where H = height of the torispherical shell
D = diameter of the cylindrical body
R_s = radius of the spherical cap
r = radius of the torus

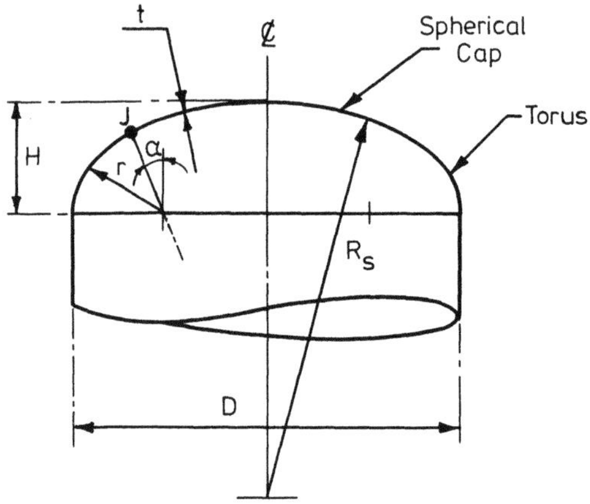

Fig. 3. A typical torispherical end closure

From Eqs. (1) and (2), it is a simple matter to show that

$$^{r}/D = \left[\frac{(2\ R_s/D - H/D)\ H/D - 1/4}{2(R_s/D) - 1}\right] \tag{3}$$

After a value of $^{H}/D$ is selected, the values of $^{r}/D$ and $^{R}s/D$ which satisfy Eq. (3) are easily determined. The results of such a calculation, over a limited range of $^{r}/D$, $^{R}s/D$ and $^{H}/D$, are shown in Fig. 4. The locus of the maximum values of $^{r}/R_s$ (shown dashed) gives the geometries of torispheres which have been called 'optimum' by some investigators [32]. The relation of the geometric ratios of the German Korbbogen head to this 'optimum' line is also indicated in Fig. 4; the geometric ratios of the 'optimum' 2:1 torispherical head (point A in Fig. 4) are $^{r}/D = 0.1726$, $^{R}s/D = 0.904$, $\alpha = 26.50^{\circ}$.

Actually, the shapes of the 'optimum' 2:1 torisphere and the 2:1 ellipsoidal shell are quite similar to the shape of an ideal Biezeno head (see Fig. 4 of [1]). The latter is a shape in which the shear stress (i.e. the difference between the two principal direct membrane stresses) in the dished end is constant and it is made equal to that in the cylindrical body. Another shape of end closure is the Cassinian head [33] and some shapes which have no compressive membrane stresses in them

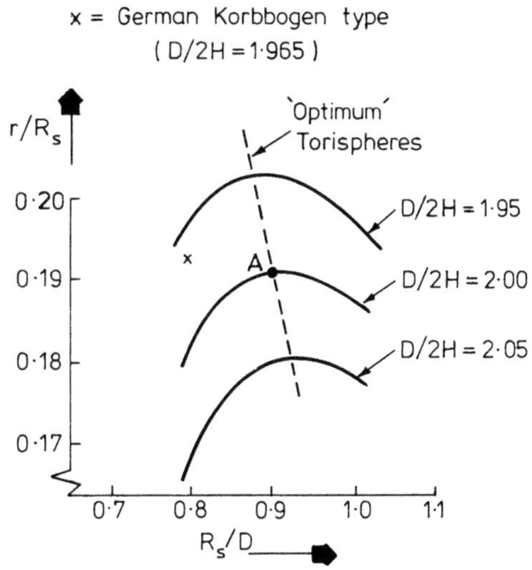

Fig. 4. Illustrating the range of torispheres which satisfy given diameter-to-head height ratios

are discussed in [34]. Of course, most of these special shapes have a varying meridional radius of curvature and, thus, are usually more difficult to manufacture than torispherical heads.

THEORETICAL PLASTIC BUCKLING RESISTANCE OF SOME PERFECT DISHED ENDS

It is not the intention here to compare the buckling pressures of many shapes of dished end but it is of some interest to compare the relative theoretical strength of a few selected heads. These are shown in the following tabulation and the plastic buckling pressures were determined with the aid of the computer program BOSOR 5. The shapes of the 2:1 ellipsoidal shell, the 2:1 'optimum' torispherical shell and the Korbbogen head are not very different but the theoretical p_{cr}'s of the 2:1 ellipsoidal shell are the highest - by at least 25% according to Table I.

The Korbbogen shell and the 2:1 'optimum' torispherical shell have similar plastic buckling resistances and the Klöpper shell is the weakest of the four. This latter is not surprising as the Klöpper shell is also the shallowest of those tabulated (its head height H is only three-quarters that of the others).

D/t	Type of Head	σ_{yp} (N/mm^2)	
		207	414
750	2:1 Ellipsoidal	No Buckling	1.09
	2:1 'Optimum' Torispherical	0.59	0.86
	Korbbogen	No Buckling	0.84
	Klöpper	0.28	0.51
1000	2:1 Ellipsoidal	0.40	0.77
	2:1 'Optimum' Torispherical	0.33	0.56
	Korbbogen	0.32	0.59
	Klöpper	0.18	0.30

Table I. Theoretical plastic p_{cr}'s for some dished ends (assumed perfect)

It would appear from this comparison that designers ought to specify the 2:1 ellipsoidal shape whenever possible. However, because of its variable meridional curvature, it is not as easy to manufacture as the torispherical types. In addition, one has to take into account the fact that welded dished ends in practice are imperfect. The Codes permit the actual shape of the manufactured ends to deviate from the specified mathematical shape by certain amounts (tolerances). This has the effect of sometimes transforming an apparent 2:1 ellipsoidal shape into a 2:1 torispherical one [32], with a consequent loss in theoretical buckling resistance [28]. This subject has not been investigated in much detail, either experimentally or theoretically. However, if the calculations are correct, then the shape of 2:1 ellipsoidal shells will have to be controlled carefully if their potential buckling resistance is to be achieved (or, if the actual shape is not controlled, then any discontinuities in meridional curvature should be).

The imperfections in shape discussed above can be relatively large (up to 1% of the diameter [32]) and are considerably greater than those normally considered when discussing the effects of initial geometric imperfections on the buckling pressures of shells. However, for this internal pressure buckling problem, the influence of the latter does not appear, fortunately, to be very significant.

SOME CHARACTERISTICS OF THE PRESENT BUCKLING PROBLEM

a. Compressive Stresses in Dished Ends

It is well-known, from the membrane theory of shells, that the circumferential stress resultants in internally-pressurised dished ends can become compressive [3] if the ratio between the two principal radii of curvature, R_θ/R_φ, is greater than 2.0.

The ratio is greater than two for many ellipsoidal and torispherical shells and buckling in the circumferential direction becomes a possibility. Whether it will occur depends on various factors, such as the diameter-to-thickness ratio, the yield point of the material, whether any strain-hardening occurs, etc. In the buckling analysis, bending stresses and large deflections have to be considered.

b. Buckle Formation

When internally-pressurised torispheres buckle, the dimples which occur may be inward or outward. For the small machined aluminium alloy models the dimples are often outward [15] but not always [19]; for the plastic (PVC, etc.) models they are normally inward [11]. With the Nottingham stainless steel heads, both types of buckle were observed and they were called 'gradual outward' or 'snap inward'. The two types of buckle are shown in Fig. 1 and further discussion on them may be found in [35]. The buckles in the stainless steel heads in [4] were all inward.

The theoretical number of buckles can vary between wide limits (say, 20 to 160), depending on the geometry. However, the author has not seen more than twelve in a test.

c. Load-Deflection Curves

A typical plot of the internal pressure versus the axial deflection at the crown of the head is shown in Fig. 5. After the occurrence of the first one or two buckles, with a resulting small drop in the pressure, it is possible to increase the pressure until another buckle forms. This sequence can be repeated and the saw-tooth curve in Fig. 5 is the result. Thus, the overall slope of the pressure-deflection curve is positive

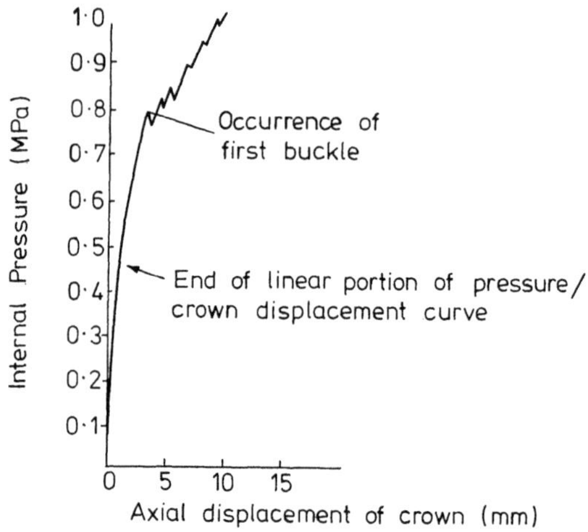

Fig. 5. Typical pressure versus crown displacement curve for a thin dished end

after the first buckles form. Buckling problems in which positive post-buckling slopes occur do not seem to be very sensitive to initial geometric imperfections (cf. Koiter's theory). Experimenters agree that this seems to be the case with this internal pressure buckling problem.

d. The Need for a Finite-Deflection Shell Theory in the Analysis

In general, it is necessary to use a finite-deflection shell theory, even for the axisymmetric prebuckling stresses in the elastic region. Supporting evidence for this statement is given in Fig. 6, in which the circumferential stresses on the outer surface of a 245 mm. diameter machined aluminium torisphere are plotted against theoretical predictions [36]. It is quite clear from Fig. 6 that the theoretical stresses as given by the non-linear (large-deflection) shell theory agree much better with the experimental results than do those given by the linear (small-deflection) theory.

It may be seen from Fig. 6 that the extreme fibre stresses predicted by the large-deflection theory are smaller than those predicted by the linear theory. It may also be shown that the prebuckling hoop direct stress resultants predicted by linear theory are larger than those given by the non-linear theory [37]. This could result in the linear theory predicting that

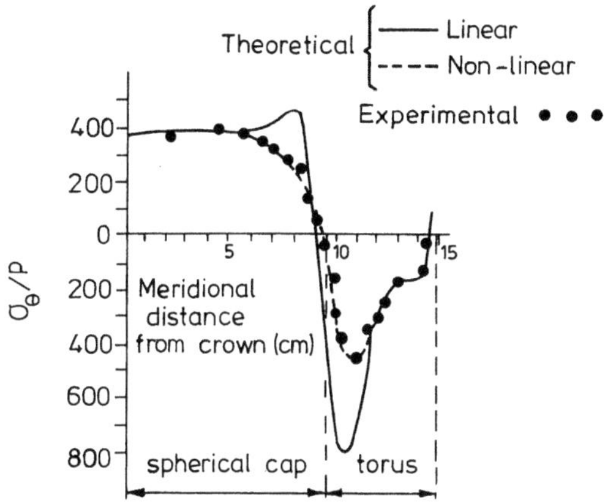

Fig. 6. Experimental and theoretical hoop stresses on the outside surface of a machined aluminium shell subjected to internal pressure (D/t = 645, R_s/D = 1.18, r/D = 0.173 and p = 0.41 N/mm^2.

buckling will occur at lower values of $^D/t$ than the non-linear theory does.

An illustration of the foregoing, for the elastic buckling of internally-pressurised 2:1 ellipsoidal shells, is given in Fig. 7. It may also be seen from this Figure that, for the larger values of $^D/t$ (say > 1500), the p_{cr}-predictions of the large-deflection and the small-deflection shell theories [38] are in good agreement. This is an interesting point which has not been fully explained so far.

e. Thinning and Residual Stresses in Fabricated Heads

Some of the smaller spun heads (≈ 2 m diameter or less) can be spun from a single sheet. Any residual welding stresses due to welding the head onto the cylindrical body will thus be axisymmetric. However, the larger spun heads will have meridional welds and accompanying non-axisymmetric residual welding stresses.

The thickness of the spun heads is sensibly constant in the hoop direction (due to the method of manufacture) but it undergoes thinning along the meridian. This reaches a maximum in the toroidal portion of the head. If the head is made from a strain-hardening material, e.g. an austenitic stainless steel,

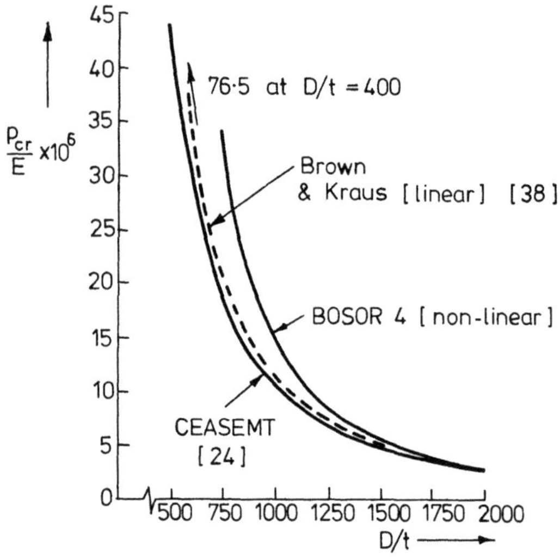

Fig. 7. Elastic p_{cr}'s for internally-pressurised 2:1 ellipsoidal shells. Predictions of linear and non-linear shell theories

then the mechanical properties of the head will vary with the amount of thinning.

Crown and segment heads have very little thinning in the knuckle region but they have more welding stresses than spun heads and the distribution of the residual stresses is not axisymmetric.

There is not a great deal of information available on residual stresses in dished ends, either experimental or theoretical. What has been done in most buckling analyses so far is to ignore them.

f. Stress-Strain Properties in Fabricated Ends

When attempting to predict the buckling pressure of a fabricated head, one needs information about the mechanical properties in the toroidal portion of the head (as a minimum). For spun heads, which undergo considerable thinning, the 'as-fabricated' properties can be quite different from the 'as-received' plate properties, especially for heads made from strain-hardening materials. The question, therefore, arises as to how these properties should be determined or estimated.

One suggestion (applicable to laboratory models) is to take coupons from the knuckle portion of the head (after it has buckled) and then determine the stress-strain curves from them. However, the coupons will be curved in two directions in general and trying to straighten them may introduce more problems. Another possibility is to take strips from the 'as-received' plate and roll them down to various thicknesses. These latter can then be used to determine stress-strain properties (e.g. the 0.2% proof stresses) and curves of reduction in thickness versus these properties can be drawn. When the actual as-fabricated thickness is known, an estimate can then be made of the relevant mechanical properties.

However, designers of these heads need to know, before manufacture commences, the approximate amount of thinning likely to occur in practice and the resulting mechanical properties. From the head manufacturer's prior experience (or otherwise), it should be possible to estimate the reduction in thickness. However, the question of how to decide the mechanical properties is still an open one.

COMPUTER PROGRAMS AVAILABLE TO ANALYSE PERFECT DISHED ENDS

The shell buckling programs which have been used by the author to analyse perfect dished ends are BOSOR 4 and BOSOR 5. These programs incorporate large-deflections and are written for general shells of revolution, i.e. they are one-dimensional and are very efficient. They have been verified on a number of different buckling problems and are believed to be reliable.

A large-deflection finite element program which has been developed recently is that of the Bochum group. Their program ROT B has been used to analyse some torispherical models which buckled plastically and which had been previously analysed using BOSOR 5. The agreement between the theoretical predictions of these two programs was very good [39].

Another finite element program is one developed in the U.S.A. and known as FESIA [23]. It has recently been used by Unger and Mang [40] to study means for reducing the discontinuity in meridional curvature in torispherical ends. However, the author has not seen any results obtained with it for the plastic buckling of dished ends.

As mentioned earlier, the CEN at Saclay, France have a large-deflection finite element program known as CEASEMT, which can be used to analyse dished ends. However, it does not seem to have been checked thoroughly against programs like BOSOR 5. One or two checks made by the author [5] tend to indicate that BOSOR 5 agrees better with the experimental results than does CEASEMT. Also, some elastic buckling results on ellipsoidal shells under internal pressure obtained using the CEASEMT program [24] agree better with the small-deflection results than the large-deflection ones [41] - see Fig. 7. Presumably, this discrepancy has a simple explanation, e.g. perhaps the small-deflection option in the CEASEMT program was used by error in the calculations.

SOME APPROXIMATE EQUATIONS FOR THE BUCKLING AND COLLAPSE OF PERFECT STEEL TORISPHERES

The author and his colleagues have used the BOSOR 4 and BOSOR 5 programs to carry out limited parametric surveys on perfect ellipsoidal and perfect torispherical shells subjected to internal pressure. Both elastic and plastic buckling were considered, as were first yield and axisymmetric plastic collapse. The computer results have been given in the form of curves and these may be found, for torispherical shells, in Refs. [3, 5, 18, 26, 27, 42]. In general, the flow theory option was used in BOSOR 5 when calculating the plastic behaviour of torispheres.

In addition to curves of computer results, simple approximate equations for the various failure modes were obtained which should be of assistance to designers. These empirical equations were determined by curve-fitting methods applied to the computer results and some examples, for steel torispherical shells, follow:

a. Elastic Buckling

For $R_s/D = 1$, the equation given in [42] assumes the simpler form

$$p_{cr}/E = 100\ [3.7\ r/D + 0.68](t/D)^{2.45} \tag{4}$$

For the ranges $500 < D/t < 2000$ and $0.06 < r/D < 0.15$, Eq. (4) predicts torispherical p_{cr}'s which are within 12% (on the safe side) of the BOSOR program predictions. Also, the experimental results of Adachi and Benicek agree with the predictions of Eq. (4) to within 10%, the theoretical results being the higher - see [42].

b. Elastic-Plastic Buckling (No Strain-Hardening)

For steel torispherical shells, the simple equation suggested in [26] is

$$p_{cr}/\sigma_{yp} = \frac{285[1-125\sigma_{yp}/E](r/D)^{0.84}}{(D/t)^{1.53}\ (R_s/D)^{1.1}} \tag{5}$$

For $R_s/D = 1$, the percentage differences in p_{cr} obtained using Eq. (5), instead of using BOSOR 5, are given in Table 1 of [5]. An alternative to Eq. (5), but which was determined only for $R_s/D = 1$, is as follows:

$$\ln\left(\frac{p_{cr}}{100}\right) = -0.60 + 0.89\ (A + B) - 1.49C + 0.42B^2 + 0.33A^2B + 0.305\ AC + 0.78ABC \tag{6}$$

where

$$A = \ln\left(\frac{\sigma_{yp}}{60000}\right)\ ;\quad B = \ln\left(\frac{r/D}{0.125}\right) \text{ and } C = \ln\left(\frac{D/t}{1000}\right) \tag{7}$$

In Eqs. (6) and (7), p_{cr} and σ_{yp} are in lbf/in^2.

Table 2 of [5] gives the percentage differences in p_{cr} obtained from Eq. (6) and BOSOR 5. The errors are somewhat smaller than when Eq. (5) is used; however, more numerical values of p_{cr} were available (for different values of σ_{yp}) when Eq. (6) was derived.

c. Plastic Collapse Pressure (No Strain-Hardening)

By plastic collapse is meant a failure mode involving large axisymmetric deformations. The pressure at plastic collapse is similar to the limit pressure but a large-deflection elastic-plastic axisymmetric shell analysis is needed to determine it. The definition of plastic collapse is rather arbitrary but in Ref. [27] it was taken as the internal pressure needed to cause twice the yield point deflection. The deflections at the crown of the shell were used in the calculations.

For steel torispherical shells having $R_s/D < 1$, the simple equation given in [27] for the axisymmetric plastic collapse pressure is

$$p_c/\sigma_{yp} = \frac{12.6[1 + 240\sigma_{yp}/E](r/D)^{1.04}}{(D/t)^{1.09}\,(R_s/D)^{1.1}} \qquad (8)$$

The predictions of Eq. (8) agree with the computer results to within 10%.

d. Some Comments on the Empirical Equations

The simple approximate empirical equations given in this section for steel torispheres were determined so that designers could make quick estimates of buckling and collapse pressures. Similar equations are available for aluminium torispheres in Ref. [3]. However, it is recognised that it might be possible to derive simpler equations than the foregoing, if different dimensionless variables were chosen or if the results were plotted in a different manner. Taking either of these options might also increase insight into the problem and this is certainly a desirable goal.

It could also be argued that some of the exponents on the approximate equations could be simplified (e.g., as r/D usually only varies from 0.06 to 0.20, perhaps the 0.84 exponent could be replaced by 1.0); also, that it is not absolutely necessary to have separate equations for steel and aluminium dished ends. While these points may have some merit in them, they have yet to be studied in detail.

e. The Drucker-Shield Limit Pressures

Another equation for predicting the onset of appreciable axisymmetric plastic deformations in torispheres subjected to internal pressure is given in [2]. It is as follows:

$$p_{DS}/\sigma_{yp} = (0.33 + 5.5\,r/D)(t/R_s) + 28(1 - 2.2r/D)(t/R_s)^2 - 0.0006 \qquad (9)$$

These limit pressures were derived using small-deflection shell theory and, thus, the predictions of Eq. (9) are likely to be lower than those of Eq. (8), especially for the larger values of D/t. It should also be noted that the highest value of D/t used in the calculations in [2] was 500. Thus, some caution is needed when employing Eq. (9) for values of D/t higher than this.

f. Effect of Strain-Hardening on the Predicted Plastic p_{cr}'s
There is no difficulty in including strain-hardening when using the BOSOR 5 program to compute buckling pressures. However, in relation to its incorporation into the simple approximate equations for p_{cr} given earlier, there are not many results available in the literature. Some, for 5% linear strain-hardening, are given in [26] and [41] for 2:1 ellipsoidal and torispherical shells. On the basis of this very limited study, it seems that the effect of strain-hardening on the plastic p_{cr} decreases with increasing D/t and is largest for the smaller values of σ_{yp}. However, more work needs to be done on this aspect of the problem.

FAILURE MODES IN STEEL DISHED ENDS

It was mentioned earlier that elastic buckling is not likely to occur in steel torispherical shells which have values of D/t and σ_{yp} that are common in practice. The reason for this statement is apparent from the following tabulation, which refers to a steel torispherical shell having r/D = 0.10 and R_s/D = 1.0:

	Asymmetric Buckling Pressures p_{cr} (lbf/in^2)			Axisymmetric Collapse Pressures p_c (lbf/in^2)	
		Plastic			
D/t	Elastic	σ_{yp} = 30000 lbf/in^2	σ_{yp} = 60000 lbf/in^2	σ_{yp} = 30000 lbf/in^2	σ_{yp} = 60000 lbf/in^2
500	770	80.2	138	48.9	117
1000	141	27.8	47.6	23.0	54.8
1500	52.1	14.9	25.6	14.8	35.2
2000	25.7	9.62	16.5	10.8	25.7

Table II. A comparison of the asymmetric p_{cr}'s and axisymmetric p_c's for a steel torispherical shell with r/D = 0.10 and R_s/D = 1.0. (No strain-hardening).

The numerical values in Table II were obtained from the approximate Eqs. (4), (5) and (8) and it may be seen that all the elastic p_{cr}'s are higher than the corresponding plastic

p_{cr}'s. Except for D/t = 2000, σ_{yp} = 60,000 lbf/in^2, the elastic p_{cr}'s are also higher than the plastic p_c's (see also Table 2 in [42]).

The plastic p_c's and p_{cr}'s are also plotted in Fig. 8 from which it can be seen that sometimes it is p_{cr} which is the lower of the two and sometimes it is p_c. The cross-over point of the two curves depends on the value of σ_{yp} and occurs at higher values of D/t for the lower values of σ_{yp}.

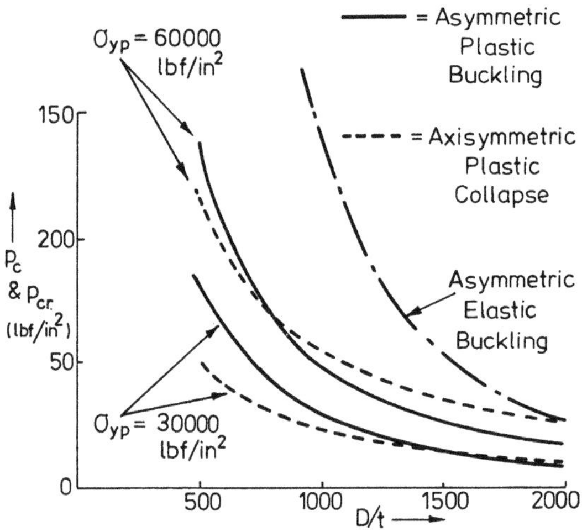

Fig. 8. Curves of p_{cr} and p_c (from Eqs. (4), (5) and (8)) for a steel torispherical shell and two values of σ_{yp}

Thus, for perfect steel torispheres subjected to internal pressure, the controlling mode of failure will be either plastic buckling or plastic collapse. It is only for steel heads with large values of D/t, or with high yield points, that elastic buckling needs to be considered. Also, over the range of D/t's considered, plastic collapse is more likely to occur than plastic buckling if the yield point is low.

For aluminium heads, the situation is a little different from that outlined above. For one thing, elastic buckling is more likely to occur with aluminium shells than with steel shells - due to the lower value of the modulus of elasticity, E. This point is illustrated in Fig. 4 of [26] where elastic buckling occurs for D/t = 1000 and σ_{yp} > 50,000 lbf/in^2 (see also [4] for

2:1 ellipsoidal shells). Another difference is that the plastic p_{cr}'s for 2:1 ellipsoidal shells appear to be lower than the plastic collapse pressures, p_c, over the range of D/t's normally encountered in practice [3, 43].

Although Fig. 8 was obtained using the approximate Eqs. (4),(5) and (8), similar curves would have been obtained if the computer program BOSOR 5 had been used. Such curves, for 2:1 ellipsoidal shells, are given in [44].

Some vessels in service experience cyclic operating conditions and, thus, their structural design should be checked to see if shakedown occurs. For ellipsoidal heads, and using small-deflection shell theory, this problem has been examined by Fox et al. [45]. However, the conclusion in that case was that shakedown is not the controlling criterion in design [34]. Whether this conclusion needs modifying for shells which undergo large deflections is not known at the moment. However, in this connection, it should be noted that the prebuckling deflections in the toroidal part of the torisphere (i.e. where the buckling occurs) are very much smaller than those at the crown of the head (see Fig. 11 of Ref. [3]).

Ductile fracture occurring simultaneously with the formation of a buckle has also been observed on one or two models [15]. However, it does not seem to have been a problem with vessels in service.

COMPARISON OF THEORY AND EXPERIMENT

a. Small Machined Models

While the main concern of this paper is with fabricated torispheres, it is of interest to know how the buckling theory for perfect shells agrees with experimental results on some small (0.137 m diameter) 'near-perfect' machined torispheres. These models were machined from solid billets (15 cm. diameter) of an aluminium alloy and the wall thicknesses of the models were about $^1/4$ mm and $^1/8$ mm. Most of the models were $^1/4$ mm thick and had a D/t ratio of ≈ 540; also, a constant thickness in the hoop direction was not achieved with the $^1/8$ mm models.

The UMIST (University of Manchester Institute of Science and Technology) tests [14], [19] have been analysed by Lagae and Bushnell [22] using the BOSOR 5 program. While there were some discrepancies between theory and the test results, the agreement overall was reasonably good (p_{expt}/p_{theo} varied from 1.05 to 0.82). The experimental/theoretical ratios found in the Liverpool tests on small machined aluminium alloy models were broadly similar [20] (i.e. 0.98 to 0.90) although rather higher values were found for some mild steel models.

If, instead of using the BOSOR 5 program, the simple approximate equation for p_{cr} is used (similar to Eq. (5) but for aluminium models) then the ratio p_{expt}/p_{theo} for the UMIST DA-models ranged from 1.19 to 0.90 [27]. When a similar procedure was adopted for the Liverpool aluminium models, the ratio rose to 1.40. This seems a little high and indicates that further study of the approximate equations for aluminium shells is needed.

b. The Nottingham One-Piece Spun Torispheres

Torispheres which are pressed and spun from one sheet of material do not have any residual welding stresses in them which have to be taken into account. Some will arise at the welded junction between the head and the cylindrical body and these will be axisymmetric (as will any residual forming stresses). If the magnitude and distribution of these residual stresses were known, and the effect of the forming operation on the mechanical properties was also known, then it is possible that accurate stress and buckling analyses of the head could be undertaken. A suitable computer program for this task would probably be BOSOR 5, as it has already been utilised for a similar problem with ring-stiffened cylinders [46], [47].

Even though some measurements have been made of residual stresses in spun heads, there is still not enough information available to enable one to proceed with such analyses. What will be done in the analyses which follow is, therefore, to ignore all residual stresses. Some account will, however, be taken of the effect of thinning on the stress-strain properties of materials which strain-harden (e.g. austenitic stainless steel). It should be noted that it is not yet known whether residual stresses are important in this buckling problem; an analysis of buckling results for spun heads made from mild steel might help to shed light on the topic.

In the following, buckling results for four 1.37 m diameter austenitic stainless steel spun heads which were tested at Nottingham University will be considered. The four heads were one-piece torispheres, their D/t-ratios were almost the same (≈ 450) and other details are given in Table III (t_T is the average thickness in the torus). The test results were first reported in 1975 [13] and two papers summarising this work have just appeared [35a, b].

There is a wealth of experimental information in refs. [13] and [35] but not, unfortunately, much detailed data on the stress-strain properties of the as-fabricated shell wall in the region of the knuckle. What these authors did was to take strips of the 'as-received' plate, roll them down to various thicknesses and then measure their properties (hardness, 0.2% proof stress, etc.) in this state. Complete stress-strain curves for one or two of the reduced thickness strips were also determined.

As mentioned before, the amount of thinning experienced in a spun head varies along the meridian, reaching a maximum in the knuckle region. The mechanical properties of the stainless steel will also vary along the meridian but by how much is not known for the Nottingham heads. However, since buckling is confined to the knuckle region, it seemed reasonable to proceed on the assumption that the mechanical properties for this portion of the head are the relevant ones for a buckling analysis. Due to the lack of detailed information, it was also assumed in the analyses that the stress-strain properties of each head were constant and were those appropriate to the average thickness (and thinning) in the toroidal portion of the torispheres. More details may be found in [5].

Using the foregoing in the BOSOR 5 program, the predictions of p_{cr} shown in Table III were obtained. As may be seen, the agreement between theory and experiment (bearing in mind the various assumptions made) is good for heads 3, 4 and 5, with theory being about 18% higher than the test results. The predicted result for head 6 was no buckling, when the flow theory of plasticity option was used in BOSOR 5 (although use of the deformation theory of plasticity predicted buckling at 236 lbf/in^2). However, with a more realistic modelling of the thickness variation in this head (and possibly the mechanical properties) it is possible that the flow theory option in BOSOR 5 would then predict buckling.

Head No.	D/t_T	r/D	R_s/D	Internal Buckling Pressures, lbf/in^2		Theory / Expt.
				BOSOR 5, using stress-strain curve	Expt.	
3	458	0.111	1.0	288	248	1.16
4	450	0.074	1.0	233	198	1.18
5	446	0.074	0.83	327	278	1.18
6	443	0.074	0.78	No Buckling	278	-

Table III. The 54 in. diameter Nottingham stainless steel spun heads. A comparison of experimental and theoretical results (all residual stresses neglected).

From the above, it seems reasonable to conclude (if only tentatively) that use of the BOSOR 5 program, together with the strain-hardened mechanical properties corresponding to the average thickness in the knuckle, satisfactorily predicts the buckling pressures of one-piece torispheres subjected to internal pressure. Since all axisymmetric residual stresses were neglected in the analysis, it could be the case that the agreement between theory and experiment is fortuitous. Clearly, more work on the subject is needed. In the interim, the more optimistic outlook will be adopted.

c. The Other Nottingham Heads

In all, there were 17 torispherical stainless steel heads tested at Nottingham University. Eleven of these were pressed and spun and six were crown and segment heads. Four 1.37 m diameter spun heads were discussed in the preceding section and, of the other seven, five were 2.74 m in diameter and two were 2.06 m. With the six crown and segment ends there were two each of three diameters, viz. 1.37 m, 2.06 m and 2.74 m.

If one is willing to neglect all the residual welding stresses, then it would be possible to analyse the other 13 Nottingham heads using the procedure outlined in the previous section (with no strain-hardening for the crown and segment ends). However, instead of doing this, a simpler method of analysing the ends will be outlined which should be of interest to designers.

A SIMPLE PROCEDURE FOR ESTIMATING THE PLASTIC p_{cr}'s OF FABRICATED TORISPHERES

It was shown above that use of the BOSOR 5 program, together with the stress-strain curve appropriate to the average thickness (and thinning) in the torus, gave reasonable predictions of p_{cr}. However, instead of using the program, it would be much simpler for designers if an approximate equation like Eq. (5) could be used to determine p_{cr}. As that equation was derived for perfect, constant-thickness steel torisphere/cylinder combinations made from elastic-perfectly plastic materials, it is necessary to decide what should be used for t and σ_{yp} in that equation when applying the equation to fabricated torispheres. The author has suggested [5] using t_{avg} in the torus for t and the 0.2% proof stress for σ_{yp}. Since there is not much thinning in crown and segment ends, it would seem appropriate to use the 0.2% proof stress of the 'as-received' plate material (in the fully annealed condition) for them. As there is considerable thinning in spun ends, a reasonable suggestion for these cases would be to use the strain-hardened 0.2% proof stress corresponding to the average reduction in thickness in the knuckle.

Utilising the above proposals, the author analysed all the Nottingham heads and the results are shown in Table IV (see [5] for more details).

As may be seen from Table IV, the experimental p_{cr} was greater than the theoretical value predicted by Eq. (5) for all eleven pressed and spun heads and for four (out of six) crown and segment ones.

It can thus be seen that if, for design purposes, the theoretical p_{cr}'s given by Eq. (5) were to be divided by 1.50 for spun ends and 1.85 for the crown and segment ones, then the two lowest ratios of experiment/design would be 1.545 (for Head 8) and 1.52 (for Head 7). If this minimum ratio of 1.5 could always be guaranteed, then one would have a satisfactory design procedure for the prevention of buckling in internally-pressurised torispherical shells.

Spun Heads

Head No.	3	4	5	6	8	9	10	11	12	16	17
$\frac{\text{Expt. } p_{cr}}{\text{Theo. } p_{cr}}$	1.27	1.43	1.67	1.59	1.03	1.28	1.32	1.23	1.72	1.32	1.22

Crown and Segment Heads

Head No.	1	2	7	13	14	15
$\frac{\text{Expt. } p_{cr}}{\text{Theo. } p_{cr}}$	1.19	1.27	0.82	1.03	1.00	0.90

Table IV. The 17 Nottingham stainless steel heads. A comparison of the experimental results with the predictions of Eq. (5). (Strain-hardening included approximately).

Incidentally, it should be noted that the ratio expt. p_{cr}/ theo. p_{cr} was less than 1.0 for the spun heads 3, 4 and 5 when the BOSOR 5 program plus the stress-strain curve was used for the analysis - see Table III.

A POSSIBLE DESIGN PROCEDURE FOR THE PREVENTION OF BUCKLING IN INTERNALLY-PRESSURISED TORISPHERES

A designer will know the nominal geometric ratios of any proposed dished end, as well as knowing its minimum mechanical properties and the proposed method of manufacture, i.e. spun or crown and segment. It would be very useful if an approximate method for predicting the internal plastic buckling pressure of fabricated torispheres were available which involved only a knowledge of the foregoing items. This objective will be discussed in the following paragraphs.

As was the case for the Nottingham heads in the previous section, the starting point will be an approximate equation for p_{cr} obtained for perfect constant-thickness torispheres made from elastic-perfectly plastic material, e.g. Eq. (5), which is for steel shells.

In the proposed method it is assumed that:

(i) any out-of-circularity caused by welding the petals of crown and segment ends may be neglected,

(ii) any residual welding or forming stresses may be ignored,

(iii) the materials property to be used as σ_{yp} for strain-hardening materials is the minimum specified 0.2% proof stress in the fully-annealed condition (it may turn out subsequently that another materials property is more appropriate, e.g. the 0.1% proof stress),

(iv) any variation in thickness along a meridian of a spun end is neglected and the relevant thickness is the thickness of the toroidal portion. This latter thickness is taken to be the nominal thickness x 0.9, i.e. a 10% reduction in thickness is assumed to occur in the torus (the factor of 0.9 is an approximate one which may be modified after further investigation; from Table 10 in [5], a factor of 0.9 seems reasonable for the reduction in thickness of the 108 in. dia. models),

(v) the relevant thickness for crown and segment ends is the nominal thickness (since there is very little thinning during manufacture),

(vi) to allow for the increase in proof-stress caused by the forming operation with spun ends, a factor of 1.5 is applied to the 0.2% proof stress in (iii), and

(vii) the p_{cr} obtained using Eq. (5), in conjunction with the foregoing items, is divided by 1.75 to arrive at the design pressure.

The above factors 1.5 and 1.75 were suggested in the written discussion of [5]. From the figures given in the preceding section of this paper, the magnitudes seem reasonable (for spun ends, at least).

As the manufacturing processes, and the residual stresses, are different for spun heads and crown and segment ones, it is not necessarily the case that exactly the same design procedure will apply to both types of head. Thus, in what follows, the two types will be tabulated separately.

The vessels which will be analysed using the above procedure are:

i. The 17 stainless steel heads tested by Stanley et al. at Nottingham University [35] and which have been given the prefix S in Table V, and

ii. The 10 stainless steel torispherical designs reported by Kemper [4] and which have been given the prefix K in Table V. Actually, buckling occurred in only five of the ten designs reported and, even for these, 90% of the heads manufactured to those designs did not buckle.

For both sets of heads, the geometric ratios used in the calculations which follow are the nominal ones (the thickness being modified according to (iv) above where necessary). However, the geometries of the heads in [35] were measured and the authors intend publishing the information in due course. When this is done, it will be interesting to find out (i) how the nominal and actual geometric ratios compare for these vessels and (ii) how the results in Table V are changed. The data should also be useful in assessing the manufacturing tolerances which should be permitted for dished ends.

The austenitic stainless steel used in both the Kemper and the Stanley heads was high-proof 304 S 65 (except for K4 which was 321 S 12). The minimum specified 0.2% proof stress for the material in the fully-annealed condition is 42,500 lbf/in^2 and this was taken as σ_{yp} for the crown and segment ends.

In addition to determining the internal buckling pressures, p_{cr}, of the vessels via Eq. (5), it was also decided to calculate the plastic collapse pressures, p_c, (from Eq. (8)) and the Drucker-Shield limit pressures, p_{DS}. These axisymmetric collapse pressures will be discussed after the asymmetric buckling pressures.

DISCUSSION OF RESULTS

The results of the calculations for p_{cr}, p_c and p_{DS} are shown in Table V. The margins of safety against buckling (Y/X) are shown in the last column).

a. <u>Internal Asymmetric Buckling Pressures</u>

Head No.	D/t	r/D	R_s/D	Theoretical (lbf/in^2) p_{DS} Eq.(9)	p_c Eq.(8)	p_{cr} Eq.(5)	$X = \frac{p_{cr}}{1.75}$ lbf/in^2	Y = Expt. p_{cr} lbf/in^2	Y/X
PRESSED AND SPUN HEADS (σ_{yp} taken as 1.5 x 42,500 lbf/in^2)									
S3	467	0.111	1.0	96	152	173	98.9	248	2.52
S4	467	0.074	1.0	69	100	123	70.3	198	2.82
S5	467	0.074	0.83	93	122	151	86.3	278	3.22
S6	467	0.074	0.78	102	131	162	92.6	278	3.00
S16	700	0.074	1.0	32	64	66	37.7	95	2.52
S17	700	0.074	0.83	47	79	82	46.9	107	2.28
S8	933	0.111	1.0	38	71	60	34.3	70	2.04
S9	933	0.074	1.0	29	47	43	24.6	62	2.52
S10	933	0.074	0.83	35	57.5	52.5	30.0	78	2.60
S11	933	0.074	0.72	43	67	61.5	35.1	86	2.45
S12	933	0.056	1.0	25	35	34	19.4	66	3.40
K5	1160	0.083	1.0	22	41.5	34	19.4	33	1.70
K9	994	0.095	0.93	36	61.5	52	29.7	NBO(34)	-
K10	811	0.103	0.82	54	96	87	49.7	NBO(53)	-
CROWN AND SEGMENT HEADS (σ_{yp} taken as 42,500 lbf/in^2)									
S1	420	0.167	1.0	105	154	215	123	280	2.28
S2	420	0.167	1.0	105	154	215	123	278	2.26
S14	630	0.167	1.0	61	99	116	66.3	120	1.81
S15	630	0.167	1.0	61	99	116	66.3	107	1.61
S7	840	0.167	1.0	43	72.5	74.5	42.6	60	1.41
S13	840	0.167	1.0	43	72.5	74.5	42.6	82	1.92
K1	825	0.159	0.91	47	78	81.5	46.6	53	1.14
K2	880	0.163	1.0	40	67	68	38.9	46	1.18
K3	915	0.166	1.0	39	65.5	65	37.1	42	1.13
K4	730	0.162	0.89	55	93	102	58.3	70	1.20
K6	575	0.146	0.83	79	117	146	83.4	NBO(53)	-
K7	730	0.166	1.0	51	84	92	52.6	NBO(50)	-
K8	915	0.159	0.91	43	69.5	69.5	39.7	NBO(40.5)	-

Table V. Fabricated stainless steel heads from Refs. [4] and [13]. A comparison of experimental first buckle pressures with some theoretical predictions.

Note: NBO = no buckling observed. Numbers in parentheses are maximum test pressures.

(i) Spun Heads

As may be seen from Table V, the use of the proposed procedure resulted in margins of safety which were greater than 1.5 for all the stainless steel spun heads which buckled. This can be considered as being very satisfactory, even though some margins of safety were rather high (e.g. 3.40 for S12).

(ii) Crown and Segment Heads

For these heads, only one of the S-heads had a margin of safety which was less than 1.5, i.e. S7. For the K-heads which buckled, the average margin of safety was about 1.16. Thus, the procedure is almost satisfactory for the S-series of crown and segment heads but it gives safety margins for the K-series which are a little low.

(iii) Spun Versus Crown and Segment Heads

It is clear from Table V that the average safety margin for the spun heads is greater than that for the crown and segment heads. This is the case for both the S-series and the K-series. It is not known why this is so, although the crown and segment heads have more non-axisymmetric welds in them (at the petal junctions). The circularity of these heads around the various latitudes may also not be as good as the circularity of the spun heads.

One way of increasing the margin of safety for the crown and segment heads would be to divide the theoretical p_{cr} by a factor greater than 1.75. If 2.25 were to be chosen, then the safety margin for the K-series would become $2.25/1.75 \times 1.16 \approx 1.5$. Another approach would be to use the 0.1% proof stress for σ_{yp} instead of the 0.2% proof stress. This would have the effect of decreasing the theoretical p_{cr} and, hence, X ($=p_{cr}/1.75$).

(iv) The S-Series Versus the K-Series

The stainless steel heads for the K- and S-series were made by the same manufacturer (i.e. A.P.V., Crawley). The K-heads were production vessels whereas the S-heads were one-off heads made for Nottingham University. It is possible that the quality control checks made during the manufacture of the S-heads were more stringent than those made during the manufacture of the K-heads (for which between 10 and 50 vessels were made to each design). However, the S-heads were supposed to be manufactured to normal practice.

Another point, which has not been investigated so far (although it would not be difficult to do using BOSOR 5), is how the thickness of the cylindrical body affects the magnitude of the theoretical p_{cr}'s. Eq. (5) assumes a constant wall thickness throughout the shell combination and all the K- and S-heads may not satisfy this requirement.

Also, as noted earlier, 90% of the vessels built to the K1 to K5 designs did not buckle. If these latter tests had been continued until buckling occurred, the average margins of safety for K1 to K5 would have been higher than those shown in Table V.

However, at this time, it is not known with any certainty why the margins of safety for the S-series are higher than those of the K-series.

(v) Other Torispherical Shells Which Have Buckled

In addition to the stainless steel torispheres listed in Table V, there are a few others which have buckled in service. These were mentioned in the written discussion of [5]. When the design procedure outlined herein was applied to them, it was found that the margin of safety was less than 1.5 only for the carbon steel 18.3 m dia. vessel described in [30]. The vessel was a crown and segment one and, thus, this finding is similar to what was found before for the other crown and segment heads. If, as was suggested for them, the factor of 1.75 is changed to 2.25, then the margin of safety for the 18.3 m dia. shell rises to 1.77.

b. <u>Axisymmetric Plastic Collapse Pressures and Limit Pressures</u>

In this paper the topic of principal interest is internal pressure plastic buckling of steel torispheres. However, in the structural design of these heads, account also has to be taken of the axisymmetric mode of failure in which large displacements and yield circles occur. In the literature and various Codes, the Drucker-Shield limit pressures for this failure mode are already available. In addition, the Liverpool group has recently suggested Eq. (8) for the plastic collapse pressures of steel torispheres. Having the experimental internal buckling pressures available in Table V gives a good opportunity to assess the accuracy of these other failure pressures.

Another investigation which should be mentioned in connection with limit pressures of torispheres is that of Calladine [48]. This is a novel analysis of the problem which also produced a simple equation for the limit pressures. The numerical results agreed with those given in [2].

The calculated values of p_{DS} and p_c for the K- and S- heads are given in Table V. For values of D/t greater than 700, estimates of p_{DS} were made by extrapolating the curves in [2].

As may be seen from Table V:

(a) all the axisymmetric limit pressures, p_{DS}, are lower than their corresponding plastic collapse pressures, p_c - which is not unexpected, since the limit pressures are small-deflection values,

(b) all the axisymmetric limit pressures, p_{DS}, are also lower than the experimental (and theoretical) asymmetric buckling pressures. This information should be of interest to designers and was noted in [4] for the K-heads. It should also mean that the failure modes were axisymmetric but they were not, and

(c) there is, surprisingly, fairly good agreement between p_{DS} and X ($=p_{cr}/1.75$). The ratio p_{DS}/X varied between 0.85 and 1.05 for the crown and segment heads and between 0.85 and 1.29 for the spun heads.

Turning now to the comparison of p_c and p_{cr} in Table V, it may be seen that:

(d) for the spun heads, p_c is less than p_{cr} for D/t = 467 and 700 and is greater than p_{cr} for D/t = 933, and

(e) for the crown and segment heads, p_c is less than p_{cr} for all D/t-ratios (except D/t = 915).

In view of how the controlling failure modes change with σ_{yp} (see Fig. 8) points (d) and (e) are perhaps not surprising. Nevertheless, if the values of p_c are reasonably correct, then there should have been a number of heads which failed axisymmetrically; this did not, however, happen. As p_c was selected (arbitrarily) as the internal pressure to cause twice the yield point deflection, this means it will be a conservative prediction and may explain part of the discrepancy. In addition, the measured value of the 0.2% proof stress of the 'as-received' plate material for the S-heads was 49,100 lbf/in^2. If this value is used for σ_{yp} in the calculations for Table V, it then turns out that $p_c > p_{cr}$ for the spun heads with D/t = 700. This is now in accord with experiment but there are still a number of heads for which $p_c < p_{cr}$. If these discrepancies are to be pursued any further, one would need to know the measured values of the geometric ratios and it might be necessary to resort to a computer program like BOSOR 5. Such a study will have to be left to another time.

With respect to the predictions of the simple approximate equations for p_c and p_{cr}, and comparisons with experimental results, the only other information known to the author refers to small (0.137 m. dia.) machined models in [27]. For those models, p_c was always greater than p_{cr} and this agreed with the test results.

CODE RULES FOR THE PREVENTION OF BUCKLING

As far as the author is aware, there are no Codes at present which cover the internal pressure buckling of torispheres made from strain-hardening materials. Even if attention is restricted to carbon steels, there are only one or two Codes which consider the buckling problem.

One of these guides is the German AD-Merkblatt B3 [49], which considers the Korbbogen and Klöpper types of torispherical head. Curves for determining the elastic buckling pressures of both types of head are given in the AD-Merkblatt but plastic buckling does not seem to be treated. It is also not stated whether the small- or the large-deflection theory of shells was used in determining the elastic p_{cr}'s.

The Americans also have rules covering the design of large, welded, low-pressure storage tanks [50]. Buckling does not seem to be considered explicitly in these rules but the allowable design stress is affected by the ratio of the tensile meridional stress to the compressive hoop stress.

It is also understood (see the discussion of [5]) that the new version of the French Pressure Vessel Code (CODAP) will contain formulae for the internal pressure buckling of torispherical heads. The formulae are supposed to be similar to Eq. (5) but the author has not seen them as yet.

The British Standards Institution (BSI) are also considering the problem and it is hoped that rules for preventing internal pressure buckling in torispheres will be incorporated in the forthcoming update of BS 5500.

CONCLUSIONS

1. Use of the stress-strain curve appropriate to the average thinning in the torus, together with the BOSOR 5 computer program, enabled the plastic buckling pressures of several stainless steel internally-pressurised one-piece spun torispheres to be predicted satisfactorily.
2. A simple method of predicting the internal buckling pressures of fabricated torispheres, via an approximate equation, was also studied. This involved the estimation of the average thickness (or thinning) in the torus and the associated 0.2% proof stress. The procedure worked reasonably well for all 11 of the Nottingham spun heads and for 4 (out of 6) of the crown and segment ones. If the 0.1% (instead of the 0.2%) proof stress was used in the calculations for the crown and segment heads, then the procedure was satisfactory for all six of these heads.
3. A variation of the method mentioned in 2. was also investigated. The theoretical p_{cr} given by the simple equation was divided by 1.75 and strain-hardening (if relevant) was accounted for via a constant factor. The procedure was tried on the 17 Nottingham torispherical heads and on about 10 others which had failed in service. For the spun heads, the margins of safety were all greater than 1.5, which is satisfactory from a Code point of view. For the crown and segment heads, some of the safety margins were as low as

1.15. For these cases, the safety margins can be increased by dividing the theoretical p_{cr} by, say, 2.25 instead of 1.75 (another way of increasing the safety margins would be to use the 0.1% proof stress instead of the 0.2% proof stress).

As a design method, the above procedure has promise and it should now be checked (by head manufacturers, etc.) on other vessels which have failed by internal pressure buckling.

4. For the heads analysed herein (and using nominal geometries) some of the theoretical axisymmetric plastic collapse pressures, p_c, were lower than the theoretical buckling pressures, p_{cr}. This should not be the case and more study of why this occurred is needed - using measured radii of curvature, etc.
5. The axisymmetric limit pressures predicted by the Drucker-Shield analysis were all lower than the plastic collapse pressures, p_c. They were also lower (sometimes considerably lower) than the experimental pressures at which the first non-axisymmetric buckles occurred.

ACKNOWLEDGEMENTS

The photographs in Fig. 1 are reproduced with the permission of Professor P. Stanley. Fig. 2 is reproduced from Ref. [7].

REFERENCES

1. Struble, R.A.: Biezeno pressure vessel heads. J. Appl. Mech., Trans. ASME, 78 (1956) 642-645.
2. Shield, R.T.; Drucker, D.C.: Design of thin-walled torispherical and toriconical pressure-vessel heads. J. Appl. Mech., Trans. ASME, 83 (1961) 292-297.
3. Galletly, G.D.: Buckling and collapse of thin internally-pressurised dished ends. Proc. Inst. Civ. Engrs. Part 2 67 (1979) 607-626.
4. Kemper, M.J.: Buckling of thin dished ends under internal pressure. Proc. Symp. on Vessels Under Buckling Conditions I. Mech. E. London (1972) 23-32.
5. Galletly, G.D.: Plastic buckling of torispherical and ellipsoidal shells subjected to internal pressure. Proc. Inst. Mech. Engrs. 195 (1981) 329-345.
6. Galletly, G.D.: Stress failure of large pressure vessels - Recommendations resulting from studies of the collapse of a 68 ft. high x 45 ft. dia. pressure vessel. Tech. Rpt. No. 45-57 Shell Development Corp., Emeryville, California, March 1957.

7. Harding, A.G.; Ehmke, E.F.: Brittle failure of a large pressure vessel. Proc. Am. Petrol. Inst. Section 3, 42 (1962) 107-117.

8. Galletly, G.D.: Torispherical Shells - A caution to designers. J. Engng. Ind., Trans. ASME 81 (1959) 51-66. Also published in Pressure Vessel and Piping Design - Collected Papers 1927-1959. ASME, New York (1960).

9. Drucker, D.C.; Shield, R.T.: Limit analysis of symmetrically loaded shells of revolution. J. Appl. Mech., Trans. ASME, 81 (1959) 61-68.

10. Mescall, J.: Stability of thin torispherical shells under uniform internal pressure. NASA Tech. Note D-1510, Collected Papers on Instability of Shell Structures (1962).

11. Adachi, J.; Benicek, M.: Buckling of torispherical shells under internal pressure. Expt. Mech. 4 (1964) 219-222.

12. Thurston, G.A.; Holston, A.A.Jr.: Buckling of cylindrical shell end closures by internal pressure. NASA CR 540 (1966).

13. Stanley, P.; Campbell, T.D.; Cooper, D.: Strains and deformations in very thin pressure vessel ends. Tech. Rpt. Dept. of Mech. Eng. University of Nottingham (1975).

14. Kirk, A.; Gill, S.S.: The failure of torispherical ends of pressure vessels due to instability and plastic deformation - An experimental investigation. Inter. J. Mech. Sci. 17 (1975) 525-544.

15. Galletly, G.D.: Internal pressure buckling of very thin torispherical shells - A comparison of experiment and theory. Proc. 3rd Inter. Conf. Struct. Mech. Reactor Technol. (SMIRT), London (1975) Paper G2/3.

16. Bushnell, D.: Stress, stability and vibration of complex branched shells of revolution. Comput. Struct. 4 (1974) 399-435.

17. Bushnell, D.: BOSOR 5 - Program for buckling of elastic-plastic shells of revolution including large deflections and creep. Comput. Struct. 6 (1976) 221-239.

18. Aylward, R.W.; Galletly, G.D.; Mistry, J.: Large deflection elastic buckling pressures of very thin torispherical and ellipsoidal shells subjected to uniform internal pressure. Proc. 3rd Inter. SMIRT Conf., London (1975) Paper G2/4.

19. Patel, P.R.; Gill, S.S.: Experiments on the buckling under internal pressure of thin torispherical ends of cylindrical pressure vessels. Inter. J. Mech. Sci. 20 (1978) 159-175.

20. Galletly, G.D.: Some experimental results on the elastic-plastic buckling of thin torispherical and ellipsoidal shells subjected to internal pressure. Prelim. Rpt. 2nd Inter. Colloq. Stability Steel Struct., Liège (1977) 619-626.

21. Bushnell, D.; Galletly, G.D.: Stress and buckling of internally-pressurised elastic-plastic torispherical vessel heads - Comparisons of test and theory. J. Press. Vess. Tech., Trans. ASME 99 (1977) 39-53.

22. Lagae, G.; Bushnell, D.: Elastic-plastic buckling of internally-pressurized torispherical vessel heads. Nucl. Eng. and Des. 48 (1978) 405-414.

23. Kanodia, V.L.; Gallagher, R.H.; Mang, H.A.: Instability analysis of torispherical pressure vessel heads with triangular thin-shell finite elements. J. Press. Vess. Tech., Trans. ASME 99 (1977) 64-74.

24. Canton, B.; Hoffman, A.; Roche, R.L.; Trochet, C.: Elastic and elastic-plastic buckling of vessel heads. Computation by the CEASEMT system. Proc. 4th Inter. SMIRT Conf., San Francisco (1977) Paper G7/4.

25. Rensch, H.J.; Wunderlich, W.: A semi-analytical finite element procedure for non-linear elastoplastic analysis of arbitrarily loaded shells of revolution. Proc. 6th Inter. SMIRT Conf., Paris (1981) Paper M4/1 North Holland Publishing Co.

26. Galletly, G.D.; Radhamohan, S.K.: Elastic-plastic buckling of internally-pressurised thin torispherical shells. J. Press. Vess. Tech., Trans. ASME 101 (1979) 216-225.

27. Radhamohan, S.K.; Galletly, G.D.: Plastic collapse and the controlling failure modes of internally-pressurised thin torispherical shells. J. Press. Vess. Tech., Trans. ASME 101 (1979) 311-320.

28. Galletly, G.D.: A comparison of the plastic buckling behaviour of 2:1 ellipsoidal and 2:1 torispherical shells subjected to internal pressure. Proc. IUTAM Symp. on Physical Non-Linearities in Structural Analysis, Senlis France May 1980.

29. Galletly, G.D.: Elastic-plastic buckling of dished ends subjected to internal pressure. Proc. 6th Inter. SMIRT Conf. Paris (1981) Paper G4/6.

30. Fino, A.; Schneider, R.W.: Wrinkling of a large thin code head under internal pressure. Bull. Weld. Res. Coun. 69 June 1961 11-13.

31. Stennett, R.: Gummed-up valve causes vessel collapse. Chart. Mech. Eng. October 1970 404.

32. Esztergar, E.P.: Development of design rules for dished pressure vessel heads. Bull. Weld. Res. Coun. No. 215 May 1976 1-34.

33. Stanley, P.: Photoelastic study of Cassinian pressure vessel end closures. Proc. 1st Inter. Conf. on Press. Vessel Tech., Delft (1969) 355-368 publ. by ASME, New York.

34. Mansfield, E.H.: An optimum surface of revolution for pressurised shells. Int. J. Mech. Sci. 23 (1981) 57-62.

35a. Stanley, P.; Campbell, T.D.: Very thin torispherical pressure vessel ends under internal pressure: Test procedure and typical results. J. Strain Anal. 16 (1981) 171-186.

b. Ibid. ....: Strains, deformation and buckling behaviour. J. Strain Anal. 16 (1981) 187-203.

36. Adachi, J.: Stresses and buckling in thin domes under internal pressure. U.S. Army Materials and Mechanics Research Center, Watertown (1968) Rpt. MS68-01.

37. Bushnell, D.: Non-symmetric buckling of internally-pressurised ellipsoidal and torispherical elastic-plastic pressure vessel heads. J. Press. Vess. Tech., Trans. ASME 99 (1977) 54-63.

38. Brown, K.W.; Kraus, H.: Stability of internally-pressurised vessels with ellipsoidal heads. J. Press. Vess. Tech., Trans. ASME 98 (1976) 157-161.

39. Wunderlich, W.: See written discussion of Ref. [5].

40. Unger, C.; Mang, H.A.: Zum spannungs- und stabilitäts-problem von kesselböden unter innendruck. Der Stahlbau, 12/1980, 373-379.

41. Galletly, G.D.: Elastic and elastic-plastic buckling of internally-pressurized 2:1 ellipsoidal shells. J. Press. Vess. Tech., Trans. ASME 100 (1978) 335-343.

42. Aylward, R.W.; Galletly, G.D.: Elastic buckling of, and first yielding in, thin torispherical shells subjected to internal pressure. Int. J. Press. Vess. Piping 7 (1979) 321-336.

43. Galletly, G.D.; Aylward, R.W.: Plastic collapse and the controlling failure pressures of thin 2:1 ellipsoidal shells subjected to external pressure. J. Press. Vess. Tech., Trans. ASME 101 (1979) 64-72.

44. Galletly, G.D.: Elastic-plastic buckling and collapse of ellipsoidal and torispherical shells subjected to uniform internal pressure. Proc. 3rd IUTAM Symp. on Shell Theory, Tbilisi August 1978. North Holland Publishing Co. (1980).

45. Fox, J.D.; Kraus, H.; Penny, R.K.: Shakedown of pressure vessels with ellipsoidal heads. Proc. Inst. Mech. Engrs. 186 (1972) 431.

46. Bushnell, D.: Effect of cold bending and welding on buckling of ring-stiffened cylinders. Comput. Struct. 12 (1980) 291-307.

47. Odland, J.: On the strength of welded, ring-stiffened cylindrical shells primarily subjected to axial compression. Dr.Ing. thesis, University of Trondheim, Norway June 1981.

48. Calladine, C.R.: Lower bound analysis of symmetrically-loaded shells of revolution. Proc. 1st Inter. Conf. on Press. Vessel Tech., Delft (1969) 335-344. Publ. by ASME, New York.

49. AD-Merkblatt B3: Dished ends subjected to internal or external pressure. Verein der Tech. Uberwachungs-Vereine e.v. (VdTUV), Essen, 1977.

50. API: Recommended rules for design and construction of large, welded low-pressure storage tanks. Standard 620 6th ed., Rev. 2 December 1980.

BUCKLING STRESSES OF SHELLS HAVING NEGATIVE GAUSSIAN CURVATURE

W. ZERNA and I. MUNGAN

Institut für Konstruktiven Ingenieurbau
Ruhr-Universität Bochum
Bochum, West-Germany

Summary

In addition to geometric parameters buckling of shells of negative Gaussian curvature depends mainly on the membrane stress state present. According to buckling tests the effect of boundary conditions is relatively low whereas the numerical results are highly dependent on the boundary conditions. Again according to tests the hyperboloidal shell has a higher buckling resistance than a cylindrical shell having the same throat radius. The buckling resistance of hyperboloidal shells against circumferential compression can be increased by arranging stiffening rings of adequate size and number. Due to nonlinear behavior of reinforced concrete the buckling load factor drops to the half or less of the value obtained assuming a linear elastic behavior. This high reduction is mainly due to orthotropy and descending tangent modulus at the near of ultimate load of reinforced concrete cooling tower shells.

Introduction

Hyperbolic natural-draught cooling towers are already the largest reinforced concrete shell structures being built. Today there is a tendency to ever increasing heigth and diameter of the cooling towers keeping the wall thickness of the shell as small as possible for economy. For this reason stability behavoir is a vital concern in design of the cooling tower shell having negative Gaussian curvature.

Although the buckling of cylindrical an spherical shells has been investigated from the beginning of this century the history of the buckling investigations on shells of negative Gaussian curvature is only 15 years young. One of the early researchs is done by Stein and McElman [1] who calculated the buckling load of bowed-in and bowed-out toroidal segments

Buckling of Shells, Ed.: E. Ramm
Proceedings of a State-of-the-Art Colloqium

under lateral or hydrostatic external pressure. On the basis of classical buckling analysis the buckling load of the bowed-out shells was several orders of magnitude larger than that of the bowed-in configuration. Hutchinson[2] studied the imperfection sensitivity of the same shells within the framework of Koiter's general theory of initial post-buckling behavior [3]. It turned out that the bowed-in shell has a lower imperfection-sensitivity associated with its considerably lower classical buckling load.

The catastrophic collapse of some of the cooling towers in Ferrybridge/England in November 1965 gave impulse to the intensive resarch on the buckling of hyperbolic cooling tower shells at the end of the sixties. Till today the most extensive tests on cooling tower shaped models in wind tunnel have been carried out by Der and Fidler [4] . Using both electroformed copper and molded PVC models Der and Fidler measured the dynamic head required to buckle their models and derived from the test results Formula (1) for the buckling pressure of a hyperboloid with proportions similar to the models

$$q_{cr} = c \cdot E \left(\frac{t}{R_T}\right)^n \qquad (1)$$

In Formula (1) E is the modulus of elasticity, T wall thickness of the shell and R_T throat radius of the hyperboloid. Der and Fidler give for n the value 2.3 and for c an average value of 0.064. Ewing [5] treated the stability problem of cooling tower shells analytically for axisymmetric pressure and non-axisymmetric wind load and found the value of $7/3$ for n .

Veronda and Weingarten [6] have tested a large number of hyperboloidal shell models made of PVC sheets by bonding along a longitudinal seam. Due to thermoforming the models were getting thinner with increasing distance from the throat. On the whole the models were very likely afflicted with initial imperfections in the shape due to relatively small wall thickness which was only 0.25 mm. The buckling began with a single dimple which mostly shapes near the upper boundary, where the models were thinnest [6] . The conclusion drawn from the tests and the theoretical analysis carried out by the authors is that,shells of negative Gaussian curvature exhibit low grade imperfection sensitivity. Values predicted applying the bifurcation theory are remarkably close to test results.

The goal of the research work carried out in Bochum was to obtain a more detailed information about the buckling states of shells of negative Gaussian curvature. Both theoretical studies and model tests on nearly perfect and relatively large models were carried out in order to obtain results which can be generalized and used in design of large cooling tower shells having 160 m or more height, nearly 140 m base diameter and standing under the action of axisymmetric dead weight together with non-axisymmetric wind load. Due to its limited frame the present work will focus on the results obtained without going into detail in the method of theoretical analysis and testing. For the theory References 7, 8, 9, 10, 11, 12, 13 and for the experimental work References 11, 14, 15, 16 are recommended.

Buckling Stress States

The advantage of operating in terms of "Buckling Stress States" instead of "Buckling Loads" is the more general character of the first. Shells are supposed to be subjected to various loading conditions inducing a large variety of biaxial stress states in the shell during their life. Having obtained the buckling stress states of a shell its stability can be checked independent of the loading conditions. This is the main idea behind the new approach to handle the shell buckling problem [15, 17, 18].

Although the buckling theory itself is straight forward its numerical realisation includes some deficiency caused through the assumptions made [6, 12]. In most cases a perfect shell geometry and an axisymmetric loading together with a periodic buckling configuration are assumed. In fact shells exhibit irregular imperfections. If there are no restrictions with respect of the shell displacements the buckling configuration may be localized at the region having the worst imperfection that initiates the first dimple. Buckling tests of Berke and Carlson [19] on complete spherical shells demonstrates the effect of the deformation control. The first inward dimple gets deeper until the deepest point of the dimple contacts a wax mandrel placed into the shell. At this moment further inward motion of the first dimple is prevented and a large number of small dimples develop rapidly at other sites, resulting in a nearly periodic buckling configuration.

The boundaries of the buckling patterns are flexible; i.e. the mode shapes developed are not confined to immovable boundaries. By changing the distance between the shell and the wax mandrel the size and number of the dimples change progressively. This transition behavior is in full agreement with the observations done in our tests on axisymmetrically loaded hyperboloidal models, only with exception the tests under uniaxial compression in meridional direction $\sigma_{11} \approx 0$ where again the single dimple got always deeper and no other dimples could be observed. From these observations it can be concluded that, after the first dimple shapes the development of the buckling configuration depends substantially on the deformation capability present. The deformation capability of the shell structure or model itself is influenced through the type of loading and boundary conditions or method of testing, respectively. For example a hyperbolic cooling tower shell under wind loading alone can not shape a periodic buckling configuration because the first inward dimple initiating the buckling at the wind pressure side gets always deeper due to external suction at the flanks and at the back side.

Fig. 1 shows a cooling tower model in wind tunnel. Without a stiffening ring at the top the buckling configuration consists of only one dimple caused through compressive stresses in circumferential direction. Fig. 2 shows the same model after attaching a very stiff steel ring to the top. In this case, after a certain size and depth of the first dimple which is again caused through circumferential compression two smaller dimples at the lower part of the flanks can be observed. These two dimples are placed symmetrically with respect of the wind direction and are caused through the meridional compression acting. An extensive research on the buckling of shells of revolution in wind tunnel is planed.

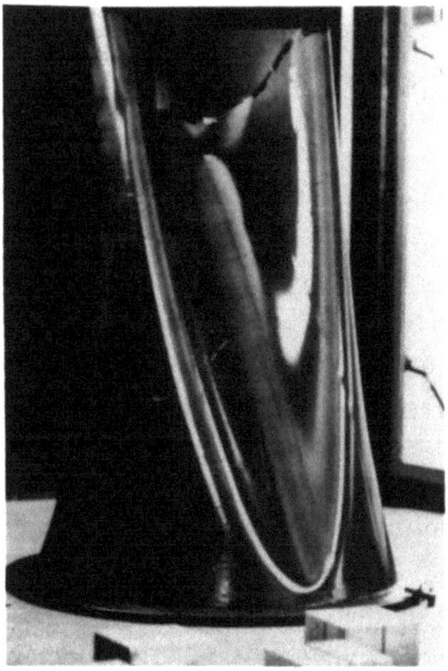

Fig. 1 Wind-Buckling of a Cooling Tower Model without Upper Edge Beam

Fig. 2 Wind-Buckling of a Cooling Tower Model with a Stiff Upper Edge Beam

Fig. 3 shows the results obtained on hyperboloidal models subjected to axisymmetric loading. In this figure also the shape and size of the models tested are given. The wall thickness of the models was nearly constant for each model. The wall thicknesses considered in the analysis of the buckling tests on 8 different models were 1.28 mm, 1.33 mm, 1.39 mm, 1.50 mm, 1.55 mm, 1.65 mm, 1.77 mm and 1.95 mm [15, 16] .

6 models were tested first in symmetrical shape S with respect of the throat. Afterwards the height of the part above the

throat was reduced from 600 mm to 220 mm to obtain cooling tower shaped models C. One model was tested only in symmetrical shape S and another one directly in cooling tower shape C. In Fig. 3 the interaction diagrams are given in terms of the circumferential and meridional stresses σ_{11} and σ_{22}, respectively, at the point where the first dimple shapes. This point was situated in each test at the region of the throat where the model was thinnest. Test results on the symmetrical models and cooling tower shaped models were of the same range. The only parameter influencing the variation in the buckling stress states was the value of the wall thickness at the region of the first dimple. For this reason test results obtained on 8 different models and 2 different shapes are extrapolated for a shell model having shape S or C, a wall thickness of 2 mm and made of a material having modulus of elasticity E = 3 450 MN/m^2 and Poissons's ratio ν = 0,38. After this extrapolation interaction diagram E is obtained, which is valid both for symmetrical and cooling tower shapes, S or C, respectively.

Additionally in Fig. 3 other 4 interaction diagrams are given in dashed lines. These are the computed interaction diagrams applying the conventional bifurcation theory. The first capital letter stands for the shape and the second stands for the boundary condition at the upper boundary considered in the numerical analysis. Boundary condition A corresponds to built-in or fixed end in adjacent state of equilibrium. In case B the boundary is still built-in, however, displacements in vertical direction are permitted. In tests the boundary condition of the top of the model was between boundary conditions A and B. Boundary condition A is stiffer than the condition present in the test. The top of the model was attached to a rigid steel plate which could sink parallelly and rotate about an axis freely if a single dimple shapes. Boundary condition A does not allow both kinds of movement in adjacent states. On the other hand boundary condition B is weaker than the test condition.According to B a vertical displacement in shape of Cos $n\Theta^1$ along the circumferential ordinate Θ^1 with any integer value of n is possible whereas according to test conditions n may be equal either to 0 or 1.

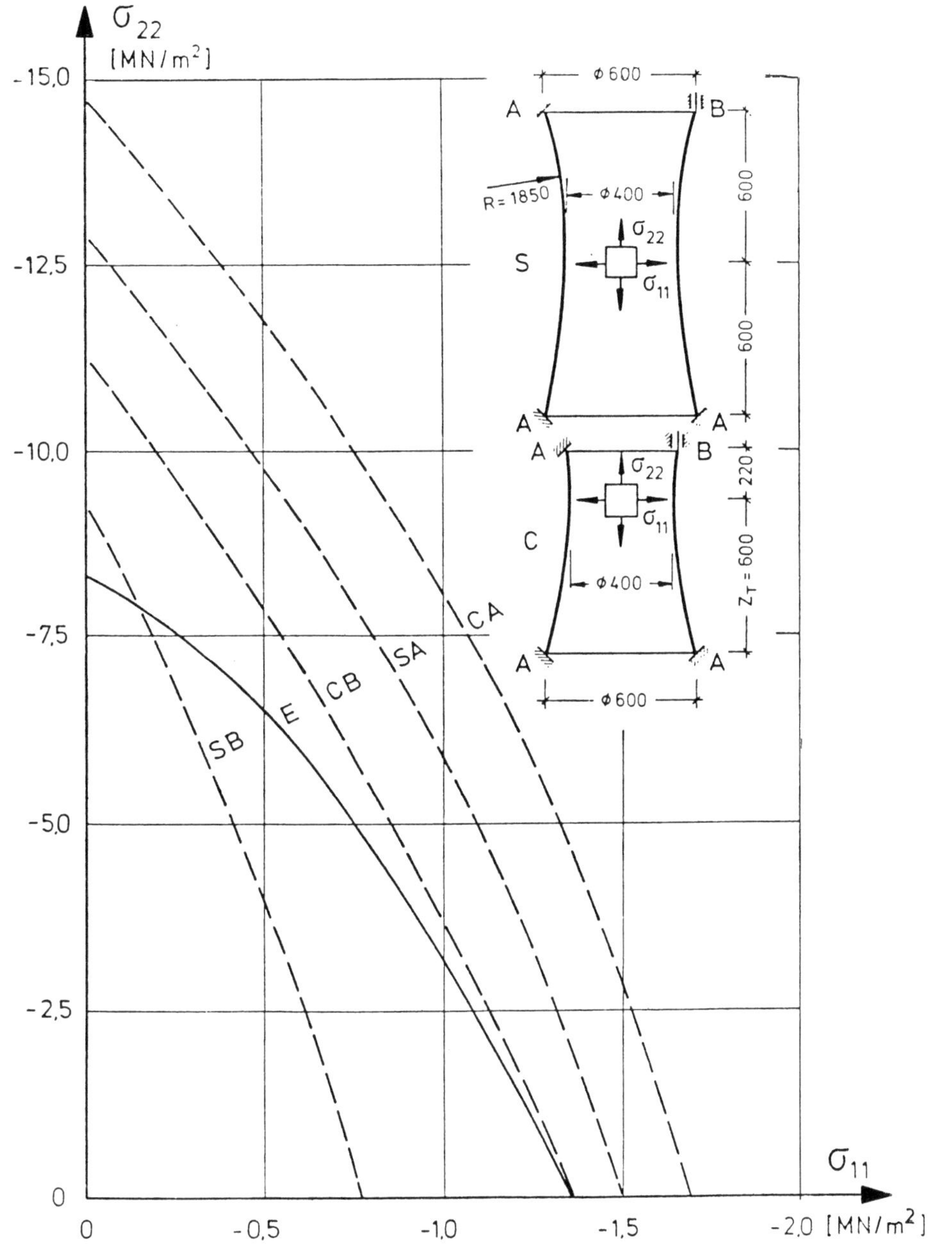

Fig. 3 Experimental and Theoretical Interaction Diagrams for the Buckling Stresses of the Models Tested.

In fact symmetrical models S can be considered as cooling tower shaped models C supported elastically at a distance 220 mm above the throat. This assumption is allowable because the first dimple shaped always away from the boundaries and at the throat. That means the computations yield four strongly deviating interaction diagrams for the same model if the boundary condition at the upper edge is altered. That is surprising because for shells of positive Gaussian curvature calculation models A and B yield the same interaction diagrams [7]. In addition, for cylindrical shells of average length there is no effect of the shell length and of the boundary conditions on the buckling stress under axial compression [14].

Comparison with Cylindrical Shells

In Fig. 4 interaction diagram E obtained experimentally on models having negative Gaussian curvature is given together with 4 interaction diagrams which are applicable to cylindrical shells C_1 to C_4. C_1 and C_2 are cylindrical shells having a radius equal to throat-radius and a length equal to the length of the cooling tower shaped models C or symmetrical models S tested, respectively. C_3 and C_4 have a radius equal to the basis-radius of the hyperboloidal models tested. Interaction diagrams C_1 to C_4 are based on the results obtained in an experimental work by applying the same testing method to cylindrical shells [14].

It is interesting to notice that in the stress state $\sigma_{11} = 0$, uniaxial compression in meridional direction, the buckling stresses of the hyperboloidal shells and of cylindrical shells C and C are equal. In the shells of negative Gaussian curvature this stress state is obtained under axial compression together with internal lateral pressure. The first dimple forms at the throat and has the shape of a flat ellipse with its long axis in the circumferential direction. For this reason the whole dimple is restricted to an area which only slightly differs from a cylinder having the throat-radius. In addition, in both cases the buckling process was initiated by means of the slight dynamic disturbance applied in the tests to check the stability in sense of Liapunov's definition.

In presence of circumferential compression the shell of negative Gaussian curvature has a higher buckling resistance than the cylindrical shells C_1 and C_2. With increasing length of the cylindrical shell the buckling resistance decreases if circumferential compression is acting. Under lateral compression the negative curvature stiffens the shell compared to cylindrical shells.

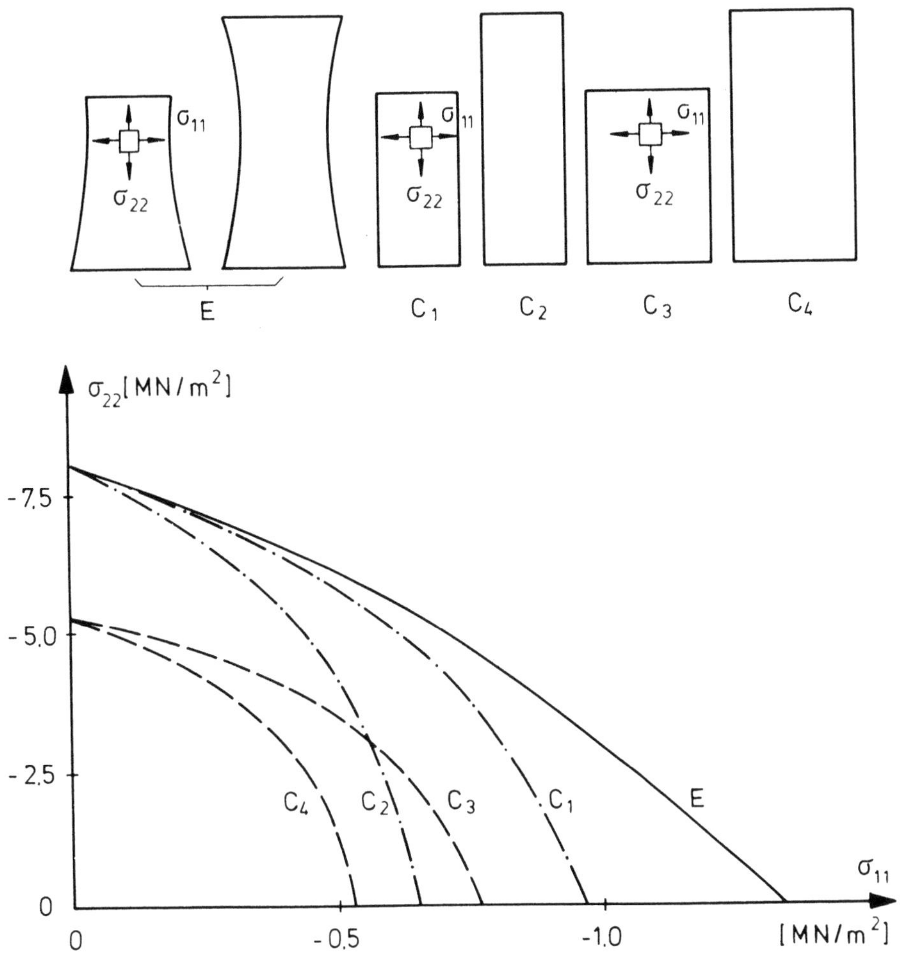

Fig. 4 Experimental Interaction Diagrams for Hyperboloidal and Cylindrical Shells

On the other hand cylindrical shell C_3 and C_4 having a radius equal to base radius of the shell of negative Gaussian curvature are in no way representative for the buckling stresses of the hyperboloidal shell because of the large deviation between interaction diagram E and each of the diagramms C_3 and C_4.

Effect of Stiffening Rings

Buckling tests show that the buckling resistance can be increased by arranging stiffening rings. Important parameters influencing the stiffening effect are the number, size and position of the rings. To study these effects some numerical and experimental investigations were carried out. In Fig. 5 the influence of the width of stiffening ring is given [9]. A 160m high cooling tower shell is stiffened by means of two rings having 0,20 thickness, one placed at the throat and the other in the middle of the distance between the throat and base. If the width of the rings is increased proportional to the local wall thickness of the shell the augmentation of bifurcation results given in diagrams of Fig. 5 is obtained. The full or

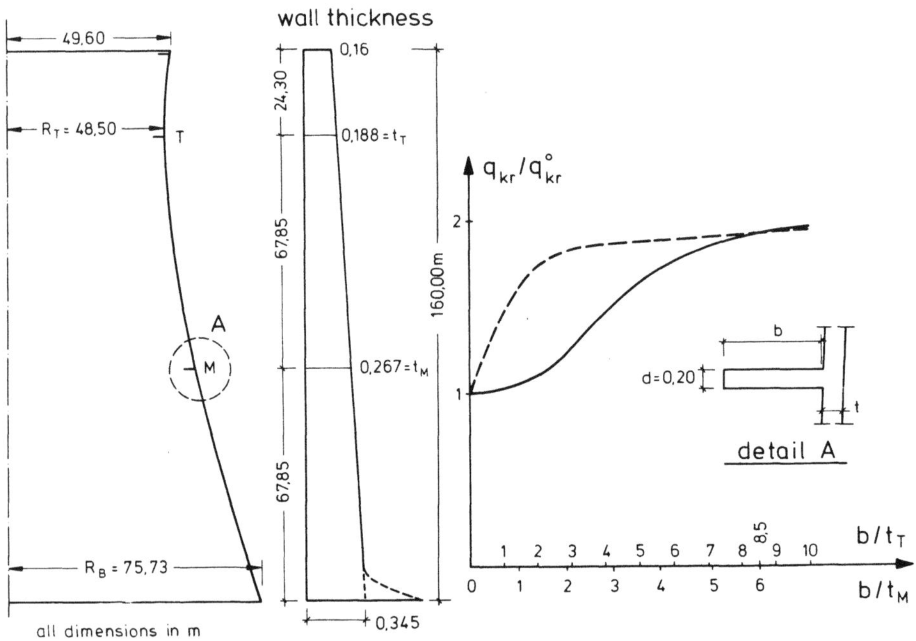

Fig. 5 Effect of Ring Width on Buckling Load According to Benz[9]

dashed curves are valid if the rings buckle in the same harmonics as the shell or never at all, respectively. For a ring width equal to 6 to 9 times the wall thickness both curves intersect, i.e. the ring does not buckle, even if it is allowed.

For cooling tower shell given in Fig. 6a the curve b) shows the slight augmentation attainable if the ring thickness is increased twofold or threefold [10] . Effect of the ring width in case of three, five or ten rings is shown in c). Again a width equal to at least 6 times of the shell thickness is optimal. According to numerical results the most effective position of a stiffening ring coincides with the peak point of the buckling configuration for the unstiffened shell. The positions of the second and third rings can be obtained by iteration successively. For a larger number of rings than five the equidistant positioning is efficient as well.

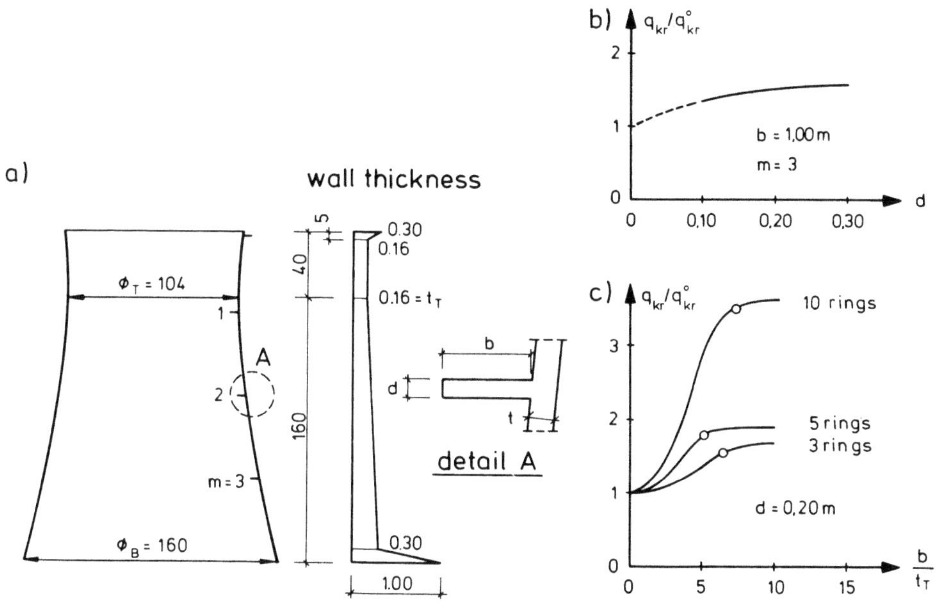

Fig. 6 Effect of Thickness, Width and Number of Rings on Buckling Load According to Lehmkämper [10].

In addition to the foregoing parametic studies some buckling tests are carried out on models stiffened by means of 5,9 or 19 equidistant rings [16]. The stiffened models have the same geometry as given in Fig. 3. The rings had 7,5mm width and 2,0,mm thickness giving a width to shell thickness ratio nearly equal to 6. The models are tested first without stiffening rings. Results obtained in this case carry index 0. Afterwards the models are tested with the arrangement of 5 rings and the results carrying index 5 are obtained. After arranging first 4 and at last 14 additional intermediate rings models with 9 and 19 rings are obtained, respectively. First the rings were glued to the shell along their entire circumference. In additional tests the rings were glued only along 40% of their circumference, i.e. alternately 4 cm were glued to the shell and 6 cm were left free. In both test series the same buckling loads and the same buckling patterns were observed.

Interaction diagrams for the buckling loads obtained in the tests (E) and calculated applying the buckling theory (T) are given in Fig. 7. T_{19}, the theoretical interaction diagram of the shell stiffenened by means of 19 rings, is omitted in Fig. 7 because in this case the numerical results were highly dependent on the assumption of global or local loss of stability and in the second case on where the local instability takes place. It is interesting to notice that three experimental interaction diagrams for models without and with 5 or 9 stiffening rings intersect in one point denoting axial compression with internal lateral pressure. For the models stiffened by means of 19 rings the buckling load is twice as high . The deviation between theoretical and experimental results under this loading is increasing with increasing number of the stiffening rings.

As stated at the beginning a better understanding of the buckling behavior is possible, if the buckling stresses are considered. In the case of ring stiffened shells this becomes im perative because the position of the first dimple changes according to the number of stiffening rings. The first dimple is shifted from the throat to the upper or lower parts with

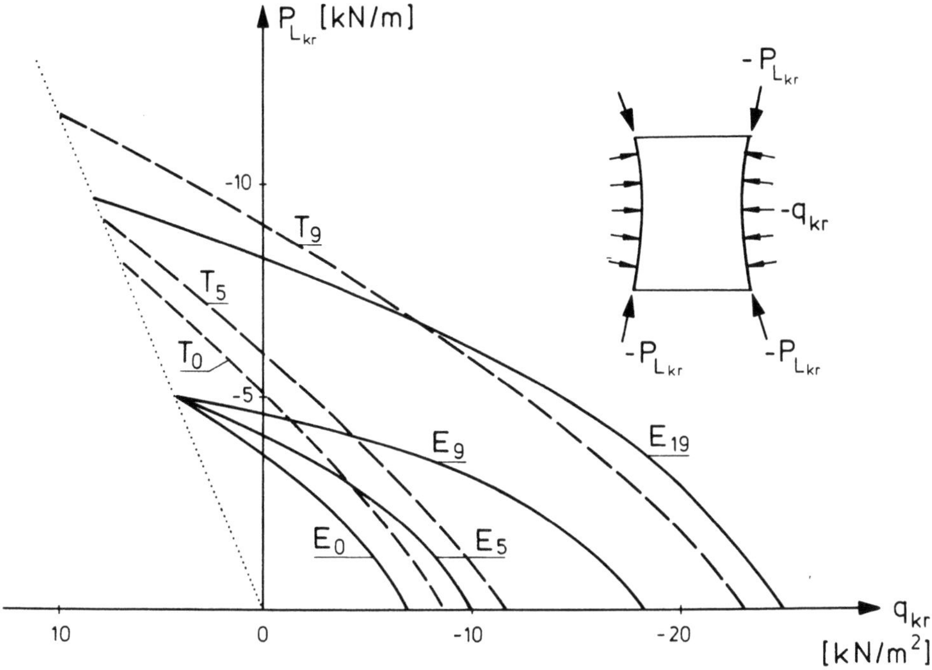

Fig. 7 Experimental and Theoretical Interaction Diagrams for the Buckling Loads of Stiffened Models

greater radius of rotation. Fig. 8 shows the position and the extension of the first buckling wave observed in tests on different models under different stress states. The buckling waves due to external radial pressure are usually larger than those due to axial compression. If the test results are analysed in terms of the stress states initiating the buckling the interaction diagrams of fig. 8 are obtained. The uniaxial buckling stress in circumferential direction σ_{110} increases nearly 1.4, 3.5 or 6.4 times after stiffening the models by means of 5,9 or 19 equidistant rings, respectively. On the other hand uniaxial buckling stress in the meridional direction σ_{220} is not affected by stiffening rings. If the interaction diagrams given in Fig. 9 are expressed in terms of the nondimensional stress states $\sigma_{11}/\sigma_{110}, \sigma_{22}/\sigma_{220}$ for all four diagrams the same equation (2) is obtained:

$$0.80\left(\frac{\sigma_{11}}{\sigma_{110}}+\frac{\sigma_{22}}{\sigma_{220}}\right)+0.20\left[\left(\frac{\sigma_{11}}{\sigma_{110}}\right)^2+\left(\frac{\sigma_{22}}{\sigma_{220}}\right)^2\right]=1 \qquad (2)$$

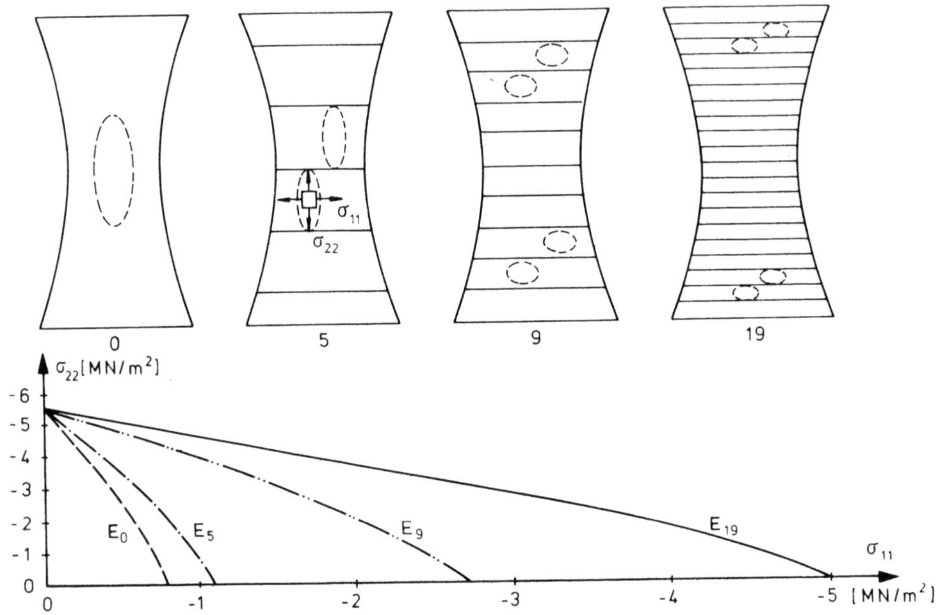

Fig. 8 Position of the First Dimple and the Experimental Interaction Diagrams for the Buckling Stresses of the Stiffened Models

Effect of Material Law

The results refered to till now are obtained on calculation and test models behaving according to Hooke´s law. An extension of these results on reinforced concrete shells is not realistic without considering the effect of the nonlinear behavior of concrete and the cracking. To demonstrate the effect of concrete nonlinearity the cooling tower shell shown in Fig. 9 is investigated under the action of an axisymmetric stress state up to failure [13, 21]. The service loading is increased in steps and in each step the bifurcation parameter is calculated. Reinforced concrete sections are divided into 11 layers the second and tenth of which being the smeared reinforcement. The material law for reinforcing steel is assumed to be bilinear. For concrete a nonlinear material law is considered depending on the biaxial stress state. Fig. 10 shows this material law. As failure criterion for concrete the secant modulus E_S is decisive. Concrete crushes if the secant modulus drops to 30% of the initial tangent modulus E_o. Any concrete layers crushed are omitted in calculating the stiffness matrices of the finite elements in following iterations.

height [m]	amount of reinforcement [%]	
	ring direction	meridional direction
153,0 - 180,0	0,4	0,4
60,0 - 153,0	0,4	0,4
15,0 - 60,0	0,4	0,3
9,0 - 15,0	0,6	0,3
0,0 - 9,0	1,0	0,3

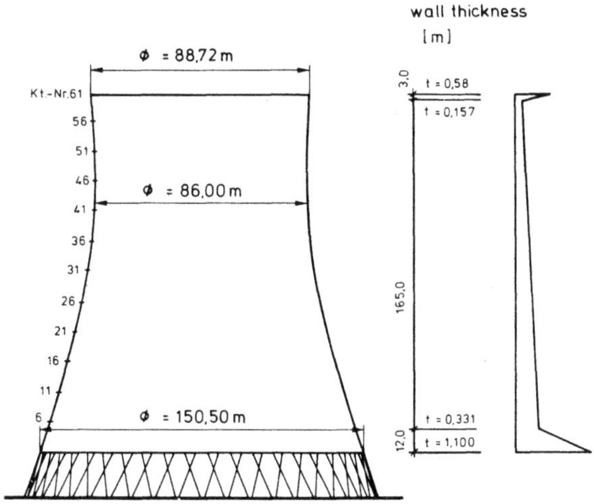

Fig. 9 Cooling Tower Shell Investigated

Bifurcation results are given in Fig. 11 in terms of Load Factor LF vs. Safety Factor γ_B against Buckling. Deviation between the elastic behavior and nonlinear behavior begins first at the load factor LF = 3.00. For higher loads the material nonlinearity becomes effective and causes the reduction of the bifurcation results. The full curve in Fig. 11 is valid for the elastic approach. The other three dashed or dotted curves are for the cases that the secant modulus E_S, the tangent modulus E_T for layers with increasing stress and the inital tangent modulus E_O for unloaded concrete layers, as suggested by Engesser-Karman, or the stress dependent tangent modulus E_T in each layer according to Shanley is put in the bifurcation analysis. Fig. 11 shows how the buckling safety γ_B decreases to half of its value in the vicinity of the ultimate strength of the cooling tower shell if instead of the linear elastic behavior the nonlinear behavior with decreasing tangent modulus is considered [13].

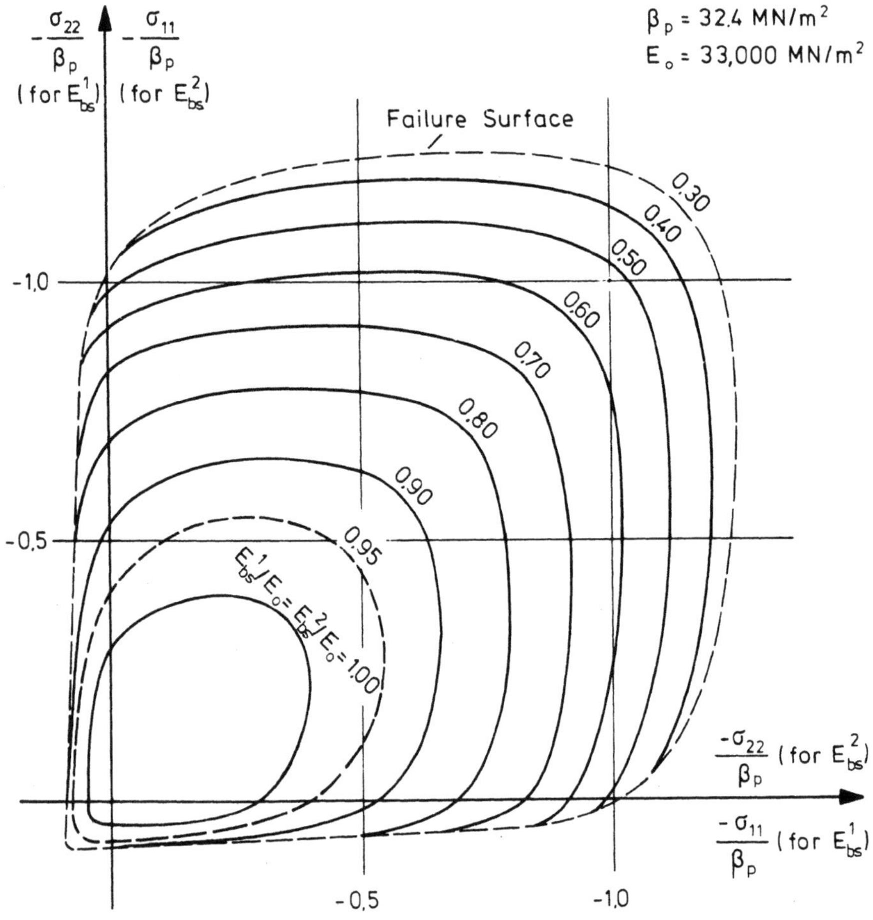

Fig. 10 Biaxial Material Law for Concrete

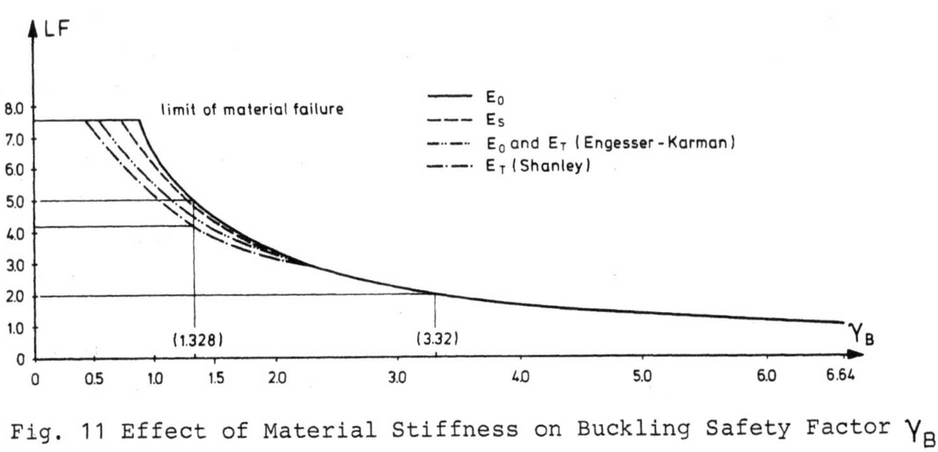

Fig. 11 Effect of Material Stiffness on Buckling Safety Factor γ_B

Conclusions

The present survey on some of the research projects carried out in Bochum points out various aspects of the buckling of hyperboloidal shells. Although not so highly imperfection sensitive as in case of axially compressed cylinders or radially loaded spheres the buckling behavior of shells of negative Gaussian curvature has its own complications. The main handicap lies in the effect of the boundary conditions as obtained from the numerical analysis. Even slight changes in the boundary conditions result in substantial deviations among the bifurcation parameters computed. In contradiction to the analysis in tests buckling starts locally and is influenced by the local imperfections. The assumption of regular and periodic buckling configurations is expected to be the reason for the discrepancy between the experimental and numerical results and among the computed values for different boundary conditions. The negative Gaussian curvature imparts a higher buckling resistance against compression in circumferential direction compared with a cylindrical shell having a radius equal to throat radius of the hyperboloidal shell. On the other hand the uniaxial meridional buckling stress is not influenced by the negative curvature of the shell. That is also the case if the hyperboloidal shell is stiffened by means of rings. Only the uniaxial circumferential buckling stress can be raised by arranging rings.

Finally the buckling resistance may drop substantially after a certain level of stress, if the material nonlinearity is taken into account. An engineering approach to apply this information in the buckling design of hyperboloidal cooling tower shells is explained in References 13, 17, 21, 22 and 23.

References

1. Stein, M. and McElman, J.A.,"Buckling of Segments of Torodial Shells," American Institute of Aeronautics and Astronautics Journal, Vol. 5, No. 2, Febr. 1968, pp.315-361.

2. Hutchinson, J.W., "Initial Post-Buckling Behavior of Torodial Shell Segments," International Journal of Solids and Structures, Vol. 3, No. 1, Jan. 1967, pp. 97-115.

3. Koiter, W.T., "Elastic Stability and Postbuckling Behavior," Proceedings of the Symposium on Non-linear Problems, edited by R.E. Langer, University of Wisconsin Press, 1963, pp. 257-275.

4. Der, T.J. and Fidler, R., "A Model Study of the Buckling Behavior of Hyperbolic Shells," Proceedings, Institution of Civil Engineers , Vol. 41, London, England, Sept. 1968, pp. 105-118.

5. Ewing, D.J.F., "The Buckling and Vibration of Cooling Tower Shells," Part I: Linearized Buckling Theory, Laboratory Report No. RD/L/R 1763, Central Electricity Research Laboratories, Leatherhead, England, Nov. 1971.

6. Veronda, D.R. and Weingarten, V.I.,"Stability of Hyperboloidal Shells," Journal of the Structural Division, ASCE Vol. 101, No. ST7, Proc. Paper 11446, July 1975, pp.1585-1602.

7. Almannai, A. Başar, Y. and Mungan, I., "Systematische Untersuchungen an Rotationsschalen: Theorie und Versuch," Bauingenieur 54 (1979), H. 6, S. 205-211.

8. Almannai, A., Başar, Y. and Mungan, I., "Basic Aspects of Buckling of Cooling Tower Shells," Journal of the Structural Division, ASCE, Vol. 107, No. ST3, Proc. Paper 16121, March 1981, pp. 521-534.

9. Benz, H.J., "Linearisierte Stabilitäts- und Schwingungsberechnung bei versteiften Rotationsschalen," Institut für Konstruktiven Ingenieurbau, Ruhr-Universität Bochum, Techn.-Wiss. Mitt. 76-10, 1976.

10. Lehmkämper, O., "Versteifte Kühlturmschalen aus Stahlbeton," Institut für Konstruktiven Ingenieurbau, Ruhr-Universität Bochum, Techn. Wiss. Mitt. 78-6, 1978.

11. Zerna, W., Başar, Y., Mungan, I. and Thiemann, W., "Beuluntersuchungen an hyperbolischen Rotationsschalen," Forschungsberichte des Landes Nordrhein-Westfalen, Nr. 2439, 1974.

12. Zerna, W., Almannai, A., Başar, Y. and Mungan., "Randbedingungen und Beulverhalten von Kühlturmschalen," Beton- und Stahlbetonbau 74 (1979), H. 8, S. 200-204.

13. Zerna, W., Mungan, I. und Steffen,W., "Bestimmung der Beulsicherheit von Schalen aus Stahlbeton unter Berücksichtigung der physikalisch-nichtlinearen Materialeigenschaften," Schriftenreihe des Deutschen Ausschusses für Stahlbeton, H. 315, Berlin: Verlag W. Ernst & Sohn 1980.

14. Mungan, I. "Buckling Stres States of Cylindrical Shells," Journal of the Structural Division, ASCE, Vol. 100, No. ST11 Proc. Paper 10965, November, 1974, pp. 2289-2306.

15. Mungan, I., "Buckling Stress States of Hyperboloidal Shells," Journal of the Structural Division,ASCE, Vol. 102, No. ST10, Proc. Paper 12465, October, 1976, pp. 2005-2020.

16. Mungan, I., Buckling Stresses of Stiffened Hyperboloidal Shells,," Journal of the Structural Division, ASCE, Vol. 105, No. ST8, Proc. Paper 14775, August, 1979, pp. 1589-1604.

17. Mungan, I., "Experimentelle Beuluntersuchungen und Stabilitätsnachweise für Kühlturmschalen," Konstruktiver Ingenieurbau Berichte 1977, Heft 29/30, S. 75-80.

18. Mungan, I., "Schalenbeulung am Beispiel der Rotationsschalen" Der Stahlbau 49 (1980) , H. 2, S. 41-45.

19. Berke, L. and Carlson, R.L., "Experimental Studies of the Postbuckling Behavior of Complete Spherical Shells," Experimental Mechanics, Vol. 8, December, 1968, pp. 548-553.

20. Mungan, I., and Lehmkämper, O., "Buckling of Stiffened Hyperboloidal Cooling Towers," Journal of the Structural Division, ASCE, Vol. 105, No. ST10, Proc. Paper 14917, October, 1979, pp. 1999-2007.

21. Zerna, W., Mungan I. and Steffen, W., "A Wind-Buckling Approach for RC Cooling Towers," Paper presented at joint ASME/ASCE Mechanics Conference, June 22-24, 1981 in Boulder, Col. Preprint No. EM5.4.

22. Mungan, I.: "Ein Konzept für die Beulbemessung der Kühlturmschalen aus Stahlbeton," Beton- und Stahlbetonbau 76 (1981), H. 10, S. 229-233.

23. Zerna, W. und Mungan, I.,"Über das Beulen von Kühlturmschalen mit Versteifungsringen," Beton- und Stahlbetonbau 76 (1981), H. 2, S. 33-36.

BUCKLING OF HYPERBOLIC PARABOLOID SHELLS

G. HEES

Fachgebiet Statik der Baukonstruktionen
TU Berlin
Berlin-West, Germany

Summary

Theoretical and experimental buckling loads of HP shells having rectangular and parallelogram shape in plan are reported. In the case of the parallelogram shell there is a large discrepancy between experiment and theory.

Introduction

Theoretical as well as experimental investigations of the buckling problem of HP shells are relatively seldom compared to other shells. The majority of these investigations was done for shells with rectangular, in particular square plan.

The typical characteristics of HP shells are straight generators and straight boundaries, Fig. 1. By definition diagonal directions the sections

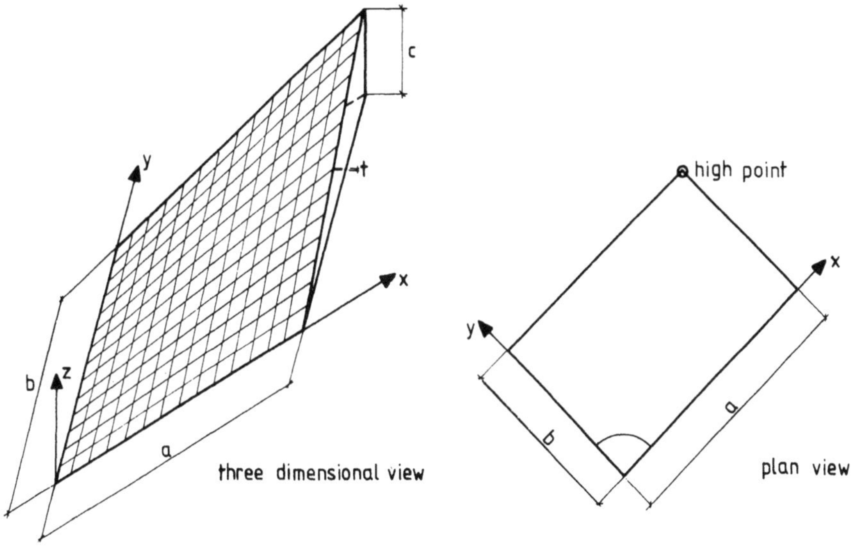

Fig. 1: HP shell with rectangular plan

Buckling of Shells, Ed.: E. Ramm
Proceedings of a State-of-the-Art Colloqium

are parabolas. The main geometrical properties of the shell are

a length of the edge in x-direction

b length of the edge in y-direction

c height at the corner $x = a$, $y = b$

t thickness of the shell

In most cases the shell has simply supported boundaries.

1. Shells with rectangular plan

1.1 Theoretical results

Early investigations of the stress state were based on the membrane analysis. A load only in the z-direction results in shear stresses in the x- and y-direction, Fig. 2. The principal stresses are oriented in the direc-

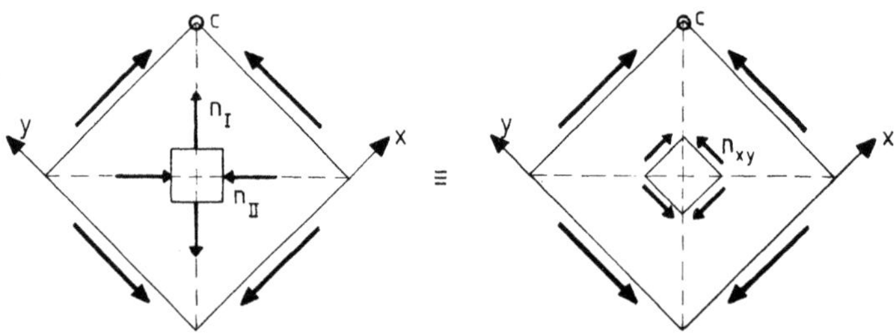

Fig. 2 : Membrane forces caused by a load p in -z-direction

tions of the diagonals: tension in the diagonal from point 0 to the high point and compression in the arch direction perpendicular to the first diagonal.

In the field of buckling the first theoretical work was described by E. Reissner [1]. He assumed that there are several buckling waves in the compression diagonal and one half wave in the tension diagonal. The critical load given by Reissner is

$$p_{cr} = \frac{2\,E}{\sqrt{3\,(1-\mu^2)}}\;\frac{t^2\,c^2}{a^2\,b^2} \qquad (1)$$

with E modul of elasticity

μ Poissons ratio.

Ralston [2] generalized the solution for the shell having a square plan with $b = a$. He extended equation (1) introducing a coefficient ρ :

$$p_{cr} = \rho \frac{2\,E}{\sqrt{3\,(1-\mu^2)}} \frac{t^2\,c^2}{a^4} \quad . \qquad (2)$$

Fig. 3 shows the coefficient ρ as a function of t/c. The curve shows the envelop of the numerical results. For a ratio $\frac{t}{c} < 0.04$ ρ is equal to 1, for $\frac{t}{c} = 1$ ρ is equal to 14.

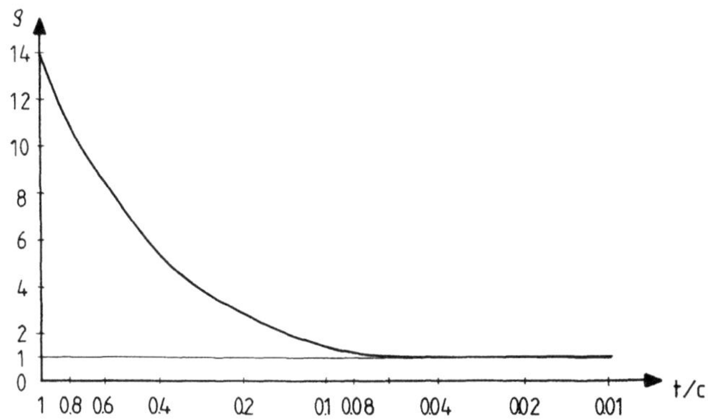

Fig. 3 : ρ as a function of $\frac{t}{c}$

Kollár [3] compared the extreme cases of the shell with the circular cylindrical shell and the plate, respectively. With

$$n_{xy} = -n_{II} = \frac{a^2}{2\,c}\,p \qquad (3)$$

and the radius R of the tension diagonal, Fig. 4:

$$R = \frac{b^2}{c} = \frac{a^2}{c} \qquad (4)$$

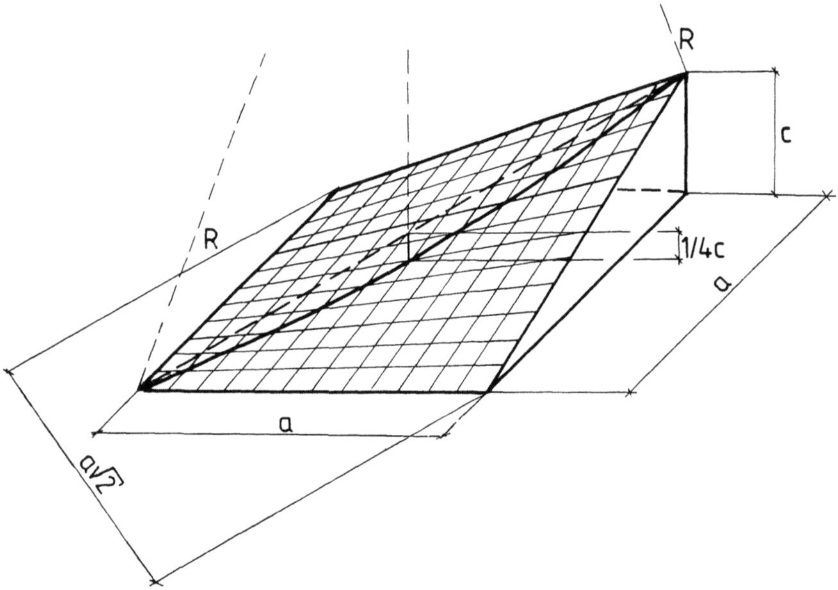

Fig. 4 : Radius R of the tension diagonal

eq. (1) describes the critical load of an axially loaded cylindrical shell [4]:

$$n_{II} = - \frac{E\ t^2}{R\sqrt{3\ (1-\mu^2)}} \qquad (5)$$

This is the value for $\rho = 1$ in Fig. 3.

The specialization to a plate with the vertical load is not as obvious [1]. Since the vertical load in a HP shell leads to shear stresses, Kollár compared the HP shell with a shear loaded plate [4] :

$$\text{plate:}\quad n_{xy,cr} = 9.7\ \frac{\pi^2\ E\ t^3}{12\ (1-\mu^2)\ a^2} = 8\ \frac{E\ t^3}{a^2\ (1-\mu^2)}\ . \qquad (6)$$

For the HP shell with the shear stresses equal to eq. (3) and with eq. (2) is:

$$n_{xy,cr} = \rho \frac{E}{\sqrt{3(1-\mu^2)}} \frac{t^2 c}{a^2}$$

$$= \frac{\rho}{\sqrt{3}} \frac{E t^3}{a^2 \sqrt{1-\mu^2}} \frac{c}{t} \quad . \tag{7}$$

The value $\frac{c}{t} = 1$ corresponds to a plate with a maximum imperfection amplitude equal to the plate thickness. Then the coefficient ρ in Fig. 3 is equal to 14. This value is introduced into eq. (7) :

$$n_{xy,cr} = 8 \frac{E t^3}{a^2 \sqrt{(1-\mu^2)}} \quad . \tag{8}$$

Equations (6) and (8) are equivalent when $\mu = 0$.

1.2 Experimental results

Leet [5] made experimental studies on HP shells having a square plan. He found that Reissners assumption - ripples in the compression direction - is correct. The buckling load of shells having very small imperfections is reduced tc approximately 70 % of Reissner's solution. If there are larger imperfections the value decreases to approximately 50 %. There was no important influence of the edge members on the buckling load except in a range with $\frac{t\,E}{A_\varepsilon\,E_\varepsilon} > 0.2$, Fig. 5. ($A_\varepsilon$ is the area and E_ε the modul of elasticity of the edge members.) In this range the buckling load depends on the axial stiffness of the edge members. - In the prebuckling phase the stress state corresponds to that of the membrane theory except the edge regions. With the beginning of buckling the load carrying capacity is not exceeded. If the edge members are stiff enough a tension field in the direction of the tension diagonals is established. The ultimate load depends on the moment of inertia of the edge beams and on the support conditions.

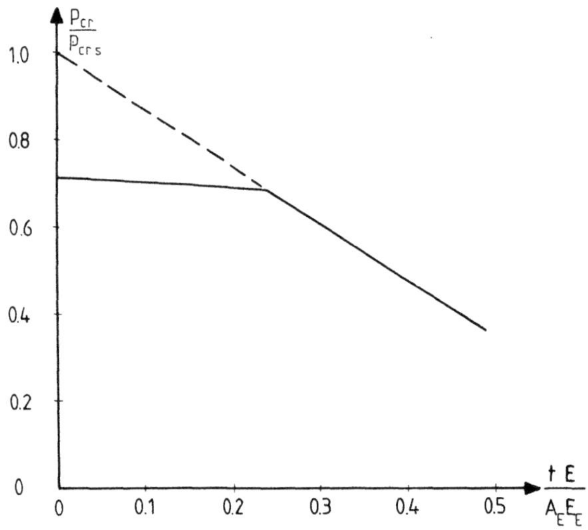

Fig. 5 : Variation of shell buckling load

2. Shells having a parallelogram shape in plan

2.1 Theoretical results

The membrane theory of HP shells with non rectangular plan was derived by Candela [6]. Assuming that there is only a load in the z-direction and that the shell is simply supported allowing only for shear forces at the boundaries the principal forces in the diagonals as shown in Fig. 6 are given by

$$\frac{n_{II}}{n_I} \approx \tan^2 \frac{\omega}{2} \tag{9}$$

The ratio is shown in Fig. 7 as a function of the skew angle ω of the shell.

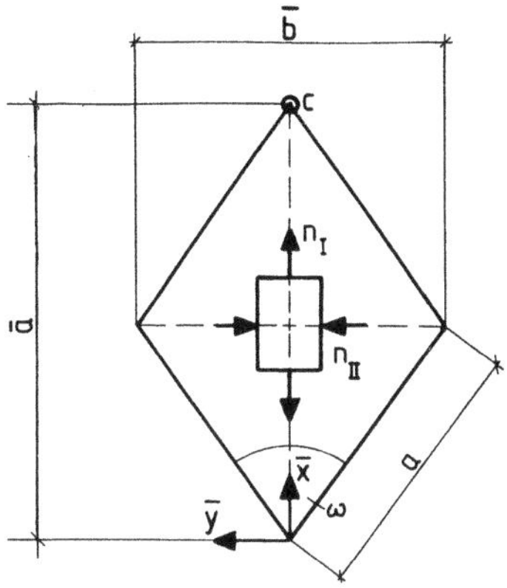

Fig. 6 : Plan view of a non rectangular HP shell

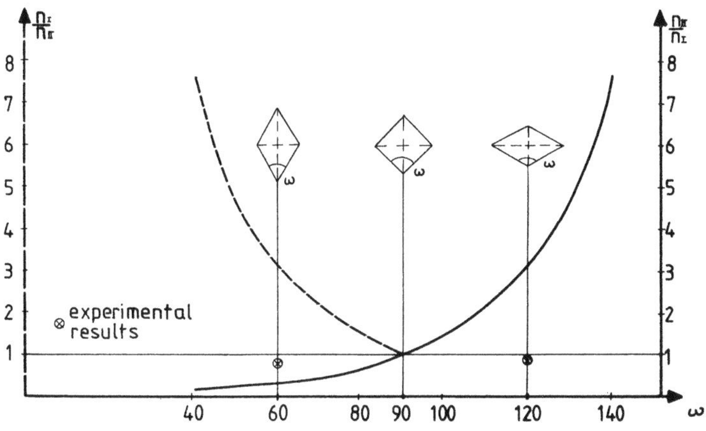

Fig. 7 : Ratio of the principal membrane forces versus the angle ω

Vosshans [7] determined the buckling load of HP shells with non rectangular plan using the same assumptions as Reissner. Using different geometrical parameters as given by Vosshans the following results:

$$p_{cr} = \frac{2\,E}{\sqrt{3\,(1-\mu^2)}} \quad 4\,\frac{t^2\,c^2}{\bar{a}^4} \quad \alpha \qquad (10)$$

with

$$\alpha = \frac{1}{2}\left[1+\left(\frac{\bar{a}}{\bar{b}}\right)^2\right]. \qquad (11)$$

The coefficient α is plotted in Fig. 8 as a function of ω and the length of the diagonals, respectively. If ω is a right angle α is equal to 1.

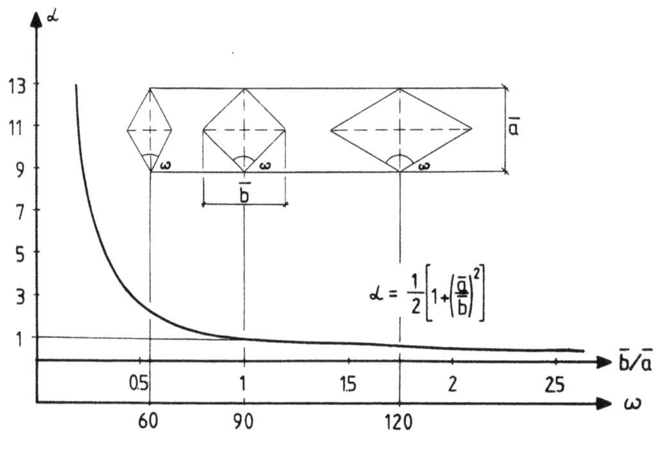

Fig. 8 : α as a function of ω and $\frac{\bar{b}}{\bar{a}}$

Then the buckling load agrees with that given in eq. (1) if

$$a = b = \frac{\bar{a}}{\sqrt{2}}\ . \qquad (12)$$

If $\bar{a} >> \bar{b}$ then α becomes infinity. In this case the shell represents a cable system with a tension stress state only. On the other side if $\bar{b}$ increases to infinity α becomes 0.5.

In the next section the theoretical results will be compared to experimental results. The tests were performed using shells with a constant side length. Using the notation of Fig. 6

$$\bar{a} = 2 \cos \frac{\omega}{2} \, a$$
$$\bar{b} = 2 \sin \frac{\omega}{2} \, a \qquad (13)$$

and introducing these values into equations (10) and (11) the buckling load is

$$p_{cr} \quad \frac{2\,E}{\sqrt{3\,(1-\mu^2)}} \; \frac{t^2\,c^2}{a^4} \; \beta \qquad (14)$$

$$\beta = \frac{1}{8\,\cos^4 \frac{\omega}{2}} \left(1 + \cot^2 \frac{\omega}{2} \right) \qquad (15)$$

The coefficient β is shown in Fig. 9 as a function of the skew angle ω.

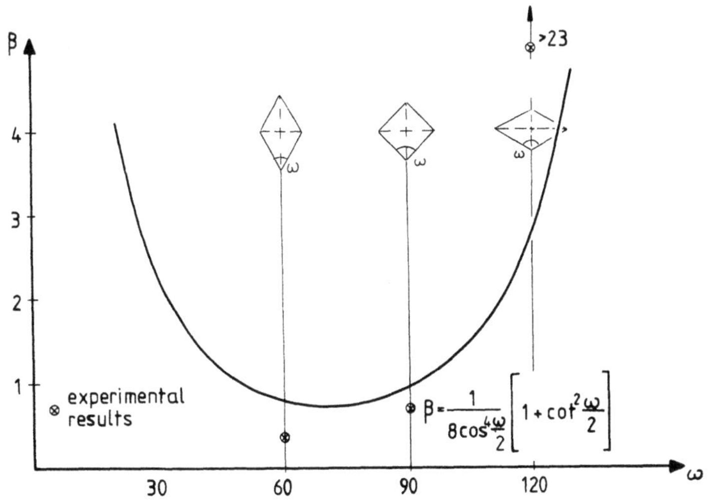

Fig. 9 : β as a function of ω

2.2 Experimental results

To verify the theoretical results, Vosshans [7] tested HP shells with angles ω of 60°, 90° and 120°. The shells made out of PVC plates were manufactured with high precision. In the test the edge members have been made rigid. The test results are added to Fig. 9. For the shell with a square plan the experimental buckling load is in agreement with the result of Leet [5], that is 70 % of the theoretical buckling load. For $\omega = 60^{\circ}$ the buckling load is reduced to approximately 40 % of the theoretical value. For the shell with $\omega = 120^{\circ}$ no buckling was observed even for a load ten times of the theoretical value. The experimental loading device did not allow a further increase of the load. To explain the discrepancy between experiment and theory the membrane stresses in the middle region of the shell were measured. The experimental results added to Fig. 7 show a ratio of $n_{II}/n_{I} \approx 1$. This does not agree with the theoretical value.

If the diagonals are regarded as an independent cable/arch system the stresses due to the membrane theory indicate that half of the vertical load is carried by either direction; that is the load ratio $p_I : p_{II}$ is 1 : 1 independent of the size of ω as defined in the first row of Fig. 10. If in addition the cable and the arch have to satisfy the compatibility condition of equal displacements at the center point the load ratio changes to the values given in the third row of Fig. 10.

	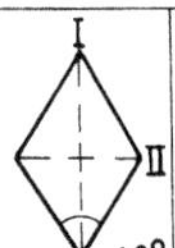 60°	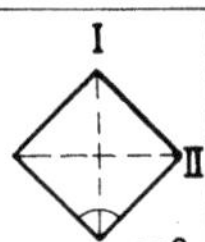 90°	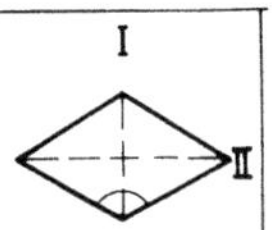 120°
independent caple/arch system	1 : 1	1 : 1	1 : 1
experimental results	1 : 2,56	1 : 1	1 : 0,30
caple/arch system connected at the center point	1 : 9,13	1 : 1	1 : 0,12

Fig. 10 : Load distribution in the principal directions p_I/p_{II}

The membrane stresses determined in the experiment allow to conclude that the ratio of the load which is carried either by the cable or by the arch is just in between the above defined values, see second row of Fig.10. Or in other words in the real system the shorter diagonal system carries more load than indicated by the membrane theory. The experimental buckling loads given in Fig. 9 do confirm this situation.

The work on the buckling problem of HP shells will be continued in Berlin together with Prof. Dr. Dierks, who initiated and supervised the work.

References

1. Reissner, E.: On some aspects of the theory of thin elastic shells. J. of the Boston Society of Civil Engineers, 42 (1955) 100-133

2. Ralston, A.: On the problem of buckling of a hyperbolic paraboloidal shell loaded by its own weight. J. Math. Phys. 35 (1956) 53 - 59.

3. Kollár, L.: Dulácska, E.: Schalenbeulung. Düsseldorf: Werner-Verlag 1975

4. Pflüger, A.: Stabilitätsprobleme der Elastostatik. Berlin, Heidelberg, New York: Springer-Verlag 1975

5. Leet, K.M.: Study of stability in the hyperbolic paraboloid. Proc. ASCE, EM 1, 1966, 121 - 142

6. Candela, F.: Structural applications of hyperbolic paraboloid shells, J. ACI 1955, 397 - 415

7. Vosshans, O.: Zum Beulverhalten von hyperbolischen Paraboloidschalen über parallelogrammförmigen Grundriß. Berlin: Universitätsbibliothek der TU Berlin, Abtlg. Publikationen 1982

Part V :

Experiments - Dynamics

THE STATUS OF EXPERIMENTAL BUCKLING INVESTIGATIONS OF SHELLS

J. SINGER

L. Shirley Tark Professor of Aircraft Structures
Department of Aeronautical Engineering
Technion-Israel Institute of Technology
Haifa, Israel

Abstract

The recent developments in shell buckling experiments are surveyed and related to a review of the progress in the seventies. Model fabrication, imperfection measurements, boundary conditions, nondestructive testing, combined loading, postbuckling behavior, composite shells and other aspects of shell buckling tests are discussed. The motivation for experiments and the conclusions drawn in the previous review are reassessed.

List of Symbols

A_1, A_2	cross sectional area of stringer and ring, respectively.
a, b	distance between centers of stringers and rings respectively.
e	eccentricity of loading (distance from shell middle surface to the point of application of load).
f	frequency.
f_{SS4L}	frequency predicted by linear theory for SS4 B.C.'s.
L	length of shell.
P_{exp}	experimental buckling load.
P_{extrap}	buckling load obtained by direct prediction from vibration tests.
P_{sp}	calculated axial buckling load for shell with effective axial or rotational restraint.
R	radius of middle surface of cylindrical shell.
t	thickness of shell.
Z	$(1-\nu^2)^{1/2}(L/R)^2(R/h)$, Batdorf shell parameter.
ρ_{sp}	P_{exp}/P_{sp}.
ρ_{th}	$P_{SS3\ imp}/P_{SS3}$.
ρ_{extrap}	P_{exp}/P_{extrap}.

Notation for Boundary Conditions:

SS3	:	$w = M_x = N_x = v$	$= 0$
SS4	:	$w = M_x = u = v$	$= 0$
C4	:	$w = w_{,x} = u = v$	$= 0$

1. Introduction

Two years ago the author reviewed developments in shell buckling experiments in the previous decade [1]. In addition to updating the

Buckling of Shells, Ed.: E. Ramm
Proceedings of a State-of-the-Art Colloqium

earlier review of Babcock [2], that paper also examined the motivation for experiments in the computer era and outlined trends and problems. The present paper will survey the very recent developments in buckling experiments on shells, reexamine their role in the light of recent developments in numerical techniques, and reassess the trends and problem areas.

Excellent reviews on buckling, and on shell buckling in particular, have been published since 1980 (see for example [3]-[6]). As in the past these surveys mostly relegate experiments to the secondary task of verification of theory, but recently more prominence has sometimes been given to certain aspects of experimental work, as for example imperfection measurements in [5]. Furthermore, some recent surveys focus on experimental studies, like [7] which discusses work carried out at the Institute for Aerospace Studies of the University of Toronto; [8] which summarizes buckling tests of fabricated steel cylindrical shells in the USA; [9] which outlines buckling research performed at Det norske Veritas in Oslo, or [10] in which empirical design rules for unstiffened cylindrical, conical, and spherical shells are proposed as a result of a very comprehensive literature survey. However, experimental techniques are scarcely discussed, and therefore their progress and the wider impact of experiments on the development of analysis and design of thin shells deserves special attention.

Recent developments apear to follow the main trends observed in 1980. Efforts are again primarily devoted to stiffened shells, combined loading and composite shells. Offshore structures, aerospace applications, and also to some extent containment shells continue to provide the prime motivation. In this paper the topics will be discussed in the same order as in [1] to facilitate comparisons, but the emphasis will be on reassessment of the conclusions drawn in Delft in 1980.

The main emphasis will again be on elastic static buckling, but in view of the extensive use of shells which buckle in the plastic region, in particular in marine and offshore structures, recent plastic static buckling tests will also be briefly discussed. Dynamic buckling experiments will not be reviewed as they usually differ significantly in scope and purpose. References already cited in [1] and [2] will usually not be repeated, except for special purposes.

2. Motivation for Experiments

Before proceeding to the review of recent experimental work, it may be of interest to study again the primary motives for buckling experiments on shells proposed by the author in [1] and examine their validity. The eight main reasons detailed in [1] are briefly recapitulated and examined here:

a. Better Understanding of Buckling Behavior and the Primary Factors Affecting it.

In addition to the buckling loads, careful experiments in which the parameters are varied one at a time yield the behavior of the shell just before, at and after buckling, and accentuate the main parameters affecting this behavior. Such a philosopy of "research type experimental programs" has been strongly advocated by Sechler [11] for many years, and has been implemented in some test programs, for example in [12] or [13]. Based on these observed parameters numerical schemes can be developed, verified, and can also be employed for "experiments on the computer" to extend the range of the parameters tested.

b. To Find New Phenomena.

This reason is a direct extension of the first one. In shell buckling experiments, the new phenomena are likely to be unexpected behavior patterns or mode interactions.

c. To Obtain Better Inputs for Computations.

The mathematical models employed in modern large multi-purpose computer programs can simulate real shell structures fairly closely, but the simulation depends very much on the input of correct boundary conditions, in particular joints or bonds, on imperfections and load applications. This has been emphasized by recent experience. Better inputs can be provided by appropriate nondestructive tests: for example, boundary conditions by vibration correlation techniques, imperfection shapes and amplitudes by imperfection scans, load transfer and eccentricities by strain measurements and vibration correlation techniques, etc. Here automated recording in experiments has just begun and much closer interaction between test and computation is developing.

d. To Obtain Correlation Factors Between Analysis and Test and for Material Effects.

Even when large powerful programs, are employed, test results still differ considerably from predictions. These differences are partly due to inaccuracies of inputs and partly to variations in buckling behavior of the mathematical model and the shells tested. They can all be lumped in a "correlation factor". The advantage of such a correlation factor is the overall correlation it provides for the designer, but its weakness is that it is completely reliable only for the shells tested. One can statistically evaluate a large number of tests to obtain overall lower bound correlation factors, so called "knock-down" factors (see for example [10] for a very recent effort in this direction), but this results in very conservative design. Hence "correlation factors" should be more specialized. Since many experiments are on laboratory scale shells, extensive studies comparing the results of laboratory scale and large scale tests are needed to reassure the experimenter and to guide the designer. Correlation type experiments will therefore continue to be a major task of research and industrial labaoratories for quite some time to come, as they provide the designer with essential correction factors which include the effects of new materials and manufacturing techniques and to some extent bridge the gap between the buckling behavior of the computation model and his realistic shell structures.

e. <u>To Build Confidence in Multipurpose Computer Programs</u>.
Extensive experimental verification is an essential element for confidence in a large computer program. This is therefore a primary motive for shell buckling experiments, which becomes more important, as the programs become more sophisticated and ambitious. Though some developers of shell programs have promoted and applied extensive experimental confirmation (for example to the BOSOR program [14]), more correlations of the results obtained from computer programs with test results are required, as pointed out in [15].

f. <u>To Test Novel Ideas of Shell Construction or Very Complicated Shell Elements</u>.
Exploratory tests of new concepts have been used extensively by aeronautical, civil, mechanical and ocean engineers, and will continue to be an important tool. Furthermore, if the shell structure is elaborate and has many openings with complicated stiffening and load diffusion elements, model testing may be less expensive and faster than computation with a large multipurpose program, even in the detail design stage [16].

g. <u>For Buckling Under Dynamic Loading and in Fluid Structures Interaction Problems</u>.
These are areas where computation is cumbersome, expensive, and difficult to interpret reliably. Experiments are therefore preferable though they too present many difficulties.

h. For Certification Tests of Full Scale Shell Structures.
This is the typical industrial task, which will continue till model experiments are sufficiently advanced and integrated with computation to eliminate the necessity for them.

Examination of these motives, proposed in 1980, reconfirms them, and recent experience reinforces the conclusion that the computer does not replace tests. The experiment remains an essential link in the analysis also in the computer era, and its scope and usefulness are even greater today.

3. Model Fabrication

a. Isotropic Shells

Very few tests on laboratory scale isotropic shells have been carried out since 1980, and no new developments in model fabrication have appeared. As pointed out in [1], much effort was invested in the seventies in making accurate specimens from Mylar and similar polyester films (like Melinex, Mitsubishi Diafoil, etc.). With the improved quality of Mylar, in particular with thickness variations not exceeding 5% and of about 10% anisotropy, these foils have become the favorite of many investigators for specimens with developable surfaces and (R/t) = 400-1000 (see for example those cited in [1] or [17]-[19]). However, the other techniques of fabricating shell models of plastic materials (PVC, PMMA, epoxy etc.) developed in the sixties and seventies, mainly for spherical and hyperbolidal shells [1], also continue to be used.

One simple method, originated by Horton at Stanford University, USA, in the early sixties, for obtaining accurate isotropic cylindrical shell specimens inexpensively, is the use of seamless beer or soft drink cans. Since these thin-walled cans have to be within the close tolerances demanded by the automated filling and sealing process, they represent a source of specimens with accurately repeated dimensions. Recently this idea has been revived by Arbocz and Elishakoff at Delft University of Technology in a test program related to stochastic stability analysis [20] and at Technion Aircraft Structures Laboratory in a test program on buckling of shells under axial impact. A special production run of empty steel beer cans without paint and without closing lids has been made, yielding specimens with (R/t) = 300. The shells have, however, an integral bottom end closure, which has to be removed, some axisymmetric thickness variations with height and quite significant residual stresses, resulting from the multiple stage deep drawing fabrication process.

These have to be studied further for more precise evaluation, but even now–though undefined – they are practically identical for the many specimens made by the process.

The alternative path, of building models by the same fabrication technique as for full scale shells (cold rolling and welding), is again being rigorously followed by civil and ocean engineers. Though their main interest has definitely shifted to stiffened shells, they also continue testing isotropic steel shells, often under combined loading [8], [21]–[23], sometimes also with simulated damages [24]. Some of the specimens in these tests could also be classified as large shells discussed below.

Plastic buckling, which was excluded from the earlier reviews [1] and [2], has received much attention in recent years, and has been the subject of many extensive test programs, primarily bending and combined loading of circular tubes [25]–[31], and some tests on spherical, torispherical and ellipsoidal shells [32], [33]. For plastic buckling tests, tubes made by standard manufacturing processes are usually employed. For example, in [25] and [27] the shells of (R/t) = 9–51 were hot formed seamless tubes, or had the custumary electric resistance welded seams. In [26] smaller diameter seamless steel and aluminum tubes were tested virtually in the "as supplied" condition, or after having been drawn down to thinner wall thicknesses to yield specimens of (R/t) = 17–39. The 88.9 mm outside diameter 304 stainless steel cylindrical shells, of (R/t) = 18–19, tested in [28] were also cut from standard stock tubes. On the other hand, the 27 shells tested in [30] were machined from 200 mm outside diameter aluminum alloy tubes to yield accurate specimens of (R/t) = 45–50, since the aim was to study the effect of different load paths on plastic buckling under combined loading. Shells for plastic buckling of shapes other than cylindrical are usually machined from solid bar, [32] or spun [33].

b. Stiffened Shells.

In the model fabrication of stiffened shells there have been further interesting developments since 1980. The group at the Technion has continued to develop their machined 7075–T6 alluminum alloy specimens, initiated more than a decade ago (see for example [13], [34] or [35]), and the tests continue to focus on end conditions, which will be discussed later. In order to verify the vibration correlation technique for definition of boundary conditions, developed on these integrally machined shells (see for example [34] or [35]), on shells of welded and riveted construction, a series of tests on such shells has been initiated, in the framework of a joint research project by RWTH Aachen - Technion [36]. Spot welded and riveted stringer-stiffened aluminum shells have been fabricated, with geometrical parameters intentionally similar to those of the smaller integrally stiffened shells tested earlier, (R/t) = 490 and $(A_1/bt) = 0.7$, to facilitate comparison. Figure 7 in section 4 shows one of these spot welded shells AAC-1 in the testing.

The frequent use of stiffened cylindrical shells as components of offshore structures and other marine structures, has motivated extensive experimental work and led both to many tests on large welded shells, to be discussed separately and to the development of special model fabrication techniques already mentioned in [1]. Walker and his associates at University College, London, developed a manufacturing and welding technique of small scale specimens with (R/t) = 200-360 representing typical offshore elements on a 1/20 scale. Their cylindrical shells were built up segmentally by rolling and machining thin steel sheet with material characteristics closely similar to those employed in offshore structures, into curved panels, and then welding these together with the stiffeners [37]. The work has been continued by Walker at Surrey University, and ring-stiffened shells with (R/t) = 150 have been made by a similar process [38]. Similar 1/20 scale stringer-stiffened shells with (R/t) = 190 were made at Imperial College by Dowling and his associates, using a different fabrication technique [39], [40]. There the cylinder itself is cut in one piece, allowing for contractions due to welding distortions, and the stiffeners are then welded into the unrolled cylinder sheet. After welding this stiffener sheet is formed around a segmented mandrel and a closing tee butt weld is then run over one stiffener. Imperfection and residual stress measurements on the U.C. and I.C. models yielded similar relative levels to those obtained on large shells. The results on these small scale tests could therefore be taken as a reliable complement to the large scale tests performed in the same program [40], [41]. An additional test program of 12 single bay stringer-stiffened cylindrical 320mm diameter shells has been initiated at Imperial College for Det norske Veritas. The shells of $(R/t) \simeq 190$ have internal stringers with $(b/t) \simeq 27.5$ and are being tested under combined axial compression and external pressure.

In a test program on the buckling strength of ring-stiffened cylindrical shells under axial compression at the Norwegian Institute of Technology, Trondheim [42], a comparison was made between machined and fabricated specimens. The tests focused on buckling in the elastic-plastic range. Twelve shells were fabricated from steel plates by cold rolling and welding, and 9 shells were machined out of a stress-relieved seamless steel tube. All the shells had $(R/t) \simeq 200$ and the other geometry parameters were also similar, though the fabricated shells were 3 times larger (with $R = 600$ mm they could already be considered as large shells).

a) b)

Fig. 1. Steel ring-stiffened cylindrical shells of $(R/t) \simeq 200$ tested at DnV under axial compression: (a) Typical welded shell, (b) Typical 1/3 scale machined shell, from [42].

Figure 1 shows two typical specimens of the program: (a) a welded shell, showing the staggered fillet weld and the axisymmetric buckling pattern; (b) a similar machined shell but with rings at half the spacing, and again axisymmetric buckling is observed. The buckling behavior of the two types of specimens was generally similar, but the buckling load (related to the classical one) of the fabricated shells was usually about 20% below that of the machined shells. When related to the yield stress, the collapse stress was sometimes even 30% lower for the fabricated shells, probably due to weld induced distortions (wrap-up) and residual stress.

Miller continued the "realistic specimens approach" he advocated earlier for unstiffened and stiffened shells [43]. For example in [44] 20 hydrostatic external pressure tests on 16-48 inch outside diameter ring-stiffened shells, made from commonly used platform steels by routine platform fabrication procedures, are reported. The shells, with (R/t) ≃ 16-63, buckled partly in the elastic and partly in the inelastic range. The geometric initial imperfections for most specimens were found to be larger than the allowable out-of-roundness tolerances. The specimens of many of Miller's tests are now even more "realistic" by being larger, and are therefore discussed in the following subsection.

Techniques developed in the seventies for fabrication of accurate plastic models for stiffened shells [1] have since been widely used and perfected. For example spin casting of liquid epoxy plastic, developed earlier for isotropic shells, has been applied to fabrication of stringer-stiffened cylindrical shells [7], [45] of (R/t) = 100-150. Figure 2 shows such a stringer-stiffened shell and the photoelastic fringe pattern associated with an initial local panel buckling mode.

Fig. 2. Small UTIAS spin-cast epoxy stringer-stiffened cylindrical shell showing local panel buckling (the patlers are photoelastic isochromatics), by courtesy of R.C. Tennyson.

Mylar and other plastic films such as lexan also continue to be employed extensively for stiffened shell models. One example is a 15 inch (381 mm) diameter ring-stiffened Mylar cylinder from a recent test program on buckling of shells with openings [46] shown in Fig. 3.

Fig. 3. 15 inch diameter Mylar ring-stiffened cylinder with reinforced opening tested under axial compression, by courtesy of C.D. Miller.

c. Large Shells.

As pointed out in [1], the ocean engineering industry has motivated extensive buckling test programs on large models of unstiffened and stiffened shell structures, since, in order to provide reliable data for welded marine and offshore structures, it was thought advisable to fabricate rather large specimens which would represent the practical geometric imperfections and residual stresses accurately. These efforts yielded many high quality specimens in the second half of the last decade and has since produced many more (see for example [8], [9], [40] or [41]).

The first phase of the U.K. Department of Energy Stiffened Shell Study for Offshore Structures, which also involved tests of large shells, has been completed [37]-[41]. The Glasgow University specimens have already been discussed in [1]. Three of the Imperial College large shells were ring-stiffened [39], [40]. In these shells, the center and end plates were butt welded while flat and rolled together, then longitudinally butt welded and rerolled. The ring-stiffeners were then added with light intermittent fillet welds and finally the end rings were welded on. The center bays of (R/t) = 157-261 were the actual test bays (see for example that of IC2 in Fig. 4). The fourth large shell IC4 was stringer-stiffened [40] and aimed at complementing the 3 Glasgow stringer-stiffened shells [41]. Its fabrication was similar to the

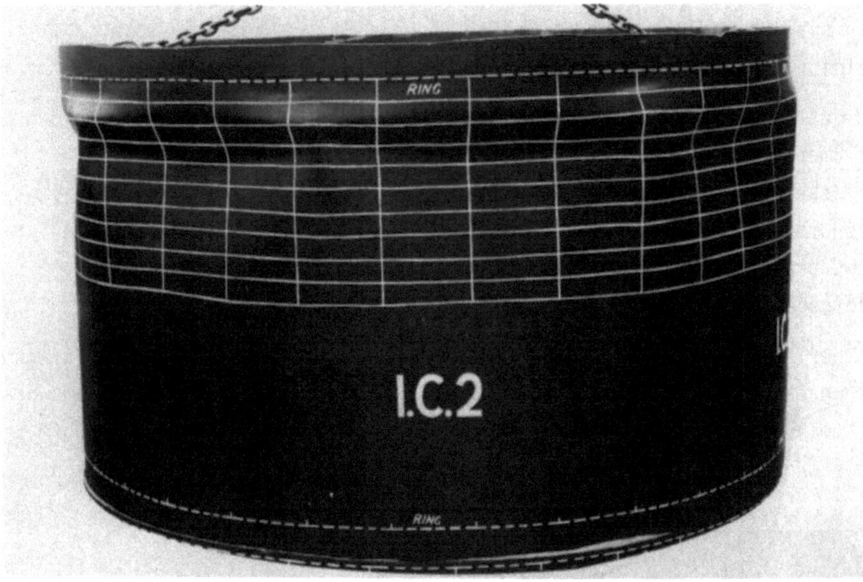

Fig. 4. Central bay buckle of large 1.5 m diameter Imperial College ring-stiffened shell I.C.2., from [40].

ring-stiffened shells, except that here the end flanges were attached before the stiffeners. Ring-stiffeners (in 4 sections) were then added to separate the bays, and finally the stringers were attached with relatively light intermittent welds. Extensive residual strain readings were taken after each major fabrication stage.

In the USA extensive test programs of large stiffened shells have been initiated by the American Petrolium Institute (API), the American Bureau of Shipping (ABS) and the offshore industry. Some of these programs have been completed [22], [23], and more are in progress. For example, a major program sponsored by Conoco and ABS involves 58 ring- and ring- and stringer-stiffened cylindrical shells of 12 different geometries, to be tested under combined axial compression and hoop compression. The shells are welded steel cylinders of $(R/t) = 150\text{-}500$, $(a/R) = 0.4\text{-}1.0$, $(a/\sqrt{Rt}) = 1.73\text{-}17.9$ and for stringers $(b/t) = 30\text{-}131$. Most of the tests will be conducted at the Chicago Bridge and Iron Co. and the largest cylinder will be 75 inches in diameter. Sixteen of the smaller shells, of 36 inch diameter, will be tested at the University of Glasgow. The results, which should soon be available, will be used by ABS to develop new strength criteria for offshore structures. As the results of these extensive large scale tests become available, more detailed comparisons with smaller shells, such as the UC, IC shells, will be possible, yielding more reliable correlation.

The details of the Det norske Veritas tests on 12 aluminum alloy welded 2 m diameter spherical shell segments mentioned in [1] have since been published [47]. These tests are good examples of modern well instrumented tests on large shells, and the fabrication of specimens shows the possibilities of technology transfer, in this case from aerospace to marine structures.

Another type of large stiffened shells, 1.8 m diameter stringer stiffened corrugated cylindrical shells are being tested in Aachen in the framework of the joint research project RWTH Aachen-Technion [36], [48]. Figure 5 shows shell QW S 8-1 of this test series, which extends the vibration

Fig. 5. Vibration correlation test on 1.8 m diameter stringer-stiffened corrugated cylindrical shell QW5 8-1 under axial compression, from [48].

correlation studies [35] [36] to large shells. The shell has circumferential corrugations customary in agricultural silos and a nominal (R/t) ≈ 590, the corrugations providing effective closely spaced ring stiffening.

d. Curved Panels.

Since curved panels and open shells raise additional loading and boundary condition problems, the experimenters preferred, as pointed out in [1], where possible, to test complete shells which, on account of their rotational symmetry, can be defined more precisely. Sometimes they even resorted to the device of testing a number of curved panels together, effectively forming a complete cylindrical shell. The scarcity of buckling tests on curved panels, mentioned in [1], therefore continued. Some preliminary tests carried out in connection with vibration correlation studies [36] reemphasize the problems of curved panel testing. It appears that much work is still needed to develop methods for satisfactory load distributions and reliable definition of boundary conditions in curved panel tests.

e. Composite Shells.

The widespread introduction of composite structures in many engineering applications in the late seventies was accompanied by many buckling experiments. The first wave of tests on composite shells has been reviewed in [1]. Rapid growth in the application of advanced composites, in particular in the aerospace industry, would suggest a similar growth in composite shell tests. However, surprisingly few such tests have been performed since 1980. The spectacular NASA Langley Tests on a large 3.00 m diameter ring-stiffened graphite epoxy corrugated cylindrical shell have been completed, and a detailed report is in preparation, but as yet no follow up tests are in progress. Some test programs on buckling and postbuckling of curved composite panels under axial compression and combined loads have been initiated in the aerospace industry. For example, at the McDonell Aircraft Company in St. Louis a program is in progress in connection with V/Stol fuselage applications. The test specimens for axial compression are very thin graphite/epoxy stiffened curved panels of $(R/t) \simeq 1000$ and $(b/t) \simeq 200$, each having 3 relatively heavy hat stiffeners. The skin is an 8-ply quasi-isotropic lay up and the stiffeners were cocured with the skin. Emphasis in the tests was on the cocured skin-to-stiffener joints in the postbuckling region and hence two test series were performed, one with baseline stiffeners and one with beefed-up stiffeners. Specimens for the combined shear and axial compression tests recently initiated are similar widely spaced panels, with $(R/t) \simeq 500$ and $(b/t) \simeq 150$, though of slightly different configuration dictated by the loading system. These two typical examples and other tests show that at present the main problem area seems to be the stiffener-skin relation, which is essentially similar in flat and curved panels. The scarcity of composite shell and curved panel tests is in contrast to many extensive test programs on stiffened flat panels (see for example [49]-[51]). It appears that the investigators have decided first to clarify buckling and postbuckling behavior of stiffened composite flat panels before embarking on further extensive shell tests.

The scarcity of composite shell buckling tests may also be due to the difficulty of producing good specimens. One group that continues its test programs in spite of these difficulties is that at the University of Toronto [1], [7], using primarily preimpregnated tape wound glass/epoxy cylinders. Stringer-stiffened cylinders were also made by this technique and used for combined loading tests [7]. Figure 6 shows one of their graphite/epoxy test specimens from a different recent study on asymmetric layups, in its postbuckled shape.

Fig. 6. Small graphite/epoxy shell, with asymmetric layup for higher buckling load, in its postbuckled state, by courtesy of R.C. Tennyson.

Another recent composite shell buckling test program dealt with dynamic buckling under impact, motivated by crashworthiness considerations [52]. The composite specimens (whose performance was compared with that of similar metal cylinders) were stringer-stiffened shells of fiber/glass epoxy, graphite/epoxy and Kevlar 49/epoxy in combination with graphite/epoxy. Woven or tape prepregs were used, and the hat-sections stringers were cocured.

4. Initial Imperfections

As pointed out in [1], probably the most important change in buckling experiments on shells in the previous decade was the extent of geometric imperfection measurments carried out by the experimenters. Since 1980 the trend has continued and today extensive geometric imperfection measurements are considered an integral part of a properly carried out shell buckling test, be it on a laboratory scale or on a large scale, (see for example [5], [7], [9], [35]-[42], [47]).

With the realization of the importance of the shape and amplitudes of initial imperfections, measurement techniques have been developed both for small and large shells as discussed in detail in [1]. Two recent additions to these measurement systems, however, deserve further discussion.

One is a unified multipurpose scanning and measurement system for imperfections and vibrations developed at Technion jointly with Instrument and Control Ltd., Haifa [53]. As pointed out in [1] and [54], [55], deficiency of some of the imperfection measurement systems in use today is that imperfections are measured before the shell is fixed in its final boundary conditions. This was particularly emphasized in relation to vibration correlation studies [35]. The new system [53] has been designed to overcome this drawback.

A closed loop noncontact probe is used to measure the vibrations and imperfections of cylindrical shells. Measurements include natural frequencies, modes of vibration, and mapping of imperfections. All measurements and mapping are carried out by the same probe inside the closed shell, which is fixed in its final boundary conditions, and an electronic control permits automatic execution of the different modes of operation. Figure 7 shows the complete system in operation on a spot welded shell AAC-1, of the joint RWTH Aachen-Technion project: (a) without the shell and (b) with the shell in position. This system can accommodate specimens of diameter 240-500 mm and length up to 470 mm. In the imperfection mode the digital data is recorded on a Nova 2 minicomputer connected to the experimental system, and analog plots of the imperfection shapes of the shell are also obtained for real time control. The digital results are stored on a computer disk which enables further numerical analysis. The scan can be taken at different load levels. In the vibration mode, the test procedure is essentially similar to that employed on the smaller integrally stiffened shells (see for example [34], [35] or [61]) though operation of the new system differs considerably. The vibration plots obtained are very clear.

The second system is the "STONIVOKS" system [20] developed at Delft University of Technology for statistical research of the influence of initial imperfections. The test apparatus is designed for the beer can specimens, discussed in Section 3, and the emphasis is on reproducible end conditions. The test apparatus is built round a stable rotating base platform and has been produced to high accuracy. It uses an LVDT contact probe of very small contact force (about 5 gr.) and the measuring procedure and data acquisition are highly automated.

a)

b)

Fig. 7. Test set-up for vibration correlation test on spot-welded shell AAC-1, with multipurpose scanning and measurement system for vibrations and imperfections. a. The scanning and measurement system, b. Test system in operation, (from [36]).

As mentioned in [1], an International Imperfection Data Bank, with branches in Delft and Haifa, has been established in order to facilitate evaluation of imperfection measurements and correlation studies. Since the following paper discusses this Data Bank in detail, it will only be very briefly dealt with here. The "standard" representation of the measured imperfection data proposed by Arbocz and Williams [56] has been adopted not only in the Delft and Haifa reports (for example [54]-[58]), but also by other investigators (for example [37], [40], [41]). A uniform form of data presentation in 5 files for data storage has been agreed upon [55], [57] and evaluation of stiffened shell characteristics from the data collected at both branches is in progress [55], [58].

As mentioned earlier [1], once the initial imperfection data is available and the boundary conditions are well defined, buckling loads can be computed for the shell. Such computations have been continued for more shells of the vibration correlation studies at the Technion [35], [55] as in Table 1 of [1]. The comparison between the theoretical buckling load ratios ρ_{th} and the experimental buckling load ratios (adjusted by vibration correlation) ρ_{sp} remains between very good and fair [35], [55].

It should be mentioned that imperfections other than geometric, such as material defects or residual stresses, are also of importance, but for the buckling of shells the geometric imperfections occupy the dominant position except in the plastic range. In the case of welded shells, however, the residual stresses may also be of similar importance in the elastic range. However, though considerable efforts have been devoted to their measurement, the integrated effect of residual stresses on the buckling of shells can still not be adequately predicted.

5. Boundary Conditions and Loading.

As pointed out in [1] the influence of boundary conditions has been intensively studied in the seventies, in particular for stiffened shells. For closely stiffened shells (as prescribed by optimization), governed largely by general instability of the entire shell, the effects of imperfections on buckling behavior have been found to be less pronounced, but the influence of boundary conditions becomes predominant. This fact has motivated closer studies of the boundary effects and also of the correlation between vibration and buckling, discussed in detail in [1] and briefly summarized in the next section.

Trends in shell mountings noted in [1] have continued. Nominal clamping, by casting the end of the shell in some material, such as a low melting temperature alloy (Cerrobend) or epoxy is usually employed for small model tests (for example [35], [40] or [46]). This boundary condition is assumed to be complete clamping, though vibration correlation tests have cast doubts on this assumption. The additional imperfections introduced by clamping (see [54]) are still ignored, though more recent tests reconfirm their occurrence (see for example [55] or [59]) and underline the need for further study. For large shells the 3-bay specimens, with the middle bay representing the test shell, continue to be used in some experiments [40], but this method is very expensive and does not assure complete simulation of actual boundary conditions. Some significant advances in shell mounting procedures are therefore still needed. The vibration correlation technique, which is also being applied now to larger shells [48], is one promising direction.

The problems of load nonuniformity are still present in recent shell buckling experiments, though more attention is given to load distribution measurements (see for example [40]). Very recent tests at the Technion also reaffirm the importance of load distribution measurement even in closely stiffened shells.

Eccentricity of loading, usually defined as the radial distance between the line of axial load application and the shell mid-skin, has been shown in the previous decade to have considerable influence on the buckling load of stringer-stiffened shells [1]. Since load eccentricity is usually not well defined, as it depends on the detailed behavior of the joint under load, its nondestructive determination would significantly improve buckling load predictions. Efforts in this direction are continuing [36].

6. Nondestructive Testing

As emphasized in [1], nondestructive methods in shell buckling experiments emerged in the previous decade as an important tool. The two groups of nondestructive test methods, those for determination of boundary conditions and those for direct determination of buckling loads, have been discussed in detail in [1] and will only briefly be recapitulated here with emphasis on advances since 1980.

The vibration correlation technique for determination of boundary conditions consists essentially of a nondestructive experimental determination of the lower natural frequencies for a loaded shell, and evaluation of equivalent elastic restraints representing the actual boundary conditions. It is based on the similarity between the strong influence of axial and rotational restraints on free vibrations of stiffened shells, in particular for the lower natural frequencies whose mode shapes resemble the buckling modes, and that observed for buckling

[35], [59], [62].

The technique was applied to shells of different Technion test series for axial compression loading. Figure 8 (reproduced from [35]) presents the results for 31 shells, 9 recent test results having been added to Fig. 28 of [1], and again shows the significant reduction in scatter as a result of the experimental determination of the boundary conditions pointed out

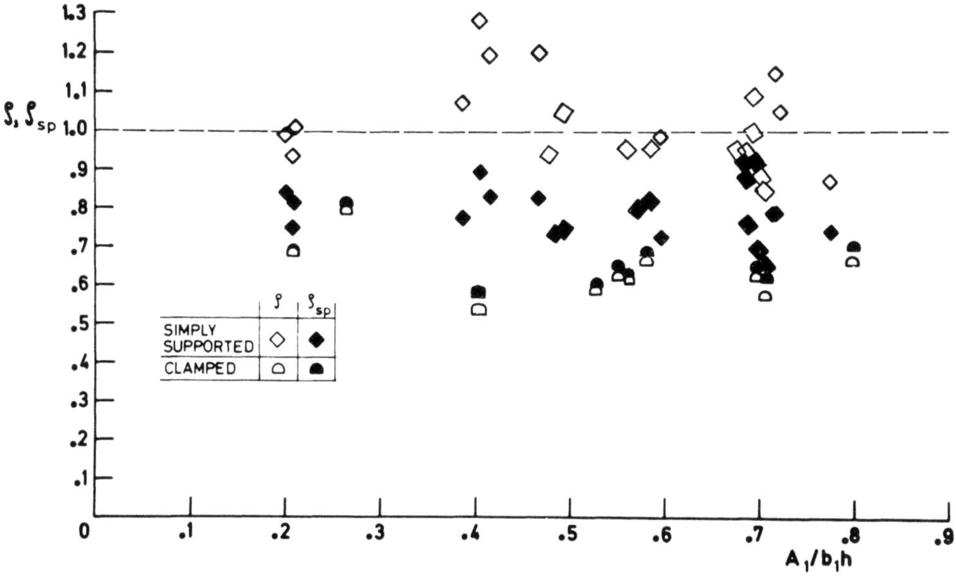

Fig. 8. "Knock-Down Factor" of 31 simply supported and clamped stringer-stiffened cylindrical shells corrected for experimentally (non-destructively) determined boundary conditions. ρ_{sp}, the corrected "Knock-Down Factor" is compared to the original ρ_{SS3} and ρ_{C4} values (from [35]).

in [1]. The scatter of the knockdown factor is reduced from 0.6-1.3 to 0.6-0.9, the low values of ρ_{sp} relating to clamped shells (probably on account of the additional imperfections introduced by clamping, which have already been mentioned in Section 4).

Since in the vibration tests an actual imperfect shell is measured, initial imperfections are indirectly included or "lumped" in the correlation. Theoretical studies of the influence of imperfections on vibrations of unstiffened and stiffened cylindrical shells have shown that imperfections have a strong influence on the frequency of vibration, similar to that on buckling (see for example [63]). The influence of boundary conditions and that of imperfections are therefore similar for vibrations and buckling and can be "lumped". However, when realistic end supports are employed they may also have load eccentricity, which may not be well defined a priori, and depends on the tolerances and behavior of the joints under load. The effects of load eccentricity differ from vibrations and buckling, and could therefore obscure the correlation. This difficulty was overcome when careful evaluation of test data revealed a salient property which distinguishes vibrations in the presence of significant load eccentricities [34]. This important property is the large increase in the frequency ratio squared $(f/f_{SS4L})^2$ with the number of circumferential waves n of the vibration pattern, which does not occur in the absence of load eccentricity. If one now plots the corresponding theoretical frequency ratio squared for some likely values of load eccentricity (for SS4 boundary conditions in the presence of load eccentricity $\bar{e}$) the same property becomes evident. One can easily find a load eccentricity which has the same slope as the experimental one.

This frequency slope property presents a tool for nondestructive identification of significant unknown load eccentricity, and once the load eccentricity has been identified a modified vibration correlation method is available [34]. Since this slope property is a crucial element of the method, it was verified on shells of prescribed load eccentricity [64], [65].

The vibration correlation method for definition of boundary conditions has been applied to integrally stringer-stiffened cylindrical shells on a variety of boundary conditions, both laboratory type edges (simple supports, clamped edges or edges with prescribed load eccentricity [35], [60]-[65]) and "practical" boundary conditions, which simulate actual joints [34]. Recently another test series with "practical" boundary conditions, the DK shells designed for variation of boundary conditions during the test, has been initiated. Figure 9 shows the postbuckling pattern and end rings of a typical shell of the series, shell DK2.

A series of tests of larger 500 mm diameter stringer-stiffened, spot-welded or riveted, cylindrical shells (see Fig. 7) has also been initiated in cooperation with RWTH Aachen [36], as well as a series of tests of large 1.8 m diameter stringer-stiffened corrugated shells [48] (see Fig. 5), in order to increase confidence in the vibration correlation technique and transform it into an industrial tool.

Vibration correlation can also be employed for direct prediction of buckling loads. At Technion, the same experimental technique used to

Fig. 9. Technion integrally stringer-stiffened 240 mm diameter cylindrical shell DK3 on realistic boundary conditions, buckled under axial compression after vibration correlation test.

define boundary conditions was extended to direct prediction of buckling loads. The method is essentially curve fitting to the experimental points of the frequency squared versus axial load, but using only those points below 50-60% of the buckling load, to make the procedure truly nondestructive. The curve fitting is actually carried out with respect to a straight line, representing the experimental points to a certain exponent q, obtained from previous vibration and buckling tests on similar shells [66]. This exponent was related to the dominant geometric properties of stringer-stiffened cylindrical shells, the Batdorf shell parameter Z and the stringer area ratio (A_1/b), and a functional relation with two empirical constants was found [66]. Being empirical, this relation is obviously limited to the range of geometries tested, and requires many more tests for wider applicability. In [66] and [35] the buckling load ratios for direct prediction ρ_{extrap} of all the shells studied are summarized and compared the the theoretical predictions for effective boundary conditions found with the vibration correlation technique ρ_{sp}. The scatter of ρ_{extrap} is found to be about 2/3 that of ρ_{sp}. Similar comparisons for additional shells tested since confirm this further reduction in scatter.

7. Combined Loading

Since combinations of external pressure and axial compression and tension are very important loading conditions for unstiffened and stiffened shells, in particular for offshore applications, renewed interest and considerable research effort have been directed to combined loading, as already pointed out in [1]. In tests too more emphasis has been put on combined loading, in particular for stiffened and composite shells, for which no reliable interaction curves were available, and this emphasis has greatly increased in recent years (see [7]-[10], [22], [23], [30], [31], [36], [45], [59], [62]).

For example, a test program on a large fabricated steel shell, 6.35 m in diameter and 3.14 m long, with $(R/t) \approx 500$, was carried out at the Chicago Bridge and Iron Company [8], [23] for buckling under various combinations of axial compression, external pressure and concentrated loads. Tests were performed on the shell without and with ring-stiffeners, all the tests being in the elastic range. The different stiffener arrangements (shown in Fig. 7 of [8]) were obtained by progressively adding rings to the shell as the tests advanced and reducing the size of the central ring. The tests are an excellent example of the "multiple tests on the same specimen" approach, also discussed in [1]. This approach was advocated and extensively used on thin-walled aluminum alloy conical shell in the sixties (see for example [67], [68]) and for composite shells in the seventies (for example [69]), sometimes also with an internal mandrel to prevent damage due to repeated buckling [70]. With the obvious advantage of reduced scatter the repeated buckling on the same specimen nonetheless introduces minor permanent damages which slightly reduce the buckling capacity of metal or composite shells, even with $(R/t) \approx 500$ [8], in contrast to Mylar (or similar polyester) shells [18] which may be repeatedly buckled many times without any deterioration. For metal cylindrical shells with $(R/t) \ll 500$, and certainly for shells which buckle in the plastic range, separate specimens must therefore be employed for each point to obtain reliable interaction curves (see for example [22], [31]).

The need for more experimental data under combined loading, especially in the inelastic range, outlined in [8] or [9], has resulted in the extensive test programs of large stiffened shells under combined loads in the USA, discussed in Section 3c.

At the Technion the importance of combined loading motivated the initiation of a program extending the vibration correlation technique to combined loading in order to obtain improved interaction curves. The first test program [62] involved 4 steel shells and 6 aluminum alloy shells, of geometries similar to the stringer-stiffened shells tested undser axial compression, and established the applicability and usefulness of the technique as a nondestructive method for improved definition of boundary conditions also for external pressure and combined loading. The results were also used to evaluate theoretical interaction curves, and to obtain improved curves by referring to the effective boundary conditions found with the aid of vibration correlation. In order to obtain better interaction curves, an additional series of tests [59], with nominally clamped or simply supported integrally stringer-stiffened shells was carried out.

Significant improvement in the interaction curve is obtained by the vibration correlation technique. It should be pointed out that these improved interaction curves are obtained by a series of definitely nondestructive tests on the same shell at different combinations of axial compression and external pressure, taking care not to exceed about thalf the theoretical (linear theory) buckling load at each load combination. In the more recent tests [59], the use of a single specimen for the entire range of load combinations also for buckling tests, which was found difficult to achieve in the earlier tests on stiffened shells [62], has been successfully performed. Additional tests of stiffened shells under combined loading are in progress.

8. Postbuckling Behavior

As pointed out in [1], the previous decade has seen many careful experiments on cylindrical and conical shells extending into the far postbuckling range. Some of these, for example the experiments of Yamaki and his associates at Tohuko University, Sendhai, noteworthy for the detailed measurements of postbuckling radial displacements and their circumferential and axial distributions, have recently been continued [19] for pressurized circular cylindrical shells of $(R/t) \approx 400$ under torsion, a typical case of combined loading. Again polyester foil specimens, which are capable of relatively large elastic deformation, are used, and the results are therefore not directly applicable to metal shells, for which large deformations would be mostly plastic, or to composite shells where material damages would appear much earlier.

Closely stiffened metal shells, as pointed out in [1], show significant improvements over unstiffened shells in postbuckling behavior even under axial compression. Furthermore, eccentricity of loading which is usually present in realistic joints [34], though reducing the buckling load, is accompanied by large improvements in postbuckling behavior [64], [65]. Hence closely stiffened shells may be advantageous for some applications that require a considerable postbuckling load carrying capacity for safety in unlikely extreme loading cases.

Finally, as in [1], it should be emphasized that on account of the large displacements involved, a proper study of the postbuckling behavior of shells should include the appropriate stress-strain behavior, and therefore specimens should simulate the material properties of the full scale structure.

9. Plastic Buckling

Whereas theoretical and numerical studies of plastic buckling differ materially from the elastic case, the overall setup of a plastic buckling test is nearly identical to that of an elastic buckling test. Usually the transfer from the elastic regime to the plastic regime is practically undefined, unless the test is a priori designed so that buckling shall initiate well in the plastic region. The differences appear primarily in the specimens, discussed earlier, in the interpretation of the results and sometimes in measurement techniques.

The recent plastic buckling tests of unstiffened shells [25]-[33], [71] and stiffened shells [42], [44], [72] fit this pattern and the experimental techniques they employ do not differ from those discussed in [1] and in the previous sections of this paper. The test results for cylindrical shells exhibit some general properties peculiar to plastic buckling: (a) Thick-walled cylindrical shells, which buckle essentially in an axisymmetric mode, are not sensitive to geometric imperfections [28]. This does not apply to other modes of plastic buckling occurring in thinner shells [71]. (b) Residual stresses and weld induced distortions can significantly lower buckling stresses [42], [44]. (c) The effect of rotational restraint (clamping) on the buckling load is small [28]. (d) For axisymmetric modes, buckling commences near the ends of the cylinder. This axisymmetric bulge represents a predictable "imperfection" [28], [72], [4].

However, for design purposes the borderline between elastic and plastic regions is most important, but buckling in this area has not yet been thoroughly investigated. Much experimental work is therefore still required.

10. Composite Shells

Composite structures exhibit much stronger material-structure interrelations than similar metals. This can be used to advantage by designers, but makes the task of the experimenter more difficult, as was emphasized in [1]. Our knowledge of the buckling behavior of composite shells is still very incomplete, but the experiments in the previous decade have already made it clear that the buckling behavior of laminated fiber-reinforced shells, and even more their postbuckling behaviour, differ considerably from that of similar metal shells, particularly in the violent failure of composite shells tested.

As indicated in Section 3e, the obvious need for more experimental research demonstrated in [1], should have resulted in many buckling tests of composite shells, but apparently these will have to wait till the behavior of flat panels is more completely understood.

11. Special Problems and Techniques

As the last topic, two of the subjects considered in [1] under this heading are reexamined:

a. Cutouts

The effect of cutouts on the buckling of shells continues to be a challenge to experimenters and analysts. A recent example is a test program on reinforcing of openings in shells subjected to axial compression and combined loading carried out at the Chicago Bridge and Iron Co. [46]. The test specimens were 15 inch diameter Mylar shells (see Fig. 3), and 24 inch diameter Lexan shells, and the specimens were reused with different openings and reinforcements to permit parametric studies. The most interesting aspects being investigated are the effect on buckling of adding reinforcement both normal and along the cylinder wall, the influence of stiffeners rather than pads or neck reinforcement in ring-stiffened cylinders, and the influence of combinations of external pressure and axial load on the effects of reinforcements. One may note that these parametric tests represent an excellent recent example of reason (e) for shell buckling experiments outlined in Section 2.

b. Optical Techniques

In [1] the efforts devoted to development of holographic methods, speckle interferometry and moire technique for shells were reviewed. It was concluded there that in 1980, in spite of promising results in the laboratory, these methods, except the moire techniques, had not yet reached industrial application. Very little progress has been reported since then in the application of optical methods to curved surfaces, and a literature search yielded only one paper since 1979 [73], which describes a white light speckle method for deformation measurements on a spherical ball. It appears that optical methods still require much research before they can be widely employed for measurement of shell deformations. This applies also to moire techniques which, though they are now standard method for flat or slightly curved panels, have not yet been used successfully on shells.

12. Summary and Conclusions

The present survey attempts to evaluate the recent progress in shell buckling experiments as a continuation of the trends outlined by the developments in the previous decade reviewed in [1]. Reexamination of the conclusion reached there shows that recent experiments follow those trends and corroborate the conclusions. They are therefore briefly recapitulated here with the modifications resulting from the new inputs superimposed on them.

The motivations for experiments presented in Section 2 are essentially the main conclusions. They emphasize that experiments are vital for sound physical insight, which is a prerequisite to meaningful analysis, even when performed with the aid of powerful digital computers. The physical process of shell buckling, postbuckling and collapse is a complicated phenomenon, highly nonlinear, combining geometric and material nonlinearities, evident in particular in the final collapse stage. Even the most advanced programs still use fairly crude mathematical models, and as these improve better inputs will be required. Boundary conditions, joint efficiencies, geometric and material imperfections, material behavior of new materials, are examples of such inputs that only careful experiments can provide.

In the seventies and particularly in recent years, experiments have become more precise and have begun to yield a large amount of data that characterizes more accurately the shells tested. This data, when properly measured, collected and reduced can lead to significant improvements in our buckling analysis.

The trends in shell buckling experiments of the previous decade [1] have continued, and with recent developments they can be summarized as follows:

a. Greater precision and uniformity in model fabrication, particularly in welded shells and plastic models.
b. Good correlation between small scale tests and corresponding tests on large shells, if the model fabrication processes are properly correlated to the full scale ones.
c. Universal acceptance of imperfection measurements as an essential element of a well executed test.
d. Development of nondestructive techniques for assessment of the parameters governing the buckling behavior, and their interaction with computations.
e. Renewed emphasis on combined loading tests.
f. Emergence of composites as a primary material for shell structures and the recognition of their different buckling behavior.
g. Renewed interest in the behavior of shells in the deep postbuckling region for assessment of fail-safe properties.

Finally, the author feels that, as pointed out in [1], the following areas of shell buckling experimentation deserve attention in coming years:

a. Composites. Extensive experiments are required to clarify their buckling behavior under different load conditions, load combinations and repeated buckling.
b. Load Interaction. Reevaluation of interaction curves and how they are affected by material behavior and boundary conditions.
c. Imperfection Data Bank. "Standard" presentation of geometric imperfection measurements and collection in a central data bank is essential for correlation studies.
d. Material Imperfections and residual stresses should be given more attention.
e. Load Distribution. More precise and extensive measurements are required to assess the influence of load nonuniformities on shell buckling.
f. Mode Interaction. To evaluate the theoretical studies that have shown this to be one of the dangers of structural optimization in stiffened shells.

g. Cutouts. More work is required for stiffened and composite shells to define optimal reinforcements.

h. Optical Techniques should be further developed for identification of danger regions in a shell structure.

i. Elastic-Plastic Buckling. Extensive tests are required to provide a broader foundation to the many theoretical studies in particular for stiffened shells.

j. Buckling due to Dynamic Loading. More systematic experiments are needed both in the elastic and plastic regions.

Acknowledgements

The support of the DFG in cooperation with the RWTH Aachen for part of the work is gratefully acknowledged. The author would like to thank Dr. T. Weller of the Technion for valuable discussions. He would also like to thank his colleagues in laboratories in other countries for the useful information they kindly provided. He wishes to thank Mrs. A. Goodman for typing the manuscript.

References

1. Singer, J.: Buckling experiments on shells - a review of recent developments. Presented at Euromech Colloquium No. 128 "Stability, Buckling, and Postbuckling Behavior, Foundation and Analysis". Delft, March 31 - April 2, 1980. (To be published in Solid Mechanics Archives). Also, TAE Report 403, Dept. of Aeronautical Engineering, Technion, Israel Inst. of Technology, April 1980.

2. Babcock, C.D.: Experiments in shell buckling. Thin-shell structures, theory, experiments and design. Fung, Y.C. and Sechler, E.E. (eds.), Prentice-Hall, Englewood Cliffs, N.J., 1974, 345-369.

3. Budiansky, B. and Hutchinson, J.W.: Buckling: progress and challenge. Trends in solid mechanics. Besseling, J.F. and van der Heijden, A.M.A. (eds.), Delft Univ. Press 1979, 93-116.

4. Bushnell, D.: Buckling of shells - pit-fall for designers. AIAA Journal, Vol. 19, No. 9 (1981) 1183-1226.

5. Arbocz, J.: Past, present and future of shell stability analysis. Zeitschrift für Flugwissenschaften und Weltraumforschung, Vol. 5, No. 6 (1981) 335-348.

6. Citerley, R. L.: A review of imperfection sensitivity of stiffened shells. Submitted to the subcommittee on shells of the Pressure Vessel Research Committee of the Welding Research Council, Jan. 1982.

7. Tennyson, R.C.: Interaction of cylindrical shell buckling experiments with theory. Theory of shells. Koiter, W.T. and Mikhailov, G.K. (eds.), North-Holland Publishing Co. 1980, 65-116.

8. Miller, C.D.: Summary of buckling tests on fabricated steel cylindrical shells in USA. Buckling of shells in offshore structures.

Harding, J.E., Dowling, P.J. and Agelidis, N. (eds.), Granada, London 1982, 429-471.

9. Valsgard, S. and Foss, G.: Buckling research in Det norske Veritas. Buckling of shells in offshore structures. Harding, J.E., Dowling, P.J. and Agelidis, N. (eds.), Granada, London 1982, 491-548.

10. Schulz, U.: Der Stabilitätsnachweis bei Schalen. Berichte der Versuchsanstalt fur Stahl, Holz und Steine der Universität Fridericiana in Karlsruhe, 4. Folge, Heft 2, 1981.

11. Sechler, E.E.: The role of experimentation in shell research. Mechanic Today, Vol. 5, Nemat-Nasser, S. (ed.), Pergamon Press Oxford 1980, 439-449.

12. Arbocz, J. and Babcock,C.D.: The effect of general imperfections on the buckling of cylindrical shells. Journal of Applied Mechanics, Vol. 36, No. 1 (1969) 28-38.

13. Weller, T. and Singer, J.: Experimental studies on the buckling under axial compression of integrally stringer-stiffened circular cylindrical shells. Journal of Applied Mechanics, Vol. 44, No. 4, (1977) 721-730.

14. Bushnell, D.: Thin shells. Structural mechanics computer programs, surveys, assessments and availability. Pilkey, W.; Saczalski, K. and Schaeffer, H. (eds.), University of Virginia Press, Charlottesville 1974, 277-358.

15. Buchert, K.P.: Practical applications of shell research. Buckling of shells in offshore structures. Harding, J.E.; Dowling, P.J. and Agelidis, N. (eds.), Granada, London 1982, 257-283.

16. Babcock, C.D.: Personal communication. January 1980.

17. Toda, S.: Experimental investigation on the effects of elliptic cutouts on the buckling of cylindrical shells loaded by axial compression. Japan Society for Aeronautical and Space Sciences, Transactions, Vol. 23 (1980) 57-63.

18. Foster, C.G.: Interaction of buckling modes in thin-walled cylinders. Experimental Mechanics, Vol. 21, No. 3 (1981) 124-128.

19. Kodama, S.; Otomo, K. and Yamaki, N.: Postbuckling behavior of pressurized circular cylindrical shells under torsion - 1. Experiment. International Journal of Non-Linear Mechanics, Vol. 16, No. 3/4 (1981) 337-353.

20. Verduyn, W.D. and Elishakoff, I.E.: A testing machine for statistical analysis of small imperfect shells. To be presented at the 7th International Conference on Experimental Stress Analysis, Haifa, Israel, August 23-27, 1982.

21. Ostapenko, A. and Grimm, D.F.: Local buckling of cylindrical tubular columns made of A-36 steel. Lehigh University, Fritz Engr. Lab. Report No. 450.7. February 1980.

22. Marlow, R.S.: Collapse performance of fabricated cylinders under

combined axial tension and external pressure. SWRI Project No. 03-6139, Southwest Research Institute, Final Report, January 1981.

23. Miller, C.D.: Buckling design methods for steel structures, a state of the art. Presented at Integrity of Offshore Structures Symposium, Glasgow, July 1-3, 1981.

24. Smith, C.S.: Strength and stiffness of damaged tubular beam columns. Buckling of shells in offshore structures. Harding, J.E., Dowling, P.J. and Agelidis, N. (eds.), Granada, London 1982, 1-23.

25. Sherman, D.R.: Tests of circular tubes in bending. ASCE Journal Structural Div., Vol. 102, ST 11, (1976) 2181-2195.

26. Reddy, B.D.: An experimental study of the plastic buckling of circular shells in pure bending. International Journal of Solids and Structures, Vol. 15 (1979) 669-683.

27. Sherman, D.R.: Inelastic local buckling of circular tubes. Buckling of shells in offshore structures. Harding, J.E.; Dowling, P.J. and Agelidis, N. (eds.), Granada, London 1982, 365-392.

28. Sobel, L.H. and Newman, S.Z.: Plastic buckling of cylindrical shells under axial compression. ASME Journal of Pressure Vessel Technology, Vol. 102 (1980) 40-44.

29. Bushnell, D.: Elastic-plastic bending and buckling of pipes and elbows. Computers and Structures, Vol. 13 (1981) 241-248.

30. Gudramovich, V.S.; Gaiduchenko, A.P.; Demeshko, M.F. and Konovalenkov, V.S.: Plastic buckling of cylindrical shells in the case of complex loading paths. Soviet Applied Mechanics, Vol. 16, No. 1 (1980) 19-24 (Translated from Prikladnaya Mekhanika, Vol. 16, No. 1 (1980) 26-31.)

31. Miller, C.D.: Buckling of cylindrical shells under combined axial tension and hoop compression. Chicago Bridge and Iron Co. Research Report, June 1980.

32. Kitching, R.; Houlston, R. and Johnson, W.: A theoretical and experimental study of hemispherical shells subject to axial loads between flat plates. International Journal of Mechanical Sciences, Vol. 17 (1975) 693-703.

33. Galletly, G.D.: Plastic buckling of torispherical and ellipsoidal shells subjected to internal pressure. Proc. Institution of Mechanical Engineers, Vol. 195, No. 26 (1981) 329-345.

34. Singer, J., and Abramovich, H.: Vibration techniques and definition of practical boundary conditions in stiffened shells. AIAA Journal, Vol. 17, No. 7 (1979) 762-769.

35. Singer, J.: Vibration correlation techniques for improved buckling predictions of imperfect stiffened shells. Buckling of shells in offshore structures. Harding, J.E.; Dowling, P.J. and Agelidis, N. (eds.), Granada, London 1982, 285-329.

36. Singer, J.: Die Bestimmung der Randbedingungen von

orthotrop-versteiften Schalen und Teilschalen anhand von Schwingungsversuchen. Final Report 1.10.79-30.9.81, Joint Research Project RWTH Aachen-Technion, ASL Report 121, Technion, Israel Inst. of Technology, Dept. of Aeronautical Eng., Haifa, Israel, October 1981.

37. Walker, A.C.; Andronicou, A. and Sridharan, S.: Experimental investigations of the buckling of stiffened shells using small scale models. Buckling of shells in offshore structures. Harding, J.E.; Dowling, P.J. and Agelidis, N. (eds.), Granada, London 1982, 45-71.

38. Sridharan, S.; Walker, A.C. and Andronicou, A.: Local plastic collapse of ring-stiffened cylinders. University of Surrey, Dept. of Mechanical Engineering Report, August 1980.

39. Dowling, P.J. and Harding, J.E.: Experimental behaviour of ring and stringer stiffened shells. Buckling of shells in offshore structures. Harding, J.E.; Dowling, P.J. and Agelidis, N. (eds.), Granada, London 1982, 73-107.

40. Dowling, P.J.; Harding, J.E.; Fahy, W. and Agelidis, N.: Report on the testing of large and small scale stiffened shells under axial compression. Imperial College, London, Dept. of Energy, Report No. OT-R-8107, August 1981.

41. Green, D.R. and Nelson, H.M.: Compression tests on large-scale, stringer-stiffened tubes. Buckling of shells in offshore structures. Harding, J.E.; Dowling, P.J. and Agelidis, N. (eds.), Granada, London 1982, 25-43.

42. Oldland, J.: An experimental investigation of the buckling strength of ring-stiffened cylindrical shells under axial compression. Norwegian Maritime Research, Vol. 4, No. 9 (1981) 22-39.

43. Miller, C.D.: Buckling of axially compressed cylinders. Journal Structural Div. ASCE, Vol. 103, ST3 (1977) 695-721.

44. Miller, C.D. and Kinra, R.K.: External pressure tests of ring-stiffened fabricated steel cylinders. 13th Annual Offshore Technology Conference Houston, May 1981, paper No. OTC 4107.

45. Tennyson, R.C.; Booton, M. and Hui, D.: Changes in cylindrical shell buckling behavior resulting from system parameter variations. Proceedings 15th UTAM Congress, Toronto 1980, 417-430.

46. Miller, C.D.: Experimental study of the buckling of cylindrical shells with reinforced openings. To be presented at the ASME/ANS Nuclear Engineering Conference, Portland, Oregon, July 25-28, 1982.

47. Odland, J.: Theoretical and experimental buckling loads of imperfect spherical segments. Journal of Ship Research, Vol. 25, No. 3 (1981) 201-218.

48. Öry, H. and Hoffmann, H.: Bericht zum Forschungsvorhaben DFG II D 1-Or 44/1, 1.10.80-30.11.81, Institut fur Leichtbau, RWTH Aachen, Nov. 1981.

49. Foreman, C.R.: Design concepts for composite fuselage structures.

Fibrous composites in structural design. Lenoe, E.M.; Oplinger, D.W. and Burke, J.J. (eds.), Plenum Press, New York 1980, 103-123.

50. Renieri, M.P. and Garrett, R.A.: Postbuckling behavior of flat stiffened graphite/epoxy shear panels. Mcair Report No. 81-015, McDonnell Aircraft Company, 1981.

51. Peddie, G.D. and Spier, E.E.: Design, fabrication and compression testing of cocured hat-stiffened panels. To be presented at 12th ICAS Congress, Seattle, August 22-27, 1982.

52. Foye, R.L.; Swindehurst, C.W. and Hodges, W.T.: A crashworthiness test for composite fuselage structure. Fibrous composites in structural design. Lenoe, E.M.; Oplinger, D.W. and Burke, J.J. (eds.), Plenum Press, New York 1980, 241-257.

53. Rosen, A.; Singer, J.; Grunwald, A.; Nachmani, S. and Singer, F.: Unified noncontact measurement of vibrations and imperfections of cylindrical shells. To be presented at the 7th International Conference on Experimental Stress Analysis, Haifa, Israel, August 23-27, 1982.

54. Singer, J.; Abramovich, H. and Yaffe, R.: Initial imperfection measurements of stiffened shells and buckling predictions. Proceedings 21st Israel Annual Conference on Aviation and Astronautics, Israel Journal of Technology, Vol. 17 (1979) 324-338.

55. Abramovich, H.; Yaffe, R. and Singer, J.: Evaluation of stiffened shell characteristics from imperfection measurements. Collection of Papers, 23rd Israel Annual Conference on Aviation and Astronautics, Feb. 1981, 114-122.

56. Arbocz, J. and Williams, J.G.: Imperfection surveys on a 10-ft. diameter shell structure. AIAA Journal, Vol. 15, No. 1 (1977) 949-956.

57. Arbocz, J. and Abramovich, H.: The initial imperfection data bank at the Delft University of Technology - part I. Report LR-290, Dept. of Aerospace Eng., Delft, The Netherlands, December 1979.

58. Abramovich, H.; Singer, J. and Yaffe, R.: Imperfection characteristics of stiffened shells - group 1. TAE Report 406, Dept. of Aeronautical Eng., Technion, Israel Inst. of Technology, September 1981.

59. Abramovich, H.; Singer, J. and Grunwald, A.: Nondestructive determination of interaction curves for buckling of stiffened shells. TAE Report 341, Technion, Israel Inst. of Technology, Dept. of Aeronautical Eng., Haifa, Israel, Dec. 1981.

60. Singer, J. and Rosen, A.: Design criteria for buckling and vibration of imperfect stiffened shells. ICAS Proceedings 1974. Dexter, R.R. and Singer, J. (eds.), Weizmann Science Press of Israel, Jerusalem, 1974, 495-517.

61. Singer, J.: Recent studies on the correlation between vibration and buckling of stiffened cylindrical shells. Zeitschrift für Flugwissenschaften und Weltraumforschung, Vol. 3, No. 6 (1979) 333-344.

62. Abramovich, H. and Singer, J.: Correlation between vibration and buckling of stiffened cylindrical shells under external pressure and combined loading. Israel Journal of Technology, Vol. 16, Nos. 1-2 (1978) 34-44.

63. Singer, J. and Prucz, J.: Influence of imperfections on the vibrations of stiffened cylindrical shells. Journal of Sound and Vibration, Vol. 80, No. 1 (1982) 117-143.

64. Rosen, A. and Singer, J.: Vibrations and buckling of eccentrically stiffened cylindrical shells. Experimental Mechanics, Vol. 16, No. 3 (1976) 88-94.

65. Singer, J. and Segal, Y.: Further experimental studies on vibrations and buckling of eccentrically loaded stiffened cylindrical shells. TAE Report 331, Technion, Israel Inst. of Technology, Dept. of Aeronautical Eng., Haifa, Israel, Dec. 1978.

66. Singer, J. and Segal, Y.: Direct prediction of the buckling load of stiffened cylindrical shells from vibration tests. TAE Report 405, Technion, Israel Inst. of Technology, Dept. of Aeronautical Eng., (to be published).

67. Singer, J.: On experimental technique for interaction curves of buckling and shells. Experimental Mechanics, Vol. 4, No. 9 (1964) 279-280.

68. Berkovits, A.; Singer, J. and Weller, T.: Buckling of unstiffened conical shells under combined load. Experimental mechanics, Vol. 7, No. 11 (1967) 465-467.

69. Wilkins, D.J. and Love, T.S.: Combined compression-torsion buckling tests of laminated composite cylindrical shells. Journal of Aircraft, Vol. 12, No. 11 (1975) 885-889.

70. Booton, M. and Tennyson, R.C.: Buckling of imperfect anisotropic circular cylinders under combined loading. AIAA Journal, Vol. 17, No. 3 (1979) 278-287.

71. Bornscheuer, F.W.: Short report on the Stuttgart experiments with axially loaded circular cylinders in the plastic region. Discussion paper presented at International Conference on Buckling in Offshore Structures, Imperial College London, April 23-24, 1981.

72. Weller, T.; Singer, J. and Yaffe, R.: Axisymmetric plastic buckling of closely ring-stiffened shells under uniform axial compression. ASL Report 116, Technion, Israel Inst. of Technology, Dept. of Aeronautical Eng., Haifa, Israel, May 1981 (in Hebrew).

73. Chiang, F.P. and Asundi, A.: White light speckle method with tandem plates for 3-D displacement and deformation measurement on curved surfaces. Applied Optics, Vol. 20, No. 13 (1981) 2167-2169.

THE IMPERFECTION DATA BANK, A MEAN TO OBTAIN REALISTIC BUCKLING LOADS

J. ARBOCZ

Department of Aerospace Engineering
Delft University of Technology, The Netherlands

Summary

Despite decades of intensive research work most shells continue to be designed by the so-called Lower Bound Design Philosophy which has been in use since the turn of the century. It is shown that by collecting the data of extensive initial imperfection surveys on full scale structures in Initial Imperfection Data Banks improved design criteria can be developed. Especially the use of the Monte Carlo Method to derive reliability functions (based on measured initial imperfection distributions that are characteristic of the different fabrication processes), may finally provide the practicing structural engineers with a design procedure that incorporates the concept of imperfection sensitivity in a rational manner.

Introduction

In this paper an attempt will be made to show that the results of detailed initial imperfection measurements on full scale structures can be used effectively to develop improved design criteria for weight critical, low margin of safety, axially compressed cylindrical shells. As is known such structures are usually buckling critical. A typical design procedure used for the layout of such structures can be summarized as follows:

1. Lay-out the preliminary dimensions.
2. Select a wall construction and a stiffening concept.
3. Use one of the many shell of revolution codes[1,2] to calculate the buckling load of the "perfect structure" taking into account the appropriate boundary conditions and the prebuckling deformations.
4. Select a "knock-down" factor to account for the "imperfections" present in the finished product.
5. Apply the appropriate safety factor.

Buckling of Shells, Ed.: E. Ramm
Proceedings of a State-of-the-Art Colloqium

Currently available computer codes can usually model the "perfect" structure in great detail including discrete elastic stiffeners, realistic end attachements, various wall constructions etc. Thus the buckling load of the "perfect" structure can be calculated very accurately. The great unknown in the above procedure is the magnitude of the "knock-down" factor by which the buckling load prediction for the perfect shell (with all its 9 significant figures) should be multiplied by in order to arrive at a reliable buckling load prediction for the real structure.

All current shell design manuals, including the "DASt Richtlinie 013" from the "Deutscher Ausschuss für Stahlbau" are based on the so-called "Lower Bound Design Philosophy". That is, they recommend the use of an empirical knock-down factor, which is so chosen that when it is multiplied by the classical buckling load a "lower bound" to all existing experimental data is obtained.

For applications where the total weight of the structure is of no concern the Lower Bound Design Approach is simple to apply, and in most cases it will provide a safe buckling load prediction. However, in those cases where the design is weight critical one is usually forced to accept a smaller margin of safety and hence a more refined method of design and analysis is called for.

Extensive research over the past few decades[3,4,5] has shown that for axially compressed thin cylindrical shells the largest portion of the "knock-down" factor is due to initial imperfections in shell geometry, with thickness variations, plastic behavior, initial stresses and poor load distribution contributing the remaining part. (It is assumed here that the effect of realistic boundary conditions has been properly accounted for.)

It has also been established[6] that if the initial geometric imperfections of a shell are known then with the current generation of nonlinear structural analysis codes it is possible to predict the actual buckling load to within about 10%, a margin that is within the accuracy expected for imperfection sensitive buckling load calculations. Based on the results of this study Arbocz and Babcock[7] have proposed a new design procedure which makes an efficient use of the currently available non-linear shell analysis capabilities. It is felt that especially for applications where the total weight of the structure is one of the critical parameters (i.e. aerospace structures) this new method will provide for definitive improvements in the design process.

The difficulty in applying the theoretical findings of the imperfection sensitivity studies is related to the fact that in most cases they require some advanced knowledge of the geometric imperfections of the structure under consideration, which is rarely available. For a prototype the imperfections can be measured experimentally and then incorporated into the theoretical analysis to predict the buckling load. This approach, however, is inpracticable for predicting the structural strength of shells manufactured in normal production. The best one can hope to do for these shells is to establish the characteristic initial imperfection distribution, which a given fabrication process is likely to produce, and then to combine this information with some kind of statistical analysis of both the imperfections and the critical loads, a kind of Statistical Imperfection Sensitivity Analysis. Thus for imperfection sensitive structures any further advances in our buckling load prediction capabilities are closely dependent on the establishment of the so-called Initial Imperfection Data Banks[8,9].

To assist those that might be planning to do imperfection surveys in the near future, a few examples of how large scale or full scale imperfection measurements have been carried out are presented. For a detailed review of the state-of-the-art of imperfection surveys the reader is referred to Singer's authoritative paper from 1980[10]. Further a portable computerized data acquisition system with the corresponding data reduction procedure is described. Characteristic imperfection distributions associated with different fabrication techniques are shown. Finally the derivation of reliability functions based on these measured characteristic initial imperfection distributions is discussed. It is shown that once the reliability function for a specific fabrication process has been found the designer can select the appropriate "knock-down" factor routinely.

Imperfection Surveys on Large or Full Scale Shells

In general, it has now been accepted that in order to apply the theory of imperfection sensitivity with confidence one must know the type of imperfections that occur in practice.

Thus the initial imperfection surveys carried out on laboratory scale shells by the Caltech group[11,12] and others[9,13] has been followed by imperfection measurements on large scale and full scale structures in different parts of the world[14,15,16]. When carrying out such measurements one must always remember that Koiter's work[3,17] has shown conclusively that the "knock-down" factor depends not only on the magnitude but also on the shape of the initial imperfections. Thus it is not sufficient to spot check the shell surface for the maximum imperfection amplitude by carrying out selected circumferential and/or axial scans. One should always provide for sufficient cross-reference data, so that later the individual scans can be pieced together to a complete surface map of the measured structure via numerical techniques on a digital computer. The need for providing cross-reference data sometimes poses added complications but as will be illustrated by the subsequent examples in most cases the difficulties can be overcome rather routinely.

Imperfection Surveys at NASA-Langley

In 1977 Arbocz and Williams[15] published the results of detailed imperfection surveys carried out on a large (radius: 1534.0 mm, length: 2387.6 mm, wall thickness: 2.54 mm) integrally ring and stringer stiffened shell. Figure 1 shows the test specimen with the equipment used to carry out the imperfection scans visible on the side of the shell. The technique employed measured the deviation of the cylinder outer surface relative to an imaginary cylindrical reference surface. Physically, this was accomplished with a 3.0 m long aluminum guide rail supported on the outside diameter of the two steel end rings, which were machined so accurately that they matched to within 0.15 mm. A trolley car carrying a direct current differential transformer instrument was slowly driven along the guide rail by an electric motor. The position of the car was electronically measured using a potentiometer. Accuracy of the displacement measurements was to within $\pm$ 0.05 mm. The guide rail was moved stepwise around the cylinder circumference in 5-degree increments thus yielding a total of 72 discrete scans for the complete cylinder. Discrete digital data was recorded every 6.0 mm along the cylinder length, thus yielding 343 data points for a typical 2057.4 mm scan. This digitized data were recorded on a magnetic tape and used later for data reduction.

In all cases where one attempts to measure the exact shape of a shell, before one can determine the initial imperfection one has to define the perfect cylinder. In the above case, the radial deviations from the imaginary cylindrical reference surface, defined by the two very accurately machined steel end rings and the rigid aluminum rail connecting them, were measured. Next the data reduction program described in Reference [18] was used to find the best-fit cylinder to the measured data of the initial imperfection scan. Using the method of least-squares, this program computed the eccentricities Y_1 and Z_1, the rigid body rotations ε_1 and ε_2 and the mean radius R (see Figure 2). Finally, the measured displacements were recomputed with respect to the newly found "perfect" cylinder. These values were then used to prepare the 3-dimensional plot of the initial imperfections shown in Figure 3. The three welded seams are clearly visible in the longitudinal direction.

Imperfection Surveys at Georgia Institute of Technology

In the early 70's Horton and his co-workers carried out a series of shell buckling tests at the Georgia Institute of Technology [14]. Both small and large scale specimens were tested. The large shells (radius: 945.8 mm, length: 2743.2 mm, wall thickness: 0.643 mm) were made of 7075-T6 aluminum alloy, each consisting of 6 identical panels. On the inside they were reinforced by 312 closely spaced Z-shaped stringers (see Figure 4). One edge of each panel was joggled, and two stringers were riveted along each joint line. The remaining stringers were attached to the sheet with adhesive. The shells were held circular by two heavy end rings, which were rolled out of [-shaped extruded sections, and by 7 equally spaced Z-shaped rings on the outside. On two of the large shells complete imperfection surveys consisting of 32 equally spaced circumferential scans were carried out. For this purpose stiff end-plates with central bearings were attached to the specimen, which then was mounted in a heavy-framework, with its axis horizontal, in such a fashion that it could be rotated about its axis. A 3 meter long precision straight edge was positioned parallel to the shell and served as a reference beam. A LVDT (linear voltage differential transducer) was then attached to the straight edge at different axial positions. The shell was rotated about its axis and the variation of the profile recorded at equal intervals along the length of the shell. A block-diagram of the overall system is given in Figure 5. A 3-dimensional view of the measured initial imperfections referenced to the previously described "best-fit" cylinder is shown in Figure 6.

Imperfection Surveys at Different Aerospace Companies in the USA

In carrying out imperfection surveys on a manufacturing floor one has to be able to adjust to the realities of working on a tight schedule. Also, in most cases, one has to use the rotation hardware that is present at the manufacturing site in order to make the required axial and circumferential scans of the test specimen. This data is then later pieced together by an appropriate data reduction program using a digital computer. In order to ascertain that the data recorded is of good quality a partial data reduction should be done on site.

The Caltech group has used the two piece portable rail system shown in Figure 7 to carry out imperfection surveys on flight hardware at different aerospace companies[19]. On the same figure one of the test shells is shown mounted in a rotation device. The rail consists of a standard I-beam that is connected with tapered pins at the center. It is supported by a center stand which is movable in the vertical (z-) direction and by two end stands which have x-y-z positioning capability. The displacement transducer is mounted on a cart which slides along the rail. The cart is spring loaded so that its precision bearing rollers are pressed against the reference axis (the front edge of the top flange). The rail can be aligned optically using a theodolite and an optical target mounted on the cart. Deviations from straightness are determined using this instrument. The initial error can be determined with the present method to about $\pm$ 0.01 mm over the 7.6 m length of the beam. The exact shape of the beam is recorded and later removed from the measured data during the data reduction process. The shape of the shell generator is thus measured with respect to this rail by means of an LVDT of sufficient range to determine not only the expected initial imperfections but also any misalignment of the reference rail with respect to the "best fit perfect shell" axis system. The position of the cart with the LVDT along the rail is determined by a black and white tape attached to the rail. The tape is read by a light emitting photo-diode pair, which generates a square wave to trigger the data acquisition system. The cart and the rail are also shown in Figure 8. The data (LVDT and position signal) are carried to the data acquisition system by a flexible cable which follows the cart on a slide wire. The cart is moved by a steel cable driven by a variable speed motor and limited by switches mounted on the rail.

The portable data acquisition system was specially designed for this type of field testing. It consists of an analog to digital converter, an HP 9825A programmable calculator to control the data acquisition sequence and perform on-site date reduction and a digital x-y plotter to display the raw data as well as the reduced data.

The A/D converter is an 11-bit system giving ± 5 mV accuracy for an input signal of 10 Volts. Each input channel (4 in the present system) has a fixed gain amplifier and these signals in turn are amplified by a programmable gain amplifier controlled by the calculator. The digitized signals are stored in the calculator using direct memory access in a 2-byte word. The data is then converted to the standard calculator format (8-byte word) and stored on the cassette tape. Each cassette will store approximately 30 000 such words. Typical tests consisted of from 100 000 to 250 000 data points.

Figure 9 shows a 3-dimensional plot of the measured initial imperfections referenced to the "best fit" cylinder from a large (radius: 1212.1 mm, length: 6454.1 mm, wall thickness: 1.549 mm) integrally stiffened shell constructed out of three pieces. The lines are drawn in the circumferential direction but the data were collected from axial scans. The three welded seams are clearly visible in the imperfection plot. The disturbance at the center results from a ring frame at this location which tends to minimize the weld imperfection. The waviness is caused by the pockets of the stiffened shell. A better view of this is shown in Figure 10, where the regular pattern of the pockets is clearly seen. The large jumps at the end of the shell are a result of the end domes of the shell being forced-fit inside the cylindrical section. In all 3-dimensional plots positive imperfections are outward. An exception to this rule are Figures 9 and 10 where positive imperfections are pointing inward. This change in orientation became necessary because of the form of the measured initial imperfections near the two edges. A closer look at the axial cross-plot shown in Figure 10 reveals that the initial imperfections consist of 3 main components, namely a large step function like component due to the uniform radial displacement produced by the forced-fit end domes, a large half-wave sine component in the axial direction plus a small short wave imperfection component which accounts for the waviness. Since for stability calculations the uniform radial expansion is of no consequence the edge zones have been eliminated from the data. Thus the Fourier decomposition of the measured data was done with the initial imperfections shown in Figure 11. Here positive imperfections are pointing outward as usual. The measured initial imperfections referenced to the "best fit" cylinder from another large (radius: 1527.4 mm, length: 6047.7 mm, mean wall thickness: 2.629 mm) stringer stiffened shell made out of four pieces are shown in Figure 12. The four longitudinal welded seams are once again clearly visible. Also, since this shell has only longitudinal stiffeners the short wave waviness caused by the intersecting transverse stiffeners observed on the previous shell is absent. On the other hand this shell has more waviness in the circumferential direction.

Imperfection Surveys on Aerospace Shells in Europe

The Solid Mechanics Group at the Delft University of Technology uses the one piece portable rail system shown in Figure 13 to carry out imperfection surveys on the Ariane interstage I/II and II/III shells[16]. Here the shell is positioned up-right on a 2-piece turntable, with the reference beam placed parallel to it on an adjustable tripod.

There are 3 LVDT pick-ups installed on the reference beam. The fixed ones on the top and on the bottom are bearing against the machined end rings. The third one is installed on a carriage which is moved along the beam by an electric drive to record the shape of the corresponding shell generator. Next the shell is rotated to a new position followed by another axial scan. The process is continued until the whole surface has been surveyed and recorded. The exact shape of the reference beam has been measured optically and it is removed from the measured data during the data reduction step. The axial position of the carriage is recorded by an electro-optical device which scans a strip with equally spaced cut-outs. The resulting square shaped signal is used to digitize the data in intervals of 10 mm. Special care is taken to record possible random rigid body displacements of the shell assembly with respect to the fixed position of the reference beam during rotating the turntable to a new circumferential position. This is accomplished by monitoring the planar displacements of a calibrated circular ring placed in the center of the turntable. These displacements are then also removed from the measured data during data reduction (see Reference [16] for a detailed description).

For data acquisition a system similar to the one of the Caltech group is employed. It consists of an HP 9825S programmable calculator, a HP 6940B multiprogrammer and a digital x-y plotter. The analog/digital converter card is a 12 bit system giving ± 5 mV accuracy for an input signal of 10 Volts. The digitized signals are recorded directly in the standard calculator format (8-byte word) and are later stored on the cassette tape. Once back at the university the data is transferred via a permanent line connection to the Amdahl V7/A for further data reduction whereby the programmable calculator acts as a remote terminal.

Figure 13 shows an Ariane interstage II/III shell (radius: 1300.0 mm, length: 2730.0 mm, wall thickness: 1.2 mm) which is built-up out of 8 identical panels. The joints between adjacent panels are joggled and on the outside there are 120 equally spaced hat-shaped stringers riveted to the shell surface. The whole structure is held circular by two precision machined end rings on the outside and 5 equally spaced [-shaped rings on the inside. A 3-dimensional view of the measured initial imperfections referred to the so-called "best fit" cylinder is shown in Figure 14. It should be mentioned here that in all 3-dimensional views the measured initial radial imperfections have been normalized by the corresponding wall thickness.

Initial Imperfection Data Banks

At the moment there are two Initial Imperfection Data Banks in existence, one at the Delft University of Technology[8] and one at the Technion in Haifa[9]. The purpose of establishing these data banks is twofold. First, all the imperfection data obtained at different laboratories by different investigators is presented in identical form so that a comparison and critical evaluation is made possible. Second, with the continued advances being made in the development of powerful nonlinear shell analysis codes the need for realistic imperfection distributions to be used for collapse load predictions is expected to increase. The Imperfection Data Banks are intended to provide this information.

In all cases the measured initial imperfections are represented by the following two double Fourier series:

$$\bar{W}(x,y) = t \sum_{k,\ell=0}^{N}\sum \cos \frac{k\pi x}{L} \left(A_{k\ell} \cos \frac{\ell y}{R} + B_{k\ell} \sin \frac{\ell y}{R}\right)$$

referred to as the half-wave cosine representation, and

$$\bar{W}(x,y) = t \sum_{k,\ell=0}^{N}\sum \sin \frac{k\pi x}{L} \left(C_{k\ell} \cos \frac{\ell y}{R} + D_{k\ell} \sin \frac{\ell y}{R}\right)$$

called the half-wave sine representation. Here R, L and t are shell radius, length and wall thickness; x and y are axial and circumferential co-ordinates; k and ℓ are integers denoting the number of axial half waves and the number of full waves in the circumferential direction, respectively. In all cases the measured imperfections are referred to the so-called "best fit" cylinders.

The adoption of a standard representation for the measured initial imperfections is necessary in order to be able to compare the different imperfection distributions, in an attempt to find characteristic imperfection distributions that are associated with the different fabrication processes. It should be stressed here that in all cases the Fourier coefficients are normalized by the corresponding wall thicknesses of the shell bodies.

Shells Fabricated with Welded Seams

Many aerospace tanks are constructed out of a fixed number of curved parts joined initially by longitudinal welded seams to form a cylinder. Later the ends are closed by appropriate end pieces. The curved parts are made out of initially thick walled flat plates, which are rolled to the desired shape after the chosen stiffening pattern has been integrally machined into them. Figure 15 shows the variation of the half-wave sine Fourier coefficients as a function of the circumferential wave number ℓ (for selected axial half-wave numbers k) of the large integrally ring and stringer stiffened shell measured at Langley[15] (see Figure 3 for the corresponding initial imperfection map). The amplitudes of the Fourier harmonics with a single half wave in the axial direction have a distinct maximum at 9 circumferential waves. Only those harmonics that are integer multiples of the 3 longitudinal welds have significant amplitudes. All Fourier coefficients with more than a single half wave in the axial direction are comparatively much smaller. Also, because the heavy end rings have been machined to very close tolerances the shell exhibits practically no ovalization. (That is the amplitude of the Fourier coefficients with $\ell = 2$ is very small.)

Figure 16 shows the variation of the half-wave sine Fourier representation for a large integrally stiffened aerospace shell (see Figure 11 for the corresponding initial imperfection map), which is also assembled out of 3 curved parts. In this case the amplitudes of the Fourier harmonics with a single half wave in the axial direction have a distinct maximum at $\ell = 3$, which corresponds to the number of longitudinal welds. Only harmonics that are integer multiples of 3 have significant amplitudes. Also, the amplitudes of the Fourier coefficients decay with increasing wave numbers both in the axial and in the circumferential direction. Further since the circular end domes are forced-fit into the cylindrical part, the shell possesses a negligibly small $\ell = 2$ component (ovalization).

Finally Figure 17 shows the variation of the half-wave sine Fourier representation for the other large integrally stiffened aerospace shell (see Figure 12 for the corresponding imperfection map), which is assembled out of 4 curved parts. In this case the end domes were but-welded to the cylindrical part. This resulted in a relatively large ovalization. As a matter of fact the Fourier harmonics with a single half wave in the axial direction have a distinct maximum at $\ell = 2$ (out-of-roundness or ovalization). Besides $\ell = 2$ and $\ell = 6$ only those harmonics that are integer multiples of 4 have significant amplitudes. Also in this case the amplitudes of the Fourier coefficients decay with increasing wave numbers both in the axial and in the circumferential direction.

Shells Fabricated with Riveted Seams

The fixed number of curved panels can also be assembled into a circular cylindrical shell by means of joggled and riveted joints. One example of such a shell is the one tested by Horton[14]. Figure 18 shows the variation of the half-wave sine Fourier representation for this shell (see Figure 6 for the corresponding initial imperfection map). The amplitudes of the Fourier harmonics with a single half wave in the axial direction have 2 distinct maxima, one at $\ell = 2$ (out-of-roundness) and another at $\ell = 6$ (which corresponds to the number of panels the shell is assembled from). The Fourier coefficients with more than a single half wave in the axial direction are comparatively much smaller.

Figure 19 shows the variation of the half-wave sine Fourier representation for the Ariane interstage II/III shell measured by the Solid Mechanics Group of the Delft University of Technology[16] (see Figure 14 for the corresponding initial imperfection map). The amplitudes of the Fourier harmonics with a single half wave in the axial direction have a distinct maximum at $\ell = 8$ (which corresponds to the number of panels the shell is assembled from). There is also a sizable $\ell = 2$ (out-of-roundness) component. All other Fourier coefficients are comparatively much smaller.

Characteristic Imperfection Distributions

It has been established earlier[20], that for integrally stringer stiffened laboratory scale shells, which are machined out of seamless thick-walled aluminum alloy following the procedure developed at the Technion[21], the amplitudes of the Fourier coefficients decay with increasing wave numbers both in the axial and circumferential direction. It has been shown that the Donnell-Imbert[22] analytical imperfection model

$$\bar{\xi} = \frac{\bar{X}}{k^r \ell^s}$$

where the coefficients $\bar{X}$, r and s are determined by least-square fitting the measured data, represents the variation of the harmonic components with the axial (k) and circumferential (ℓ) wave numbers satisfactorily. Thus the above formula summarizes the characteristic imperfection distribution for the specific fabrication technique described.

It appears from the results presented in this paper that for full scale aerospace shells built-up out of a fixed number of curved panels the initial imperfection distributions will be dominated by two components only, if the joints are riveted. Using the half-wave sine representation both components will have a single half-wave in the axial direction and respectively 2 and ℓ full waves in the circumferential direction, where ℓ is the number of full length panels out of which the shell is assembled. By using accurately machined rigid end rings the $\ell = 2$ out-of-roundness component can be significantly reduced in size. Thus the variation of the measured harmonics shown in Figures 18 and 19 can be considered as the characteristic initial imperfection distribution for this particular type of fabrication process.

On the other hand, if the longitudinal seams, needed to assemble the curved parts, are welded then the resulting imperfection distributions are more complicated. The dominant imperfection harmonics continue to have a single half wave in the axial direction and the number of full waves in the circumferential direction are obviously influenced by the number of full length panels out of which the shell is assembled. However, the welding procedure used has also a strong influence on the resulting number of full waves. Thus both the Langley shell LA-1 and the first aerospace shell X-1 (radius R = 1212.1 mm) consist of 3 full length curved panels, both have negligibly small out-of-roundness (the $\ell = 2$ Fourier coefficients are very small), however, for the Langley shell LA-1 the largest Fourier coefficient has 9 full waves in the circumferential direction (3 times the number of welded seams), whereas for the first aerospace shell X-1 the largest Fourier coefficient has 3 full waves (equal to the number of welded seams). Apparently the welding procedure used by the two aerospace companies were similar because also the second aerospace shell X-2 (radius R = 1527.4 mm) has a large Fourier coefficient with the same number of full waves in the circumferential direction as the number of curved panels out of which it is assembled, namely 4. However, the initial imperfection distribution of this shell has also comparatively large harmonics with $\ell = 2$ (out-of-roundness) and $\ell = 6$ full waves in the circumferential direction, besides the harmonics that are integer multiples of 4, the number of welded seams.

Thus for shells assembled out of a fixed number of curved parts by welded seams an important question remains to be answered; namely, when is the number of full waves of the dominant Fourier coefficient in the circumferential direction equal to the number of welded seams and when will it be equal to the number of welded seams times an integer?

Stochastic Stability Analysis

As pointed out in the introduction the great unknown in any design procedure is the selection of the "knock-down" factor by which the buckling load prediction of the perfect shell must be multiplied by in order to arrive at a reliable buckling load prediction for the real structure. It has also been suggested that the existence of Initial Imperfection Data Banks makes it possible to associate statistical measures with the different methods of fabrication, which in turn can be used to derive the corresponding Reliability Functions $R(\lambda)$. Here $R(\lambda)$ is defined as the probability that the (random) buckling load of the structure Λ will exceed (or be at least equal to) the specified load level λ. Thus

$$R(\lambda) = \text{Prob}\ (\Lambda \geq \lambda)$$

where λ is the normalized load parameter ($= P/P_{c\ell}$ where $P_{c\ell} = 2 R\pi t \sigma_{c\ell}$ and $\sigma_{c\ell} = Et/cR$, $c = \sqrt{3(1 - \nu^2)}$) and Λ is the normalized random buckling load. As can be seen from Figure 20, the knowledge of the reliability function makes it possible to find the allowable load, defined as the normalized load level λ_a for which the desired high reliability (say 0.95) is achieved, for the whole ensemble of shells produced by a given manufacturing process. The corresponding "knock-down" factor is thus λ_a.

Several investigators[23,24,25] have studied the static and dynamic buckling of imperfection sensitive structures with small random initial imperfections. For an authoritative review the reader should consult Amazigo's paper from 1976[26]. However, it was not until 1979 that a method has been proposed by Elishakoff[27], which made it possible to introduce the results of the initial imperfection surveys routinely into the stochastic stability analysis.

Basically Elishakoff suggested to utilize the Monte Carlo Method for the solution of the stability problem of axially compressed cylindrical shells with random initial imperfections. When applying this method the initial imperfections are expanded in terms of the usual double Fourier series and then the Fourier coefficients are treated as random variables. Next the mean functions and the variance-covariance matrices of the experimental sample are calculated. Then by using a special numerical procedure[28] the Fourier coefficients of the desired large sample of random initial imperfection shapes are simulated. This is followed by a deterministic buckling load analysis for each of the simulated shells. Finally from the resulting histogram of nondimensional buckling loads the reliability function representing the probability (i.e. fraction of an ensemble) of the buckling load exceeding the specified load is calculated.

The relative ease with which one can derive the reliability function via the Monte Carlo Method, once a sufficiently large sample of initial imperfection measurements on nominally identical shells is available, will be illustrated by considering the case of axially compressed cylindrical shells with axisymmetric imperfections. Having N realizations of the measured imperfections

$$\bar{W}(x)^{(m)} = \frac{\bar{w}(x)^{(m)}}{t} = \sum_i A_i^{(m)} \cos \frac{i\pi x}{L} \quad (m = 1, 2, \ldots, N)$$

one calculates first, by taking "ensemble averages", the estimated mean of the Fourier coefficients $A_i^{(m)}$

$$\bar{A}_i^{(e)} = \frac{1}{N} \sum_{m=1}^{N} A_i^{(m)}$$

and then the estimated variance-covariance matrix

$$\sigma_{jk}^{(e)} = \frac{1}{N-1} \sum_{m=1}^{N} \sum_{n=1}^{N} \left[A_j^{(m)} - \bar{A}_j^{(e)}\right] \left[A_k^{(n)} - \bar{A}_k^{(e)}\right]$$

Since $\sigma_{jk}^{(e)}$ is a non-negative symmetric matrix, therefore it can be decomposed as a product of lower and upper triangular matrices by the Choleski algorithm

$$\left[\sigma_{jk}^{(e)}\right] = \left[C\right]^{\Delta} \left[C\right]^{\nabla}$$

Next the vector $\left\{A\right\}$ of the simulated initial imperfections is obtained as follows

$$\left\{A\right\} = \left[C\right]^{\Delta} \left\{r\right\} + \left\{\bar{A}^{(e)}\right\}$$

where

$\left\{\bar{A}^{(e)}\right\}$ = estimated "mean" vector

$\left\{r\right\}$ = random vector

The r's are normally distributed random numbers with zero mean and unit variance generated by the computer. Taking, for example, 1000 different r's, one gets 1000 different A's, that is, different simulated shells with the A's as the Fourier coefficients of the initial imperfections. For each of the "created" initial imperfections one then carries out a deterministic buckling load calculation generating the buckling load histogram of the group of shells under consideration shown in Figure 21. Having defined the reliability function $R(\lambda)$ as the probability that the buckling load Λ will exceed the prescribed value λ, one then proceeds to calculate $R(\lambda)$ from the histogram of the buckling loads by the frequency interpretation (i.e. fraction of an ensemble) yielding the dots shown in Figure 22. The accuracy of the Monte Carlo Method can be seen on the close coincidence with the solid curve which represents a closed form solution in terms of error functions for the same case published by Roorda and Hansen[25].

The feasibility of using the Monte Carlo Method to derive reliability functions using the data on experimental imperfection surveys assembled in the Imperfection Data Bank at the Delft University of Technology[8] has been demonstrated by Elishakoff and Arbocz for both axisymmetric[29] and general asymmetric imperfections[30].

Conclusions

Because it is simple to apply and since for many cases it provides a safe and reliable buckling load prediction the Lower Bound Design Philosophy will continue to be widely used also in the future, especially for standard applications where the total weight of the structures is of no concern.

However, since structureal optimization invaryingly seems to lead to thin-walled configurations whose buckling loads are affected by initial imperfections, therefore one can expect that the interest in developing improved design procedures will continue to occupy the scientific community. It is this author's opinion that new approaches leading to possible improvements in our buckling load prediction capability, for imperfection sensitive structures will require in all cases extensive, and the more detailed the better knowledge of the actual imperfections that are present in the real structures. It is encouraging to see that the need for detailed imperfection surveys on full scale and laboratory scale structures and the usefulness of establishing Initial Imperfection Data Banks is being recognized by more and more investigators[31,32,33].

The idea of generating reliability functions via the Monte Carlo Method, which display the degrading effect on the buckling load of the expected initial imperfection distribution characteristic of a given fabrication process, seems to offer the means of combining the Lower Bound Design Philosophy with the notion of Goodness Classes. Thus shells manufactured by a process, which produces inherently a less damaging initial imperfection distribution, will not be penalized because of the low experimental results obtained with shells produced by another process, which generates a more damaging characteristic imperfection distribution.

Looking into the future, it is to be expected that the existence of extensive data on characteristic initial imperfection distributions classified according to fabrication processes, the availability of improved versions of the present generation nonlinear structural analysis codes[34], the stochastic stability approach via the reliability functions and the greatly increased computational speed offered by the so-called supercomputers will finally result in a series of improved design recommendations which will incorporate the latest theoretical findings and make them routinely accessible to the designers.

Acknowledgement

The author wishes to express his sincere thanks to Miss Henny Schreuders for the skilful typing. The fine artwork by Mr. Ronald P. Notenboom and Mr. René H.M. van Boheemen is also much appreciated.

References

1. Bushnell, D.: Stress, stability and vibration of complex branched shells of revolution: Analysis and user's manual for BOSOR-4. NASA CR-2116 (1972).

2. Cohen, G.A.: User document for computer programs for ring-stiffened shells of revolution. NASA CR-2086 (1973).

3. Koiter, W.T.: On the stability of elastic equilibrium. Ph.D. thesis, in Dutch, TH Delft, the Netherlands, H.T. Paris, Amsterdam (1945). English translation issued as NASA TT F-10, 833 p. (1967).

4. Budiansky, B.; Hutchinson, J.W.: Dynamic buckling of imperfection sensitive structures. Proceedings 11th IUTAM congres, Springer Verlag, Berlin (1964), pp. 636-651.

5. Hutchinson, J.W.; Amazigo, J.C.: Imperfection-sensitivity of eccentrically stiffened cylindrical shells. AIAA Journal, vol. 5, no. 3, (1967), pp. 392-401.

6. Arbocz, J.; Babcock, C.D. Jr.: Computerized stability analysis using measured initial imperfections. Proceedings 12th Congress of the International Council of the Aeronautical Sciences, October 12-17, 1981, München, pp. 688-701.

7. Arbocz, J.; Babcock, C.D. Jr.: The buckling analysis of imperfection sensitive shell structures. NASA CR-3310 (1980).

8. Arbocz, J.; Abramovich, H.: The Initial Imperfection Data Bank at the Delft University of Technology - part I. Report LR-290, Delft University of Technology, the Netherlands (1979).

9. Singer, J.; Abramovich, H.; Yaffe, R.: Initial imperfection measurements of integrally stringer-stiffened shells. TAE report no. 330, Technion, Haifa, Israel (1978).

10. Singer, J.: Buckling experiments on shells - a review of recent developments. TAE report no. 403, Technion, Haifa, Israel (1980).

11. Arbocz, J.; Babcock, C.D. Jr.: The effect of general imperfections on the buckling of cylindrical shells. Journal of Applied Mechanics, vol. 36, no. 1 (1969), pp. 28-38.

12. Singer, J.; Arbocz, J.; Babcock, C.D. Jr.: Buckling of imperfect stiffened cylindrical shells under axial compression. AIAA Journal, vol. 9, no. 1 (1979), pp. 68-75.

13. Walker A.C.; Stridharan, S.: Buckling of compressed, longitudinally stiffened cylindrical shells. BOSS '79, Proceedings 2nd International Conference on the Behaviour of Off-shore Structures, Imperial College, London (1979), vol. 2, pp. 341-356.

14. Horton, W.H.: On the elastic stability of shells. NASA CR-145088 (1977).

15. Arbocz, J.; Williams, J.G.: Imperfection surveys on a 10-ft diameter shell structure. AIAA Journal, vol. 15, no. 7 (1977), pp. 949-956.

16. Sebek, R.W.L.: Imperfection surveys and data reduction of ARIANE interstages I/II and II/III. ir. thesis, TH Delft, Department of Aerospace Engineering (1981).

17. Koiter, W.T.: Elastic stability and postbuckling behavior. Proceedings Symposium Nonlinear Problems, University of Visconsin Press, Madison (1963), pp. 257-275.

18. Arbocz, J.: The effect of general imperfections on the buckling of cylindrical shells. Ph.D. thesis, California Institute of Technology, Pasadena (1968).

19. Babcock, C.D. Jr.; Arbocz, J.: Imperfection measurements of large scale shells. GALCIT report SM 78-7, California Institute of Technology, Pasadena (1978).

20. Arbocz, J.: Past, present and future of shell stability analysis. Zeitschrift für Flugwissenschaften und Weltraumforschung, vol. 5, no. 6 (1981), pp. 335-348.

21. Weller, T.; Singer, J.: Recent experimental studies on the buckling of integrally stringer-stiffened cylindrical shells. TAE report no. 100, Technion, Haifa, Israel, April 1970.

22. Imbert, J.: The effect of imperfections on the buckling of cylindrical shells. Aeronautical engineer thesis, California Institute of Technology (1971).

23. Bolotin, V.V.: Statistical methods in the nonlinear theory of elastic shells. NASA TT F-85 (1962). (Translation of a paper presented at a seminar in the Institute of Mechanics of the Academy of Sciences USSR on October 3, 1957.)

24. Amazigo, J.C.: Buckling under axial compression of long cylindrical shells with random axisymmetric imperfections. Quart. Appl. Math., vol. 26, no. 4 (1969), pp. 537-566.

25. Roorda, J.; Hansen, J.S.: Random buckling behavior in axially loaded cylindrical shells with axisymmetric imperfections. Journal of Spacecraft, vol. 9, no. 2 (1972), pp. 88-91.

26. Amazigo, J.C.: Buckling of stochastically imperfect structures. Proceedings IUTAM Symposium Buckling of Structures, edited by B. Budiansky, Springer Verlag, Berlin (1976), pp. 172-182.

27. Elishakoff, I.: Buckling of stochastically imperfect finite column on a nonlinear elastic foundation - a reliability study. Journal of Appl. Mech., vol. 46, no. 2 (1979), pp. 411-416.

28. Elishakoff, I.: Simulation of space-random fields for solution of stochastic boundary-value problems. Journal Acoustical Society of America, vol. 65, no. 2 (1979), pp. 399-403.

29. Elishakoff, I.; Arbocz, J.: Reliability of axially compressed cylindrical shells with random axisymmetric imperfections. Report LR-306, Delft University of Technology (1980).

30. Elishakoff, I.; Arbocz, J.: Buckling of shells with general random imperfections. Paper presented at the 15th IUTAM Congress in Toronto, 17-23 August (1980). (Also Memorandum M-401, Delft University of Technology, Department of Aerospace Engineering, 1981.)

31. Grove, T.; Didriksen, T.: Buckling experiments on 4 large axial stiffened and 1 ring stiffened cylindrical shells. Det Norske Veritas report no. 76-432 (1976).

32. Green, D.R.; Nelson, H.M.: Compression tests on large-scale stringer stiffened tubes. Proceedings International Conference Buckling of Shells in Off-shore Structures, held 23-24 April 1981 at Imperial College, London.

33. Odland, J.: Theoretical and experimental buckling loads of imperfect spherical shell segments. Journal of Ship Research, vol. 25, no. 3 (1981), pp. 201-218.

34. Almroth, B.O.; Brogan, F.A.; Miller, E.; Zele, F.; Peterson, H.T.: Collapse analysis for shells of general shape. II user's manual for the STAGS-A computer code. Air Force Flight Dynamics Lab., Wright Patterson AFB, AFFDL-TR-71-8 (1973).

Figure 1 Initial imperfection survey instrumentation at NASA Langley [15]

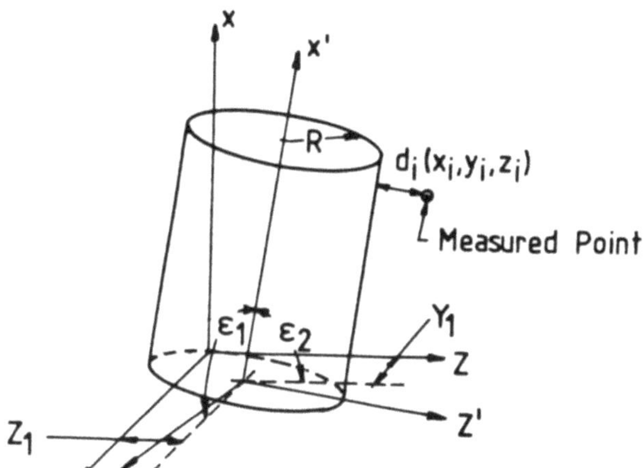

X,Y,Z	Reference axis of traversing pick-up
X',Y',Z'	Reference axis of best fit cylinder
d_i	Normal distance from measured point to best fit cylinder

Figure 2 Definition of the "perfect" cylinder

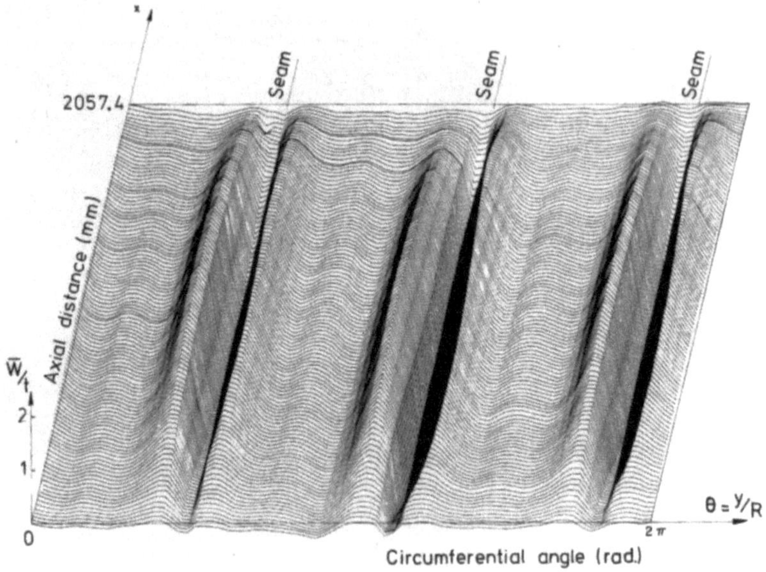

Figure 3 Measured initial shap of Langley shell LA-1 [15]

Figure 4 Construction of Horton's shell HO-1 [14]

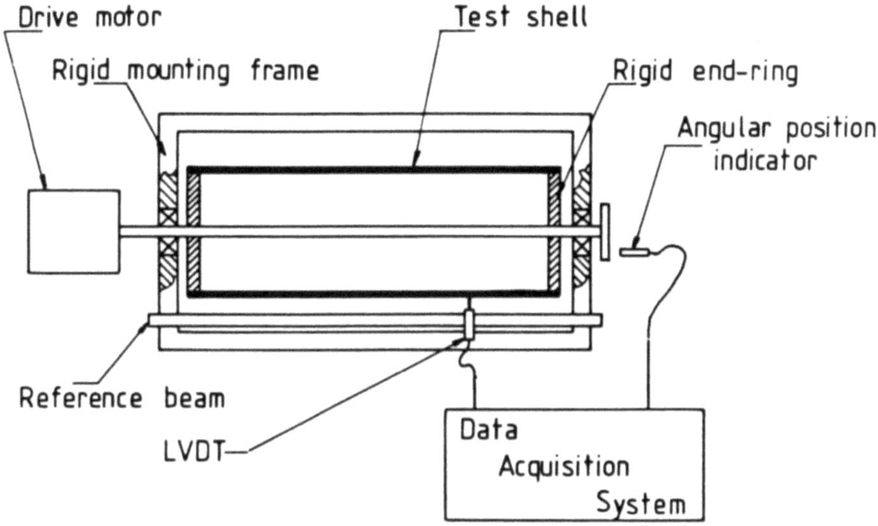

Figure 5 Block diagram of Horton's imperfection survey instrumentation

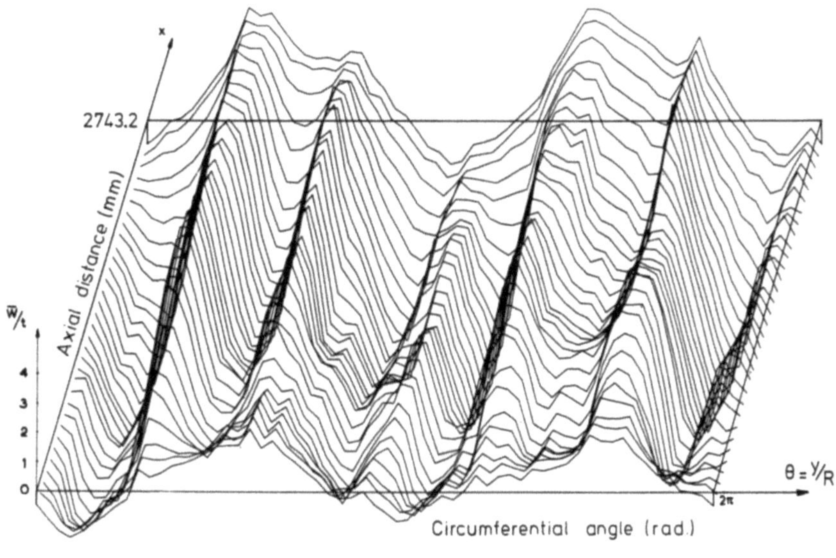

Figure 6 Measured initial shape of Horton's shell HO-1 [14]

Figure 7 Measurement of a full scale aerospace shell [19]

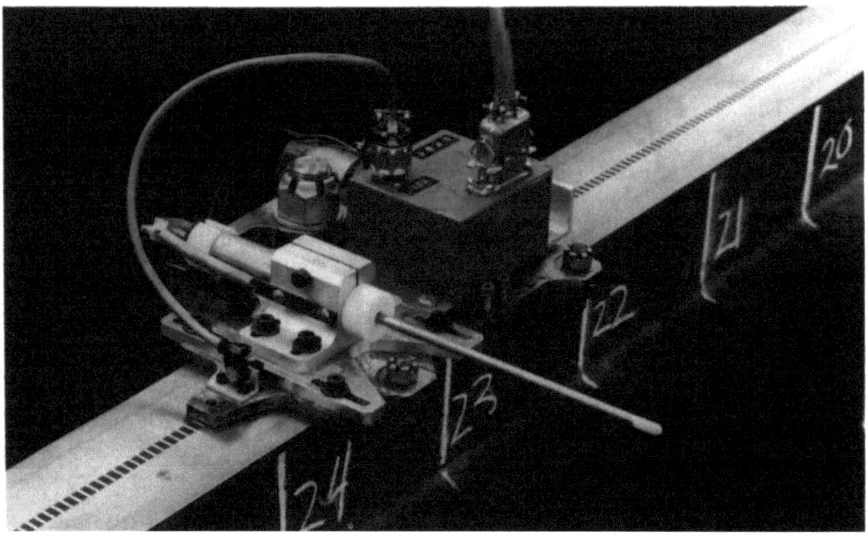

Figure 8 Instrument card and position tape on reference beam

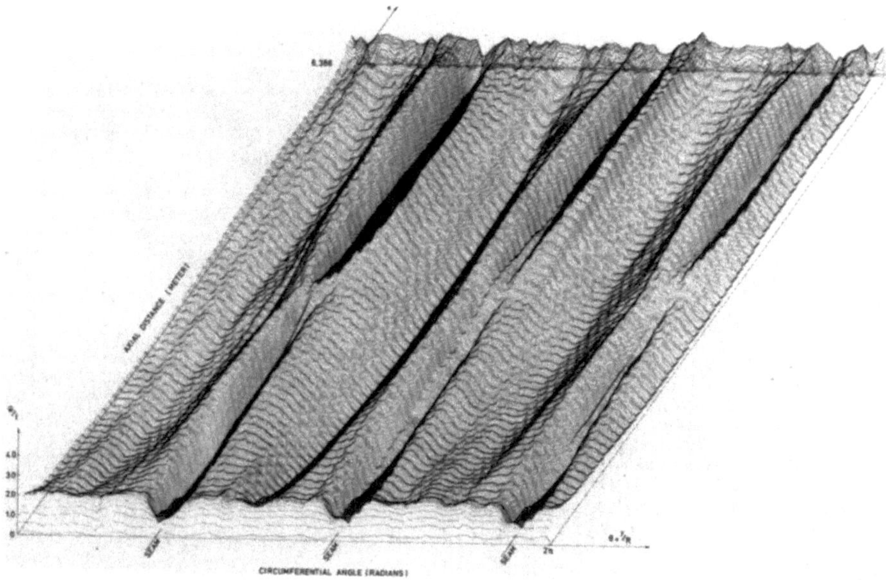

Figure 9 Measured initial shape of aerospace shell X-1

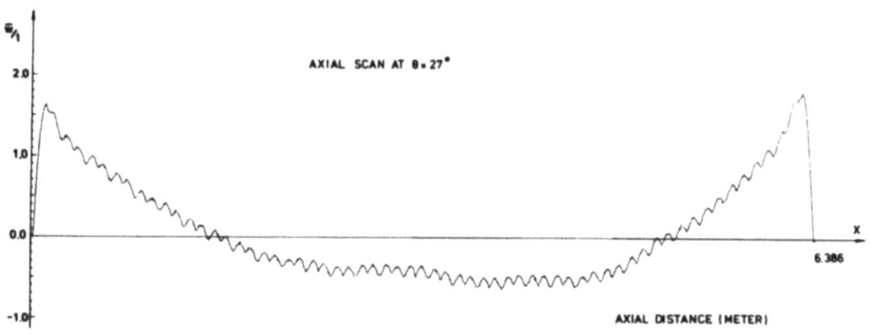

Figure 10 Typical meridional shape of aerospace shell X-1

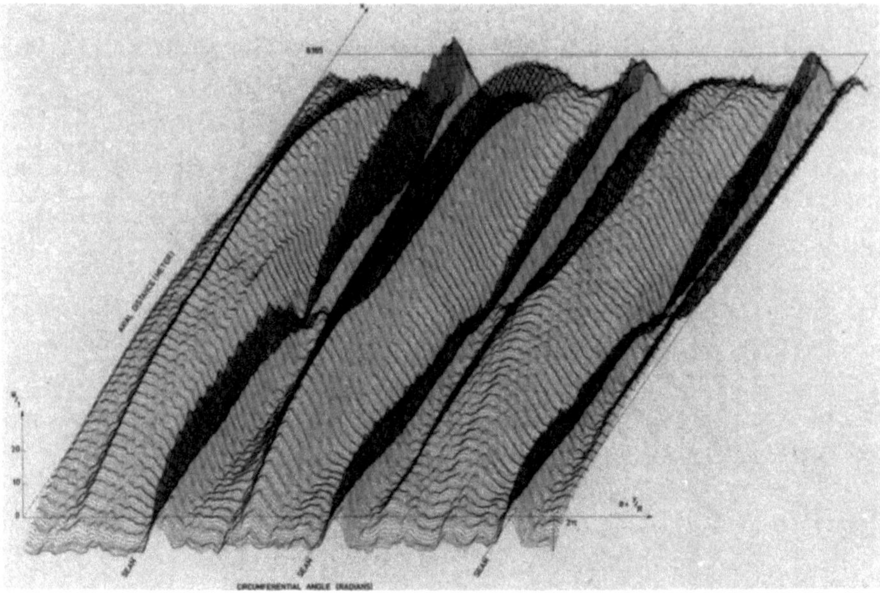

Figure 11 Measured initial shape of aerospace shell X-1 with edge zones removed

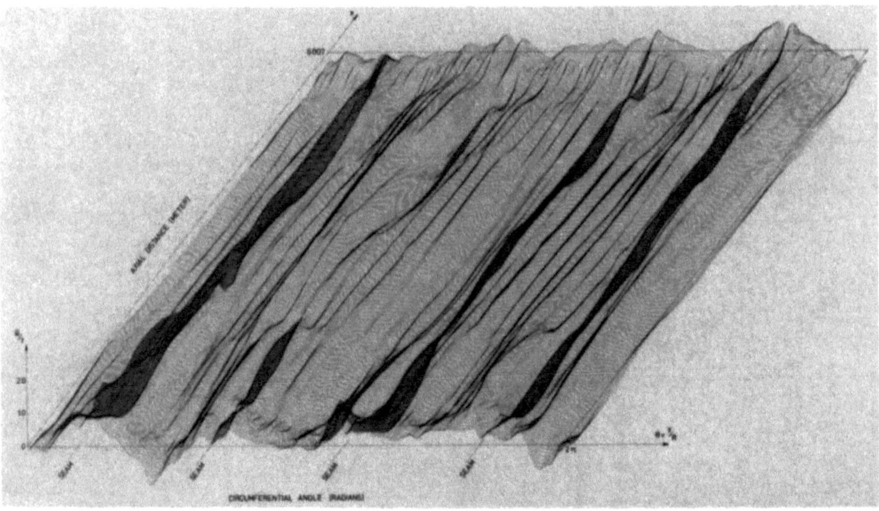

Figure 12 Measured initial shape of aerospace shell X-2

Figure 13 Initial imperfection survey system from the TH Delft

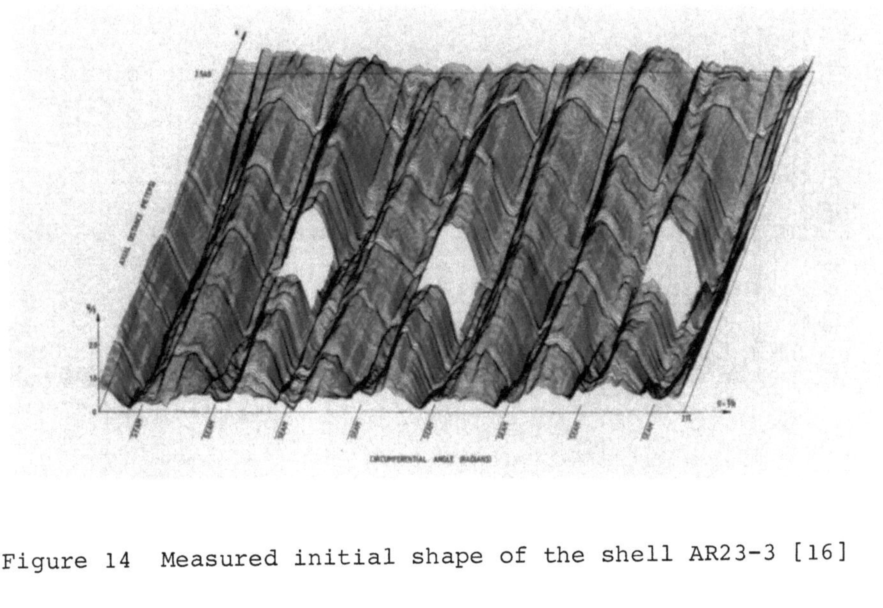

Figure 14 Measured initial shape of the shell AR23-3 [16]

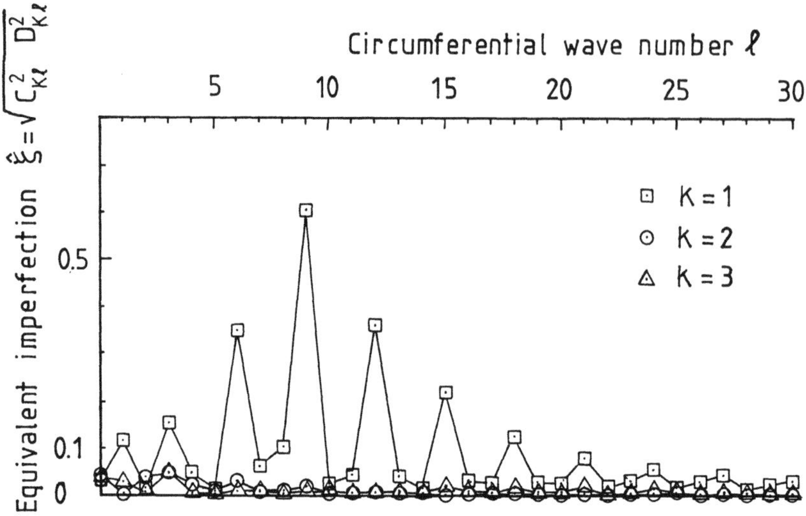

Figure 15 Circumferential variation of the half-wave sine Fourier representation (shell LA-1 [15])

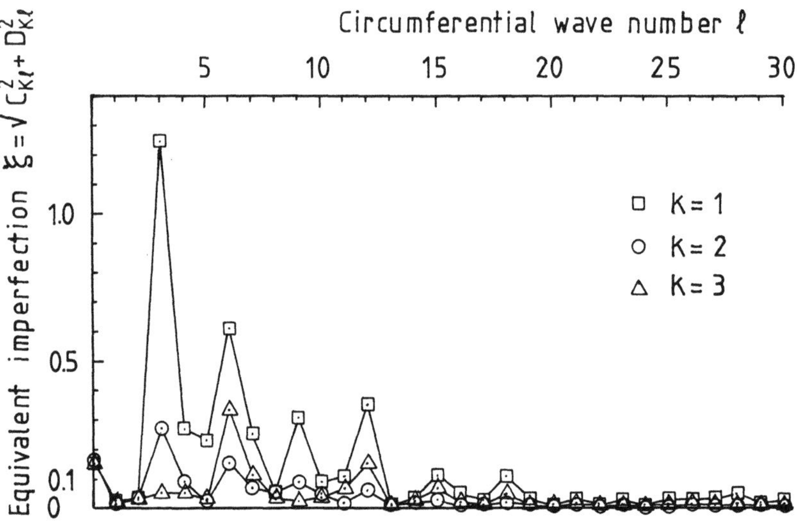

Figure 16 Circumferential variation of the half-wave sine Fourier representation (aerospace shell X-1)

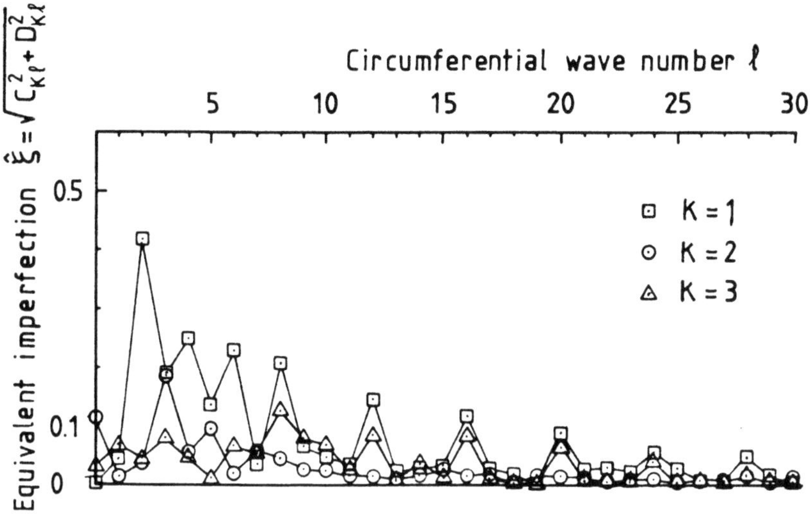

Figure 17 Circumferential variation of the half-wave sine Fourier representation (aerospace shell X-2)

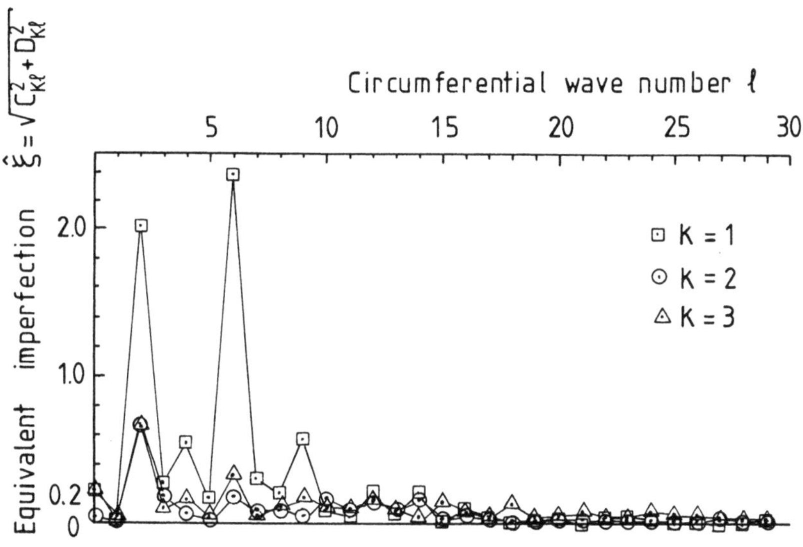

Figure 18 Circumferential variation of the half-wave sine Fourier representation (Horton's shell HO-1 [14])

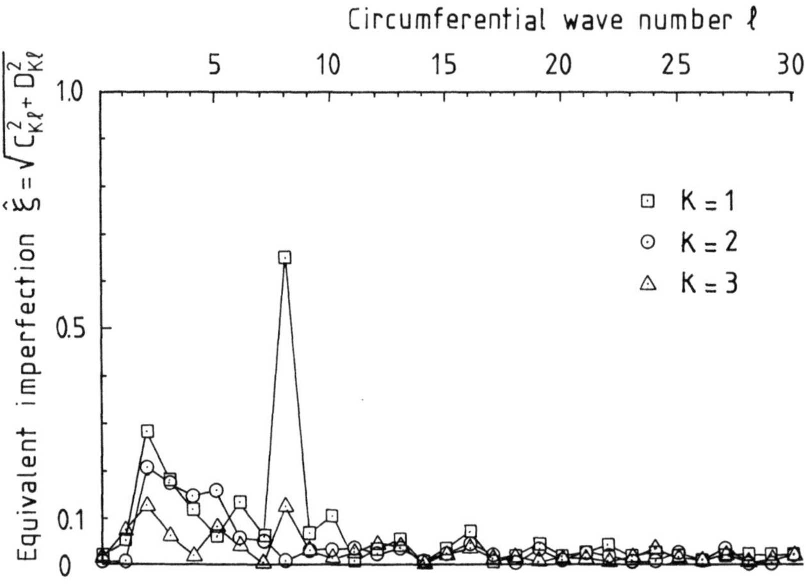

Figure 19 Circumferential variation of the half-wave sine Fourier representation (aerospace shell AR23-3 [16])

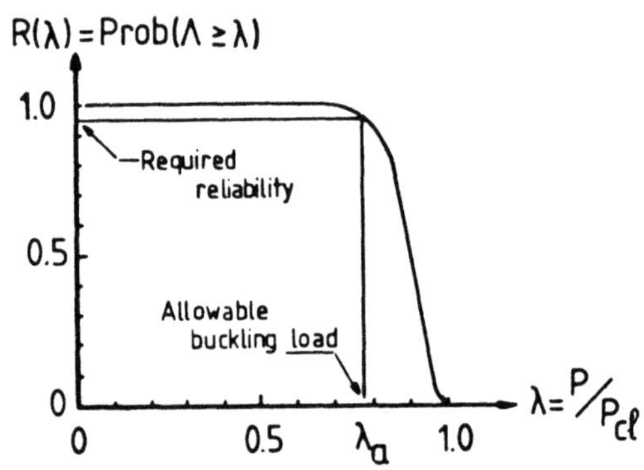

Figure 20 Determination of the "knock-down" factor λ_a by means of a reliability function $R(\lambda)$

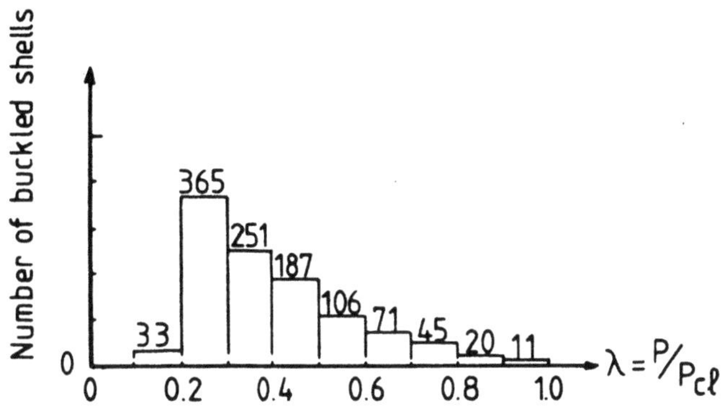

Figure 21 Histogram of the nondimensional buckling loads [29]

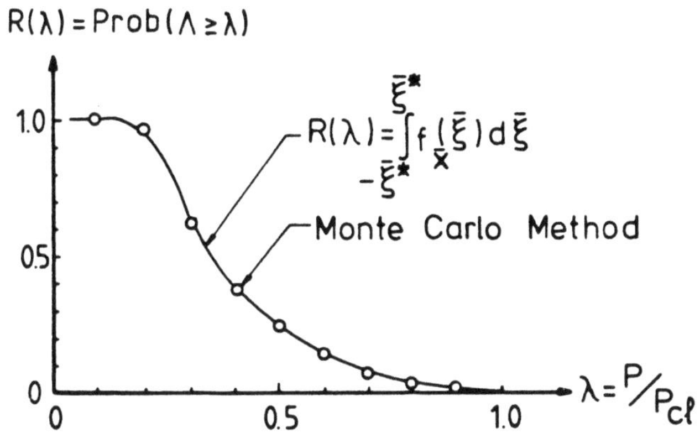

Figure 22 Comparison of analytical reliability function [25] with results of Monte Carlo simulation [29]

THE STABILITY OF LIQUID-FILLED CYLINDRICAL SHELLS UNDER DYNAMIC LOADING

D.F. FISCHER, F.G. RAMMERSTORFER

Technical and Scientific Computing Div. FAT
VOEST-ALPINE AG Linz, Austria

1. Introduction

In the paper we are looking at cylindrical shells fixed at the bottom and free at the top. Let us assume that these shells are filled with a viscous, compressible fluid of a mass density, ρ_L, up to height, H. For definitions see Fig. 1. The shell is accelerated at the bottom, i.e. at position x = 0, by a horizontal acceleration, A(t), (t = time).

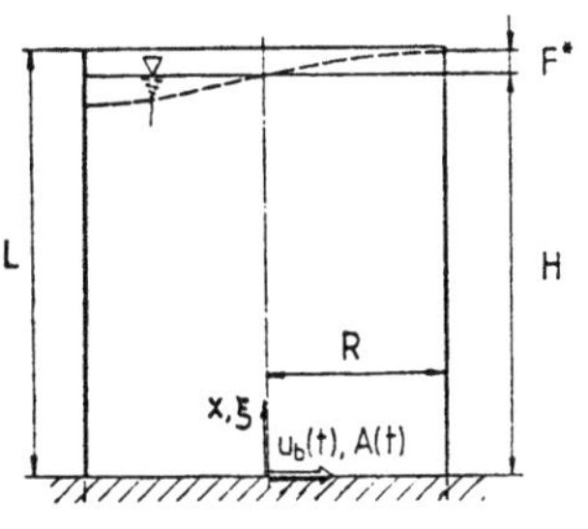

Fig. 1 Definitions

The topics of a vertical acceleration and rocking of the base are not treated in this paper; they are the object of further research. The maximum value of A(t) is the maximum bottom acceleration, A:

$$\max A(t) = A. \tag{1}$$

Such a horizontal acceleration is caused e.g. by an earthquake, an explosion in or near the ground, or by a plane crash, and is exerted on a building surrounding the shell. Such hazards can activate stresses in the cylindrical shell which lead to a loss of stability ("shell crippling" or "elephant footing" as e.g. at accidents in Alaska and Japan and on wine storage tanks in

Buckling of Shells, Ed.: E. Ramm
Proceedings of a State-of-the-Art Colloqium

Livermore, Calif.).
The analysis of the interaction of the shell, the fluid, the soil and a potential construction fixed to the shell shows a complex fluid-structure-soil-interaction problem and can be treated by the finite-element-method and direct time integration, e.g. [1]. An extensive survey of the literature is contained in [2]. Direct time integration, however, in practice can be circumvented by use of the response spectrum method. This method which has found broad acceptance recently, takes approximate dynamic behaviour of a construction into account by looking at individual modes of the structure [3]. The maximum values of the stresses and deformations are estimated for each mode and superposed in accordance with recognized superposition rules.

Therefore we are going to look at a cylindrical shell which at the time t experiences a horizontal bottom displacement, $u_b(t)$, i.e. a parallel displacement of the shell, and which in addition vibrates in an eigen-mode with the relative displacements - in relation to the base circle - $w(x,\varphi,t)$, $v(x,\varphi,t)$. w stands for the relative radial displacement, v for the relative tangential displacement. ω in the following means the circular frequency according to the first eigen-mode. Here we have to take into account that during horizontal ground acceleration in perfect circular cylindrical shells only eigen-modes with the circumferential wave number m = 1 are being activated, [4]. As a matter of principle, when applying the response spectrum method, all, i.e. an infinite number of eigen-modes for the circumferential wave number m = 1 would have to be taken into account. For the sake of simplicity, however, this paper only deals with the first eigen-mode, a procedure which seems justified for many practical applications. (The action of higher eigen-modes is negligible in many cases; on the other hand, for the higher eigen-modes we can proceed in an analogous way and superimpose in accordance with the superposition rule.)

2. Determination of the pressure distribution in the fluid

2.1 Subdivision of the components of the dynamically activated fluid pressure

The dynamically activated pressure distribution can be divided into three parts:

- The convective fluid pressure, p_{bc}, attributed to the parallel displacement, $u_b(t)$, (rigid body displacement) of the cylindrical shell and created by surface vibrations ("sloshing"). For sloshing, the cylindrical shell can be treated as if rigid. The sloshing period, T_j^*, as well as the approximate amplitude at the freeboard, F^*, are demonstrated e.g. in [2] as follows

$$T_j^* = 2\pi \left(\frac{R}{\eta_j \quad g \tanh(\eta_j \frac{H}{R})} \right)^{\frac{1}{2}} , \quad f_j^* = \frac{1}{T_j^*}, \quad j = 1,2 \; \ldots (2)$$

$$F^* = 0.831 \frac{A_1^*}{g} R, \qquad (3)$$

η_j is the j-th zero of the derivative of the Bessel function J_1 of the first kind (η_1=1.841, η_2=5.331, η_3=8.536, ...).

g being the acceleration of gravity. For deformable shells, see [11]. A_j^* is the spectral acceleration value to the frequency, f_j^*, and can be taken from a response spectrum for a maximum base acceleration, A, and a small modal damping value (ζ < 0.5 %).

- The impulsive fluid pressure, p_{bi}, attributed to the parallel displacement, $u_b(t)$, (rigid body displacement) of the cylindrical shell and generated by the parallel displacement of the "fluid column". The attributable maximum acceleration is A, see (1).

- The impulsive fluid pressure, p_d, activated by the relative motion of the shell and hence the fluid as against the base circle. The eigen-frequency, $f = \omega/2\pi$, of this vibration almost always ranges an order of magnitude above the sloshing frequency, f^*. The convective fluid pressure activated by the common vibration between shell and fluid is negligible compared to the other pressure portions. Due to the eigen-frequency, f, a spectral acceleration value, a, is taken from a response spectrum due to the maximum ground acceleration, A, and a damping coefficient, ζ (see appendix 2). The maximum acceleration of the deformable shell results from the product of "a" with the mode participation factor, γ, (see appendix 1). It seems now appropriate (see [2], section 3.3) to attribute the maximum relative acceleration

$$A_d = \gamma(a-A) \tag{4}$$

to the impulsive fluid pressure, p_d, despite the fact that the maximum ground acceleration, A, and the maximum relative acceleration of the shell usually do not occur at the same time and in the same direction.

2.2 Calculation of the amplitudes of the individual components of the fluid pressure

As explained in [2], section 2, when defining the pressure distribution, viscosity and compressibility can be neglected in most cases. The following consideration is therefore based on the potential theory of incompressible fluids.

2.2.1 Determination of p_{bc}

The extreme distribution of the convective fluid pressure, p_{bcj}, corresponding to the j-th sloshing mode, can be taken from [5] and is as follows

$$\begin{aligned} p_{bc_j} &= R\rho_L \; A_j^* \frac{2}{\eta_j^2-1} \frac{\cosh(\eta_j \frac{H}{R}\xi)}{\cosh(\eta_j \frac{H}{R})} \cos\varphi \\ &= R\rho_L \; \bar{P}_{bc_j}(\xi) \cos\varphi A_j^*, \; j = 1,2,3 \ldots \end{aligned} \tag{5}$$

The actual extreme fluid pressure, p_{bc}, results from a superposition (see chapter 2.2.4). In most cases only the first sloshing mode is essential.

2.2.2 Determination of p_d

An exhaustive derivation of the amplitude distribution of the impulsive fluid pressure, p_d, is demonstrated in [6].
Here the two relative displacement components, v, w, are described as

$$w = C\ f(x)\ \cos\varphi\ \sin\omega t,$$
$$v = D\ g(x)\ \sin\varphi\ \sin\omega t. \tag{6}$$

f(x), g(x) are assumed as explicitly known functions to which the following applies

$$|f(x)|,\ |g(x)| \leq 1.0. \tag{7}$$

When introducing with $\xi = \frac{x}{H}$ instead of (f(x), g(x)) the dimensionless functions $(\hat{f}(\xi), \hat{g}(\xi))$ we arrive at

$$\bar{p}_d = R\rho_L \sum_{j=1}^{\infty} \frac{2I_1(\lambda_j)}{\lambda_j\ I'_1(\lambda_j)} \left(\int_0^1 \hat{f}(\xi)\cos\left(j\frac{\pi}{2}\xi\right) d\xi\right) \cos\left(j\frac{\pi}{2}\xi\right) \cos\varphi$$
$$= R\rho_L\ \bar{P}_d(\xi)\ \cos\varphi,\quad \lambda_j = j\frac{\pi R}{2H},\ j = 1,3,5... \tag{8}$$

I_1 is the modified Bessel function of the first kind and first order, I'_1 its derivative.
The actual relative acceleration amplitude, $C\omega^2$, is now applied according to the considerations contained in 2.1 and follows with (4)

$$C\omega^2 = \gamma(a-A). \tag{9}$$

The actual fluid pressure, p_d, results by multiplying $\bar{p}_d$ with $C\omega^2$. Thus one has to know the deformation figure, f(x), (eigen-mode of vibration of the shell filled with fluid), circular frequency, ω, and the participation factor, γ. A method for the iterative determination of these values is found under [2], section 3.2, or appendix 1 of the present paper.

2.2.3 Determination of p_{bi}

This impulsive fluid pressure distribution is attributed to a parallel displacement of the rigid tank, $u_b(t)$. If we divide this parallel displacement into a radial component, $w_b(t)$, and a tangential component, $v_b(t)$, we arrive at

$$\begin{aligned} w_b(t) &= u_b(t)\cos\varphi, \\ v_b(t) &= -u_b(t)\sin\varphi. \end{aligned} \qquad (10)$$

When comparing equation (10) with (6) or (7), respectively, we find

$$\begin{aligned} C &= u_b(t), \\ f(x),\ g(x) &\equiv 1.0 \end{aligned} \qquad (11)$$

We can use equation (8) taking (11) into account when calculating $\bar{p}_{bi}$ and find

$$\begin{aligned} \bar{p}_{bi} &= H\rho_L \sum_{j=1}^{\infty} \frac{8}{\pi^2} \frac{(-1)^{\frac{j-1}{2}} I_1(\lambda_j)}{j^2 I'_1(\lambda_j)} \cos\left(j\frac{\pi}{2}\xi\right)\cos\varphi \\ &= H\rho_L \bar{P}_{bi}(\xi)\cos\varphi, \quad j = 1,3,5,\ \ldots \end{aligned} \qquad (12)$$

The actual fluid pressure, p_{bi}, results by multiplying $\bar{p}_{bi}$ with the maximum ground acceleration, A. λ_j is given with equation (8).

2.2.4 Determination of the resultant pressure distribution

The maximum convective fluid pressure, p_{bc}, generally does not occur simultaneously with the two impulsive fluid pressure parts, p_{bi}, p_d. For superposition we use the SRSS method, [3], p.562. The maximum values of the resultant pressure distribution result from equations (5), (8), (12) with a maximum ground acceleration, A :

$$
\begin{aligned}
p(\xi,\varphi) &= R\rho_L\{\sum_j [A_j^*\bar{P}_{bc}(\xi)]^2+[\gamma(a-A)\bar{P}_d(\xi)+A\frac{H}{R}\bar{P}_{bi}(\xi)]^2\}^{\frac{1}{2}}\cos\varphi \\
&= P(\xi)\cos\varphi.
\end{aligned}
\tag{13}
$$

The amplitude function, $\frac{R}{H}\bar{P}_{bc_1}(\xi)$, can be seen in Fig. 2 for various values of $\frac{H}{R}$. The amplitude function, $\bar{P}_{bi}(\xi)$, is represented in Fig. 3.

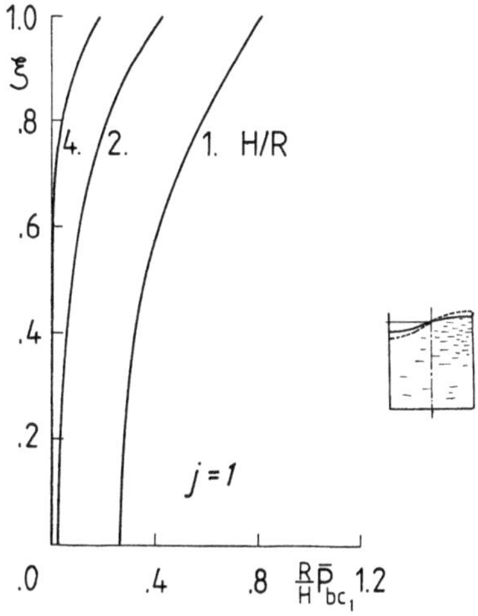

Fig. 2 Convective Pressure, $\bar{P}_{bc}(\xi)$, due to "Sloshing"

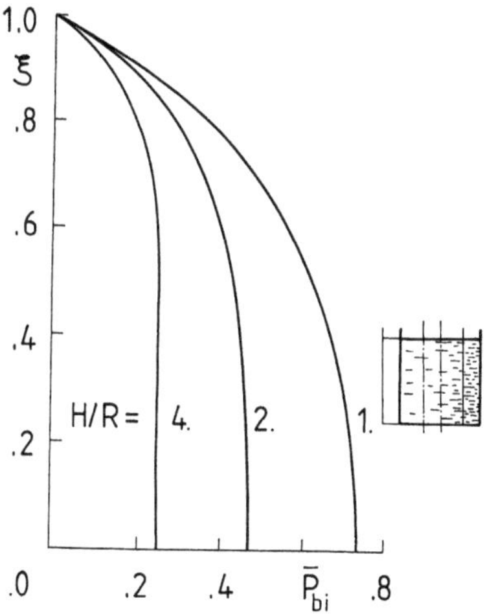

Fig. 3 Impulsive pressure, $\bar{P}_{bi}(\xi)$, due to rigid body displacement of the shell

The values of the amplitude function, $\bar{P}_d$, are calculated for different courses for $\hat{f}(\xi)$ and represented in Fig. 4.1, 4.2 and 4.3.

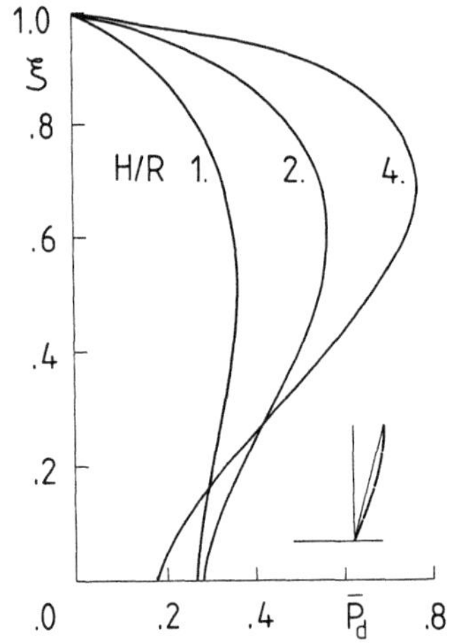

Fig. 4.1 Impulsive pressure, $\bar{P}_d$, due to the vibration of the shell with $\hat{f}(\xi) = \sin \frac{\pi}{2}\xi$

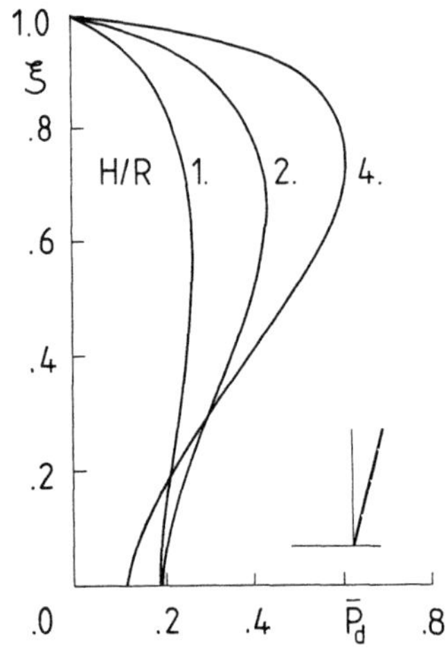

Fig. 4.2 Impulsive pressure, $\bar{P}_d$, due to the vibration of the shell with $\hat{f}(\xi) = \xi$

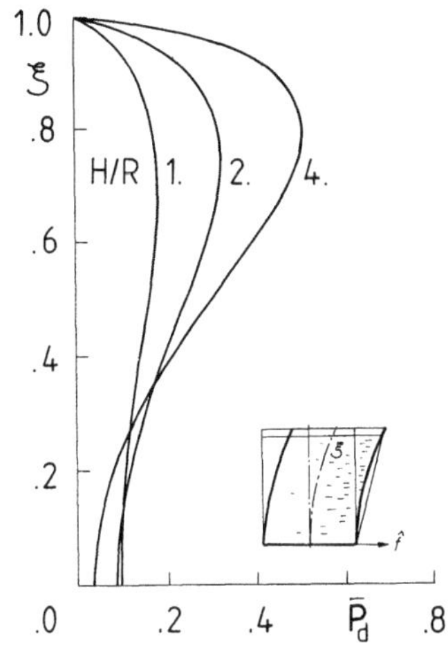

Fig. 4.3 Impulsive pressure, $\bar{P}_d$, due to the vibration of the shell with $\hat{f}(\xi) = 1-\cos\frac{\pi}{2}\xi$

2.2.5 Determination of p_d in an imperfect cylinder

As mentioned in the introduction vibration-modes with a circumferential wave number $m > 1$ cannot be excited in a perfect cylinder by a horizontal rigid base motion. However, fabrication tolerances in civil engineering tanks permit a deviation from a nominal circular cross section and this tends to excite those modes. Little can be found in the literature about the importance of the modes with $m > 1$ in an earthquake response analysis. Turner [10] calculated the impulsive pressure, p_{bi}, in a rigid cylindrical shell with a radius of the surface of the unloaded tank, $\tilde{R}(\varphi) = R(1+\varepsilon\cos n\varphi)$; R is the nominal tank radius and ε is a small number much less than unity. The pressure, p_{bi}, consists now of three components with the circumferential wave numbers $m = 1$, $n-1$, $n+1$. But he could show that the components with $m = n-1$, $n+1$ are much smaller than the pressure components due to $m = 1$.

Haroun [1] performed an analysis of the impulsive pressure, p_d, in a flexible cylindrical shell with an above mentioned shape function of the shell surface numerically and found also the three components with $m = 1$, $n-1$, $n+1$. He also found that the components with $m = n-1$, $n+1$ are much smaller than that due to $m = 1$ for practical imperfections.

Therefore it may be concluded for practical reasons, that the pressure components with m = n-1,n+1 may be neglected in an earthquake analysis. But it seems to be necessary to take into account the imperfections for the calculation of the membrane forces and the stability of the shell; see following chapters.

One of the most recent analytical investigations of vibrating, liquid-filled cylindrical shells can be found in ref. |12|.

3. Determination of the membrane stresses in the full shell generated by the dynamically activated fluid pressure

3.1 Perfect Shell

As shown in |8,13|, for the current type of vessels the influence of the shell mass forces can be neglected when calculating the stresses activated by the fluid pressure in the cylindrical shell and analyzing the stability of the shell, i.e. the stress- and stability analysis can - as an approximation - be performed for a quasi-static system. The quasi-static stability approach may be justified by the fact that the frequency of the time dependent pressure loading is considerably lower than the fundamental natural frequency of the shell. Furthermore the buckling mode is completely different from the vibration mode due to earthquake excitation. Now let us look at a cylindrical shell fixed at the base, free above, under a static load, $p(\xi,\varphi)$. For a stability analysis of the shell we need the membrane forces n_x, n_φ, $n_{x\varphi}$. The definition of these terms is contained in Fig. 5.

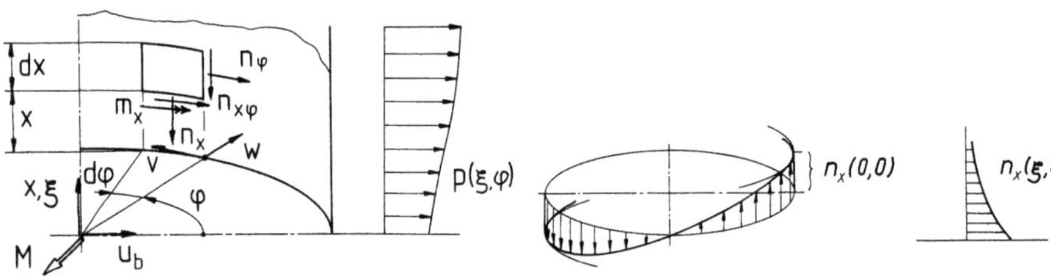

Fig. 5 Definitions

According to the membrane theory of the circular shells [9], p. 397 ff. for a pressure distribution according to equation (13) we get

$$n_\varphi(\xi,\varphi) = R(P(\xi)\cos\varphi + H\rho_L g(1-\xi)) \qquad (14)$$

Hence

$$\frac{\partial n_{x\varphi}}{\partial \xi} = -\frac{H}{R}\frac{\partial n_{\varphi}}{\partial \varphi} = H\,P(\xi)\,\sin\varphi \tag{15}$$

The integration of equation (15) with the boundary condition $n_{x\varphi}(1,\varphi)\equiv 0$ results in

$$n_{x\varphi}(\xi,\varphi) = H\int_{\xi}^{1} P(\tilde{\xi})\,d\tilde{\xi}\,\sin\varphi. \tag{16}$$

Furthermore, the membrane theory together with (16) yields

$$\frac{\partial n_x}{\partial \xi} = -\frac{H}{R}\frac{\partial n_{x\varphi}}{\partial \varphi} = -\frac{H^2}{R}\int_{\xi}^{1} P(\tilde{\xi})\,d\tilde{\xi}\,\cos\varphi. \tag{17}$$

The integration of equation (17) with the boundary condition $n_x(1,\varphi)\equiv 0$ and applying partial integration yields

$$n_x(\xi,\varphi) = -\frac{H^2}{R}\int_{\xi}^{1} P(\tilde{\xi})(\tilde{\xi}-\xi)\,d\tilde{\xi}\,\cos\varphi \tag{18}$$

Now we shall examine if $n_x(0,0)$ is described sufficiently accurate by the membrane theory of cylindrical shells. We use the bending theory of the circular shells [9], p. 465. One can show that both the stresses and the deformations are functions of $\cos\varphi$ or $\sin\varphi$, respectively, assuming a pressure load according to equation (13).

The longitudinal membrane force calculated by means of the bending theory is defined as $n_x^b(\xi,\varphi)$.

The resultant moment of the pressure load corresponding to the axis ($\xi=0$, $\varphi=\pi/2$) is

$$M = H^2R\pi\int_{0}^{1} P(\xi)\,\xi\,d\xi. \tag{19}$$

If only membrane forces $n_x(0,\varphi)$ were active, the following would apply for the resultant moment of these membrane forces corresponding to the axis ($\xi=0$, $\varphi=\pi/2$)

$$M = R^2\pi n_x(0,0). \tag{20}$$

Using the bending theory and assuming linear elastic material ten different stress-resultants varying with $\sin\varphi$ or $\cos\varphi$, respectively, in circumferential direction are applicable. If the resultant moment of these stress-resultants is also calculated corresponding to the axis ($\xi=0$, $\varphi=\pi/2$), only the longitudinal membrane force $n_x^b(0,\varphi)$ and the bending momentum $m_x(0,\varphi)$ furnish a contribution. Hence

$$M = R^2\pi \; n_x^b(0,0) + R\pi m_x(0,0). \tag{21}$$

If we now introduce the stress-resultant deformation relations [9], p. 469, taking into account the boundary conditions at the position $\xi = 0$,

$$u = v = w = 0, \quad \frac{\partial w}{\partial x} = 0, \tag{22}$$

we get

$$n_x^b(0,\varphi) = \frac{Eh}{H(1-\nu^2)} \left(\frac{\partial u}{\partial \xi} - k\frac{R}{H}\frac{\partial^2 w}{\partial \xi^2}\right), \tag{23}$$

$$m_x(0,\varphi) = \frac{Eh^3}{12(1-\nu^2)H^2} \left(\frac{\partial^2 w}{\partial \xi^2} - \frac{H}{R}\frac{\partial u}{\partial \xi}\right). \tag{24}$$

E being the Young's modulus of the shell, ν the Poisson's ratio, and h the thickness of the shell's wall. The shell parameter, k, is defined by

$$k = \frac{h^2}{12R^2}; \tag{25}$$

for thin shells $k \ll 1$ applies.
If we now insert the relations (23) and (24) into (21), we get

$$M = \frac{R^2\pi}{H}\,\frac{Eh}{1-\nu^2}\,\frac{\partial u}{\partial \xi}\,(1-k). \tag{26}$$

In the immediate ground surroundings there will be hardly any ovalisation of the shell. Hence

$$w(0,0) = -v(0,\tfrac{\pi}{2}) \tag{27}$$

or for the amplitudes of the displacement functions, respectively,

$$W(0) = -V(0). \tag{28}$$

If the boundary conditions (22) are inserted into the second of Flügge's equations of the isotropic circular cylindrical shell [9], p. 471, taking (28) into account, we get the following results for $\xi = 0$:

$$\frac{R}{H}\frac{\partial^2 w}{\partial \xi^2} = -\frac{(1+\nu)}{1-\nu(1-2k)}\frac{\partial u}{\partial \xi}. \tag{29}$$

Using equation (23) and expressing $\frac{\partial u}{\partial \xi}$ by $n_x^b\,(0,\varphi)$ and $\frac{\partial^2 w}{\partial \xi^2}$, eliminating $\frac{\partial^2 w}{\partial \xi^2}$ by equation (29), after some calculations we arrive at

$$M = R^2 \pi n_x^b (0,0)\,(1-\frac{2k}{1-\nu}). \tag{30}$$

After comparing equation (30) with (20), we see that the membrane longitudinal force, n_x^b, determined according to the bending theory is only slightly larger than the axial membrane force, n_x, determined according to the membrane theory. The relation is as follows

$$n_x^b\,(0,0) = n_x(0,0)\,(1+\frac{2k}{1-\nu}). \tag{31}$$

The membrane forces n_x, $n_{x\varphi}$, n_φ determined by means of the membrane theory therefore describe the membrane forces within the cylindrical shell with sufficient accuracy and will be employed for the following consideration.

In addition we note that when neglecting k against 1, the result is a simple relationship between the maximum longitudinal bending stresses, σ_{mx}, and the maximum longitudinal membrane stresses, σ_{nx}, occuring at $\xi = 0$. Using (29), both $n_x^b\,(0,\varphi)$ and $m_x(0,\varphi)$ are expressed by $\frac{\partial u}{\partial \xi}$. Hence

$$|\sigma_{mx}| = \frac{h}{R(1-\nu)}\,|\sigma_{nx}|. \tag{32}$$

4. Estimation of the safety factor for buckling due to earthquake

As already mentioned the stability of the liquid filled shell can be approximatively investigated by considering the shell to be quasi-statically loaded by the pressure, $p(\xi,\varphi)$, corresponding to equation (13).

Following the concept of several design standards, as for example [14,15,16], let us calculate the safety factor regarding stability limit by comparing the classical critical axial membrane stress of a perfect circular cylindrical shell under a uniformly distributed axial load [17]

$$\bar{\sigma}_x^{cr} = \bar{n}_x^{cr}/h = \frac{1}{\sqrt{3(1-\nu^2)}}\,\frac{Eh}{R}\,, \tag{33}$$

with the maximum value of the axial membrane compressive-stress $-n_x(0,0)/h$, corresponding to equation (18). In order to take account for imperfections a "knockdown" factor, α, is applied and we estimate the safety factor, s, by

$$s = \alpha\,\frac{\bar{n}_x^{cr}}{-n_x(0,0)}\,. \tag{34}$$

Experimental investigations, as for example shown in ref. [8], leads to the conclusion that this procedure represents a conservative estimation even if no "knockdown" is used to account for imperfections. Also from these experiments [8] it can be concluded that the imperfection is balanced by the nonuniformity effect as discussed at the end of this chapter.

Niwa and Clough [18] propose for freebase tanks a formula which is based on a single experiment by application of one specific earthquake to one free base model tank. This estimation of the critical maximum axial membrane force leading to diamond-shaped buckling during earthquake is given by:

$$\sigma_x^{cr} = n_x^{cr}(0,0)/h = 0.373\,\frac{Eh}{R} \tag{35}$$;

h is the wall thickness at the lowest shell part.
This equation renders α values of about 0.6.
However, also this σ_x^{cr} for the specific free base tank is considerable higher than the critical axial stress, $\alpha\bar{\sigma}_x^{cr}$, which has been adopted in the standards [14,15,16], i.e. in these standards a knockdown value $\alpha = 0.2 \div 0.4$ is proposed which is too small and, hence, rather conservative.

The knockdown factor, α, in these standards is based on static buckling tests of small cylinders under uniform axial compression. The axial compressive stress of the earthquake loaded shell is, however, concentrated rather locally in the area of small values of φ and ξ. Hence, a certain stiffening effect may be contributed by the surrounding shell. In addition to this the over-all imperfections are less relevant than the local imperfections near the base which are usually smaller than in the remaining cylinder. Furthermore the axisymmetrical static fluid pressure, p_s, leads (except a vary narrow area at the base) to circumferential tensile stresses which give a certain stabilizing effect. These facts are the reason why the standards are rather conservative in case of earthquake loading.

5. Numerical analysis of the behaviour of the liquid filled shell during earthquake

The estimation of the stability limit by the procedure outlined in the previous chapters contains some uncertainties:

- the ovalization is neglected in the stress calculation,
- the non-uniform axial membrane stress is replaced by a uniform distribution of the maximum value of this compressive stress,
- the influence of nonlinearities is desregarded in the stress calculation as well as in the stability estimation,
- the influence of imperfections is not sufficiently taken into account.

In order to overcome these uncertainties we propose the application of the finite element method for the calculation of the behaviour - the quasi-static behaviour - of the liquid filled shell. It should be emphasized that - contrarily to some other numerical approaches to related problems [19,20] - we apply the analytically obtained pressure distribution, equ. (13), to the "empty" shell which is modelled by finite shell elements. In refs. [19,20] the solid shell as well as the liquid are modelled by finite elements and the fluid-structure interaction is regarded by modelling the coupling criteria.

The main object of our numerical computations is to demonstrate the influence of nonlinearities and imperfections on the stress distribution and on the stability as recently shown for wind loaded shells in [21,22].

5.1 Description of the model tank under consideration

Akira Niwa performed an extensive experimental work on the dynamic behaviour of liquid filled circular cylindrical shells [23]. Since he reports on results of imperfection measurements we will demonstrate the numerical calculation using one of Niwa's model tanks, see Fig. 6

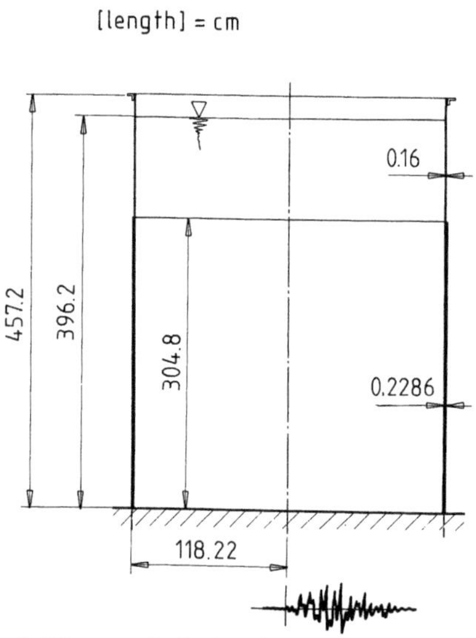

aluminium shell:
material properties

$E = 6.86 \times 10^{6}\ N/cm^2$

$\nu = 0.3$

$\rho_S = 2.7\times10^{-5}\ Ns^2/cm^4$

liquid (water):

$\rho_L = 1.0\times10^{-5}\ Ns^2/cm^4$

Fig. 6 The model tank under consideration (according to [23])

Niwa [23] used a modified ElCentro earthquake with a maximum ground acceleration of

$A = 0.5$ g.

Application of equ. (2) leads to a fundamental sloshing vibration period of

$T^* = 1.161$ s,

which, using the relevant earthquake response spectrum, renders a spectral acceleration

$A^* = 0.32$ g.

With the aid of the procedure outlined in chapter 2.1 and appendix 1 we find the mode participation factor, γ, and the spectral acceleration, a, for the tank wall vibrations to be:

$\gamma = 1.81$

$a = 0.929$ g.

The fundamental natural frequency of the liquid filled shell is $\omega = 77.79\ s^{-1}$; with a mode shape almost exactly represented by a mode as used in fig. 4.2 (see chapter 2.2.4). According to equ. (4) the corresponding maximum relative acceleration amounts to

$A_D = 0.777$ g.

With the aid of this information and equs. (5), (8) and (12), respectively, the individual components of the dynamically activated fluid pressure amplitudes are calculated. By equ. (13) we obtain the resultant maximal dynamic pressure distribution shown in Fig. 7.

As explained in chapter 2.2.5 the pressure distribution obtained for the perfect cylinder is as a good approximation also applicable to the imperfect shell.

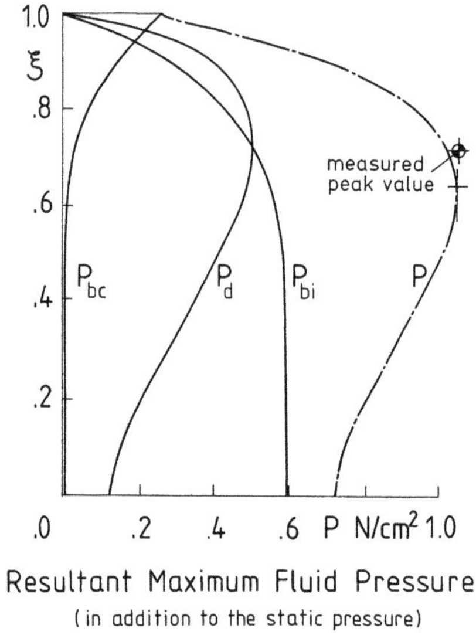

Fig. 7 Pressure amplitudes - components and SRSS-resultant pressure

5.2 An analytical estimation of the stability

Application of the procedure outlined in chapter (3) to the specific example under consideration leads to the maximum value of the axial compressive stress (according to equation (18) and a pressure distribution as shown in Fig. 7):

$$-n_x(0,0) = \frac{H^2}{R} \int_0^1 \xi\, P(\xi)\, d\xi = 566.4 \text{ N/cm}$$

If we use equation (33) for calculating the critical buckling stress, as justified by Ref. [8], we obtain

$$\bar{n}_x^{cr}(0,0) = 0.605\, \frac{Eh^2}{R} = 1\,834.6 \text{ N/cm}$$

Hence we find an approximative value of the safety factor concerning stability which is

$$s = \frac{\bar{n}_x^{cr}(0,0)}{-n_x(0,0)} = 3.24$$

This estimation will be compared with the result of the non-linear stability analysis described in the following chapters.

5.3 The finite element solution procedure

The theoretical background of the nonlinear finite element shell analysis using the computer program NISA80 [24] is described in detail in [21,25,26]. We will focus on the description of the procedure.

It should be mentioned that the sequence of pressure load application is essential, i.e. before applying the dynamically activated pressure the static fluid pressure of the system at rest

$$p_s(\xi,\varphi) = \rho_L gH(1-\xi) \tag{36}$$

has to be incremented up to the final value and furthermore held constant. This is of particular importance in the case of geometrically imperfect cylinders for which the static pressure has a smoothening effect, i.e. a reduction of the imperfections relevant for the stability under axial compressive stresses due to the subsequently applied dynamic pressure.

Therefore a non-proportional load incrementation has to be carried out:

$${}^{i}p_{res}(\xi,\varphi) = {}^{i}\lambda_1 \, p_s(\xi,\varphi) + {}^{i}\lambda_2 \, p(\xi,\varphi) \tag{37}$$

(i is the increment number)

with p_s and p corresponding to equs. (36) and (13), respectively. The incrementation histories of the load multiplyers, ${}^{i}\lambda_1$ and ${}^{i}\lambda_2$, are shown schematically in Fig. 8.

The history ${}^{i}\lambda_1$ of the static pressure is defined in advance and modified Newton Raphson iteration [27] is applied during the increments $i = 1,2 \ldots i^0$. However the incrementation and iteration strategy for the dynamic pressure application follows a concept described in [28] based on Riks' & Wempner's algorithm [29].

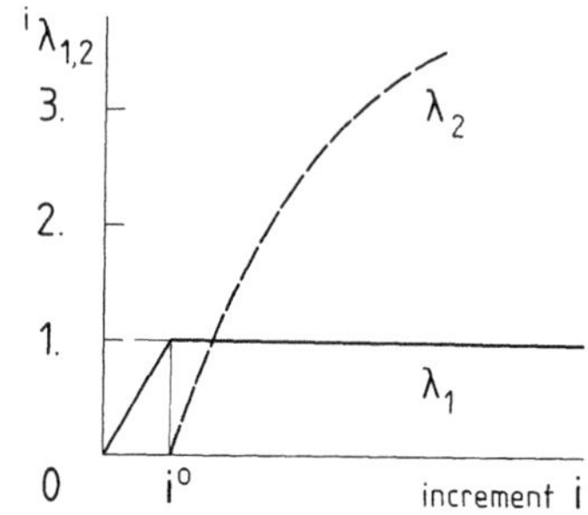

Fig. 8 Load incrementation for the different pressure contributions

Since these automatic incrementation algorithms were developed for proportional load only a modification for non-proportional loading is necessary, i.e. p_s acts as a "dead load".

Due to the non-proportionality of the load the interpretation of the eigenvalues of the accompanying eigenvalue analysis, see for instance [21], is rather difficult. The approximative linearized eigenvalue problem

$$\{{}^{o}\underset{\sim}{K}_e + {}^{m}\eta\,[\underset{\sim}{K}_u({}^{m}\lambda_1, {}^{m}\lambda_2) + \underset{\sim}{K}_g({}^{m}\lambda_1, {}^{m}\lambda_2)]\}\,{}^{m}(\delta\underset{\sim}{u}) = 0 \qquad (38)$$

(${}^{o}\underset{\sim}{K}_e$, ${}^{m}\underset{\sim}{K}_u$, ${}^{m}\underset{\sim}{K}_g$ are the linear elastic stiffness matrix, the initial displacement matrix and the geometric matrix, respectively, at increment m), $m > i^0$

for the estimation of the critical load leads to eigenvalues, ${}^{m}\eta_j$, which have the following meaning:
Load combinations

$${}^{m}\eta_j\;{}^{m}\lambda_1\;p_s + {}^{m}\eta_j\;{}^{m}\lambda_2\;p = {}^{m}p^*_{res_j}\,,\quad m > i^0,\quad {}^{m}\lambda_1 = 1.0 \qquad (39)$$

represents approximate critical loads. But due to the practical reality the full static pressure, p_s, must not be varied. This condition would require to multiply only contributions due to ${}^{m}\lambda_2$ in ${}^{m}\underset{\sim}{K}_u$ and ${}^{m}\underset{\sim}{K}_g$ with ${}^{m}\eta$:

$$\{{}^{o}\underset{\sim}{K}_e+\underset{\sim}{K}_u^1({}^m\lambda_1)+\underset{\sim}{K}_g^1({}^m\lambda_1)+{}^m\tilde{\eta}\left[\underset{\sim}{K}_u^2({}^m\lambda_2)+\underset{\sim}{K}_g^2({}^m\lambda_2)\right]\}\,{}^m(\delta\underset{\sim}{u})=\underset{\sim}{0} \qquad (40)$$

$${}^m\lambda_1 = 1.0$$

with approximative critical load combinations

$$p_s + {}^m\tilde{\eta}_j\ {}^m\lambda_2\ p = {}^m\tilde{p}^*_{res_j}\,. \qquad (41)$$

On the other hand one should hold in mind that the eigenvalues ${}^m\eta_j$ as well as ${}^m\tilde{\eta}_j$ become as more meaningful as more the load situation m, i.e. ${}^m\lambda_1 = 1.0$, ${}^m\lambda_2$, approaches the critical situation, m*, for which ${}^{m^*}\eta_1$ as well as ${}^{m^*}\tilde{\eta}_1$ approach the value 1. This leads to the conclusion that the application of the eigenvalue procedure (38), (39) is also useful provided the considered equilibrium state, m, is near the critical one, denoted by m*.

5.4 The finite element model and results of the analysis

The cylindrical shell is modelled by (4x4)-noded degenerated nonlinear shell elements [24]; the ring at the top of the shell is modelled by nonlinear beam elements coupled with the shell by the translational degrees of freedom.

The geometric imperfection is chosen to be

$$\tilde{R}(\xi,\varphi) = R\left[1+\varepsilon(a_1\xi+a_2\xi^2+a_3\xi^3)\cos n\varphi\right] \qquad (42)$$

with

$\varepsilon = 4.23 \times 10^{-3}$

$a_1 = 2.275$
$a_2 = 0.875$
$a_3 = -3.15$
$n = 4$

This is in good correspondence with the information given in [23] about the shape of the model tank under consideration.

Fig. 9 shows the finite element mesh in the undeformed and in the deformed configuration immediatly before buckling appears, respectively.

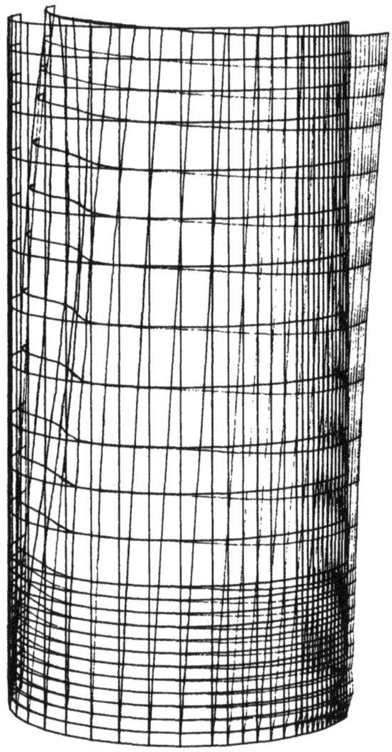

Fig. 9 Undeformed and deformed mesh (immediately before buckling)

It should be emphasized that this Fig. shows the essential fact that the initial imperfections are smoothened at the front side (around $\varphi = 0$) however they are reactivated at the rear side (around $\varphi = \pi$) since the static pressure is decreased due to dynamic activation. From this one can conclude that not all out-of-round deformations observed during earthquake excitation are contributions of higher order vibration modes, i.e. the alternating appearance and disappearance of the geometric imperfections lead also to an imagination of vibrations with $m > 1$.

The distribution of the axial membrane forces at the base, $n_x(0, \varphi)$, is shown in Fig. 10.

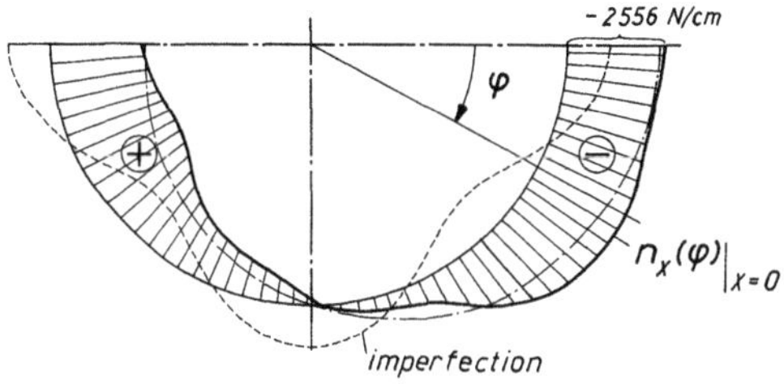

Fig. 10 Axial membrane force at the base, $n_x(0, \varphi)$

A certain deviation from the cos φ-distribution according to the membrane theory of the perfect cylinder, equ. (18), can be observed. There is a significant correlations between this deviation and the distribution of the geometric imperfections. This correlation agrees well with measurements as shown in ref. [23].

Carrying out an eigenvalue analysis corresponding to equation (38) at the load level λ_2 = 4.056 leads to an eigen value of $\tilde{\eta}_1$= 0.967 (which is near 1.0; see remarks above). This means that the finite element analysis results in a critical load multiplice of $\lambda_2^* \approx 3.92$ i.e. safety factor s ≈ 3.92. This result for the specific example under consideration is a further justification for the conservativeness of the application of equ. (33).

Fig. 11 shows the fundamental buckling mode, $\delta \underset{\sim}{u}_1$, calculated at the above mentioned load level.

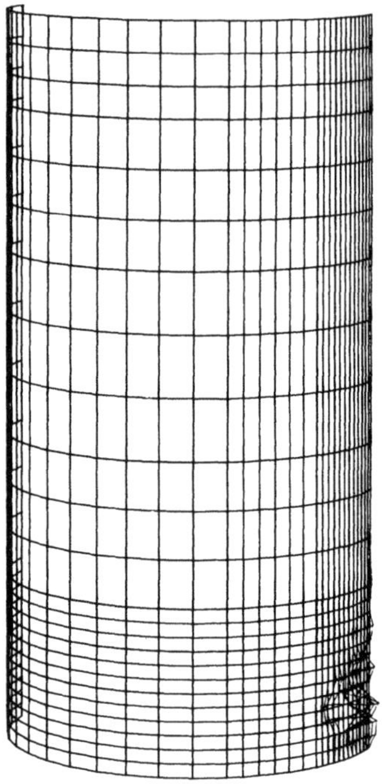

Fig. 11 Fundamental buckling mode

This buckling mode shows local buckling at the front near the base. This mode corresponds very well with shell crippling modes observed at real tanks and model tanks [18] during earthquake.

Acknowledgement

The authors would like to appreciate the contributions given by Mr. Stegmüller of Stuttgart University and Mr. Gruber of VOEST-ALPINE AG Linz who have carried out most of the numerical work. Furthermore we thank Prof. Ramm at the University of Stuttgart for his encouraging and helpful discussions.

References

1. Haroun, M.A.: Dynamic Analysis of Liquid Storage Tanks. Report No. EERL 80-2, California Institute of Technology, Pasadena, California, 1980.

2. Fischer, D.: Ein Vorschlag zur erdbebensicheren Bemessung von flüssigkeitsgefüllten zylindrischen Tankbauwerken, Der Stahlbau, 1, 13 - 20, 1981.

3. Clough, R.W., Penzien, J.: Dynamics of Structures, McGraw Hill, 1975.

4. Kalnins, A., Godfrey, D.A.: Seismic Analysis of Thin Shell Structures, Nucl. Engng. Design, 27, 68 - 76, 1974.

5. Yang Jong Yih: Dynamic Behaviour of Fluid Tanks, PhD-Thesis, Rice University, Houston, Texas, 1976.

6. Fischer, D.: Dynamic Fluid Effects in Liquid-Filled Flexible Cylindrical Tanks. J. Earthquake Engng. & Struct. Dyn., 7, 587 - 601, 1979.

7. Welded Steel Tanks for Oil Storage, American Petroleum Institute, API Standard 650, 7th edition, Nov. 80, Appendix E.

8. Shih, C.-F. and C.D. Babcock: Scale Model Buckling Tests of a Liquid Filled Tank Under Harmonic Excitation, ASME 80-C2/PVP-66, 1980.

9. Girkmann, K.: Flächentragwerke, 5th edition, Springer-Verlag, Wien, 1959.

10. Turner, J.W.: Effect of Out-of-Roundness on the Dynamic Response of Liquid Storage Tanks, M.S. thesis, Rice University, Houston, Texas, May 1978.

11. Parkus, H.: Überschwappwahrscheinlichkeit für einen Flüssigkeitsbehälter unter Erdbebeneinwirkung, Ing.-Archiv, 49, 179 - 185 (1980).

12. Parkus, H.: Modes and Frequencies of Vibrating Liquid-Filled Cylindrical Tanks, Int. J. Engng. Sci., 20, 319 (1982).

13. Shih, C.F., Babock, C.D.: Buckling of Cylindrical Tank Under Earthquake Excitation, ASCE Proc. 3rd Engng. Mech. Conf.,81-84, 1979

14. ÖNORM B4650, Teil 4: Stahlbau, Beulung von Kreiszylinderschalen, 1977.

15. DASt Richtlinie 013: Beulsicherheit von Schalen, Juli 1980.

16. API Standard 650, Sixth edn., Rev. 3: Welded Steel Tanks for Oil Storage, Amer. Petroleum Inst., Washington, 1979.

17. Timoshenko, S.P., Gere, J.M.: Theory of Elastic Stability, 2nd edn., McGraw-Hill, New York, 1961.

18. Niwa, A., Clough, R.W.: Buckling of Cylindrical Liquid-Storage Tanks Under Earthquake Loading, J. Earthquake Engng. & Struct. Dyn., 10, 107 - 122 (1982).

19. Liu, W.K.: Finite Element Procedures for Fluid-Structure Interactions and Application to Liquid Storage Tanks, Nucl. Engng. Design, 65, 221 - 238 (1981).

20. Ma, D.C., Liu, W.K., Chang, Y.W.: Seismic Response of the Flexible Fluid-Tank Systems - a Numerical Study, to be presented at the ASME Portland Conf., June 27 - July 7, 1982

21. Brendel, B., Ramm, E., Fischer, D.F., Rammerstorfer, F.G.: Linear and Nonlinear Stability Analysis of Thin Cylindrical Shells under Wind Loads, J. Struct. Mech., 9(1), 91 - 113 (1981).

22. Ramm, E., Sättele, J.M.: Elasto-Plastic Large Deformation Shell Analysis Using Degenerated Elements, Symp. Nonl. Finite Element Analysis of Shells, ASME-Winter Annual Meeting, Washington DC., 1981.

23. Niwa, A.: Seismic Behaviour of Tall Liquid Storage Tanks, Earthquake Engng. Research Center Report UCB/EERC-78/04, Univ. of Berkeley, 1978.

24. Brendel, B., Häfner, L., Ramm, E., Sättele, J.M.: Programm-dokumentation - Programmsystem NISA, Bericht Inst. f. Baustatik, Univ. Stuttgart, 1977.

25. Brendel, B.: Geometrisch nichtlineare Elastostabilität, Bericht Nr. 79-1, Inst. f. Baustatik, Univ. Stuttgart, 1979.

26. Sättele, J.M.: Ein Finites Elementkonzept zur Berechnung von Platten und Schalen bei stofflicher und geometrischer Nichtlinearität, Bericht Nr. 80-3, Inst. f. Baustatik, Univ. Stuttgart, 1980.

27. Bathe, K.-J., Ramm, E., Wilson, E.L.: Finite Element Formulations for Large Deformation Dynamic Analysis, Int. J. Num. Meth. Engng., 9, 353 - 386 (1975).

28. Ramm, E.: Strategies for Tracing the Nonlinear Response Near Limit Points, Nonlinear Finite Element Analysis in Structural Mechanics (Eds.: W. Wunderlich, E. Stein, K.-J. Bathe), Springer-Verlag, Berlin, Heidelberg, New York, 1981.

29. Riks, E.: An Incremental Approach to the Solution of Snapping and Buckling Problems, Int. J. Solids Struct., 15, 529 - 551 (1979).

Appendix 1

Determination of f(x), ω, γ

The maximum value, $r(\xi)$, of the resultant of the pressure distribution, p_d, equ. (8), per unit length in axial direction is

$$r(\xi) = C\omega^2 R\rho_L \int_0^{2\pi} \bar{P}_d(\xi)\cos^2\varphi R d\rho = C\omega^2 \rho_L \bar{P}_d(\xi) R^2 \pi. \quad \text{(A1)}$$

We define now

$$m(\xi) := \frac{r(\xi)}{C\omega^2} = \rho_L \bar{P}_d(\xi) R^2 \pi,$$

$m(\xi)$ has the dimension of a mass per unit length. Let us now consider the resultant relative acceleration of a shell point (ξ,φ) in a plane ξ = const.

Its component in the direction of $\varphi = 0$ with (6) is as follows

$$\frac{\partial^2 w}{\partial t^2}\cos\varphi - \frac{\partial^2 v}{\partial t^2}\sin\varphi = -(C\hat{f}(\xi)\cos^2\varphi - D\hat{g}(\xi)\sin^2\varphi)\omega^2 \sin\omega t, \quad \text{(A2)}$$

its component in the direction of $\varphi = \pi/2$ with (6) is

$$\frac{\partial^2 w}{\partial t^2}\sin\varphi + \frac{\partial^2 v}{\partial t^2}\cos\varphi = -(C\hat{f}(\xi) + D\hat{g}(\xi)\sin\varphi\cos\varphi\omega^2 \sin\omega t \quad \text{(A3)}$$

Assuming that the shell hardly ovalizes, we get

$$w(\xi,0) = -v(\xi,\pi/2). \quad \text{(A4)}$$

From (A3) it follows that

$$C\hat{f}(\xi) = -D\hat{g}(\xi). \quad \text{(A5)}$$

By inserting (A5) into (A2), (A3) we see that the maximum resultant relative acceleration of one shell point (ξ,φ) in the plane ξ = const. has the value $C\hat{f}(\xi)$ and is directed parallel to $\varphi = 0$. Proceeding from (A4), we may regard the resultant of the pressure distribution, $r(\xi)$ per length also as the maximum inertia force of a mass, $m^*(\xi)$:

$$m^*(\xi) = \frac{m(\xi)}{\hat{f}(\xi)}, \tag{A6}$$

taking into account that this mass, $m^*(\xi)$, as an additional mass is fixed to the shell at the position ξ and experiences the same acceleration as the shell itself during vibration.
If the circular frequency, ω, of the shell is to be calculated under the consideration of the fluid, it is useful to assume a deformation figure $\hat{f}_i(\xi)$; i is the number of the i-th cycle in the iteration process described below. In order to take the fluid pressure into account, the shell is attributed with a fictitious additional mass density, $\rho_i(\xi)$, corresponding to (A6), (A1)

$$\rho_i^*(\xi) = \frac{m^*(\xi)}{2R\pi h} = \rho_L \frac{R}{2h} \frac{\bar{p}_d(\xi)}{\hat{f}_i(\xi)} . \tag{A7}$$

The overall density, $\rho_i(\xi)$, results as

$$\rho_i(\xi) = \rho_s + \rho_i^*(\xi), \tag{A8}$$

ρ_s ... mass density of the shell material.
The circular frequency, ω, can now be determined for a "dry" shell with mass density, $\rho_i(\xi)$, either analytically or numerically by means of a computer program. The eigen-mode found by this calculation can be used for an improved assumption of the deformation figure, $\hat{f}_{i+1}(\xi)$. This iteration process can be continued until the assumed mode corresponds to the calculated mode with satisfying accuracy. In [2], section 3.2 a simplified procedure is described with a linear fictitious additional mass density. In most cases, a few iterations are sufficient. For horizontal acceleration, the following applies for the mode participation factor:

$$\gamma = \frac{\int_\Omega \psi \rho \, d\Omega}{\int_\Omega \psi^2 \rho \, d\Omega}, \tag{A9}$$

Ω is the surface of the shell and ρ the effective mass density (A8). ψ is the first eigen-mode of free vibration of the fluid-filled shell. In (A9) only unidirectional ($\varphi=0$) horizontal deflections are taken into account. For the first eigen mode the following applies:

$$\psi(\xi,\varphi) = \psi_w(\xi,\varphi)\cos\varphi - \psi_v(\xi,\varphi)\sin\varphi. \tag{A10}$$

Assuming that the shell hardly ovalizes and considering (A4), (A5) it follows

$$\psi(\xi,\varphi) = \hat{f}(\xi) \tag{A11}$$

Neglecting ρ_s as against ρ^*, (A7), which is admissible in many practical cases, the approximative expression for γ is

$$\gamma = \frac{\int_0^1 \frac{\bar{P}_d(\xi)}{h(\xi)} d\xi}{\int_0^1 \frac{\hat{f}(\xi)}{h(\xi)} \bar{P}_d(\xi) d\xi} \tag{A12}$$

Here, the wall thickness, h, of the shell is regarded as function $h(\xi)$.

Appendix 2

Determination of modal damping ζ

In order to determine the damping during the common vibration of the shell and the fluid, there are only few experimental and theoretical hints. The question arises whether the damping of the common system is smaller than the damping of the empty shell or not. The authors are at present undertaking an extensive experimental and theoretical development program. Haroun [1] as well as API 650 take for a system steel-shell - water (or light petroleum respectively) a damping value of $\zeta = 0.02$. However, the authors recommend the use of smaller damping values, e.g. $\zeta = 0.01$.

Part VI :

Codes and Practice

To the Problem of Buckling Safety of Shells in the Plastic Range

F. W. BORNSCHEUER

Institut für Baustatik
Universität Stuttgart, West-Germany

1. Starting Point

In steel structures like vessels or offshore constructions shells are often not as thin as in aircraft and space constructions. The rather thick-walled shells which are stressed in the elastic-plastic range are of great importance. To take this fact into account the plastic range was introduced in the codes for shell buckling during the past years. Compared to the number of experiments in the elastic range there are unfortunately only a few tests performed in the plastic range. And these few ones are not even well documented. Especially there is a lack of exact values of the yield stresses of the materials being used.

The only acceptable procedure was to assume inaccurate formulas for the transition from the elastic to the plastic range. So two different linear functions were introduced into the new German rules for shell buckling, the DASt-Richtlinie 013 [1], as shown in Fig. 1. This code was released in July 1980 and will be the basis for part 4 of the new German stability standard DIN 18 800. The transition line (1. 2) applies to cases as circular cylindrical and conical shells subjected to axial compression and spherical shells under external pressure which are very sensitive to imperfections. The function (1. 3) applies to circular cylinders and conical shells under external pressure which are less sensitive to imperfections. In this diagram the normalized load carrying stresses $\bar{\sigma}_u = \sigma_u/\sigma_F$ *) are plotted versus the special shell slenderness parameter $\bar{\lambda}_S = \sqrt{\sigma_F/\alpha\,\sigma_{Ki}} = \sqrt{\sigma_F/\sigma_e}$. In contrary to the buckling of steel columns the denominator of the root includes the

*) other notations: $\bar{\sigma}_u = \sigma_u/\sigma_F \equiv \sigma_K/\sigma_r \equiv \sigma_u/ß_S \equiv \sigma_u/\sigma_y \equiv \sigma_u/f_y$

Buckling of Shells, Ed.: E. Ramm
Proceedings of a State-of-the-Art Colloqium

reduction (or knock-down) factor α known from aerospace technology which applies to purely elastic buckling. This reduction factor α reduces the ideal buckling stress $\sigma_{Ki} \equiv \sigma_{cr}$ according to the bifurcation load of the linear theory to the ultimate stress $\sigma_u = \sigma_e$ in the elastic range. In the plastic region a further reduction will be made using formulas (1.2) and (1.3) resp.

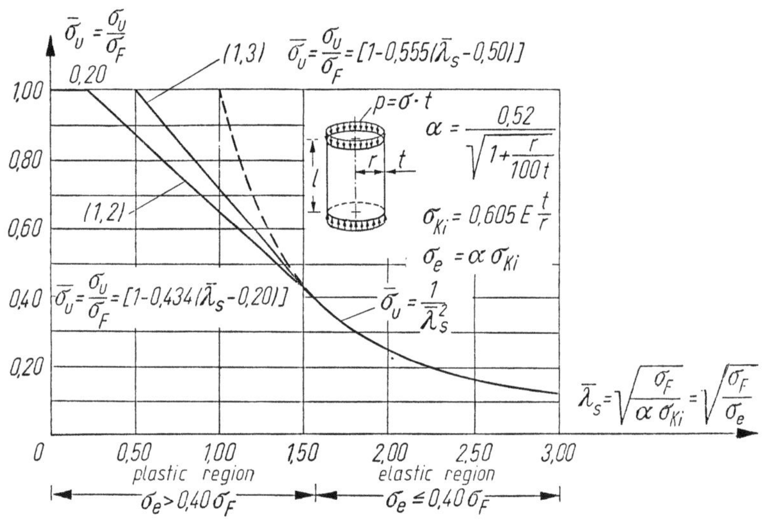

Fig. 1: Load carrying capacity curves according to the DASt-Richtlinie 013

In the European code for shell buckling, the ECCS Recommendations R 4.6 [2], which was worked out during the past years a similar procedure for the case of the circular cylinder under axial compression and pure bending was adopted. Using the transition curve plotted in Fig. 2 results in about the same shell design as if the DASt-Richtlinie 013 is applied though the reduction factors α and α_o are different from each other. This was pointed out already in a recent paper published in the "Stahlbau" [3].

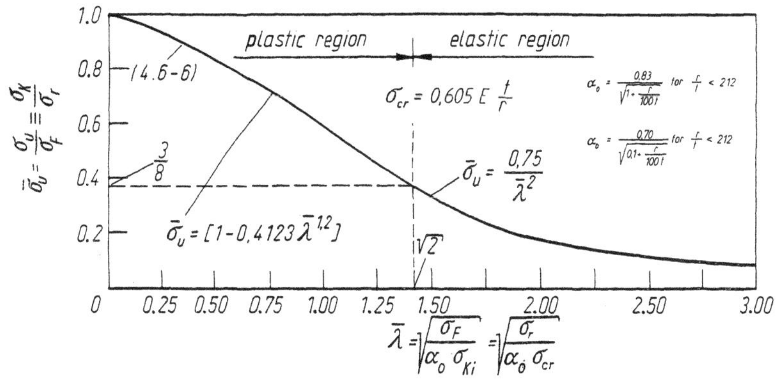

Fig. 2: Load carrying capacity curve for the axially loaded circular cylindrical shell according to the ECCS Recommendations R 4. 6

In the presentation given in the ECCS Recommendations the reduction factor α_o represents the lower bound of test values. In the DASt-Richtlinie according to Fig. 3 the reduction factor α is taken as 75 % of the

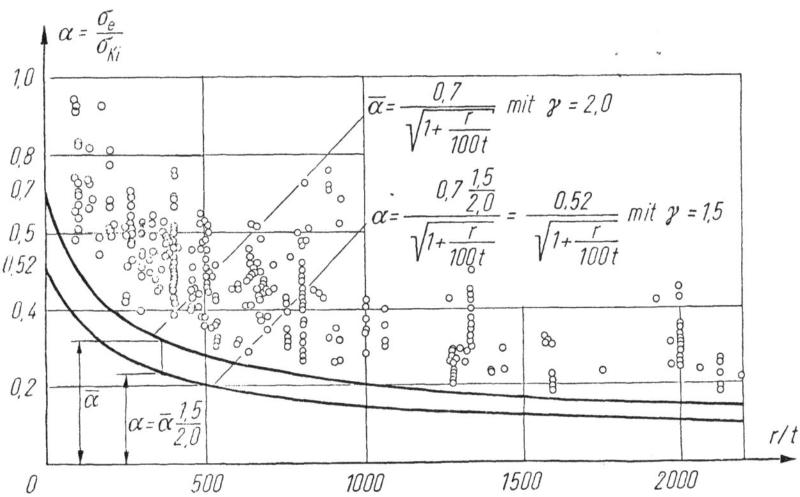

Fig. 3: Determination of the curve of the reduction factor α for the axially loaded circular cylindrical shell according to the DASt-Richtlinie

lower bound of the test results [3]. This was necessary in order to get an effective safety factor of $\nu = 2.0$ with a constant load increasing factor of $\gamma = 1.5$ (global load and resistance factor). In the ECCS Recommendations this was achieved by lowering the Euler curve from $1/\bar{\lambda}^2$ to $0.75/\bar{\lambda}^2$.

In other codes quite different curves were adopted for the elastic-plastic transition zone. In Fig. 4, taken from [3], the admissible stresses of the axially compressed cylinder of 6 different codes are plotted. For reasons

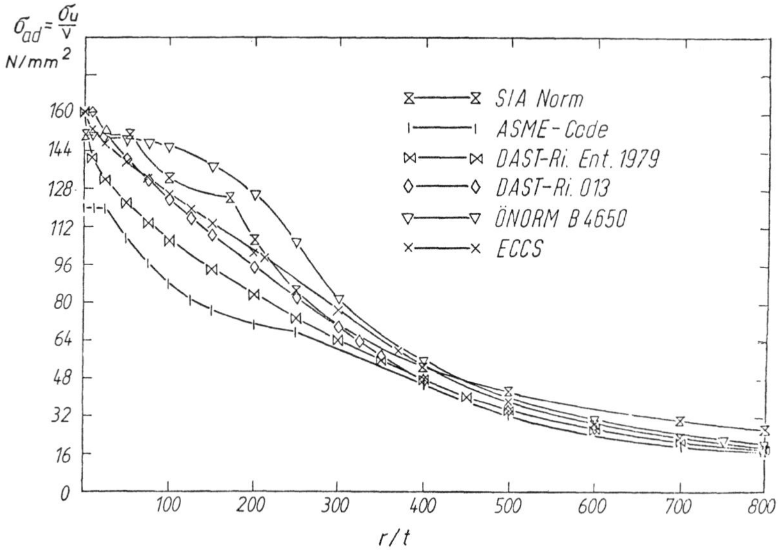

Fig. 4: Admissible stresses for the axially loaded circular cylindrical shell of medium length for steel St 37

of better comparison these curves are evaluated for main loads only. Steel St 37 (St E 240) with a yield limit of 240 N/mm² was chosen. The admissible stresses on the ordinate are the ultimate stresses σ_u divided by the safety factor ν. They are plotted as a function of the ratio of radius r to wall thickness t. For slender shells, i.e. for large values of r/t, the curves nearly coincide. It is noteworthy that the results for small r/t values in the plastic range show a remarkable scatter. In the limit of r/t = 0 one can notice different safety factors. The ECCS Recommendations R 4.6 [2] as well as both versions of the DASt-Richtlinie 013 [1] suggest $\nu = 1.5$, the ASME code [4] $\nu = 2.0$ and the Austrian ÖNORM B 4650 [5] and Swiss SIA Norm [6] $\nu = 1.6$. In the elastic-plastic region the Austrian and Swiss curves are located very high in contrast to the ASME code where the curve is very low. Slightly above the curve of ASME a curve is shown which is taken from an earlier draft of the DASt-Richtlinie 013 of the year 1979. Due to some thorough analyses from Saal [7], Vandepitte/Rathé [8] et. al. this curve was abandoned in favour of the function (1.2) which lies above the 1979 curve and is therefore more economical.

Due to the large imperfection sensitivity in the elastic range in both rules, the ECCS Recommendations as well as the DASt-Richtlinie, the effective safety factor was defined by $\nu = 2.0$ based on the lower bound of ultimate stresses obtained from test results. In the elastic-plastic transition zone the effective safety factor decreases in both rules from $\nu = 2.0$ to $\nu = 1.5$. This procedure seems to be reasonable as long as some of the extreme test results are omitted. Hence, Saal [7] left out a part of the tests from Newmark/Wilson and Robertson. Vandepitte/Rathé eliminated additionally some results of Steinhardt/Schulz [9]. Up to now it could not be cleared whether or not this procedure is justified. In this quite vague situation it is difficult to answer the question of the safety of shell buckling in the plastic range, especially being aware of the fact that many unsolved problems are still existing in the elastic region. Of course, all efforts in that field are appreciated. For the time being we are obliged to refer mainly to experiments.

2. Test Results for the Elastic-Plastic Region in the Presentation of the ECCS Recommendations R 4.6 and the DASt-Richtlinie 013

In Figs. 5 and 6 test series given in the literature are added to the diagrams of both codes [1] and [2]. Also shown are the tests carried out in Karlsruhe this year by Schulz [10]. A paper on these experiments will be published shortly in the "Stahlbau".

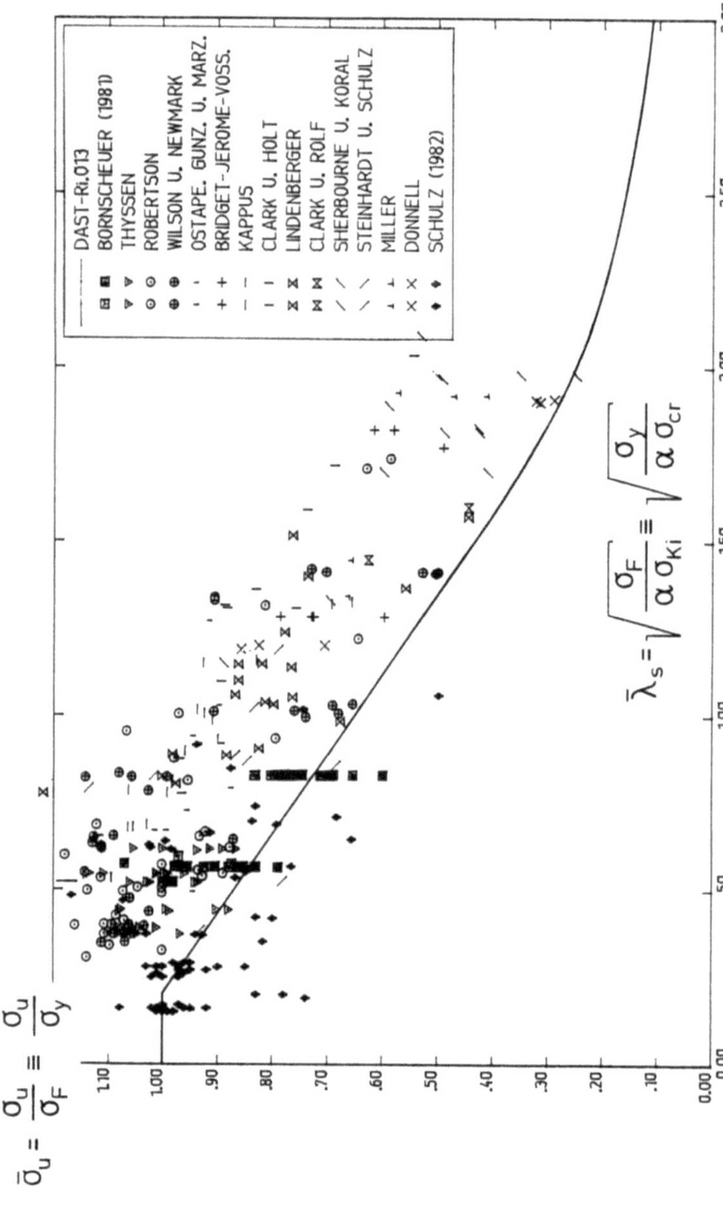

Fig. 5: Test results for the axially loaded cylindrical shell according to the DASt-Richtlinie 013

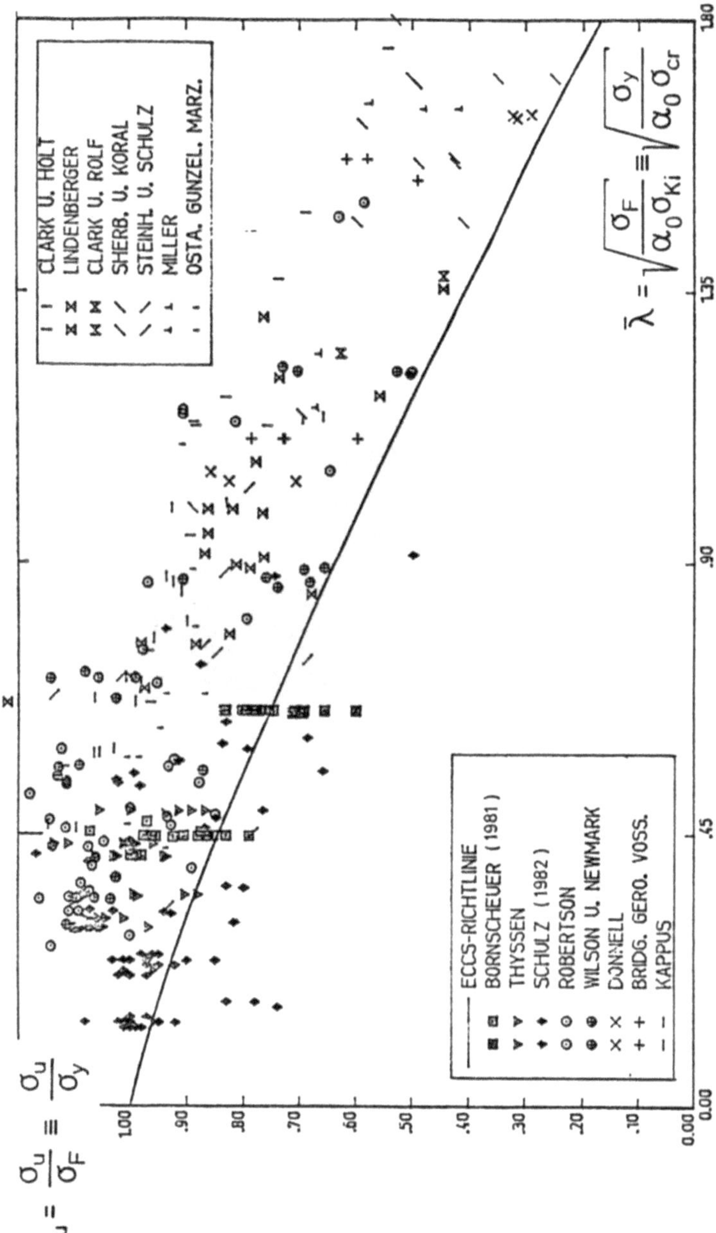

Fig. 6: Test results for the axially loaded cylindrical shell according to the ECCS Recommendations R 4.6

3. Modes of Collapse in the Elastic-Plastic Range

In order to allow a better judgement of the safety problem the specific modes of collapse in the plastic range are described in the following. In case of small slenderness (i.e. $r/t \leq 100$) axially compressed cylinders do fail in an axially symmetric hump-like mode at one end of the cylinder having a diameter of 250 mm and a longitudinal weld ($t = 2$ mm) [3].

Fig. 7: Characteristic buckling shape in the fully plastic region

This shape of buckling is already known from the earliest shell buckling test performed some decades ago. The buckles near the end are essentially caused by bending moments which come from the boundary constraint. Strictly speaking the axially symmetric buckles in Fig. 7 do not represent a real instability phenomenon but rather a collapse through a loss of strength. With increasing slenderness this buckling mode changes into a few flat buckles along the circumference. The typical collapse mode of axially loaded thin shells with a few single buckles along the circumference can be recognized in Fig. 8. With larger slenderness as $r/t > 300$ the influence of the boundary layer effect considerably decreases compared to the influence of the in practice unavoidable imperfections in the entire shell. Finally buckles of the classical chess board - like shape develop which proceed into the well-known diamond shaped buckles in the postcritical range.

Fig. 8: Characteristic buckling shape in the elastic-plast: transition zone ($r = 125$ mm; $t = 1$ mm)

4. Comparison between the Axially Loaded Column and the Circular Cylindrical Shell

Comparing the stress-strain-curves in the postcritical range of compact and slender columns and cylinders some conclusions on the required safety factors are possible.

In Fig. 9 the buckling of columns and in Fig. 10 the buckling of shells is outlined in principle. The two upper diagrams are applicable to the elastic region where the ideal buckling stress of the linear theory of stability σ_{Ki} is chosen as reference value. The lower diagrams apply to the plastic range. Here σ_F is chosen as reference value.

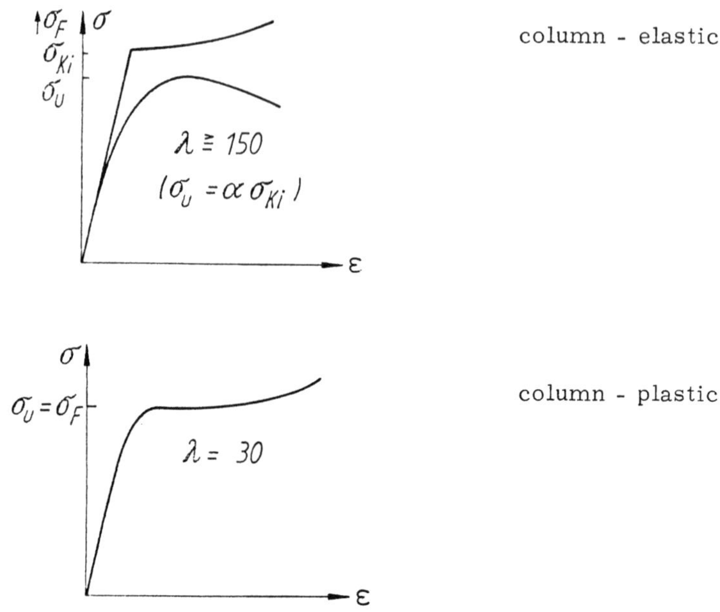

Fig. 9: Stress - strain - curves of slender and compact columns

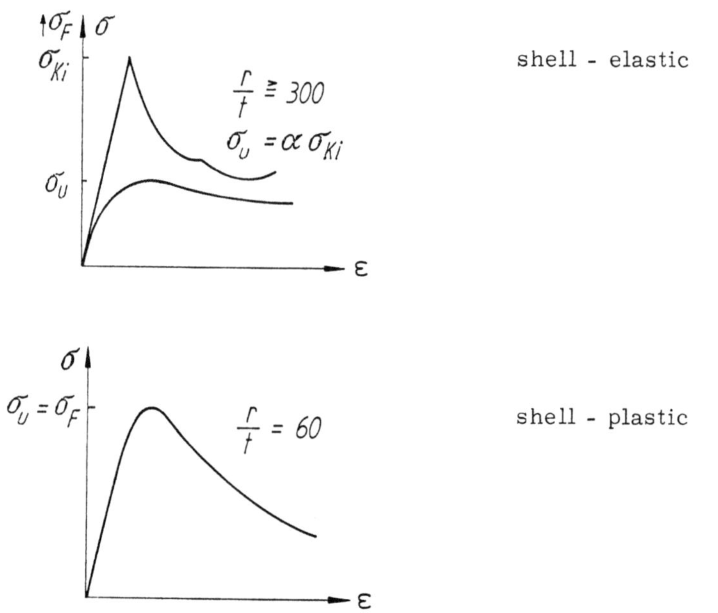

Fig. 10: Stress - strain - curves of slender and compact cylindrical shells

In the elastic range the sensitivity with respect to imperfections may be taken into consideration by a reduction factor α. In the case of the columns the reduction of the critical stress σ_u with respect to the ideal buckling stress σ_{Ki} is by far not as large as in the case of the shells which are much more sensitive to imperfections. In column buckling one attains in the elastic range reductions, being equivalent to α values, which range between 0.5 and 0.9. In shell buckling the α values decrease to as far as 0.2 or even less.

In the plastic range it was assumed in both diagrams that the ultimate stresses reach the yield limit. The postcritical regions differ fundamentally. For columns made of conventional construction steel one could notice even a hardening effect across the entire cross-section. In shell buckling the postcritical minimum is located far below the yield limit. Even if one does not assume postcritical stresses as a measure for admissible stresses one should not disregard this dramatic decrease. Anyway a safety factor of only $\nu = 1.5$ seems to be not sufficient compared to the factor used for columns. An increase to 1.6 or 1.7 would be reasonable. On the other hand one does not have to go as far as $\nu = 2.0$ as it is stated in the ASME code, classes A and B, defining the main load cases. One possibility to fix an effective safety factor ν that takes into account the behaviour in the postcritical region is shown in Fig. 11.

range	column plastic		column elastic	shell plastic		shell elastic
safety factor ν (for main loads)	1.5	<	1.6	1.6	<	2.0
knock-down factor γ/ν (for main loads with $\gamma = 1.5$)	1.00	>	0.9375	0.9375	>	0.750

Fig. 11: Safety factor for columns and axially compressed cylinders

5. Proposal for a Modification in the Elastic-Plastic Range of Buckling

The definition of a safety factor in the highly plastic region is not the only priority task. It is at least as important to find the buckling curve of the transition zone. For example the curve adopted by the ÖNORM lies too much on the unsafe side inspite of a safety factor of 1.6. The following proposals retain the concept of a reduction factor α introduced in the aerospace technology. Here either the curves in the transition zone or the reduction factor α itself may be lowered.

5.1 Lowering of the Curve for the Ultimate Stress

In Figs. 12 and 13 the version is shown in which the curves in the transition zone are lowered.

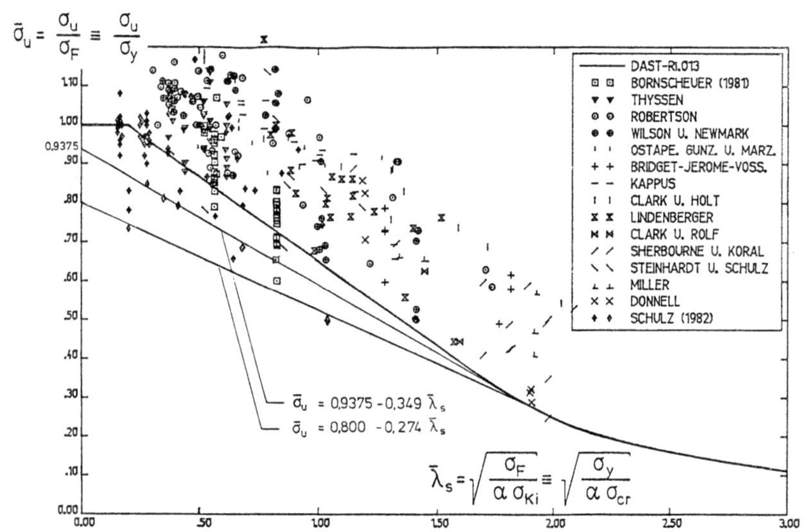

Fig. 12: Proposals for lowered ultimate stress curves according to the DASt-Richtlinie 013

Following the concept of the DASt-Richtlinie 013 two transition straight-lines are plotted in Fig. 12 in addition to those given in Fig. 5. The lower line which covers essentially all the test results leads to an ultimate stress of $\bar{\sigma}_u = 0.800$ for $\bar{\lambda}_S = 0$. For main loads with a global load and resistance factor $\gamma = 1.5$ one attains an effective safety of $\nu = 1.5/0.800 = 1.875$ for the limit case of $\bar{\lambda}_S = 0$. For the upper line an effective safety of $\nu = 1.6$ is assumed for the limit case $\bar{\lambda}_S = 0$. This results in a reduced ultimate stress of $\bar{\sigma}_u = 1.000 \cdot 1.5/1.6 = 0.9375$. So the effective safety which is $\nu = 2.0$ in the elastic range would decrease gradually with diminishing slenderness from $\nu = 2.0$ to $\nu = 1.875$ and $\nu = 1.6$ respectively. The formulas for the two transition straight lines are:

$\bar{\sigma}_u = 0.8000 - 0.274\ \bar{\lambda}_S$ for the lower straight line

$\bar{\sigma}_u = 0.9375 - 0.349\ \bar{\lambda}_S$ for the upper straight line

Here it is supposed that the elastic-plastic transition zone begins with $\bar{\sigma}_u = 0.3$. Despite the fact that the upper line does not cover all test points the author tends to vote for this latter design curve from the economical point of view. In the highly plastic range three test points of the latest test series of Schulz [10] are not covered. These points at about $\bar{\sigma}_u \cong 0.8$ are supposed to be in the vicinity of $\bar{\sigma}_u = 1.0$. Because of this large discrepancy one could think that the reference yield limit had not been properly determined. But in the meantime Schulz confirmed that the yield stress had been correctly obtained. It should be pointed out that the determination of the reference yield limit is in general a major problem. Strictly the yield stresses determined by tension tests ought not to be used as reference values - as it is often done in shell and plate buckling tests - rather the yield limits of compression tests should be chosen. This method is common in column buckling where the yield stress in compression is determined by stub column tests. This procedure is not applicable to very thin walled shell models. Therefore, only the use of tension yield limits remains. But even then considerable differences occur depending whether the upper yield limit, the static yield limit or a $\sigma_{0.2}$-limit is to be determined. The closest approach to reality is if the buckling tests and the determination of the yield stress are performed with the same loading speed. Moreover, discrepancies arise whether the buckling load is referred to either the yield stress out of a single test piece or the mean value minus twice the standard deviation. The latter procedure was used in the European column buckling curves.

Furthermore, the influence of the thickness of the structure on the yield limit is important. To simplify the application of most of the codes the essential reduction of the effective yield stress due to a large wall thickness was neglected. Instead of the thickness upper limits are introduced, e. g. 40 mm, but this is not valid for all steels. In the context of this paper this important problem how the accuracy of test results can be interpreted and how this influences the structural design cannot be discussed in further detail. In Fig. 13 representing the ECCS concept the analogous procedure to Fig. 12 is given. Also two transition straight lines are plotted in addition to those of Fig. 6 which again end with vanishing slenderness at $\bar{\sigma}_u = 0.8000$ and $\bar{\sigma}_u = 0.9375$ resp.

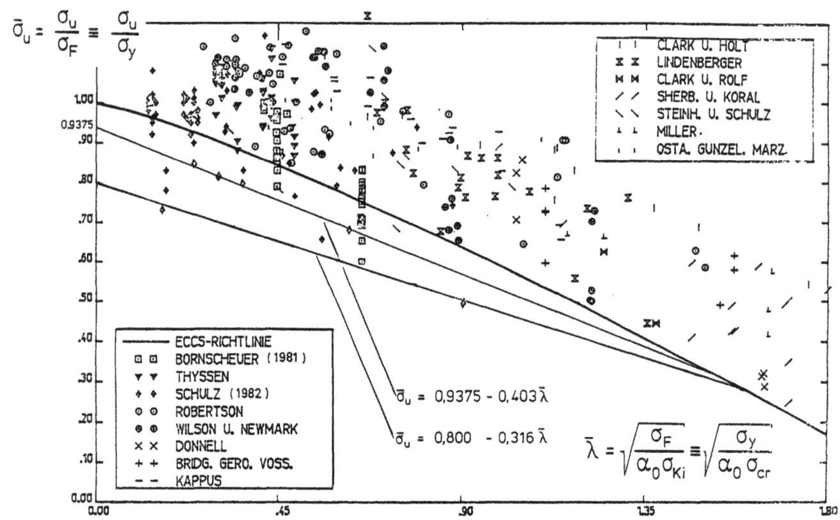

Fig. 13: Proposals for lowered ultimate stress curves according to the ECCS Recommendations R 4. 6

The formulas for both transition lines are as follows:

$\bar{\sigma}_u = 0.8000 - 0.316\ \bar{\lambda}$ and $\bar{\sigma}_u = 0.9375 - 0.403\ \bar{\lambda}$ resp.

Here, too, the effective safety in the case of the upper line decreases gradually from $\nu = 2.0$ to 1.6. Instead of straight lines also curves can be chosen as given in the ECCS Recommendations.

5.2 Reducing the Reduction Factors α

Now the second version to modify the α - concept will be shown. It is sufficient to demonstrate the application of this method to the DASt-Richtlinie 013. Smaller α values alter the scale of the abscissa in such a way that the test results move to the right of the diagram. If α is chosen small enough all the test results lie above the design curves (1.2). Fig. 14 taken from [3] shows some of the curves for the reduction factor α.

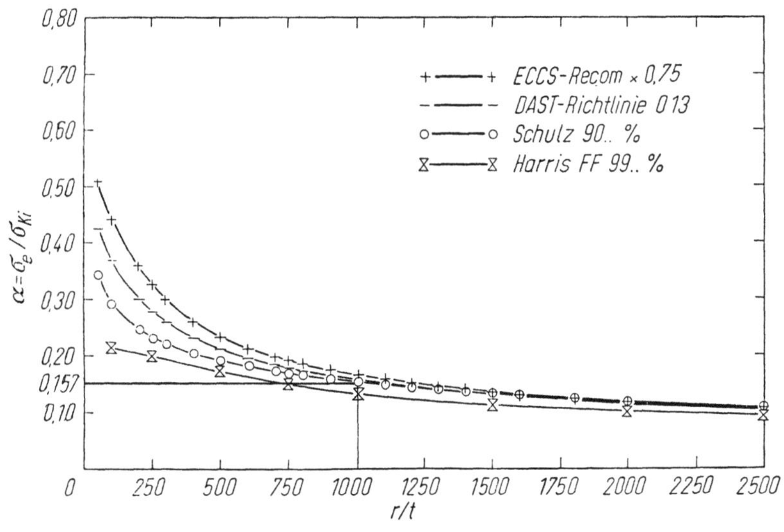

Fig. 14: Reduction factors α for the axially loaded cylindrical shell

For $r/t \leq 1000$, that is the main range of application in structural engineering for shells chiefly subjected to axial load, the smallest reduction factor according to the DASt-Richtlinie is $\alpha = 0.52/\sqrt{1 + 1000/100} = 0.157$. Taking a constant value $\alpha = 0.157$ for the entire range from $r/t = 0$ up to $r/t = 1000$ results in a horizontal shifting of the test points to the right as depicted in Fig. 15. In this figure only four test results of Schulz [10] are included which are located near the new proposed transition curve in Fig. 12.

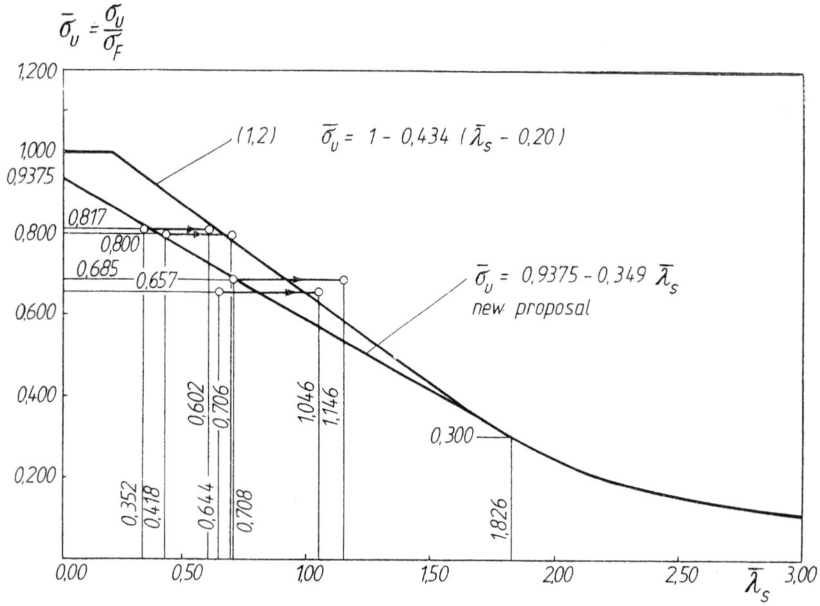

Fig. 15: Selected test results according to the DASt-Richtlinie 013 with a reduced reduction factor α

To achieve a safety of $\nu = 1.6$ for the limit case of $\bar{\lambda}_S = 0$ an upper bound of $\bar{\sigma}_u = 0.9375$ for the ultimate stress $\bar{\sigma}_u$ has to be assumed. It should be mentioned that inspite of this limitation of the ultimate stress $\bar{\sigma}_u$ this will result in a more economical design for compact tubular cross-sections with small r/t ratios compared to the present methods used in Germany. With main loads the German DIN 4114 which still has to be applied requires a safety of $\nu = 1.71$ for a slenderness $\lambda \rightarrow 0$, that means designing with $\nu = 1.6$ would lead to a more economical structure.

A constant reduction factor can be explained as follows:

It is true that in the case of the axially loaded cylinder the reduction factor α decreases with increasing slenderness. This was already shown by Pflüger [11] 20 years ago when he determined snap-through loads. In accordance with the procedure to obtain the load carrying capacity of columns in DIN 4114 he based his concept on "efficient imperfection" amplitudes. But this can only be applied to comparable buckling modes which are for example not too much influenced by boundary conditions. In the elastic-plastic and even more in the highly plastic region the influence of the boundary layer is responsible for the collapse. So it is questionable to base the calculations of reduction factors on this concept where the range of applicability is not valid any more. Constant values for reduction factors can be adopted as long as the ultimate stresses known from model tests are all covered by this concept. This is demonstrated exemplarily in Fig. 15. There is another argument which pleads for the assumption of constant α - values:

In all other load cases of both codes [1] and [2] constant reduction factors were assumed for the time being. So, in the case of the externally loaded cylinder constant reduction factors were introduced. These factors vary between 0.5 and 0.8 depending on the code. In the case of spherical caps reduction factors which vary from 0.1 to 0.2 according to the boundary conditions were adopted.

6. Summary and Outlook

6.1 Summary

The most important points of this study are summarized as follows:

The buckling behaviour of axially loaded circular cylindrical shells in the elastic-plastic region was examined:

- Comments are given on the special design curves of the ECCS- and DASt-codes.
- It was shown that the design formula of the various codes vary considerably.
- Results of tests performed recently in Stuttgart and Karlsruhe were presented. Some of these buckling loads are located below the limit curves of the ECCS- and DASt-codes.
- The collapse mode influenced by the constraint of the radial displacements at the boundary was discussed. The axisymmetric buckles in the end-zone are caused thereby.

- In contrary to the compact column which fails under yielding in compression the case of the circular cylindrical shell requires a slightly higher safety factor because of unfavourable boundary conditions.
- Based on the α - concept of ECCS- and DASt-codes a sufficiently safe design could be achieved by lowering the curves in the transition zone or by reducing the reduction factor α or by a combination of both.

6.2 Outlook

This paper only discusses the buckling problem of the axially loaded circular cylinder. Of course, other kinds of shells and load cases ought to be examined as well. Much research work will have to be done in order to base the design of a sufficiently safe and yet economical construction on easy-to-handle formulas.

Besides the usual α - concept in shell buckling it should be investigated whether a concept without explicit reduction factors α like that of the European column buckling curves is desirable. Since in the meanwhile some plate buckling codes were modified in that way a similar procedure is possible for all stability problems. Also comparative studies of the influence of different reference yield limits on the various stability problems should be performed.

7. Acknowledgement

The author wishes to thank the Institut für Bautechnik, Berlin, for the financial support of the Stuttgart and Karlsruhe model tests with circular cylindrical shells under axial compression.

Professor Schulz provided the Karlsruhe test results. This is gratefully appreciated.

References

[1] Deutscher Ausschuß für Stahlbau: DASt-Richtlinie 013: Beulsicherheitsnachweise für Schalen. July 1980, 1/16.

[2] European Convention for Constructional Steelwork (ECCS): European Recommendations for Steel Construction (ECCS Recommendations) R 4.6: Buckling of Shells. 1979.

[3] Bornscheuer, F. W.: Plastisches Beulen von Kreiszylinderschalen unter Axialbelastung. Stahlbau 50 (1981) 9, 257/62.

[4] Cases of ASME - Boiler and Pressure Vessel Code Case N-284: Metal Containment Shell Buckling Design Methods. Section III, Division 1, Class MC. 1980.

[5] ÖNORM B 4650, Teil 4: Stahlbau, Beulung von Kreiszylinderschalen. November 1977.

[6] Schweizerischer Ingenieur- und Architektur-Verein: SIA-Norm 161. With comment, 1979.

[7] Saal, H.: Buckling of Circular Cylindrical Shells under Combined Axial Compression and Internal Pressure. ECCS Stability of Steel Structures. Preliminary Report, Liège, April 1977.

[8] Vandepitte, D.; Rathé, J.: Buckling of Circular Cylindrical Shells under Axial Load in the Elastic-Plastic Region. Stahlbau 49 (1980) 369/73.

[9] Steinhardt, O.; Schulz, U.: Zum Beulverhalten von Kreiszylinderschalen. Schweizerische Bauzeitung 89 (1971) 1/14.

[10] Schulz, U.: Beulversuche mit axialdruckbelasteten Kreiszylinderschalen im plastischen Bereich.
To be published shortly in: Stahlbau.

[11] Pflüger, A.: Zur praktischen Berechnung der axial gedrückten Kreiszylinderschale. Stahlbau 32 (1963) 6, 161/65.

Practical Analysis Methods for Design of Circular Cylinders with Longitudinal Stiffeners and Subjected to Axial Compression

L. Å. SAMUELSON

IFM Akustikbyrån AB
Stockholm, Sweden

Summary

The possibilities to analyze complicated shell stability problems have improved enormously during recent years through the development of powerful computer programs. Still there is a need for simple design rules for shell geometries commonly used in practice. The present investigation forms part of the work by task group TWG 8/4 "Shell Stability" of the European Convention for Constructional Steelworks, ECCS. The purpose is to provide simplified methods of analysis for the design of circular cylindrical shells with longitudinal stiffeners and subjected to axial compression. Various methods of analysis are discussed and comparisons with tests are carried out. The design procedures proposed for the ECCS recommendations are presented. In view of the complexity of the problem it is necessary for the simplified rules to be fairly conservative in order to guarantee a safe design over the range of geometries considered. However, the designer should always have the option to use improved design methods whenever required.

Introduction

The first edition of the recommendations for design of shell structures with respect to buckling produced by TWG 8/4 of the ECCS was published in 1981, ref /1/, and covers unstiffened cylinders under axial compression, unstiffened and stiffened cylinders under external pressure, spherical shells under external pressure, and certain cases of combined loads. Meanwhile, work on other shell elements is going on, such as conical shells and stiffened cylinders. The present paper reviews the work on longitudinally stiffened cylinders which is carried out with the purpose to provide simplified design rules.

Realizing the complexity of the problem, the task group set up the following guidelines for the work:

Buckling of Shells, Ed.: E. Ramm
Proceedings of a State-of-the-Art Colloqium

* The recommendations should be simple to use
* The result should be conservative - the degree of conservatism may be reduced in the future through introduction of improved methods
* The designer should have the option of using a refined method of analysis.

In view of these requirements, methods of analysis are proposed, based on assumptions with regard to boundary conditions, initial imperfections, load excentricities etc., which realistically simulate the environment normally found in actual structures.

It is the opinion of the task group that buckling in any form is unacceptable even if the postbuckling behaviour is stable. Despite this fact it may be necessary to consider the general instability limit of a shell where local buckling of the shell panels has occured and methods of analysis for this case are discussed.

Literature review

A very large number of papers on the stability of stiffened cylindrical shells under axial compression have been published during the last two decades. Most of the work has been carried out within the aerospace industry and mainly on elastic structures. Reviews of theory and experiments are given in ref /2-4/ and no attempt will be made to duplicate these in the present paper.

During the last few years, the interest in cylindrical shells with longitudinal stiffeners has increased considerably in offshore applications and significant research work has been published by a large number of authors, ref /3, 5-9/. Characteristically the shells considered are mainly of steel and buckling occurs in the elastic-plastic range. The number of papers published at various conferences, ref /3, 10-11/, indicates the significance of the problem.

Design rules for stringer stiffened cylinders have been issued by Det Norske Veritas, (DNV), ref /12/ and by ASME, ref /13/. In some respects the proposal for the ECCS recommendations are based on conditions similar to those given in these rules. The format is slighty different in order to provide continuity with the existing ECCS recommendations on shell and plate structures.

Theory of stringer stiffened shells

Various types of buckling modes may be anticipated in a stiffened shell as demonstrated in figure 1:[1)]

* Local buckling of the stiffener
* Local buckling of the shell panels (ℓ) between the stiffeners
* Buckling of the curved stiffened panel (p)
* General instability involving buckling of shell and ring-stringer stiffener system (g).

For the time being local buckling of the stiffeners is disregarded since stiffener dimensions used in practical applications are fairly stocky and instability through tripping or web buckling does not occur.

In the design work it is necessary to consider all the different modes of instability:

L o c a l B u c k l i n g of the shell panels between the stiffeners may be predicted by the theory of plane panels when the stringer spacing is sufficiently small (narrow panel buckling). On the other hand curved panel buckling is to be considered in the case of wide panels.

1) The definition of buckling modes is that given in ASME code case No N284.

Plane panel behaviour is illustrated in figure 2 which gives the stability factor K for various types of boundary conditions. If the torsional stiffness of the stringer is reasonably small, the unloaded edges may be considered as simply supported and a factor K = 4.0 according to figure 2 is applicable. The postbuckling behaviour is stable.

Curved panels under axial compression behave like unstiffened cylinders provided the stiffener spacing is sufficiently wide. Both types of behaviour may be represented by the formula:

$$\sigma_{c\ell}^{\ell} = C\ E\ \frac{t}{r} \tag{1}$$

where

$$C = 0.605 \text{ in wide panel buckling and} \tag{2}$$

$$C = 3.6\, rt/b^2 \text{ in narrow panel buckling} \tag{3}$$

The theoretical transition point between the two cases is given by

$$b = 2.44\ \sqrt{rt} \tag{4}$$

It has been shown by several authors, compare reference /4/, that the carrying capacity at the transition point is approximately 20% higher than the value given by (1). More over, the imperfection sensitivity of the shell gradually increases with the value of $b/\sqrt{rt}$.

A design diagram for the evaluation of the buckling coefficient of (1) is given in figure 3. *The higher* of the two values of C given in diagram is used in the design analysis. The theoretical value of the buckling stress as given by (1) is multiplied by a reduction factor α. The resulting stress value is assumed to be a lower bound for the actual buckling stress of any shell that fulfills the requirements of the shell. The condition may be written:

$$\sigma_{cr}^{\ell} = \alpha\sigma_{c\ell}^{\ell} = C^{*}E\frac{t}{r} \tag{5}$$

The ECCS recommends the following values:

For plane panels a constant value $\alpha = 0.83$ is used, while the α-values of cylindrical shells vary with the r/t ratio. Figure 4 gives recommended values of the buckling coefficient $C^* = \alpha C$ for various r/t ratios and for the two imperfection tolerance limits prescribed by ECCS.

When buckling occurs in the plastic range the design stress (lower limit) calculated by use of the elastic buckling formula has to be reduced accordingly. Different methods are available, compare ref /1, 13/. The formula given by ECCS for unstiffened cylinders in axial compression is:

$$\sigma_K^{\ell} = \sigma_y \left[1-0.4123\left(\frac{\sigma_y}{\alpha\sigma_{c\ell}^{\ell}}\right)^{0.6}\right] \text{ if } \alpha\sigma_{c\ell}^{\ell} \geq \frac{1}{2}\sigma_y \tag{6}$$

which is applicable for local buckling of the shell panels between stiffeners.

General Instability, or (stiffened) Panel Buckling, if no rings are present, may be analyzed by use of the theory of orthotropic shells provided the stiffener spacing is sufficiently small. One of the earliest solutions to the problem was published by Block, Card and Mikulas Jr. ref /14/, whose work was based on results by Baruch and Singer. Similar approaches may be found in for instance ref /15-17/.

The particular theory for buckling analysis proposed for use in the ECCS recommendations is the same as that given in ref /13/. The equations given by Block et al, ref /14/ were extended by Miller /17/ to include the effect of local buckling of the shell panels on the overall buckling of the stiffened cylinder.

The bifurcation limit of the axial stress σ_x is obtained from:

$$\sigma_{cl}^{p} = \min_{\substack{n>3 \\ m\geq 1}} \left[A_{33} + \frac{A_{12}A_{23}-A_{13}A_{22}}{A_{11}A_{22}-A_{12}^{2}} A13 + \right.$$

$$\left. + \frac{A_{12}A_{13}-A_{11}A_{23}}{A_{11}A_{22}-A_{12}^{2}} A23 \right] / \left[\left(\frac{m\pi}{\ell}\right)^{2} t_s \right] \tag{7}$$

where the coefficients A_{ij} depend on the geometry, the stiffness parameters and the wave numbers m and n in the axial and circumferential directions. The coefficients are given in the appendix. The combination of wave numbers m, n which yields a minimum of the right hand side of eq. (7) represents the (classical) bifurcation buckling mode and the corresponding σ-value represents the buckling stress.

Eq. (7) was based on the assumption of simply supported edges and a buckling mode of the form; Figure 5

$$\begin{aligned} u &= u_o \cos mx \sin n\theta \\ v &= v_o \sin mx \sin n\theta \\ w &= w_o \sin mx \sin n\theta \end{aligned} \tag{8}$$

Other forms may be developed, considering a different set of boundary conditions, compare ref /18/. For practical purposes, however, the conditions proposed here should produce acceptable results as will be shown later.

The theory of orthotropic shells is strictly valid only if the stiffener spacing is sufficiently small. A condition commonly used is:

$$n_s \geq 4n \tag{9}$$

where n is the wave number of the critical buckling mode

determined from eq. (7). There is no restriction, though, on the "slenderness" of the shell panels since the concept of an effective width is included in the load range where local buckling has occured. The specific formula used in the present investigation may be written as:

$$b_e/b = 0.9\sqrt{\frac{\sigma^{\ell}_{cr}}{\sigma^{p}_{cr}}} \leq 1 \qquad (10)$$

Solution of eq. (7) with consideration of (10) involves an iterative procedure since the effective width is a function of the stress level. The algorithm is readily programmed for the computer, see for instance ref /19/.

As in case of unstiffened cylinders, a reduction factor has to be introduced to account for the influence of various types of imperfection parameters. A constant value of $\alpha = 0.65$ is proposed for the recommendations regardless of the radius to thickness ratio.

Eq. (7) is strictly valid only in the case of elastic buckling and there is no obvious extension to elastic - plastic buckling problems as in the case of for instance an Euler column. A rigorous analysis would be preferable, but it would be far too complicated at this time to introduce within the framework of the ECCS recommendations. Simplified methods of analysis should therefore be proposed. A first attempt would obviously be to use the reduction formula for unstiffened shells presented in eq. (6). Other methods are discussed in the next paragraph. Design charts according to figure 6 may be developed.

Approximate methods of analysis

Even though eq. (7) has been reduced on a fairly simple form the analysis still requires some experience with regard to the buckling behaviour of the shell. It is possible to analyze the problem by use of other methods which, in most cases,though, yield a very conservative result.

For short shells and when heavy stiffeners are used the stiffener including an effective strip of shell plate may be treated as a column. The model always yields a conservative result and in particular for long, lightly stiffened shells a very large descrepancy between the design load and the actual buckling load will result. The method is, however, used at present in the DNV rules for offshore constructions, ref /12/.

A modified approach was proposed by Walker et al, ref /18/, who included the elastic bedding effect of the shell on the column. Also it was conjectured that elastic-plastic buckling of the shell would occur when the stress in the stiffener reaches the yield limit of the material. ("First Yield" criterion). The method is being proposed for inclusion in the DnV rules, ref /8/.

Finally, the DnV proposes a simplified formula for the determination of the critical stress, ref /20/:

$$\sigma_{cr}^{p} = \frac{a^2 E i_{ef}^2}{\ell^2} + \frac{1}{1+A_s/bt} \frac{E}{\sqrt{3(1-\nu^2)}} \frac{t}{r} \tag{11}$$

The expression clearly yields a better estimate of the critical stress than the simple Euler theory. On the other hand the effect of the stiffener is marginal unless the section is very stiff. Specific results will be discussed below.

Rigorous analyses

There is a large number of computer programs available which have the capability to analyze shell stability problems, ref /21-26/. In order to treat the problem "exactly" it is necessary to consider geometrically and materially nonlinear behaviour and to include initial imperfections and other disturbances which affect the stability of the shell.

The numerical model - based on for instance the finite element technique - becomes extremely complicated and the cost of the analysis may be prohibitive. In particular, it is often impossible to carry out the parametric studies necessary during the design phase. The analysis is therefore often carried out in steps by use of a gradually refined model and only the final design is verified through a 3-dimensional comprehensive analysis.

Practical design tools are available in the form of programs for *shells of revolution*, ref /21-23/, which have been utilized by for instance Singer, ref /27/ in the analysis of stringer stiffened cylindrical shells in the elastic range. The applicability of the theory is limited to the case of narrow paneled stiffened cylinders where the theory of orthotropic shells is valid. An interesting extension of the use of the BOSOR 4 program by Bushnell is described in ref /21/ where the shell is regarded as part of a torus with a very large radius. This model may be used to study the bifurcation buckling of the shell-stiffener system, see figure 7.

The shell of revolution model can usually not be utilized for analysis of general nonsymmetric imperfections. However, Cohen, ref /24/ included the analysis of the postbuckling coefficients in his program, the values obtained may be used to estimate the imperfection sensitivity of the shell as shown by Arbocz in ref /28/. Theories for shells of revolution including nonsymmetric behaviour have been developed, compare ref /25,26/ but none of these has today the capability to treat stringer stiffened cylinders.

For elastic-plastic problems, the general theory of orthotropy has to be modified in order to provide a rigorous model of the shell stiffness. Few programs exist which can treat stability problems in the plastic range. One of these is BOSOR 5, ref /29/, which was utilized by Richards, ref /30/ in the analysis of stringer stiffened cylinders. Some results were also obtained in the present investigation based on the model given in figure 8 where the stiffeners were substituted for a continous elastic-plastic layer with stiffness in the axial direction only. The models for the material properties are included in figure 8.

The analysis of shells with random, nonsymmetric imperfections, nonuniform load distribution etc requires use of a *general shell* computer code. Some general finite element codes have capabilities for nonlinear analysis, others have been specifically developed for shell analysis such as STAGS, ref /23/. STAGS has been utilized by Arbocz in the analysis of stringer stiffened cylinders with general imperfections, ref /31/.

Comparisons with test results

Numerous experimental results are available in the literature, ref /16, 32/, and comprehensive studies of the correspondence between theory and tests have been carried out.

A summary of elastic buckling test results vs theory is given in ref /4/ and forms the basis for the choice of eq. (7) for the ECCS recommendations. Additional data are available in ref /32/ which support the design rules of ref /13/. Basically the validity of eq. (7) in the elastic range is demonstrated by figure 9 which shows the ratio of the experimental buckling stress to the classical bifurcation stress. The reduction factor α_p was determined from the lower bound of the test data in figure 9.

It was shown in ref /4/ that eq. (7) provides a reasonably good approximation for simply supported and "semiclamped" shells according to tests by Singer. The results are reproduced in table 1. It may be noted that clamped conditions in actual construction would be similar to those used in the tests by Singer. This observation supports the choice of eq. (7) as a basis for design. Similar conclusion may be drawn from the results of ref /32/.

Most of the investigations referenced above concern narrow

panel buckling in the elastic range and few results on local buckling of the shell panels between stiffeners are included. Tests reported,ref /5-7, 33/,include wide panel buckling and in some cases buckling occured well into the plastic range.

Moreover, interaction of buckling modes has been observed in some tests.

A summary of the theoretical predictions in comparison with test results is given in table 2 where the followihg notation is used:

$\bar{\sigma}_{u1}$ = Maximum stress/yield stress (global buckling).

$\bar{\sigma}_{u2}$ = Local buckling stress/yield stress.

$\bar{\sigma}^{*}_{u2}$ = Average stress at first yield in stiffener/yield stress.

$\bar{\sigma}_{Bif}$= Predicted bifurcation stress/yield stress.

The panel buckling stress, $\bar{\sigma}_{u1}$, predicted by the theory proposed for ECCS is based on eq. (7) and the plasticity reduction formula, eq. (6). In the evaluation, the local buckling of the shell panels between stiffeners is considered through eq. (10). However, it was noted in ref /5, 6/ that local buckling induced large sideway deformations of the stiffeners, in some of the tests and therefore the theory may yield a slight overestimate of the panel buckling stress. Comparison with the test values shows reasonably good, slightly conservative results.

The analysis of *the local buckling stress* of the shell panels, $\bar{\sigma}_{u2}$ was carried out by use of eq. (1) and (6). Again acceptable agreement with experiments may be noted except in a few cases where the theory overestimates the buckling stress. The ECCS recommendations require that an additional safety factor of 4/3 is used for the case of wide panel buckling. It is also possible that use of figure 4.a is unconservative.

If the design curve of fig 4.b is adopted, better agreement is obtained.

Use of *simplified methods*, such as the *Euler column theory* to predict the stability limit leads to an unacceptable underestimate of the carrying capacity as was shown in for instance ref /33/. The simplified formula for stiffened shell buckling given by eq. (11) was checked against the general expression, eq. (7) over a limited range of parameters. The result is included in figure 6 which shows that eq.(11) yields a conservative result. Checking of (11) against available test data remains to be done.

The predictions of the stability limit given by Walker, ref /18/ are included in Table 2. In particular the "*first yield criterion*" appears to give very reasonable agreement with the test results and the method should therefore be considered for design purposes.

Computer analysis by use of BOSOR 5, ref /29/ has been carried out for two shells according to the model of figure 8. The "first yield" stress $\bar{\sigma}_{u2}^{*}$ does not agree very well with the experimental results, which may be a consequence of the choice of boundary conditions. Also, the bifurcation load $\bar{\sigma}_{Bif}$ seemed to be highly overestimated by the model used.

The bifurcation stresses are those given by Richards, ref /30/. It is clear that more experience is needed in this type of analysis.

Systematic analyses of stringer stiffened cylinders have been carried out by Arbocz, ref /31/ who used the STAGS code for *general shells*, ref /23/. The results indicate that very close agreement between test results and numerical predictions are obtained when initial imperfections and load excentricities are known. Arbocz also studied the dependence of the buckling load on the initial imperfections by use of statistical methods, ref /34/. The results may become very important for future design methods for shells of revolution.

Concluding remarks

The primary purpose of the present investigation was to evaluate methods of analysis for stringer stiffened cylinders under axial load, to be included in the ECCS recommendations.

It was shown that the classical theories for a) buckling of the shell panels between stiffeners and b) stiffened panel buckling may be used for both elastic and elastic-plastic analyses. The proposed values of the design stress were found to be slightly conservative in most cases.

Simplified methods are available but it has been shown that they often give a serious underestimation of the carrying capacity.

Reduction with respect to the yield limit is done by use of an empirical formula. More rational methods are available, such as the "first yield" criterion proposed by Walker, which will be proposed as an alternative method in the recommendations.

Use of computer programs for shell analysis has been demonstrated to yield very good results in some cases. Successful use in design work appears to require long experience and may become expensive. Further research in this area is needed.

Acknowledgements

The present investigation was sponsored in part by the Swedish Work Environment Fund. under contract No 81-0797. Dr J.P. Mouton, Dr L. Dahlberg and Mr P. Vestergren have contributed to the work and their support is highly appreciated.

References

1. European Recommendations for Steel Construction. The Construction Press, London 1981.

2. Singer, J: Buckling Experiments on Shells - a Review of Recent Developments. Technion Rep. TAE No 403, April 1980.

3. Harding, J.E., Dowling, P.J., Agelidis, N.E. Conference on: Buckling of Shells in Offshore Structures. London 23-24 April 1981. Granada, London 1981.

4. Samuelson, Å., Vestergren, P.: Analysis of Axially Loaded Stringer Stiffened Circular Cylinders. FFA TN HU-2134, 1979.

5. Dowling, P.J., Harding, J.E., Fahy, W., Agelidis, N.: Report on the Testing of Large and Small Scale Stiffened Shells Under Axial Compression. Department of Energy, report No OT-R-8107.

6. Sridharan, S., Walker, A.C.: Experimental Investigation of the Buckling Behaviour of Stiffened Cylindrical Shells. Department of Energy, report No OT-R-7835. London 1980.

7. Nelson, H.M., Green, D.R., Phillips, D.V.: Buckling Studies of Large Diameter Stiffened Tubes. Department of Energy, report No OT-R-7838. Department of Civil Engineering, Glasgow University 1980.

8. Valsgard, S., Steen, E.: Simplified Strength Analysis of Narrow Panelled Stringer Stiffened Cylinders Under Axial Compression and Lateral Load. Progress report No 4. DnV Report No 80-0590.

9. Miller, C.D.: Commentary on the Nov 13, 1979 Edition of Appendix - "Metal Containment Shell Buckling Design Methods" of the ASME Boiler and Pressure Vessel Code Section III - Division 1. Chicago Bridge & Iron Co., Dec 1979.

10. Second International Conference on Behaviour of Offshore Structures. Imperial College, London 28-31 Aug 1979.

11. Annual Technical Session and Meeting of the Structural Stability Research Council 30-31 March 1982, New Orleans, LA.

12. Det Norske Veritas: Rules for the Design, Construction and Inspection of Fixed Offshore Structures. Oslo 1974 and 1977.

13. Cases of ASME Boiler and Pressure Vessel Code, Case N-284. Aug 25 1980.

14. Block, D.L., Card, M.F., Mikulas Jr, M.M.: Buckling of Eccentrically Stiffened Orthotropic Cylinders. NASA TN D-2960.

15. Esslinger, M., Geier, G.: Buckling and Postbuckling Behaviour of Discretely Stiffened Thinwalled Circular Cylinders. Z. Flugwiss. 18, Heft 7, 1970.

16. Singer, J.: Buckling of Integrally Stiffened Cylindrical Shells - a Review of Experiment and Theory Contributions to the Theory of Aircraft Structures. Delft University Press, 1972. pp. 325-357.

17. Miller, C.: Buckling Stresses of Ring and Stringer Stiffened Cylindrical Shells under Axial Compressive Load.

18. Walker, A.C., Anronicou, A., Sridharan, S.: Theoretical Analysis of Stringer and Ring Stiffened Shells. Buckling of Shells in Offshore Structures, Granada, London 1981.

19. Vestergren, P.: Buckling Analysis of Stiffened Cylindrical Shells. User's manual for BC01.

20. Dowling, P.J., Ho, T.: Effect to Initial Deformations on the Strength of Axially Compressed Cylindrical Panels. Inst. Mech. Eng. C9/79.

21. Bushnell, D.: Stress Stability and Vibration of Complex Branched Shells of Revolution. Analysis and user's manual for BOSOR, Report No LMSC-D-243605, March 1972.

22. Bushnell, D.: Stress, Buckling and Vibration of Prismatic Shells. AIAAJ. Volume 9 No 10, Oct. 1971.

23. Almroth, B.O., Brogan, F.A., Miller, E., Zele, F., Peterson, H.T.: Collapse Analysis for Shell of General Shape. II User's Manual for the STAGS-A Computer Code. Air Force Flight Dynamics Lab., Wright Patterson AFB, AFFDL-TR-71-8, March 1973.

24. Cohen, G.A.: User Document for Computer Programs for Ring-Stiffened Shells of Revolution. NASA CR-2086, March 1973.

25. Ball, R.E.: A Program for the Non-Linear Static and Dynamic Analysis of Arbitrarily Loaded Shells of Revolution. Paper presented at the Conference on Computer oriented analysis of Shell Structures, August 10-14 1970. Lockheed MSC Palo Alto Calif.

26. Stricklin, J.A., Haisler, W.E., von Riesemann, W.A.: Formulation, Computation and Solution Procedures for Material and/or Geometric Nonlinear Structural Analysis by the Finite Element Method.
SC-CR-72 3102. Sandia Laboratories, Albuguerque, New Mexico, July 1972.

27. Rosen, A., Singer, J.: Vibration of Axially Loaded Stiffened Cylindrical Shells with Elastic Restraints.
Int. J. Solids Structures, 1976, Vol. 12, pp 577-588
Pergamon Press.

28. Arbocz, J.: Past, Present and Future of Shell Stability Analysis.
Delft University. Techn. Dept. Aerospace Eng.
Report No LR-320. May 1981.

29. Bushnell, D.: BOSOR 5 - Program for Buckling of Elastic-Plastic Complex Shells of Revolution Including Large Deflections and Creep.
J. Computers & Structures, Vol. 6, pp 221-239.
Pergamon Press 1976, England.

30. Richards, D.M.: Shell Buckling Research and Design APPRAISAL PROCEDURES.
Buckling of Shells in Offshore Structures, Granada, London 1981.

31. Arbocz, J.: Utilization of STAGS to Determining Knockdown Factors from Measured Initial Imperfections.
Delft University. Techn. Dept. Aerospace Eng. Report No LR-275, Nov 1978.

32. Grove, R.B.: Summary of Buckling Tests of Stringer Stiffened Cylinders.
Chicago Iron & Bridge Co. April 16, 1981.

33. Groth, H.: Buckling of Cylinders with Longitudinal Stiffeners Subjected to Axial Compression.
FFA TN HU-2184, Dec 1979.

34. Elishakoff, I., Arbocz, J.: Reliability of Axially Compressed Cylindrical Shells with Random Axisymmetric Imperfections.
Delft University. Techn. Dept. Aerospace Eng.
Report No LR-306, Sept 1980.

EXPERIMENTAL RESULTS			THEORETICAL RESULTS			BOUNDARY CONDITIONS AT TEST
Spec. No	P_{exp} N	m/n	P_{theor} N	m/n	P_{exp}/P_{theor}	
RO 04	19 620	-	19 875	1/8	0.99	"CLAMPED"
RO 15	17 658	2/7	19 237	1/10	0.92	SIMPLE SUPPORT
RO 25	36 297	1/8	36 670	1/11	0.99	SIMPLE SUPPORT
RO 28	29 038	2/9	27 223	1/11	1.07	SIMPLE SUPPORT
RO 31	46 038	1/7	36 944	1/9	1.25	"CLAMPED"
RO 34	41 398	1/7	35 757	1/10	1.16	SIMPLE SUPPORT
RO 44	29 479	2/9	23 505	1/10	1.25	SIMPLE SUPPORT
RO 45	19 767	2/10	18 727	1/10	1.06	"CLAMPED"
RO 46	27 468	1/9	23 956	1/10	1.15	"CLAMPED"

Table 1 Comparison between test results and theoretical predictions for elastic shells with different experimental boundary conditions. From ref /4/ based on tests by Singer.

					EXPERIMENTAL			THEORY								
								ECCS PROPOSAL			REF / 18 /			BOSOR 5		
REF No	SPECIMEN ID.NO.	R/t	L/R	N_s	$\bar{\sigma}_{u1}$	$\bar{\sigma}_{u2}$	n	$\bar{\sigma}_{u1}$	$\bar{\sigma}_{u2}$	n	$\bar{\sigma}_{u1}$	$\bar{\sigma}^*_{u2}$	$\bar{\sigma}_{u2}$	$\bar{\sigma}^*_{u2}$	$\bar{\sigma}_{Bif}$	n
/33/	Unst.	357	3.04	-	.52*	.52**	-	.38	.38	-						
	8-1 0	357	3.04	8	.66	.51	(5)	.65	.38	6						
	8-2 0	357	3.04	8	.68	.44	(5)		.38	6						
	8-3 I	393	2.76	8	1.34	.53	(5)		.35	6						
	8-4 I	393	2.76	8	1.29	.51	(5-6)		.35	6						
	12-1 0	357	3.04	12	0.77	.64	6		.38	6						
	12-2 0	357	3.04	12	0.65	.46	6		.38	6						
	12-3 I	393	2.76	12	1.04	.64	6		.35	6						
	12-4 I	393	2.76	12	0.93	.47	6		.35	6						
	20-1 0	357	3.04	20	0.88	.66	5-7		.38	6						
	20-2 0	357	3.04	20	0.90	.83	6		.38	6						
	20-3 0	357	3.04	20	0.93	.71	6		.38	7						
	20-4 I	393	2.76	20	1.34	.65	6		.35	6						
	20-5 I	393	2.76	20	1.23	.55	6	.65	.35	6						
/6/	UC1	200	0.4	20	.82	.35	-	.71	.43/.55	(10)	-	-	.41	.65		(10)
	UC2	200	1.11	40	1.03	-	9	.63	.73	8	.90	.69			.97	
	UC3	200	1.11	20	.76	.40	-	.68	.54	7	-	-	.40		1.19	
	UC4	200	1.11	30	.96	.40	-	.71	.59	7	-	-			-	
	UC5	200	1.11	40	1.04	-	8	.73	.73	7	.93	.85	-		-	
	UC6	280	1.11	40	.65	-	8	.51	.58	9	.64/.73	.45	-	.85	.92	8
	UC7	280	1.11	40	.86	-	8	.62	.58	8	.83	.68	-		-	
	UC8	360	1.11	20	.51	.42	-	.44	.38	9	-	-	.25		.92	
	UC9	360	1.11	40	.66	-	9	.47	.40	9	.60	.48	-		.93	
	B1	280	1.56	40	.82	-	8	.51		7	.83	.68	-			
	B2	280	1.56	20	.54	.31	-				-	-	.25			
	B3	280	1.11	20	.60	.29	-	.54		7	-	-	.31			
	B4	280	0.78	20	.61	.29	-				-	-	.32			
	B5	280	1.56	40	.82	-	8				.81	.73				
/5/	1	190	0.41	40	.96	-	-		.75			.81				
	2	190	0.41	20	.96	-	-		.58			.92				
	3	190	1.125	40	.95	-	-	.74	.75	7		.88				
	4	190	0.41	40	1.07	-	-		.75			.85				
	5	190	0.41	20	.94	-	-	.72	.58	10		.87				
	6	190	1.125	40	1.02	-	-	.74	.75	7		.89				
	IC1S see	190	0.41	40	.97	-	-		.74						1.04	
	IC2S ref	190	0.41	20	.97	-	-		.58						1.04	
	IC3S 30	190	1.13	40	.99	-	-	.74	.74	7					1.02	
/7/	GU1	272	1.56	20	.63	.31	-		.53				.27		.98	
	GU2	95	1.33	8	.97	-	-		.76						1.07	
	GU3	196	1.11	30	.69	.23	-		.70						1.02	

Table 2 Comparison between test results and theroretical predictions for general instability and local buckling.

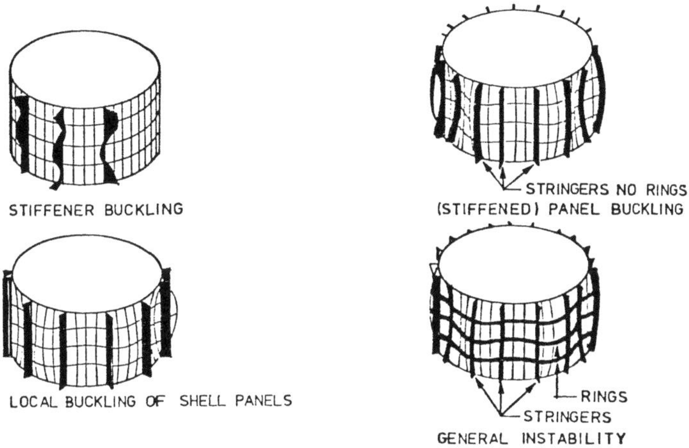

Figure 1 Possible buckling modes of a stiffened cylindrical shell.

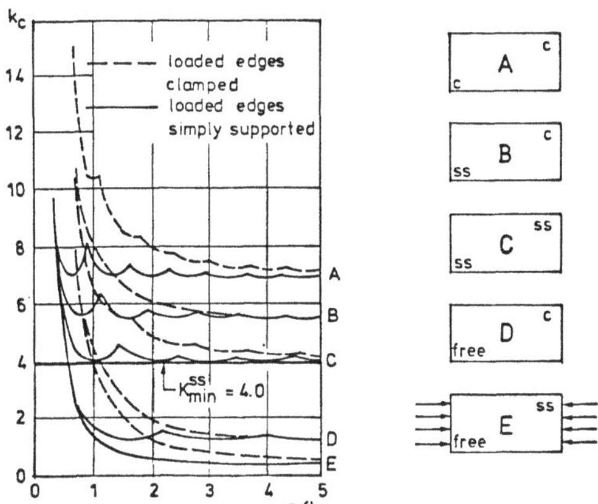

Figure 2 Buckling coefficient K_c for plane panels as a function of different edge conditions.

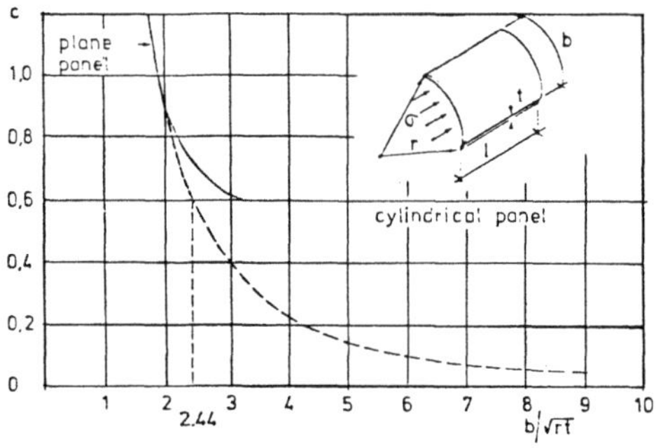

Figure 3 Theoretical buckling coefficient C, eq. (1) for a shell panel between stiffeners.

4a)

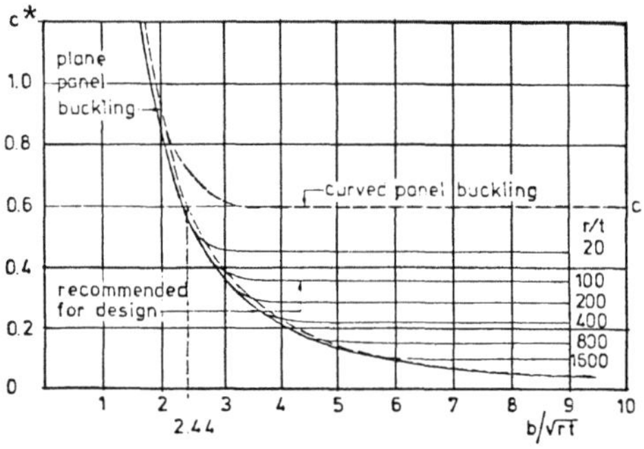

4b)

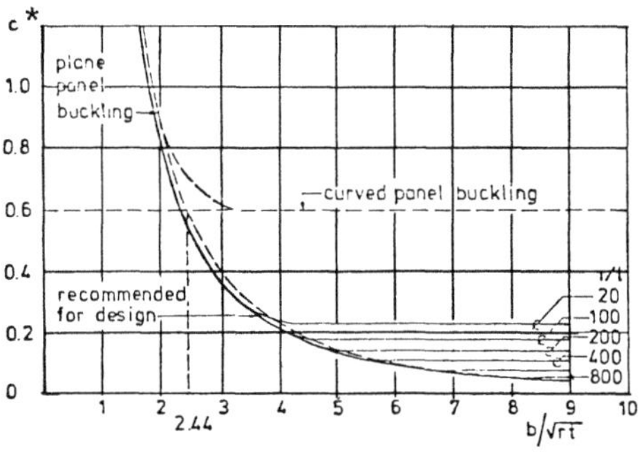

Figure 4 Proposed design values of the buckling coefficient C*, eq. (5). a) and b) correspond to the two levels of initial imperfection amplitude tolerances given by ECCS. ($\bar{w}/\ell_r \leq 0.01$ and $\bar{w}/\ell_r \geq 0.02$).

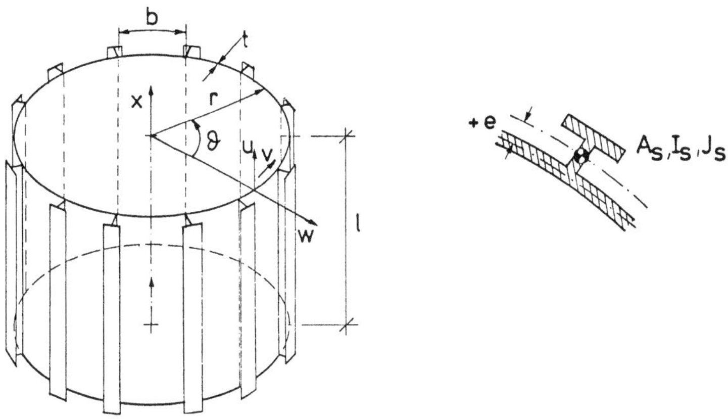

Figure 5 Geometry of stiffened cylinder.

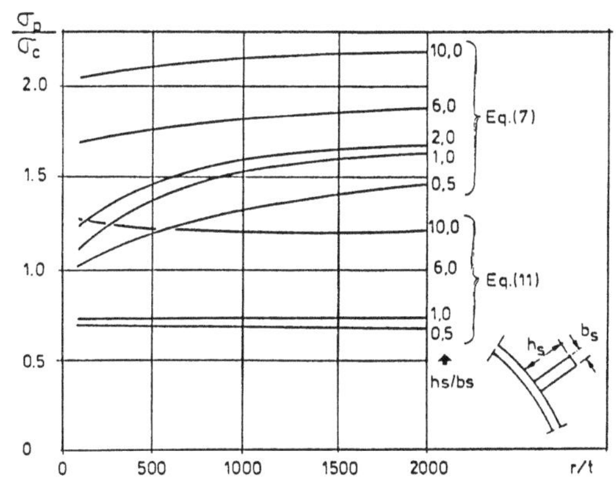

Figure 6 Buckling stresses of a stiffened cylinder with $A_s/bt = 0.5$. The classical buckling stress of the unstiffened cylinder is used as a reference.

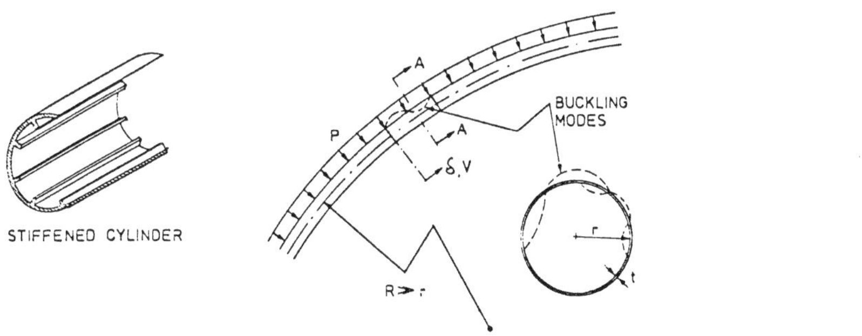

Figure 7 Model for approximate analysis of discretely stiffened cylindrical shell according to ref /21/.

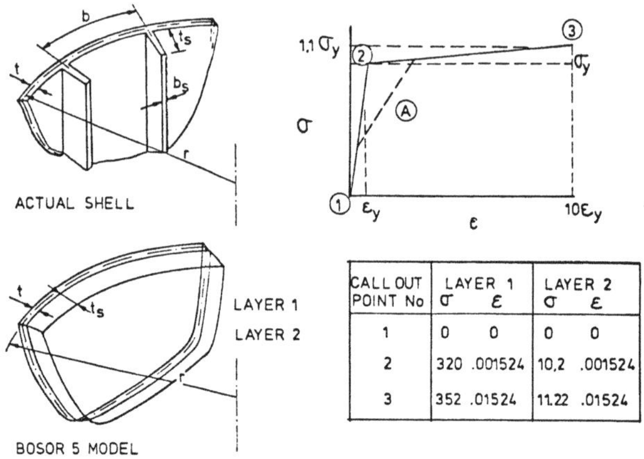

CALL OUT POINT No	LAYER 1 σ	LAYER 1 ε	LAYER 2 σ	LAYER 2 ε
1	0	0	0	0
2	320	.001524	10,2	.001524
3	352	.01524	11.22	.01524

Figure 8 BOSOR 5 model for analysis of stringer stiffened cylinder. Material data refer to test specimen UC1 of Table 2. (A) denotes a possible model for analysis when local buckling of shell panels occurs.

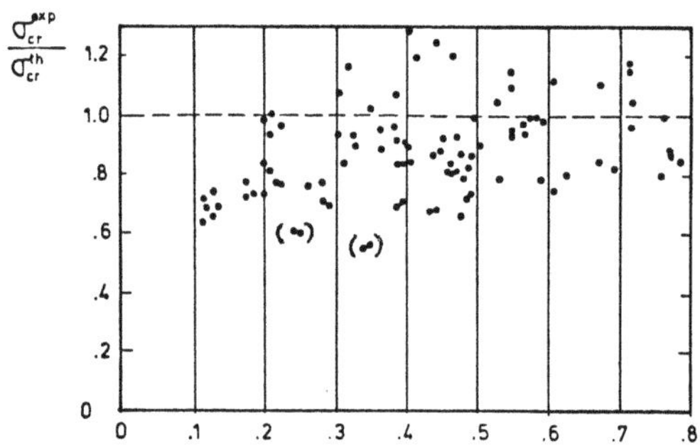

Figure 9 Comparison between test data and theoretical predictions according to eq. (7).

Appendix

The coefficients of Eq. (7) are defined as:

$$A_{11}=E_\phi\left(\frac{m\pi}{\ell}\right)^2+G_{\phi\theta}\left(\frac{n}{r}\right)^2;\quad A_{22}=E_\theta\left(\frac{n}{r}\right)^2+G_{x\theta}\left(\frac{m\pi}{\ell}\right)^2;$$

$$A_{33}=D_\phi\left(\frac{m\pi}{\ell}\right)^2+\bar{D}_{\phi\theta}\left(\frac{m.\pi}{\ell}\right)^2\left(\frac{n}{r}\right)^2+D_\theta\left(\frac{n}{r}\right)^4+\frac{E_\theta}{r^2}+\frac{2C_\theta}{r}\left(\frac{n}{r}\right)^2;$$

$$A_{12}=(E_{\phi\theta}+G_{\phi\theta})\left(\frac{m\pi}{\ell}\right)\left(\frac{n}{r}\right);\quad A_{23}=\frac{E_\theta}{r}\left(\frac{n}{r}\right)+C_\theta\left(\frac{n}{r}\right)^3;$$

$$A_{13}=\frac{E_{\phi\theta}}{r}\left(\frac{m\pi}{\ell}\right)+C_\phi\left(\frac{m\pi}{\ell}\right)^3;$$

$$E_\phi=\frac{Et}{1-\nu^2}\left(\frac{b_e}{b}\right)+\frac{EA_s}{b};\quad E_{\phi\theta}=\frac{\nu Et}{1-\nu^2}$$

$$E_\theta=\frac{Et}{1-\nu^2}\left(\frac{\ell_e}{\ell}\right)+\frac{EA_r}{\ell};\quad G_{\phi\theta}=\frac{Gt}{2}\left(\frac{b_e}{b}+\frac{\ell_e}{\ell}\right)$$

$$D_\phi=\frac{Et^3}{12(1-\nu^2)}\left(\frac{b_e}{b}\right)+\frac{EI_s}{b}+\frac{EA_s\,e_s^2}{b}$$

$$D_\theta=\frac{Et^3}{12(1-\nu^2)}\left(\frac{\ell_e}{\ell}\right)+\frac{EI_r}{\ell}+\frac{EA_r\,e_r^2}{\ell}$$

$$D_{\phi\theta}=\frac{\nu Et^3}{b(1-\nu^2)}+\frac{Gt^3}{b}\left(\frac{\ell_e}{\ell}+\frac{b_e}{b}\right)+\frac{GC_s}{b}+\frac{GC_r}{\ell}$$

$$C_\phi=\frac{EA_se_s}{b};\quad C_\theta=\frac{EA_re_r}{\ell}$$

The coefficients include terms for a ring stiffened shell. These terms are set = 0 for a stringer stiffened shell.

The coefficients of Eq. (10) are defined as:

$a = A_{ef}/A_t$

$i_{ef}^2 = I_{s_{ef}}/A_{ef}$

A_{ef} = cross sectional area of shell and stiffener

$$A_{ef} = A_s + A_t$$

A_t = cross sectional area of shell alone = bt

$I_{s_{ef}}$ = second moment of inertia of the shell-stiffener system

Ψ_1 = 1 for outside stiffeners (S)

$\Psi_1 = 1/(1+A_s/(bt))$ for inside stiffeners

THE STABILITY OF THIN CONCRETE SHELLS

HEINZ ISLER

Consulting Engineer and Research
Burgdorf, Switzerland

Summary

As a result of extensive practical experience some rules have been deduced as to how the risk of instability in thin concrete shell structures can be reduced.

1. Primary Static Task

For its primary static task a shell must have

- a good shape
- sufficient curvature at every point
- equilibrium at the supports (ties etc!)
- correct edges
- reasonable deformations and stresses
- good protection against corrosion.

2. Types of Instability

If these main conditions are fulfilled in all load cases, for both short and more extensive periods a thin shell structure must also be safe against all sorts of instabilities (pict. 1, Ref. 1.):

- snapthrough
- local buckling in the surface
- buckling of edges
- torsion of segments (windwheel effects)

Buckling of Shells, Ed.: E. Ramm
Proceedings of a State-of-the-Art Colloqium

Pict. 1 Instability dangers to be avoided in this shell: Surface buckling, buckling of the free edge, windwheel like torsion of segments. (Garden Center Paris)

Pict. 2 Good shell shape, sufficient double curvature at every point. (Mehrzweckhalle Solothurn)

3. General Buckling Equation

The general behaviour of a shell concerning stability of surface can be described by the following equation: (general buckling formula, my private "compass")

$$p_K = \underbrace{c}_{\alpha \cdot \beta \cdot \gamma} \cdot E \cdot \left(\frac{t}{r}\right)^x \geq s \cdot p_{eff}$$

p_k is the critical buckling load

α a number between 0,2 and 1,2

β practical reduction factor (inaccuracies etc.) about o,3

γ formspecific factor for every shape

E modulus of elasticity short and long period

t shell thickness

r local radius of curvature form!

x power between ca. 2 and 3
dependent on the form.
cylinder 3, sphere 2

s safety factor

p_{eff} real load

4. Design as the Main Influence

This formula proves that the main influence on stability comes from the design, namely:

Three factors depend on the form or shape which is chosen: γ, r, and the most powerful: x, the power of $\frac{t}{r}$. Cylindrical parts of a shell have about 30 (you hear right: thirty) times less buckling resistance than spherical ones because the power x becomes $2\frac{1}{2}$ up to 3 instead of 2,0 with Zoellys sphere.

Two factors depend on right or false dimensions: shell-thickness t and curvature r both also chosen in the design.

And one factor deals with the loads: p_{eff} also here design decides on the buckling safety: whether one estimates the loads correctly or insufficiently.

5. Correlation of Form and Stability

Sufficient curvature (pict. 2), sufficient shell thickness (pict. 9) and sufficient hardness (E-modul) of the material are of great importance, to avoid overall and local surface buckling. Also exact forming, accurate pouring, vibrating, smoothening (pict. 7), good curing, watering, protection against evaporation.

Free edges can be stiffened by edgemembers (pict. 3), by additional transverse curvatures (pict. 4), by transverse countercurvatures (pict. 5), or by avoiding compression in free edge zones (pict. 6).

About the correlation between designed form and stability, which is valuable for shells in any material - metal, plastics, stone, concrete - I spoke here in this room three years ago on the symposium on widespan membrane structures. Therefore now I will not go into further details about it. It can be read in the proceedings of the symposium of 1979 [Ref. 1].

6. Four Additional Problems with Concrete

This time I will look at the specific instability-problems of shells made of concrete. They are:

1st: Quality of concrete by in situ fabrication
2nd: Control of shellthickness
3rd: Creep
4th: Cracks

Pict. 3 Shells with stiffening edgebeams. (Cité scolaire Chamonix)

Pict. 4 Shell with free edges. Additional transverse curvature. (Gasoline Station Deitingen, Schweiz)

Pict. 5 Transverse countercurvature on free edge. (Sicli-building, Geneva)

Pict. 6 Free edge zone where compression is avoided due to appropriate shape. (Exhibition Hall Recherswil)

7. E and Fabrication In Situ

The quality of concrete defines E from 30 000 N/mm^2 to 10 000, good, medium or poor.

The concrete of a shell is fabricated on the building site itself (pict. 7). Sometimes the conditions are very harsh. The chemical process of hardening is strongly influenced by heat, cold,evaporation or drowning.

The quality of the concrete depends not only on the aggregates of the mixture but also on the care of treatment. The workmen have to work in all weather conditions, wind, cold, rain, sun and at night.

Most parts of a shell are steep, where vibration becomes critical (pict. 8).

For all these reasons the quality of the shell concrete varies. In most cases it is inferior to traditional concrete works such as slabs or walls.

Highest attention has to be given to obtain optimal concrete under the prevalent weather and working conditions.

8. Shell Thickness

Another problem is the thickness of the shell. It is rather difficult to control it. Slats to define the thickness every one or two metres have proved to be very uncomfortable, because they leave a great number of holes in the finished concrete surface (pict.14, a shell in South America).

Our practice uses another alternative (pict. 9): estimating the covering of the upper layer of reinforcement and highly skilled and trained work men.

Pict. 7 High quality concrete by careful usage of sledge vibration even in steep parts.

Pict. 8
Concrete work on steep shell.
(Naturtheater Stetten)

Pict. 9
Allover double-layer reinforcement and accurate concrete thickness.

The buckling load is very sensitive to incorrect shell thicknesses, because the thickness t appears in the square or third power in the equation. Here as example a spherical shell:

$$t_o = 8 \text{ cm} \qquad \left(\frac{8}{R}\right)^2 = 100\ \%$$

$$t_o = 2 \text{ cm} \qquad \left(\frac{6}{R}\right)^2 = 56\ \%$$

En error of 2 cm diminishes the buckling load to 56 %!

9. Example One

What happens if quality of concrete and shell thickness are out of control can be illustrated in the following case:

Three doubly curved shells in South America, pict. 10, collapsed because of low quality concrete, too high load (shell poured too thick) and a weak design (insufficient safety factor for the ties, etc.). Part of it really buckled (rare photo of real buckling! Pict.11) by the insufficient modulus of elasticity.

I must add that it was not my design, I was called to function as expert after a fourth and a fifth nearby collapse in the same building, two years later. Pict. 12. (Overstress of ties, because of no safety factor s = 1,0!) Pict. 13 shows the repair with additional ties.

Two other pictures of the same building: pict. 14 span 30 x 30 m, weak design:

- non ideal shell shape
- horizontal thrust partially taken by frame
- open ties, without compression reserve very susceptible to overloading
- one layer of reinforcement (pict. 15)
- bamboo scaffolding, difficult to check the shape (pict.16)

Pict. 10
Shell collapse

Pict. 11
Buckling of shell

Pict. 12
Broken tie

Pict. 13
Additional ties

Pict. 14
Open ties, very susceptible to overloading.

Pict. 15
One layer of reinforcement only. Slats to define shell thickness.

Pict. 16
Bamboo scaffolding, difficult to check shape.

10. Other Examples

It has to be stressed, that good concrete work is quite possible. But it needs additional knowledge, effort and training.

A few examples:

Pict. 17 + 18 bubble shell 1960 and stability test, 58 x 52 m span. Buckling decided the shell thickness. Rise higher as normal 1/7 of diagonal. r smaller, $\left(\frac{t}{r}\right)^2$ higher

Pict. 19 great effort: high quality of concrete, high E ! 3 days 3 nights pouring! Steep because of overhight to reduce R !

Pict. 7 sledge vibration on every part.

Pict. 20 stable scaffolding

Pict. 21 accurate formwork, sporthalls 49 m span

Pict. 22 + 23 good work under all weather conditions examples: Tessin market gunnite 5 cm in summer, Garden Centre in winter time. two extremes: dry out by heat and sun or freeze by storm and cold. In rain it is impossible to pour. Therefore a careful weather forcast is necessary. We have our own service.

After discussing the problems of fabrication of concrete on the building site I come to the next problem of concrete shells:

11. Creep and its Influence on Stability

Creep is the slow increase of deformations under loads which do not increase.

By creep a shallow arch or shell can get flatter and flatter until the angle of curvature becomes zero. At this moment even infinite reactions (normal forces) could no longer

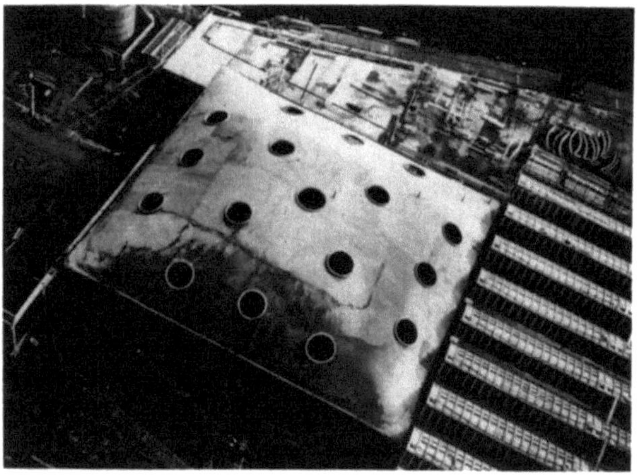

Pict. 17
Bubble shell
58 x 52 m
(Coop, Wangen)

Pict. 18
Model test for
buckling.

Pict. 19
Careful pouring of
concrete 3 days
and 3 nights.

Pict. 20
Stable scaffolding
(Tennishall Düdingen)

Pict. 21
Accurate formwork
(Sporthall Grenchen)

Pict. 22
Gunnite work in summer
heat.
(Mercato Biasca)

Pict. 23
Gunnite work in wintertime.
(Garden Center Solothurn)

hold any perpendicular force.

The only reserve lies in the bending stiffness of the buckling area. There exist buckles on real shells which have come to a stop in their inverse position. I have seen one on the same building in South America which I showed before.

In practice, creep is in a way equivalent to a reduction of the modulus of elasticity E.

In the past years several hypar shells in the United States have collapsed two, or four years after the erection of the buildings. By the action of creep. Shallow shells are very sensitive to this effect.

Extensive creep is a direct consequence of low quality of concrete and creep becomes additionally dangerous in shells of big initial deformations.

12. Small Deformations Desirable

Pict. 24 shows a shell with a high amount of deformations. The causes were a non ideal shape (segment of sphere and too flat) and low quality of concrete (pumped, little vibration, too liquid). The deformations of 4 years creep were 300 % bigger than the elastic deformations of the first days. These results are not good, although there is not yet an actual danger, because the shell is thick enough.

Pict. 25 shows shells in the same place with improved conditions: higher rise-to-span-ratio and concrete of better quality, not pumped but poured and vibrated carefully. There the results: smaller elastic (initial) deformations relative to span and creep deformations 150 % of the elastic ones.

Pict. 26 gives a good example: shells with good shape
good concrete (dry, vibrated)
small creep deformations

Pict. 24
shell with pumped concrete, high creep.

Pict. 25
shells with medium deformations (Cité scolaire Chamonix)

Pict. 26
shells with small deformations (Tennishall Burgdorf)

small initial deformations

For myself the amount of deformations - short and long period - is a measure of the quality of a shell building. I try to achieve the smallest possible, because they increase the safety of stability.

Small deformations do not cost anything. By perfect form-giving in the design, accuracy in the formwork and careful pouring of the concrete the desired small deformations can be achieved.

By keeping the deformations for continual loads as small as possible, the risk of creep-instability is reduced. How can the deformations be kept small? By a good shape of shell.

The author has observed the deformations of most of his shells over shorter or longer periods. Thereby he discovered clear differences between different shapes. [Ref. 2].

13. Influence on E by Cracking

The fourth problem typical in concrete structures is cracking.

Cracks appear where the tensional strength of the concrete is exceeded. Sometimes the cracks appear after months of use, sometimes years later. Cracking means infiltration of water, this can lead to damages by ice and corrosion. This does affect the stability only indirectly.

But stability is affected directly by the increased deformations resulting from cracking.

In the shell-buckling formula it again means a substantial reduction of the modulus of elasticity E. A reduction in my opinion to less than 50 %.

Pict. 27
Shell with tension zones (Market Bellinzona)

Pict. 28

checking deformations

Pict. 29

Finished building

14. Example

We had the opportunity to observe the influence of cracking on a shell in the South of Switzerland. Pict. 27 + 28. It is a shellshape having large tension zones.

By chance the cracks appeared only one month after prestressing and taking off the scaffolding. The tensional stresses of the load case proper weight alone were within the tensional resistence of the concrete. Cold winter weather and no snow preserved this stage for about 30 days. The elastic (initial) deformations remained constant during this period.

The intensive sun of February in this region added thermal stresses to proper weight stresses. They now exceeded the tensional resistence of the concrete: within another 30 days the complete cracking pattern appeared.

Within this second period the initial deformations increased rapidly and then came to a final stop. The increase was about 100 % of the elastic stage. It can be called a real chance to observe on a shell building the cracking influence neatly separated from the elastic range.

15. Single or Double Layer of Reinforcement

The historical quarrel regarding the question of one layer of reinforcement or two layers in a concrete shell can be resolved as follows: If instability is an actual danger a double layer reinforcement offers more stability, because it keeps bending deformations smaller.

For a given amount of reinforcement the acceptable moment is slightly larger if all the steel is put in the middle of the shell. However the deformation angle α (Fig. 1) (steel in one layer) is nearly double than β (steel in two layers). And this increase of deformation is exactly that which promotes instability. This is instability: the increase of deformation

without increase of effort. Impeding transverse deformation impedes buckling.
Buro Isler always puts a double layer of reinforcement.

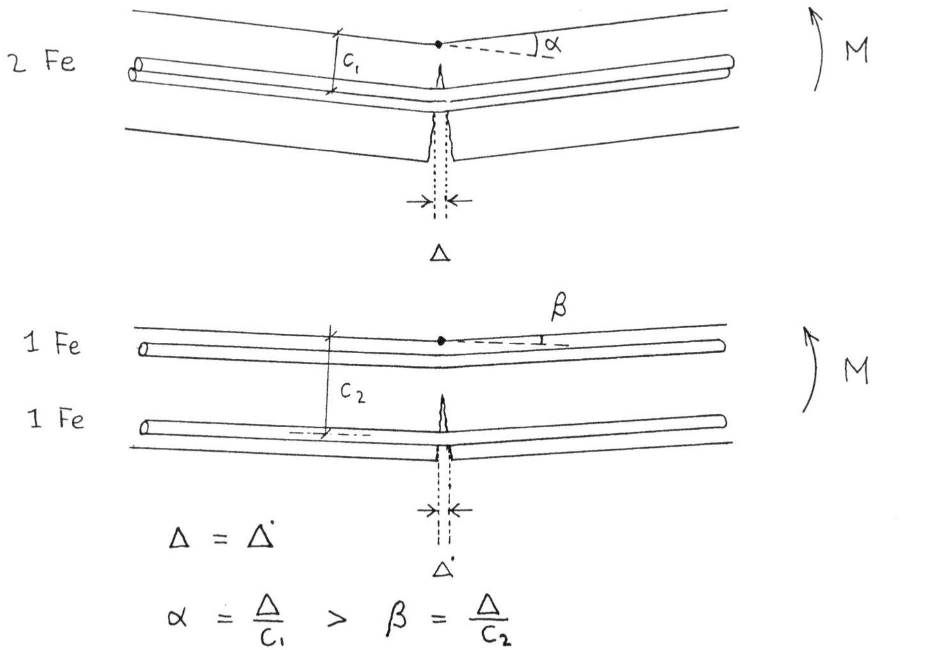

Fig. 1. Steel in two layers = smaller deformations

16. Shell Shapes Avoiding Cracking

Creep and cracks reduce the buckling safety considerably. If possible one should use shapes and techniques which avoid cracking by avoiding tensional stresses at all. That means using shells of mere compression.

Here some examples:
Pict. 30 Bubble shells for a factory. They have a pneumatic shape combined with prestressing, i.e. mere compression. This building is without coating and is still waterproof after 18 years, another proof, that there are no cracks.

Pict. 3o
Bubble shells of mere compression. After 18 years still waterproof.
(Moser Lyssach)

Pict. 31
Garage in the Swiss Alps after 16 years of use
(Epper, Engelberg)

Pict. 32
Bubble shells for factory (Langenthal)

Pict. 33

Shell in pure compression (Hotel swimming pool Lugano)

Pict. 34

asymmetric shell in pure compression

(Ballet dancing school Stetten)

Pict. 35

Compression shell on trapezoidal plan

(Laboratory Bex)

Pict. 36 Swimming pool Brugg, square shell 35 x 35 m in pure compression

Pict. 37 Swimming pool Brugg, inside view. Shellshape found by hanging membrane experiment.

Pict. 31 shows a 16 years old garage in the Swiss Alps at high altitude with exposed mountain weather conditions. It also remained watertight.

The next pictures have a shape of natural pure compression found by inversion of hanging model membranes. By experimental methods practically to every task and groundplan a shellshape can be found which has mainly compression stresses. [Ref. 3].

Pict. 33 Indoor Swimming Pool Lugano, shell in pure compression

Pict. 34 Balletdancing school Stetten, completely asymetrical shell

Pict. 35 Laboratory at Bex, compression shell on trapezoidal plan.

Pict. 36 + 37 Swimming Pool Brugg, with square shell.

In my opinion these shapes give the highest effiency,
- very small deformations, initial and longtime,
- no cracks, no corrosion,
- high safety against buckling.

In corrosive atmospheres concrete shells in pure compression are best. Cracking of the concrete can be avoided and theoretically no reinforcement is needed. Of course pure compression shells must also be reinforced (in two layers!) to cater for unforseen load cases.

17. Recapitulation

Concerning stability concrete shells have in addition to the well known factors of instability (shape, curvature, thickness, inaccuracies) additional problems:
- fabrication of formwork and concrete on the building site
- creep leading to increased deformations
- cracks

However, all these difficulties can be surmounted by a good design and careful execution. So I think the expert in shells needs not be afraid of them.

18. Overall Stability

Stability means not only buckling. There is also an overall stability, which sometimes can be extreme. To end with I will show you a short sequence of slides on accidents with shells:

19. Destruction Test

Last year two shells (bubble shells) had to give way to an other building. We profited by their destruction (in collaboration with ETH and EMPA) to simulate all sorts of severe accidents on a shell. Here they are:

Pict. 38	overloading of the edges
Pict. 39	overloading of the skylight opening
Pict. 40 + 41	punching (buckling test too expensive)
Pict. 42 + 43	meteorite
Pict. 44 + 49	settlement 20 cm, 50 cm, 1 m, 2 m, 4 m with cracking pattern
Pict. 50 + 51	destruction of 2nd column (one would still survive!)
Pict. 52 + 53	removing one by one the columns one is still protected under the shell

And finally

Pict. 54 + 55	pulverization and removing of the whole material.

Conclusions:

1. A double curved shell of this type has an extraordinary survival capacity. Even after very heavy damage it still gives high protection to inhabitants. The shell itself never breaks.

38

39

4o

41

42

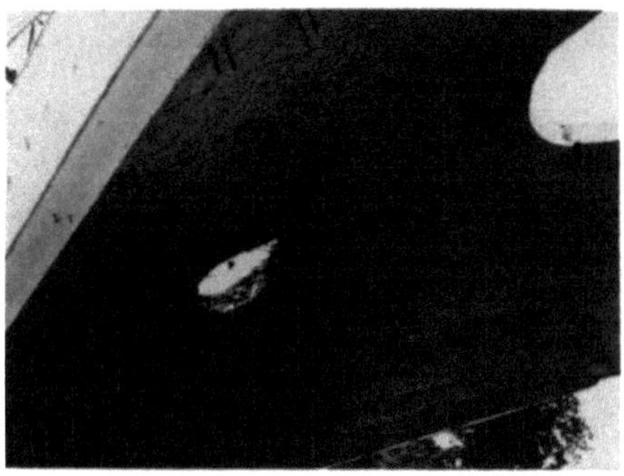

43

Pict. 38 - 43: Tests on bubble shells Uster.

Overloading of edges
overloading of skylight opening
punching tests
"meteorite"

44

45

46

47

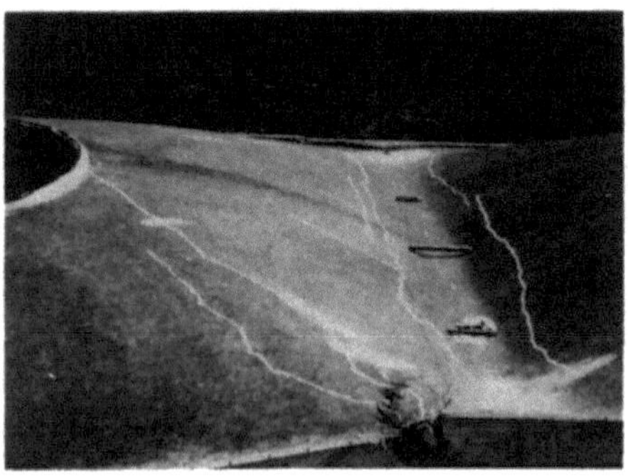

48

49

Pict. 44 - 49: Tests on bubble shells Uster.
Settlement 2o cm, 5o cm, 1 m, 2 m, 4m with final cracking pattern.

5o

51

52

53

54

55

Pict. 5o - 55: Destruction of bubble shells Uster.

Removing one by one the columns.
Even after heavy damage the shell gives high protection.

Pulverization and removal of material.

2. When pulverised systematically, a question of another one and a half hours, the remaining amount of material is so small, that it is removed in half a day.

So, what more does one want? Lets apply sound shells more frequently!

References

1. Isler, H. Zur Korrelation von Formgebung und Stabilität bei dünnen Schalentragwerken. 2. Internat. Symposium Weitgespannte Flächentragwerke, Universität Stuttgart, Mai 1979.

2. Heinz Isler as Structural Artist. The Art Museum, Princeton University, New Jersey, USA, 1980.

3. Isler, H. New Shapes for Shells - Twenty years after, IASS-Bulletin 71/72, September 1979 Madrid.

All projects by author except pict. 10 - 16.

MIX
Papier aus verantwortungsvollen Quellen
Paper from responsible sources
FSC® C105338

If you have any concerns about our products,
you can contact us on
ProductSafety@springernature.com

In case Publisher is established outside the EU,
the EU authorized representative is:
Springer Nature Customer Service Center GmbH
Europaplatz 3, 69115 Heidelberg, Germany

Printed by Libri Plureos GmbH
in Hamburg, Germany